Natural Bioactives from the Endophytes of Medicinal Plants

Endophytes from medicinal plants have garnered global attention due to their remarkable capacity to produce unique phytochemicals, pharmaceuticals and promising lead compounds. This book explores cutting-edge advancements in endophytic fungi research, encompassing a comprehensive exploration of their biodiversity, ecological dynamics, mechanisms of interaction and adaptive strategies. The book offers valuable insights into the therapeutic potential, agricultural applications, environmental impacts and commercial prospects of these fascinating organisms.

While contemporary books are limited to either taxonomic, ecological or practical discussions, this book presents a systematic compilation of biology and biotechnological applications of endophytic fungi from medicinal plants.

Key Features

1) Comprehensive exploration of the diversity, ecology and interactions of endophytes derived from medicinal plants found in their natural habitats.
2) Highlights the potential of endophytes derived as prolific producers of novel pharmaceutical and lead compounds.
3) Showcases the practical applications of endophytes in plant development and sustainable agriculture practices.
4) Includes contribution from researchers and academicians with vast experience in the field.

This book is a collection of informative illustrations, strategically integrated throughout the content, to facilitate the understanding of concepts and aid in smooth transitions between topics. Given the vast and diverse nature of the discipline, this book serves as an invaluable reference resource for students and researchers in various fields, including mycology, microbiology, biotechnology, pharmacology, botany, ecology, agronomy and molecular biology. Additionally, professionals in environmental conservation and policymakers dedicated to preserving biodiversity of medicinal plants will find this book to be a valuable asset in their work.

Exploring Medicinal Plants

***Series Editor: Azamal Husen**, Wolaita Sodo University, Ethiopia*

Medicinal plants render a rich source of bioactive compounds used in drug formulation and development; they play a key role in traditional or indigenous health systems. As the demand for herbal medicines increases worldwide, supply is declining as most of the harvest is derived from naturally growing vegetation. Considering global interests and covering several important aspects associated with medicinal plants, the Exploring Medicinal Plants series comprises volumes valuable to academia, practitioners, and researchers interested in medicinal plants. Topics provide information on a range of subjects including diversity, conservation, propagation, cultivation, physiology, molecular biology, growth response under extreme environment, handling, storage, bioactive compounds, secondary metabolites, extraction, therapeutics, mode of action, and healthcare practices.

Led by Azamal Husen, PhD, this series is directed to a broad range of researchers and professionals consisting of topical books exploring information related to medicinal plants. It includes edited volumes, references, and textbooks available for individual print and electronic purchases.

Promising Antiviral Herbal and Medicinal Plants
Edited by Nadeem Akhtar, Azamal Husen, Vagish Dwibedi, and Santosh Kumar Rath

Antimalarial Medicinal Plants
Edited by Azamal Husen

Medicinal Spice and Condiment Crops
Edited by Azamal Husen

Therapeutic Medicinal Plants in Traditional Persian Medicine
Edited by Roja Rahimi and Roodabeh Bahramsoltani

Medicinal Plants for the Management of Neurodegenerative Diseases
Edited by Jamal Akhtar, Maryam Sarwat and Fouzia Bashir

Nutraceuticals Inspiring the Contemporary Therapy for Lifestyle Diseases
Edited by Mala Trivedi, Sachidanand Singh, Parul Johri, and Pedro López-Sánchez

Plants as Medicine and Aromatics: Uses of Botanicals
Edited by Mohd Kafeel Ahmad Ansari, Mushtaq Ahmad, and Gary Owens

Exploring Traditional Wild Edible Plants
Edited by Vibhor Agarwal, Sachidanand Singh and Rhaul Datta

Propagation to Pharmacopeia: Modern Approaches in Medicinal Plants
Edited by Abha Manohar K, Gopal Shukla, Saikat Gantait, A P Das, and Sumit Chakravarty

Exploring Medicinal Orchids
Edited by Azamal Husen and Manu Pant

Symbiotic Association of Microorganisms with Medicinal and Herbal Plants
Edited by Allah Ditta, Sajid Mehmood, and Azamal Husen

Tissue Culture Techniques and Medicinal Plants: Enhancing Propagation and Production
Edited by Azamal Husen and Manu Pant

For more information about this series, please visit: www.routledge.com/Exploring-Medicinal-Plants/book-series/CRCEMP

Natural Bioactives from the Endophytes of Medicinal Plants

Edited by
Vineet Meshram, Kamlesh Kumar Shukla,
Mahiti Gupta and Nadeem Akhtar

CRC Press
Taylor & Francis Group
Boca Raton London New York

CRC Press is an imprint of the
Taylor & Francis Group, an **informa** business

Designed cover image: © Getty Images

First edition published 2025
by CRC Press
2385 NW Executive Center Drive, Suite 320, Boca Raton FL 33431

and by CRC Press
4 Park Square, Milton Park, Abingdon, Oxon, OX14 4RN

CRC Press is an imprint of Taylor & Francis Group, LLC

ISBN: 9781032781143 (hbk)
ISBN: 9781032781198 (pbk)
ISBN: 9781003486275 (ebk)

DOI: 10.1201/9781003486275

Typeset in Times
by Newgen Publishing UK

Dedication

This Book is Dedicated to
Dr. Stanley Freeman
Senior Scientist
The Volcani Institute
Agricultural Research Organization
Rishon Lezion, Israel

Contents

Contributors

Abdullah Abdullah
Department of Plant Sciences
Quaid-i-Azam University
Islamabad, Pakistan

Adil Muhammad
Pharmacology and Toxicology Section,
University of Veterinary & Animal Sciences, Lahore, Jhang Campus,
Jhang, Pakistan

Ahmad Zeeshan
Department of Plant Sciences
Quaid-i-Azam University
Islamabad, Pakistan

Ahn Young-Ho
Department of Civil Engineering
Yeungnam University
Gyeongsan, Gyeongbuk, South Korea

Alzahrani Abdulhakeem
Department of Food Science & Nutrition
College of Food and Agricultural Sciences
King Saud University
Riyadh, Saudi Arabia

Anam Giridhar Babu
Department of Biotechnology
School of Sciences
SR University
Warangal, Telangana, India

Awachar Sachin
Department of Microbiology
Sant Gadge Baba Amravati University
Maharashtra, India

Badshah Husain
Department of Plant Sciences
Quaid-i-Azam University
Islamabad, Pakistan

Bajpai Mangalam
School of Bioscience
The University of Sheffield
Sheffield, United Kingdom

Bibi Nizakat
International Joint Research Laboratory for Global Change Ecology
School of Life Sciences
Henan University
Kaifeng, Henan China

Das Jyotirmoyee
Department of Biology
University of Alabama at Birmingham
Birmingham, Alabama, United States

Dhar Meghna
Department of Life Sciences
J.C. Bose University of Science and Technology
YMCA, Faridabad, Haryana, India

Dharmappa K.K.
Inflammation Research Laboratory
Department of Studies and Research in Biochemistry
Mangalore University, JnanaKaveri PG Centre,
Kodagu, Karnataka, India

Dholpuria Sunny
Department of Life Sciences
J.C. Bose University of Science and Technology
YMCA, Faridabad, Haryana, India

Dufossé Laurent
Université de La Réunion, Laboratoire
CHEMBIOPRO (Chimie et Biotechnologie des Produits Naturels)
ESIROI Département agroalimentaire
Saint-Denis, France

Dwivedi Sneha
Laboratory of Microbiology and Plant Pathology
Department of Botany, C. M. P. Degree College
University of Allahabad
Prayagraj, India

Gade Aniket
Department of Microbiology
Nicholas Copernicus University
Torun, Poland

Giresha A.S.
Department of Chemistry and Biochemistry
School of Science
Jain (Deemed-to-be) University,
Bangalore, Karnataka, India

Goyal Soniya
Department of Biosciences and Technology
Maharishi Markandeshwar (deemed to be) University,
Mullana-Ambala, Haryana, India

Gupta Mahiti
Department of Biosciences and Technology
Maharishi Markandeshwar (deemed to be) University,
Mullana-Ambala, Haryana, India

Hande Dilip
Mahatma Fule Arts, Commerce and Sitaramji Choudhary Science College
Warud District, Amravati, MS India.

Haq Zahoor ul
International Joint Research Laboratory for Global Change Ecology
School of Life Sciences
Henan University
Kaifeng, Henan, China
Yellow River Floodplain Ecosystems Research Station
Henan University
Xingyang, China

Hassan Javed Chaudhary
Department of Plant Sciences
Faculty of Biological Sciences
Quaid-I-Azam University, Islamabad, Pakistan

Jisha M.S.
School of Biosciences
Mahatma Gandhi University
Kottayam, Kerala, India

Kamble Kapil
Department of Microbiology
Sant Gadge Baba Amravati University
Maharashtra, India

Kalsoom Maria
Institute of Food Science and Nutrition
University of Sargodha
Sargodha, Pakistan

Khan Mohsin
International Joint Research Laboratory for Global Change Ecology
School of Life Sciences
Henan University
Kaifeng, Henan, China
Yellow River Floodplain Ecosystems Research Station
Henan University
Xingyang, China

Khan Muhammad Shakeel
Department of Plant Sciences
Quaid-i-Azam University
Islamabad, Pakistan

Khan Shujaul Mulk
Department of Plant Sciences
Quaid-i-Azam University
Islamabad, Pakistan
Pakistan Academy of Sciences
Islamabad, Pakistan

Kumar Prasann
Department of Agronomy
School of Agriculture
Lovely Professional University
Phagwra, Punjab, India

Kumar Surender
Texas A&M AgriLife Research and Extension Center
Amarillo, TX, USA

Kumari Reenu
College of Horticulture and Forestry
Dr Y S Parmar University of Horticulture and Forestry
Mandi, Himanchal Pradesh, India

Mahrukh
Department of Plant Sciences
Faculty of Biological Sciences
Quaid-I-Azam University, Islamabad, Pakistan

Mazhar Rafique
Department of Plant Sciences
Quaid-I-Azam University
Islamabad, Pakistan

Mukherjee Arkadeep
Department of Civil Engineering
Yeungnam University
Gyeongsan, Gyeongbuk, South Korea

Munis Muhammad Farooq Hussain
Department of Plant Sciences
Faculty of Biological Sciences
Quaid-I-Azam University
Islamabad, Pakistan

Narale Anagha
Department of Microbiology
Rajasthan Aryans Mahavidyalaya
Washim, Maharashtra, India

Naz Sabiha
Department of Biotechnology and Microbiology
Bhilai Mahila Mahavidhyalaya
Chhattisgarh, India

Pachori Rachana
Department of Microbiology
Rajasthan Aryans Mahavidyalaya
Washim, Maharashtra, India

Pandey Shalini
Amity Institute of Biotechnology
Amity University
Chhattisgarh, Kharora, Raipur, (C.G.) India

Paker Najeeba Parre
Department of Plant Sciences
Faculty of Biological Sciences
Quaid-I-Azam University, Islamabad, Pakistan

Prabhakar B.T.
Molecular Biomedicine Laboratory
Postgraduate Department of Biotechnology
Sahyadri Science College, Kuvempu University
Shivamogga, Karnataka, India

Rajeshkumar K.C.
National Fungal Culture Collection of India (NFCCI)
Biodiversity and Palaeobiology (Fungi) Gr.
Agharkar Research Institute
Pune, Maharashtra, India

Rakshit Smita
Department of Microbiology
Sant Gadge Baba Amravati University
Near Tapovan Gate, SRPF Colony, Amravati, Maharashtra, India

Rehman Fazal ur
Department of Plant Sciences
Faculty of Biological Sciences
Quaid-I-Azam University
Islamabad, Pakistan

Saraf Ashish
School of Sciences
MATS University,
Raipur, C.G., India

Saxena Sanjai
Department of Biotechnology
Thapar Institute of Engineering and Technology
Patiala, Punjab, India

Shahwal Richa
Amity Institute of Biotechnology
Amity University
Chhattisgarh, Kharora, Raipur, (C.G.) India

Sharma Deepansh
Department of Life Sciences
J.C. Bose University of Science and Technology
YMCA, Faridabad, Haryana, India

Singh Alok Kumar
Laboratory of Microbiology and Plant Pathology
Department of Botany, C. M. P. Degree College
University of Allahabad
Prayagraj, India

Singh Joginder
Department of Botany
Nagaland University
Zunheboto, Nagaland, India

Sodhi Gurleen Kaur
Department of Biotechnology
Thapar Institute of Engineering and Technology
Patiala, Punjab, India

Sorout Kajal
Department of Life Sciences
J.C. Bose University of Science and Technology
YMCA, Faridabad, Haryana, India

Srivastava Arpita
Amity Institute of Biotechnology
Amity University
Chhattisgarh, Kharora, Raipur, (C.G.) India

Sur Arunima
Amity Institute of Biotechnology
Amity University
Chhattisgarh, Kharora, Raipur, (C.G.) India

Suradkar Kishor
Department of Botany
Indira Mahavidyalya
Kalamb, Yavatmal, MS, India

Tiwari Madhavi
Faculty of Science
Department of Applied Science
Shri Rawatpura Sarkar University,
Raipur, Chhattisgarh, India

Tiwari Pragya
Department of Horticulture and Life Science
Yeungnam University
Gyeongsan, Gyeongbuk, South Korea

Uddin Bashir
Department of Plant Sciences
Quaid-I-Azam University
Islamabad, Pakistan

Tehsin Ullah
Department of Plant Sciences
Quaid-i-Azam University
Islamabad, Pakistan

Tejaswini H.K.
Department of Microbiology
Maharani Cluster University
Bangalore, Karnataka, India

Urs Deepadarshan
Inflammation Research Laboratory
Department of Studies and Research in Biochemistry
Mangalore University, Jnana Kaveri PG Centre,
Kodagu, Karnataka, India

Varghese Sherin
School of Biosciences
Mahatma Gandhi University
Kottayam, Kerala, India

Verma Seema
Department of Botany
GD Rungta College of Science and Technology
Bhilai, Durg, Chhattisgarh, India

Wang Zixuan
International Joint Research Laboratory for Global Change Ecology
School of Life Sciences
Henan University
Kaifeng, Henan, China
Yellow River Floodplain Ecosystems Research Station
Henan University
Xingyang, China

Zeb Shakil Ahmad
Department of Plant Sciences
Quaid-i-Azam University
Islamabad, Pakistan

Zheng Junqiang
International Joint Research Laboratory for Global Change Ecology
School of Life Sciences
Henan University
Kaifeng, Henan, China
Yellow River Floodplain Ecosystems Research Station
Henan University
Xingyang, China

Editor Biographies

Vineet Meshram is Dean of the Research and Development at Anjaneya University, Raipur, India. With over 15 years of research experience in plant–microbe interactions, natural products and drug discovery, microbial secondary metabolites and management of biotic and abiotic stresses in crop plants, his work has resulted in two edited books, 23 research articles, 25 abstracts and contributions to five books chapters. He also serves as an editor and reviewer of several prestigious journals.

Kamlesh Kumar Shukla is currently serving as Associate Professor at the School of Studies in Biotechnology, Pt. Ravishankar Shukla University, Raipur, India. With over 22 years of teaching and research experience, Dr. Shukla has contributed significantly to the education and development of students in the field of Biotechnology. Dr. Shukla has a notable research background, having authored 35 research papers in peer-reviewed national and international journals as well as 12 book chapters.

Mahiti Gupta is an Assistant Professor at Maharishi Markandeshwar Deemed University, Haryana, India. She specializes in natural products, drug discovery and enzyme inhibitors. Her publications include two edited books, 19 research articles and nine book chapters.

Nadeem Akhtar is a scientist at CBS Bio Platforms Inc., leveraging biotechnology in the agri-food sector. He served as a research associate at the University of Guelph and as a postdoctoral researcher at the University of Waterloo and the University of Guelph. Dr. Akhtar has contributed to 27 published articles, 12 book chapters, four edited books and a patent, and is a reviewer for many prestigious journals.

Editor Biographies

Preface

For centuries, medicinal plants have served as vital sources of bioactive natural products offering a wide array of compounds used in traditional and modern medicines. In recent years, endophytic microorganisms residing asymptomatically inside plant tissues have emerged as a treasure trove of bioactive compounds. Endophytes establish a symbiotic relationship with their host plant that serves multiple functions including stimulating plant growth and inducing defense mechanisms against abiotic and biotic stresses. The exploration of endophytes from medicinal plants not only broadens the spectrum of bioactive compounds available for drug discovery but also enhances our understanding of the intricate interactions between the host plant and its mutualistic partner. Furthermore, endophytes can also mimic the biological properties of the host, which makes them suitable candidates for drug discovery.

The purpose of this book titled *Bioactive Naturals from the Endophytes of Medicinal Plants* is to present the frontier strategies and technologies involved in endophytic research and enable its readers to understand the significance of endophytic microorganisms in the field of pharmaceuticals and agriculture. With 22 chapters contributed by an elite group of researchers working in the forefront of bioprospecting bioactive compounds from endophytes, the book is divided into three units; the first unit discusses the taxonomy, ecology and omics approaches involved in studying endophytic microorganisms, while the second unit elaborates the bioactive compounds derived from endophytes and their possible application in drug discovery and development. The third unit highlights fitness benefits and adaptations provided by endophytes in mitigation of abiotic and biotic stresses in crop plants for increasing agricultural productivity. Additionally, the book also includes chapters on the economics and commercial opportunities of endophyte-derived products along with their environmental impacts, offering a holistic view on the roles of endophytes and their products. This book, therefore, will serve as a useful reference book for students and researchers across diverse fields, including biotechnology, microbiology, mycology, pharmacy, biochemistry, agriculture and environmental studies.

Raipur, India **Vineet Meshram**
Raipur, India **Kamlesh Kumar Shukla**
Mullana, India **Mahiti Gupta**
Calgary, Canada **Nadeem Akhtar**
2024

Unit 1

Taxonomy, Diversity and Occurrences of Endophytic Fungi of Medicinal Plants

1 Diversity and Morphotaxonomy of Endophytic Fungi from Melghat Forest, Amravati, Maharashtra, India

Dilip Hande and Kishor Suradkar*
*Corresponding author.

1.1 INTRODUCTION

The multiplicity of fungi and their natural prettiness occupy a most important place in the biological world, and India has been the cradle for such fungi. Only a fraction of the total fungal wealth has been subjected to scientific scrutiny, and mycologists have to unravel the unexplored and hidden wealth. One-third of the fungal diversity of the globe exists in India. Unfortunately, only around 5–10% of fungi can be cultured. Fungi are not only beautiful but also play a significant role in the daily life of human beings including their utilization in industry, agriculture, medicine, food industry, textiles, bio-remediation, natural cycling, as biofertilizers and in many other ways (Manoharacharya et al., 2005).

Endophytes play a significant role in establishing fungal diversity. The research interest has been increasing in ecology, biology and applications of endophytic fungi. It is believed that endophytic fungi are diverse in those areas where there is greater diversity of plants, and India is one of these areas with about 17,527 Angiosperm and 67 Gymnosperm species; however, studies on the endophytic fungal diversity from India is extremely inadequate (Singh, 2012). Tropical regions are considered to be the most important regions to study the endophytic fungal diversity, which may help to estimate the number of fungi species on Earth (Hawksworth, 2001). India is one of these tropical regions, but such studies in India are inadequate. Only a few plant hosts have been comprehensively studied for their endophytic diversity (Karthikeyan, 2009). Considering the vast plant diversity and wide range of vegetation distributed in India, it is presumed that a high degree of endophytic fungal diversity also exists, which may serve as valuable germplasm for biologically interesting antimicrobial compounds.

Several plants from temperate regions have been studied for their endophytic associations but investigations into the endophytic diversity from tropical plants are not adequate. Only preliminary studies have been carried out on endophytic fungi from the tropics in various regions of the world, such as Hong Kong (Brown et al., 1998), Brazil (Rodrigues & Samuel, 1999), Guyana (Cannon & Simmons, 2002) Thailand (Photita et al., 2001), and India (Suryanarayanan et al., 2001, 2003; Kharwar et al., 2008). Therefore, it is of utmost importance to study the endophytic fungi from India as they serve as an important component of fungal diversity.

DOI: 10.1201/9781003486275-2

1.2 MATERIALS AND METHODS

1.2.1 Selection of Plants

In this study, 12 plants were selected, namely *Adhatoda vasica* Linn, *Calotropis procera* (Ait.) R. Br., *Cissus quadrangularis* L., *Coffea arabica L.*, *Datura metel* Linn., *Dioscorea bulbiferea* L., *Helecteres isora* L., *Hemidesmus indicus* R. Br., *Murraya koenigii* Linn. Spreng, *Ocimum sanctum* L., *Vitex negundo* L. and *Withania somnifera* Linn. on the basis of their medicinal properties.

1.2.2 Study Area and Collection of Plant Samples

The samples were collected from different locations of Melghat forest, including Semadoh, Ghatang, Kolkas, Gugamal, Bori, Tarubanda and Harisal. The entire area of Melghat Tiger Reserve is covered with dry deciduous forest. The temperature of forest the ranges from 4°C to 43°C. The annual rainfall ranges between 1784–2500 mm (Central Government Water Management Section, Chikhaldara). The climate of Melghat Forest is humid, moist and cold during the monsoon. These conditions are very favorable for the growth of endophytic fungi. Plant samples were collected during different seasons, i.e., winter and summer seasons.

1.2.3 Isolation of the Endophytic Fungi

The collected plant samples were washed under running tap water to remove surface adherents. Surface sterilization was carried out according to the method described by Suryanarayanan et al. (2001). The surface-sterilized explants were then inoculated at 26 ± 2°C in Petri dishes containing water agar. The plates were periodically observed for fungal growth. The fungi obtained on WA plates were further subcultured on fresh PDA (potato dextrose agar) plates to obtain pure isolates. The purified endophytic fungal cultures were then transferred onto PDA slant and stored as stock culture for further studies. All procedures were carried out aseptically under a laminar air flow hood.

1.2.4 Identification of Endophytic Fungi

All the endophytic isolates were identified morphologically and placed in appropriate genera and species of fungi using standard taxonomic keys and monographs. Ellis (1971, 1976), Sutton (1980), Subramanian (1971), and Barnett and Hunter (1972) were referred to for the identification of endophytes. In addition, other taxonomic-related research articles describing endophytes were also referred to.

1.2.5 Data Analysis

The colonization frequencies (CF %) of fungal endophytes were calculated using the following formula (Kumaresan and Suryanarayanan 2001):

$$Colonization\,Frequency\left(CF\,\%\right) = \frac{Total\,number\,of\,segments\,colonized\,by\,fungi}{Total\,number\,of\,segments\,studied} \times 100$$

1.3 RESULTS

1.3.1 Morphotaxonomy

All the isolated endophytic fungi were studied for their morphotaxonomy. Morphological characteristics of the fungi such as color of colonies, structure of mycelium, shape, size and color of conidia or spore, structure of conidiophores, attachment of conidiophore, shape and size of pycnidia, etc. were recorded. On the basis of these microscopic and macroscopic characteristics, isolated fungi were identified morphologically. A total of 22 species of endophytic fungi belonging to 19 genera were identified morphologically from all the investigated medicinal plants. Endophytic fungi (Figures 1.1–1.16) obtained from investigated medicinal plants were identified as *Alternaria*

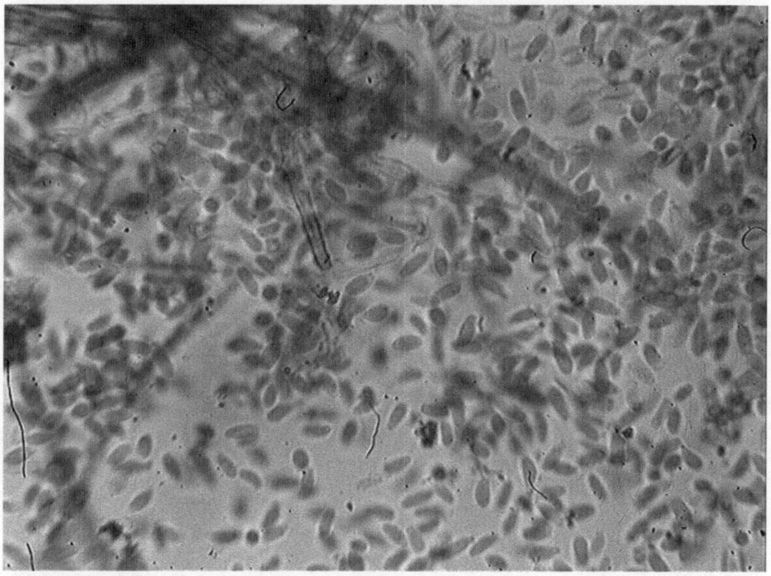

FIGURE 1.1 Mycelium with conidia of *Acremonium fusidioide*.

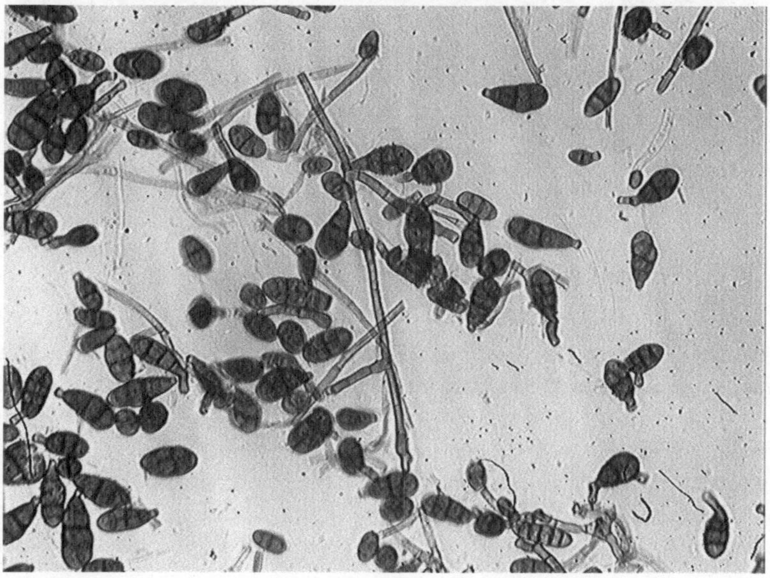

FIGURE 1.2 Mycelium with conidia of *Alternaria alternate*.

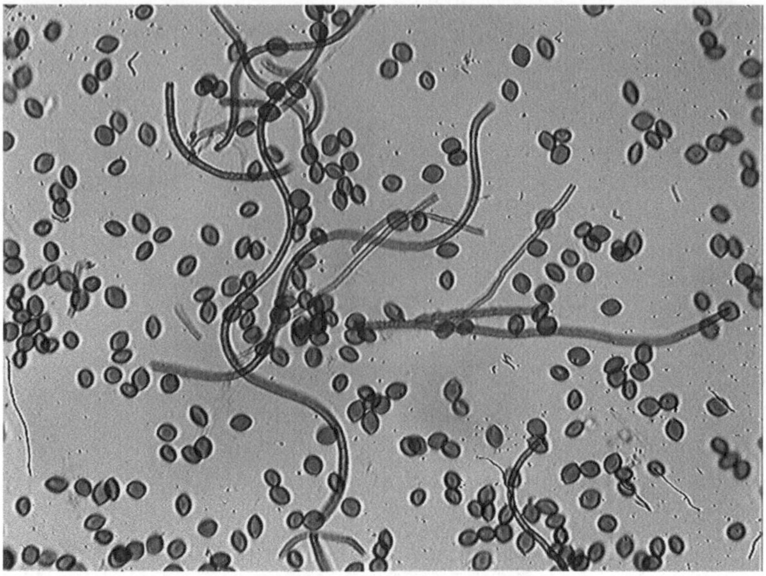

FIGURE 1.3 Mycelium with conidia of *Arthrinium phaeospermum.*

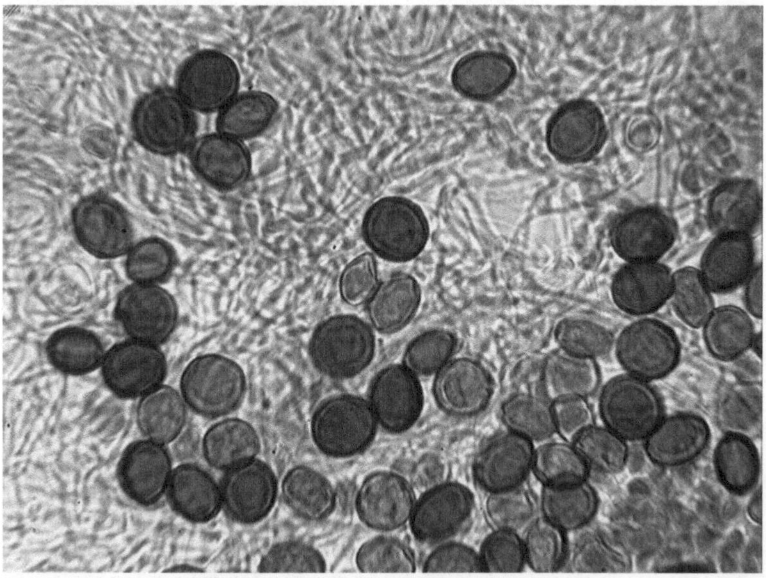

FIGURE 1.4 Conidia and conidiogenous cells of *Arthrinium hydei.*

alternata (Fr.) Keissl., *Acremonium fusidioide* (Link) ex. Fries, *Arthrinium phaeospermum* (Corda) Ellis, *Arthrinium hydei* Crous and Groenewald, *Aspergillus stellatus* Curzi, *Aspergillus nidulans* (Eidam) Winter, *Cladosporium cladosporioides* (Fres.) de Varies., *Colletotrichum gloeosporiodes* (Penz) Sacc., *Fung. agrum, Curvularia lunata* (Wakker) Boedijn, *Diplodia mutila* Fr. apud Mont., *Epicoccum nigrum* Ehrenb. ex. Schlecht., *Fusarium oxysporum* Schl. ex Fries., *Nigrospora oryzae* (Berk. and Br.) Petch., *Penicillium chrysogenum* Thom, *Pestalotiopsis funereal* Stey., *Phoma crysanthemicola* Hollos, *Ann. hist. nat. Mus. nutn. Hung.*, *Pithomyces chartarum* (Berk and Curtis)

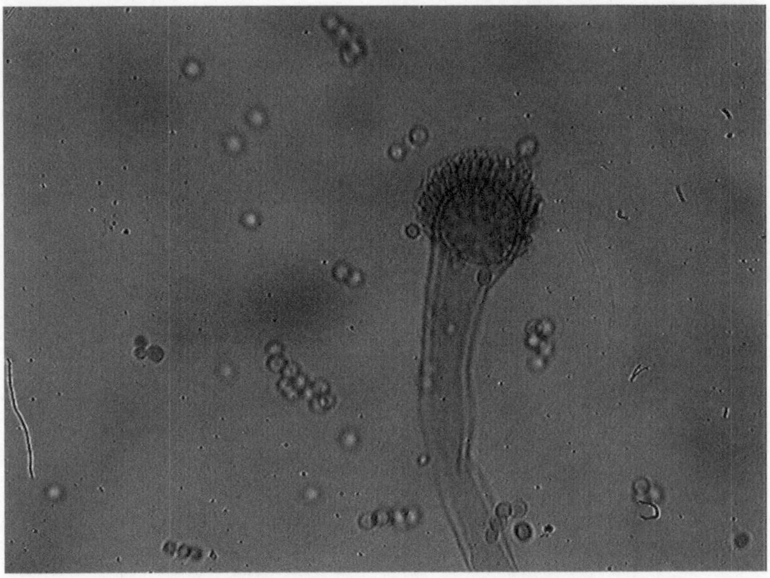

FIGURE 1.5 Conidia and conidiophore of *Aspergillus nidulans*.

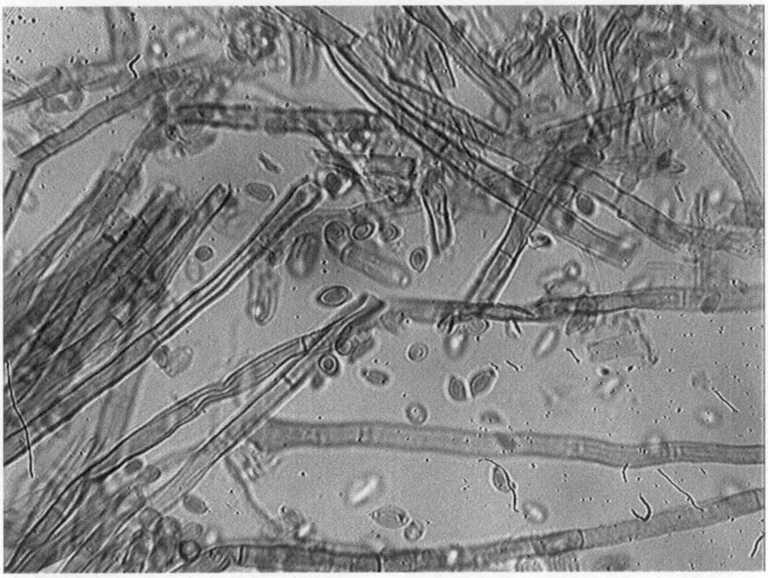

FIGURE 1.6 Mycelium with conidia of *Cladosporium cladosporioides*.

Ellis, *Stachybotrys chartarum* (Ehrenb.) Hughes, Stachybotrys *nilgirica* Subram., *Trimmatostroma hughesii* Rao and Subhedar, *Virgaria nigra* (Link) Nees., *Xylaria* sp. Hill ex. Schrank.

1.3.2 Diversity

The diversity of endophytic fungi was studied in 12 medicinal plants collected from different locations of Melghat forest. All the isolated endophytes revealed high levels of variations in occurrence and colonization in each season of the study.

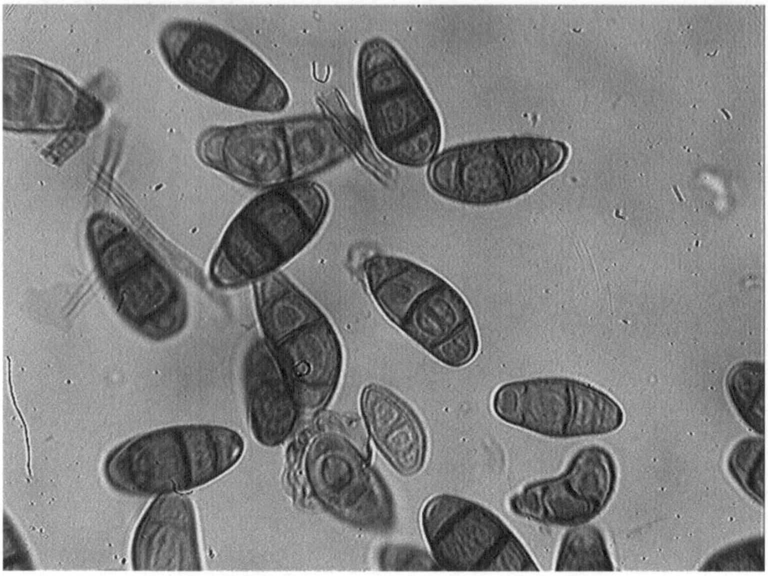

FIGURE 1.7 Conidia of *Curularia lunata*.

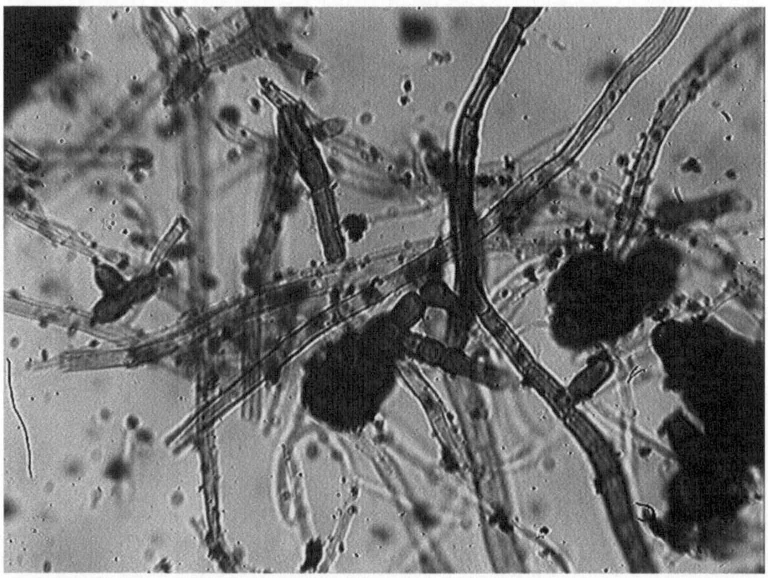

FIGURE 1.8 Mycelium with conidia of *Epicoccum nigrum*.

1.3.2.1 Diversity of Endophytic Fungi from *Adathoda vasica* Linn.

Sixteen fungal endophytes were isolated. All the isolates exhibited seasonal variations in occurrence. *Arthrinium phaeospermum* showed a higher colonization rate in leaf tissue in monsoon, whereas in winter it exhibited a high colonization frequency in stem tissue. *Aspergillus stellatus* was present throughout the study but in winter season it showed a higher colonization frequency as compared to

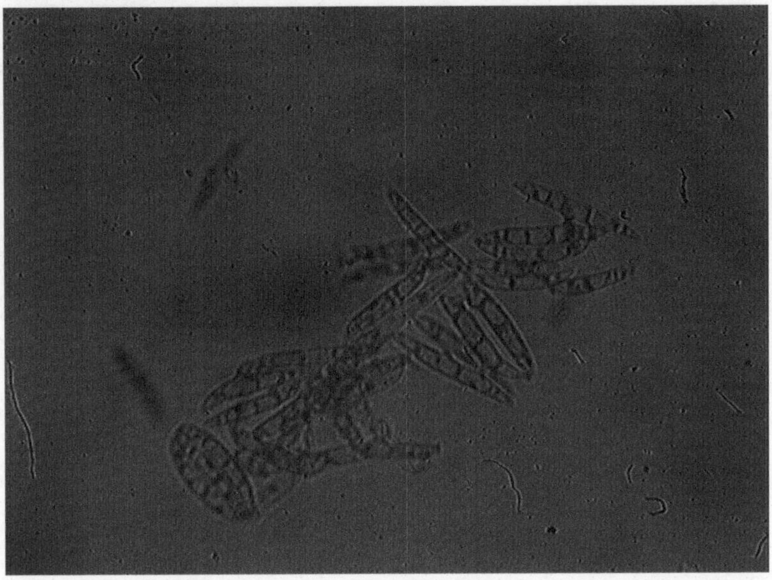

FIGURE 1.9 Conidia of *Fusarium oxysporum*.

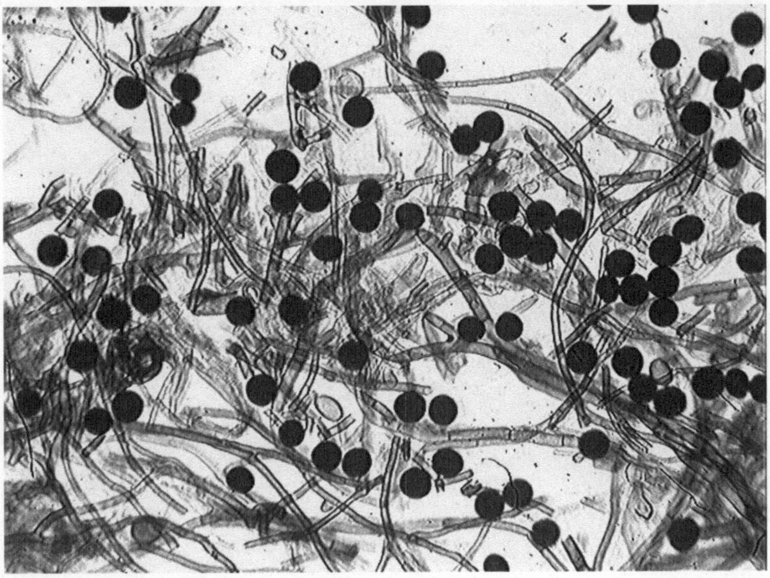

FIGURE 1.10 Mycelium with conidia of *Nigrospora oryzae*.

summer. Much less number of these fungi were recovered during the summer season with a lower colonizing frequency (Table 1.1).

1.3.2.2 Diversity of Endophytic Fungi from *Calotropis procera* (Ait) R. Br.

There was a great difference in the colonization rate and tissue preference of all the identified endophytic fungi from leaf, petiole and stem samples. Fourteen different fungal species were discovered, among these isolates *Stachybotrys nilgirica* and *Trimmatostroma hughesii* were recovered from leaf

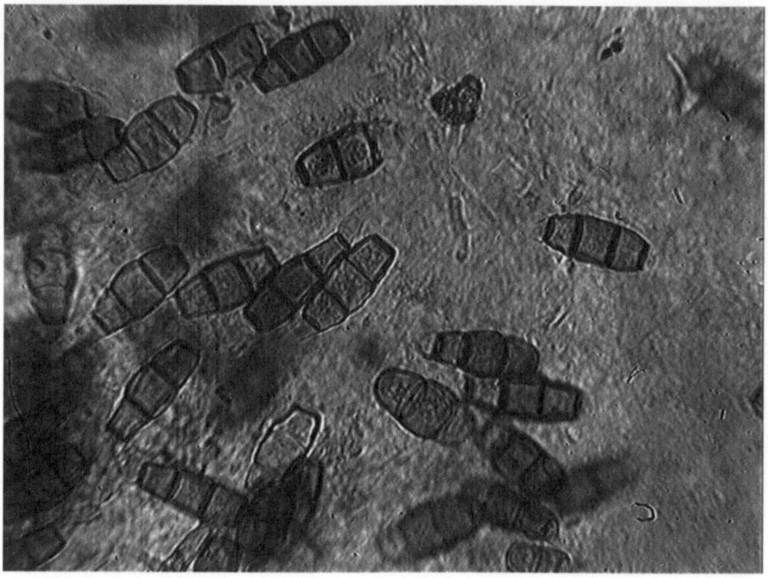

FIGURE 1.11 Mycelium with conidia of *Pestalotiopsis funereal*.

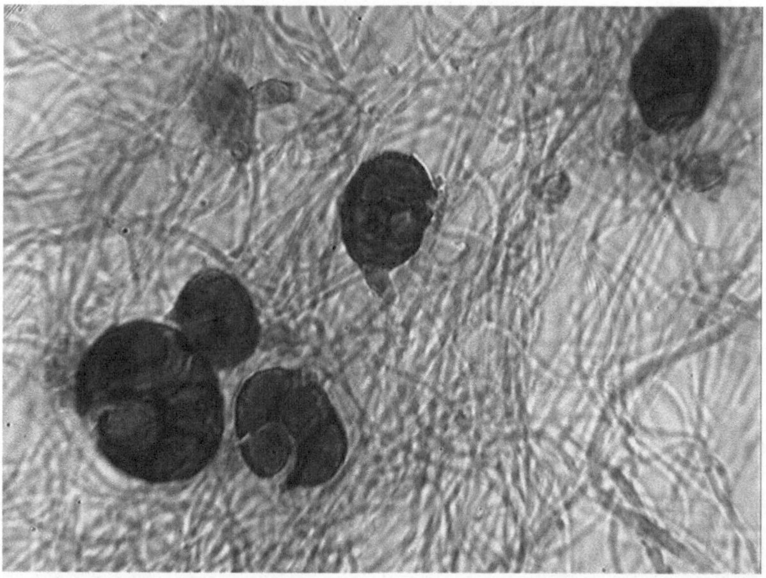

FIGURE 1.12 Conidia of *Pithomyces chartarum*.

and petiole segments in the rainy season only. *Alternaria alternata* exhibited a high colonization frequency for the stem in monsoon, while during winter and summer its colonization frequency was higher in the petiole (Table 1.2).

1.3.2.3 Diversity of Endophytic Fungi from *Cissus quadrangularis* L.

Species abundance and occurrence as well as percentage of colonization frequency were decreased from monsoon to summer. Some species like *Arthrinium hydei*, *Arthrinium phaeospermum*,

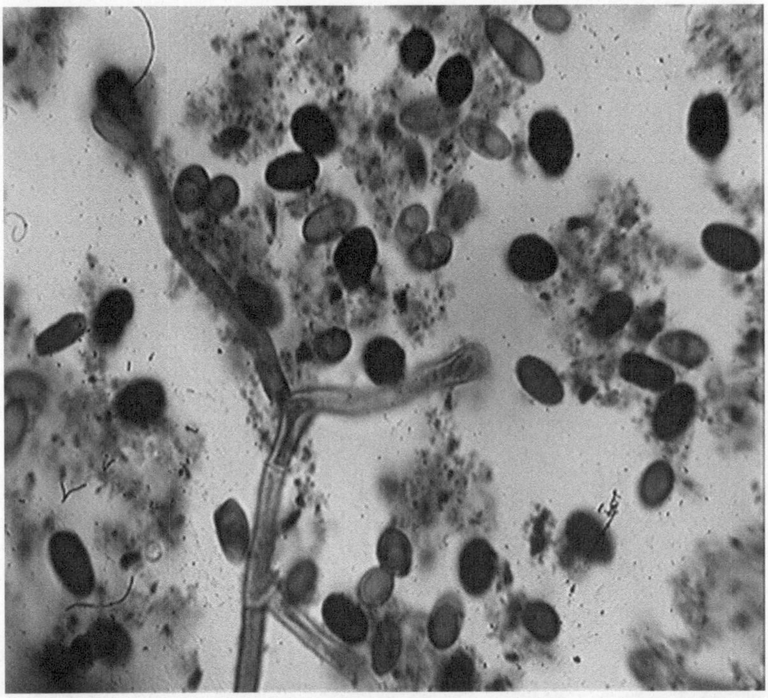

FIGURE 1.13 Conidia of *Stachybotrys chartarum*.

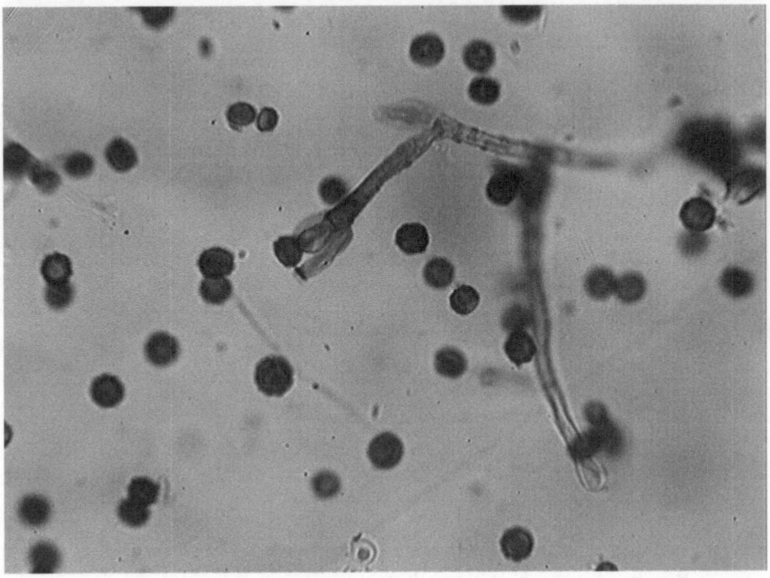

FIGURE 1.14 Conidia of *Stachybotrys chartarum*.

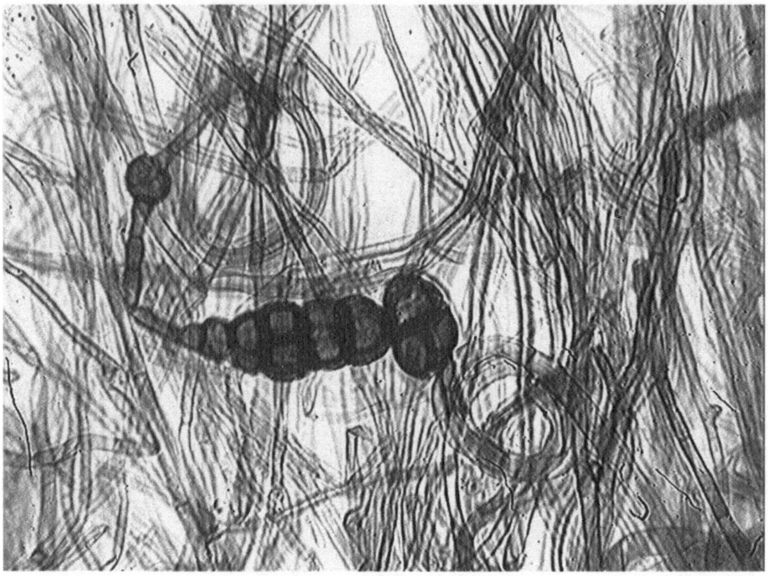

FIGURE 1.15 Mycelium with conidia of *Trimatrostroma hughesii*.

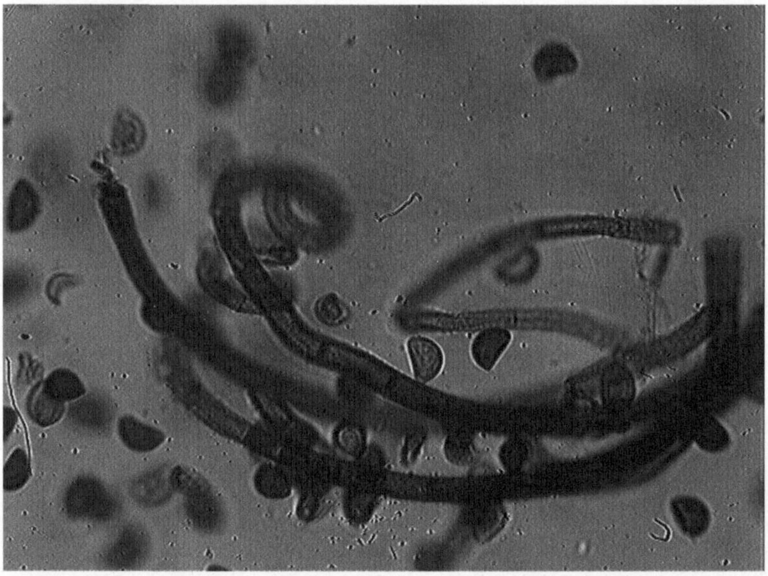

FIGURE 1.16 Mycelium with conidia of *Virgaria nigra*.

Colletotrichum gloeosporiodes, *Epicoccum nigrum* and *Stachybotrys nilgirica* may require high humidity to colonize this host, which is why they were present only in monsoon season, whereas *Acremonium fusidioide* requires low humidity for colonization, which is why it was observed in summer only (Table 1.3).

1.3.2.4 Diversity of Endophytic Fungi from *Coffea arabica* L.

Fourteen endophytes were recognized, out of which, 11 fungi were found during the rainy season with maximum colonizing frequency, while only five fungi were recovered in summer with a

TABLE 1.1
Colonization Frequency (%) of Endophytic Fungi Isolated from *Adathoda vasica*

		Colonization frequency (%) of endophytic fungi isolated from *Adathoda vasica* in three different seasons								
		Monsoon			Winter			Summer		
S. no.	Endophytes	Stem	Leaf	Petiole	Stem	Leaf	Petiole	Stem	Leaf	Petiole
1.	*Alternaria alternata*	–	–	–	–	–	–	7.78	9.70	3.58
2.	*Arthrinium hydei*	19.75	35.29	51.61	–	–	–	–	–	–
3.	*Arthrinium phaeospermum*	25.98	35.59	30.09	39.45	32.65	21.21	–	–	–
4.	*Aspergillus nidulans*	–	–	–	–	–	–	8.90	10.62	7.12
5.	*Aspergillus stellatus*	14.81	23.53	18.39	38.28	24.89	27.45	5.55	4.98	8.04
6.	*Cladosporium cladosporioides*	–	–	–	25.97	36.66	30.84	4.50	8.23	6.96
7.	*Colletotrichum gloeosporiodes*	–	–	–	18.88	22.90	28.47	–	–	–
8.	*Curvularia lunata*	–	–	–	–	–	–	9.73	5.82	6.96
9.	*Diplodia mutila*	–	–	–	22.49	26.80	18.61	–	–	–
10.	*Fusarium oxysporum*	16.44	21.88	28.31	–	–	–	10.25	8.44	5.89
11.	*Penicillium chrysogenum*	–	–	–	–	–	–	10.20	2.66	4.54
12.	*Phoma crysanthemicola*	16.25	19.99	17.42	–	–	–	–	–	–
13.	*Pithomyces chartarum*	–	–	–	14.78	19.45	23.40	–	–	–
14.	*Stachybotrys chartarum*	20.80	16.35	24.66	–	–	–	–	–	–
15.	*Stachybotrys nilgirica*	26.09	30.22	18.74	–	–	–	–	–	–
16.	*Xylaria sp.*	22.72	19.33	40.32	40.75	50.44	32.07	–	–	–

lower colonizing frequency. Few endophytes such as *Stachybotrys chartarum, S. nilgirica* and *Trimmatostroma hughesii* also present with a higher colonization rate in monsoon season but they were not isolated during the winter and summer seasons (Table 1.4).

1.3.2.5 Diversity of Endophytic Fungi from *Datura metal* Linn.

The rate of colonization was higher in monsoon season, while it was lower during summer for all the recovered fungal species. Winter isolates showed average colonization frequency. Some isolates, i.e., *Aspergillus nidulans* and *A. stellatus*, may like a low moisture in the atmosphere since they were observed only in summer. Tissue specificity and colonization frequency were also shown by a few endophytes, such as *Alternaria alternata*, in monsoon and winter when it was present in the leaf portion, whereas in summer it was found in the petiole with a higher colonization rate (Table 1.5).

TABLE 1.2
Colonization Frequency (%) of Endophytic Fungi Isolated from *Calotropis procera*

		Colonization frequency (%) of endophytic fungi isolated from *Calotropis procera* in three different seasons								
		Monsoon			Winter			Summer		
S. no.	Endophytes	Stem	Leaf	Petiole	Stem	Leaf	Petiole	Stem	Leaf	Petiole
1.	*Alternaria alternata*	30.66	19.44	25.14	16.46	19.50	22.88	6.57	9.82	10.21
2.	*Arthrinium hydei*	15.31	16.97	12.35	_	_	_	_	_	_
3.	*Arthrinium phaeospermum*	16.47	23.21	13.95	_	_	_	_	_	_
4.	*Aspergillus stellatus*	_	_	_	_	_	_	3.84	6.32	7.56
5.	*Cladosporium cladosporioides*	_	_	_	_	_	_	_	6.20	5.31
6.	*Curvularia lunata*	_	17.94	18.15	_	29.46	20.94	_	7.55	3.56
7.	*Diplodia mutila*				22.88	20.87	14.37	_	_	_
8.	*Epicoccum nigrum*	_	_	_	27.66	21.94	19.06	_	_	_
9.	*Fusarium oxysporum*	_	_	_	_	31.64	18.49	_	2.07	5.94
10.	*Nigrospora oryzae*	21.80	26.47	29.24	_	_	_	_	_	_
11.	*Penicillium chrysogenum*	_	29.35	25.44	_	13.23	18.40	_	9.21	6.74
12.	*Stachybotrys chartarum*	21.60	19.44	31.52	_	_	_	_	_	_
13.	*Stachybotrys nilgirica*	_	26.30	24.31	_	_	_	_	_	_
14.	*Trimmatostroma hughesii*	_	26.88	17.55	_	_	_	_	_	_

TABLE 1.3
Colonization Frequency (%) of Endophytic Fungi Isolated from *Cissus quadrangularis*

		Colonization frequency (%) of endophytic fungi isolated from *Cissus quadrangularis* in different seasons								
		Monsoon			Winter			Summer		
S. no.	Endophytes	Stem	Leaf	Petiole	Stem	Leaf	Petiole	Stem	Leaf	Petiole
1.	*Acremonium fusidioide*	_	_	_	_	_	_	5.06	8.56	10.08
2.	*Alternaria alternata*	16.98	21.88	26.22	17.22	20.30	23.77	3.97	5.57	4.21
3.	*Arthrinium hydei*	25.44	18.62	16.75	_	_	_	_	_	_
4.	*Arthrinium phaeospermum*	26.90	21.61	18.42	_	_	_	_	_	_
5.	*Aspergillus stellatus*	_	_	_	19.23	31.45	27.64	_	_	_
6.	*Cladosporium cladosporioides*	_	_	_	15.77	14.65	10.66	_	_	_
7.	*Colletotrichum gloeosporiodes*	30.16	19.13	29.86	_	_	_	_	_	_
8.	*Diplodia mutila*	_	_	_	16.02	19.34	21.07	_	_	_

TABLE 1.3 (Continued)
Colonization Frequency (%) of Endophytic Fungi Isolated from *Cissus quadrangularis*

S. no.	Endophytes	Colonization frequency (%) of endophytic fungi isolated from *Cissus quadrangularis* in different seasons								
		Monsoon			Winter			Summer		
		Stem	Leaf	Petiole	Stem	Leaf	Petiole	Stem	Leaf	Petiole
9.	*Epicoccum nigrum*	23.44	29.03	17.66	_	_	_	_	_	_
10.	*Fusarium oxysporum*	12.55	17.35	19.46	_	_	_	4.07	6.42	3.37
11.	*Nigrospora oryzae*	_	_	_	14.56	19.33	22.10	_	_	_
12.	*Penicillium chrysogenum*	_	_	_	22.16	26.48	34.10	2.54	4.39	5.33
13.	*Stachybotrys chartarum*	32.45	26.27	21.44	25.34	18.99	27.65	_	_	_
14.	*Stachybotrys nilgirica*	25.72	18.20	21.52	_	_	_	_	_	_
15.	*Trimmatostroma hughesii*	12.89	16.06	21.04	22.89	26.06	21.04	_	_	_
16.	*Virgaria nigra*	22.17	15.77	19.33	34.70	22.10	26.88	_	_	_

TABLE 1.4
Colonization Frequency (%) of Endophytic Fungi Isolated from *Coffea arabica*

S. no.	Endophytes	Colonization frequency (%) of endophytic fungi isolated from *Coffea arabica* in three different seasons								
		Monsoon			Winter			Summer		
		Stem	Leaf	Petiole	Stem	Leaf	Petiole	Stem	Leaf	Petiole
1.	*Alternaria alternata*	12.65	17.57	22.30	_	_	_	2.06	8.34	4.51
2.	*Arthrinium hydei*	27.85	22.41	19.04	_	_	_	_	_	_
3.	*Aspergillus nidulans*	10.88	22.64	19.42	_	_	_	4.09	2.30	1.88
4.	*Cladosporium cladosporioides*	12.54	18.73	22.31	21.33	17.54	12.99	3.06	12.55	5.26
5.	*Curvularia lunata*	_	_	_	15.32	12.08	11.87	10.02	3.24	8.21
6.	*Diplodia mutila*	35.80	17.41	25.31	_	_	_	_	_	_
7.	*Epicoccum nigrum*	13.84	19.20	26.10	_	_	_	_	_	_
8.	*Nigrospora oryzae*	29.00	21.44	18.77	22.51	18.06	16.24	_	_	_
9.	*Penicillium chrysogenum*	_	_	_	_	_	_	6.24	9.16	8.42
10.	*Phoma crysanthemicola*	22.05	18.33	24.88	16.77	18.73	38.33	_	_	_
11.	*Pithomyces chartarum*	_	_	_	23.21	20.15	29.16	_	_	_
12.	*Stachybotrys chartarum*	20.90	21.14	15.21	_	_	_	_	_	_
13.	*Stachybotrys nilgirica*	30.69	26.11	16.89	_	_	_	_	_	_
14.	*Trimmatostroma hughesii*	40.92	32.04	29.44	_	_	_	_	_	_

TABLE 1.5
Colonization Frequency (%) of Endophytic Fungi Isolated from *Datura metal*

		Colonization frequency (%) of endophytic fungi isolated from *Datura metal* in three different seasons								
		Monsoon			Winter			Summer		
S. no.	Endophytes	Stem	Leaf	Petiole	Stem	Leaf	Petiole	Stem	Leaf	Petiole
1.	*Alternaria alternata*	15.44	25.88	19.77	20.55	22.75	18.99	6.02	4.64	8.90
2.	*Aspergillus nidulans*	–	–	–	–	–	–	6.07	8.64	5.48
3.	*Aspergillus stellatus*	–	–	–	–	–	–	3.44	2.07	9.01
4.	*Cladosporium cladosporioides*	25.09	18.45	21.16	19.66	16.78	14.04	5.20	3.06	4.88
5.	*Colletotrichum gloeosporiodes*	36.10	27.38	21.00	–	–	–	–	–	–
6.	*Diplodia mutila*	16.33	33.24	12.97	–	–	–	–	–	–
7.	*Penicillium chrysogenum*	–	–	–	27.23	13.06	21.63	–	–	–
8.	*Phoma crysanthemicola*	33.20	20.48	17.41	–	–	–	–	–	–
9.	*Pithomyces chartarum*	41.09	28.31	34.11	–	–	–	–	–	–
10.	*Stachybotrys chartarum*	16.21	17.20	22.91	–	–	–	–	–	–
11.	*Stachybotrys nilgirica*	15.38	19.07	20.40	–	–	–	–	–	–
12.	*Trimmatostroma hughesii*	19.33	26.34	22.07	–	–	–	–	–	–
13.	*Xylaria* sp.	–	–	–	21.99	27.39	23.57	–	–	–

TABLE 1.6
Colonization frequency (%) of Endophytic Fungi Isolated from *Dioscorea bulbifera*

		Colonization frequency (%) of endophytic fungi isolated from *Dioscorea bulbifera* in three different seasons								
		Monsoon			Winter			Summer		
S. no.	Endophytes	Stem	Leaf	Petiole	Stem	Leaf	Petiole	Stem	Leaf	Petiole
1.	*Alternaria alternata*	23.41	14.45	19.01	–	–	–	3.86	4.45	8.12
2.	*Arthrinium hydei*	26.40	29.16	22.43	–	–	–	–	–	–
3.	*Arthrinium phaeospermum*	33.45	25.62	19.34	–	–	–	–	–	–
4.	*Aspergillus nidulans*	–	–	–	19.32	18.23	14.09	6.24	5.46	2.31
5.	*Aspergillus stellatus*	–	–	–	–	–	–	10.25	6.48	4.33
6.	*Colletotrichum gloeosporiodes*	22.90	20.13	15.56	–	–	–	–	–	–
7.	*Curvularia lunata*	–	–	–	12.88	16.24	18.50	5.79	2.83	7.36
8.	*Diplodia mutila*	37.33	29.35	24.48	–	–	–	–	–	–
9.	*Penicillium chrysogenum*	–	–	–	–	–	–	7.72	7.28	10.52
10.	*Phoma crysanthemicola*	17.64	22.24	18.18	16.94	27.08	26.66	–	–	–
11.	*Pithomyces chartarum*	–	–	–	12.74	17.65	21.28	–	–	–
12.	*Trimmatostroma hughesii*	–	–	–	24.32	19.15	14.46	2.30	1.98	6.54
13.	*Xylaria* sp.	–	–	–	25.82	35.42	35.25	5.04	8.64	3.22

1.3.2.6 Diversity of Endophytic Fungi from *Dioscorea bulbifera* L.

Out of 13 isolates, four isolates, i.e. *Arthrinium hydei*, *Arthrinium phaeospermum*, *Colletotrichum gloeosporiodes* and *Diplodia mutila*, were identified only in the monsoon season, while *Aspergillus stellatus* and *Penicillium chrysogenum* were detected only during the summer with a very low

TABLE 1.7
Colonization Frequency (%) of Endophytic Fungi Isolated from *Helecteres isora*

		Colonization frequency (%) of endophytic fungi isolated from *Helecteres isora* in three different season								
		Monsoon			Winter			Summer		
S. no.	Endophytes	Stem	Leaf	Petiole	Stem	Leaf	Petiole	Stem	Leaf	Petiole
1.	*Alternaria alternata*	12.88	26.66	18.22	25.13	28.45	14.23	2.30	10.01	6.54
2.	*Arthrinium phaeospermum*	_	_	_	_	_	_	4.66	2.31	3.82
3.	*Aspergillus nidulans*	_	_	_	_	_	_	1.65	5.45	8.04
4.	*Aspergillus stellatus*	_	_	_	_	_	_	2.08	4.64	6.01
5.	*Cladosporium cladosporioides*	20.55	24.94	12.01	18.33	14.22	22.33	6.21	8.31	4.51
6.	*Colletotrichum gloeosporiodes*	_	_	_	19.74	22.22	16.84	_	_	_
7.	*Curvularia lunata*	29.40	22.30	18.35	17.55	19.79	21.66	2.05	6.08	5.41
8.	*Epicoccum nigrum*	15.38	21.05	26.01	_	_	_	_	_	_
9.	*Fusarium oxysporum*	_	_	_	23.53	19.20	15.84	_	_	_
10.	*Nigrospora oryzae*	14.50	19.64	22.54	13.59	20.35	18.54	_	_	_
11.	*Penicillium chrysogenum*	_	_	_	_	_	_	5.32	2.06	4.33
12.	*Pestalotiopsis funereal*	_	_	_	21.04	27.45	22.77	_	_	_
13.	*Phoma crysanthemicola*	_	_	_	14.00	15.65	10.85	3.40	5.80	6.30
14.	*Virgaria nigra*	16.80	21.97	25.05	22.01	15.55	21.47	_	_	_
15.	*Xylaria* sp.	_	_	_	16.32	17.02	19.35	2.88	7.81	5.10

colonization frequency, which may due to decreased atmospheric moisture during this season (Table 1.6).

1.3.2.7 Diversity of Endophytic Fungi from *Helecteres isora* L.

The host plant was more susceptible for fungal endophytes in winter as compared to other seasons. A total of 15 fungi were isolated from the plant, out of which ten fungi occurred in the winter season. Species diversity was highest during winter. Leaf tissue was more susceptible to *Alternaria alternata* in each season (Table 1.7).

1.3.2.8 Diversity of Endophytic Fungi from *Hemidesmus indicus* R. Br.

All the endophytic fungi obtained from the present study colonized the stem, leaf and petiole segments with seasonal variations. Among the identified fungi, *Cladosporium cladosporioides*, *Fusarium oxysporum* and *Penicillium chrysogenum* were recovered during all the seasons, while *Acremonium fusidioide*, *Colletotrichum gloeosporiodes*, *Nigrospora oryzae*, *Pestalotiopsis funereal*, *Phoma crysanthemicola* and *Trimmatostroma hughesii* were observed only in the monsoon season (Table 1.8).

1.3.2.9 Diversity of Endophytic Fungi from *Murrayakoenigii* Linn. Spreng

Most of the fungi showed a greater colonization frequency for the monsoon period and it was much lower during the summer. *Arthrinium hydei*, *A. phaeospermum*, *Diplodia mutila*, *Epicoccum nigrum*, *Pestalotiopsis funereal* and *Pithomyces chartarum* were the fungi which required high humidity to colonize the host, which is why they showed their presence only in the monsoon season. Fungi like *Cladosporium cladosporioides* and *Fusarium oxysporum* were observed in all the seasons (Table 1.9).

TABLE 1.8
Colonization Frequency (%) of Endophytic Fungi Isolated from *Hemidesmus indicus*

		Colonization frequency (%) of endophytic fungi isolated from *Hemidesmus indicus* in three different seasons								
		Monsoon			Winter			Summer		
S. no.	Endophytes	Stem	Leaf	Petiole	Stem	Leaf	Petiole	Stem	Leaf	Petiole
1.	*Acremonium fusidioide*	13.48	26.12	17.08	_	_	_	_	_	_
2.	*Cladosporium cladosporioides*	19.66	13.84	17.91	16.98	14.77	25.20	3.64	1.89	2.57
3.	*Colletotrichum gloeosporiodes*	19.48	26.33	31.93	_	_	_	_	_	_
4.	*Curvularia lunata*	_	_	_	21.55	16.82	24.66	4.20	7.23	5.55
5.	*Fusarium oxysporum*	26.87	18.34	14.94	22.66	29.84	17.56	5.10	3.88	6.77
6.	*Nigrospora oryzae*	16.21	19.06	21.07	_	_	_	_	_	_
7.	*Penicillium chrysogenum*	20.45	19.55	22.33	17.80	18.94	22.17	10.30	8.01	5.31
8.	*Pestalotiopsis funereal*	29.44	21.08	24.06	_	_	_	_	_	_
9.	*Phoma crysanthemicola*	17.05	14.50	21.22	_	_	_	_	_	_
10.	*Trimmatostroma hughesii*	27.60	21.80	24.30	_	_	_	_	_	_

TABLE 1.9
Colonization Frequency (%) of Endophytic Fungi Isolated from *Murraya koenigi*

		Colonization frequency (%) of endophytic fungi isolated from *Murraya koenigi* in three different seasons								
		Monsoon			Winter			Summer		
S. no.	Endophytes	Stem	Leaf	Petiole	Stem	Leaf	Petiole	Stem	Leaf	Petiole
1.	*Alternaria alternata*	32.84	27.21	21.90	_	_	_	8.06	5.48	9.21
2.	*Arthrinium hydei*	21.55	39.55	26.95	_	_	_	_	_	_
3.	*Arthrinium phaeospermum*	19.10	22.88	16.67	_	_	_	_	_	_
4.	*Aspergillus nidulans*	_	_	_	13.23	15.05	18.91	2.05	4.94	6.14
5.	*Aspergillus stellatus*	_	_	_	_	_	_	10.20	4.65	7.25
6.	*Cladosporium cladosporioides*	19.22	23.14	16.22	12.56	23.66	17.11	3.33	4.21	5.77
7.	*Colletotrichum gloeosporiodes*	_	_	_	18.44	30.74	21.98	_	_	_
8.	*Diplodia mutila*	14.37	23.54	16.34	_	_	_	_	_	_
9.	*Epicoccum nigrum*	14.52	19.34	22.46	_	_	_	_	_	_
10.	*Fusarium oxysporum*	25.64	28.42	19.35	10.29	17.02	11.03	7.88	4.66	8.45
11.	*Penicillium chrysogenum*	_	_	_	12.75	17.56	19.24	5.46	7.72	9.55
12.	*Pestalotiopsis funereal*	19.22	23.45	16.36	_	_	_	_	_	_
13.	*Pithomyces chartarum*	17.64	33.20	23.34	_	_	_	_	_	_
14.	*Stachybotrys chartarum*	19.00	25.30	21.66	15.61	22.32	14.23	_	_	_
15.	*Stachybotrys nilgirica*	18.48	22.33	14.22	16.44	26.46	21.32	_	_	_
16.	*Xylaria sp.*	17.94	26.32	22.15	10.09	6.77	4.15	_	_	_

1.3.2.10 Diversity of Endophytic Fungi from *Ocimum sanctum* L.

Fusarium oxysporum was a common species which was found in all the seasons. The greatest abundance of endophytic species was found in the monsoon period while a very low number of fungi was recovered during the temperate season. A similar colonization frequency was revealed by *Colletotrichum gloeosporiodes* in leaf, stem and petiole tissues during the monsoon, while in winter a higher frequency of colonization was observed only for the stem (Table 1.10).

1.3.2.11 Diversity of Endophytic Fungi from *Vitex negundo* L.

A total of 13 fungal isolates were identified, and the most commonly occurring species were *Curvularia lunata* and *Fusarium oxysporum* in all the seasons. Among all the obtained fungi, *Acremonium fusidioide*, *Colletotrichum gloeosporiodes*, *Nigrospora oryzae*, *Phoma crysanthemicola* and *Xylaria* sp. needed a high moisture content in the atmosphere because they were recovered only in the monsoon season (Table 1.11).

1.3.2.12 Diversity of Endophytic Fungi from *Withania somnifera* Linn.

A few endophytes such as *Alternaria alternata*, *Fusarium oxysporum* and *Penicillium chrysogenum* colonized only leaf and petiolar tissues throughout all the seasons. The colonizing frequency of these fungi was also greater than for other isolates. Some isolates like *Colletotrichum gloeosporiodes*, *Stachybotrys chartarum*, *Stachybotrys nilgirica* and *Trimmatostroma hughesii* revealed remarkable specialization by occurring only in lower temperature periods (Table 1.12).

TABLE 1.10
Colonization Frequency (%) of Endophytic Fungi Isolated from *Ocimum sanctum*

		Colonization frequency (%) of endophytic fungi isolated from *Ocimum sanctum* in three different seasons								
		Monsoon			Winter			Summer		
S. no.	Endophytes	Stem	Leaf	Petiole	Stem	Leaf	Petiole	Stem	Leaf	Petiole
1.	*Acremonium fusidioide*	22.50	11.80	16.66	_	_	_	_	_	_
2.	*Colletotrichum gloeosporiodes*	26.21	28.63	29.32	38.42	26.43	22.09	_	_	_
3.	*Curvularia lunata*	16.89	10.57	20.90	_	_	_	2.04	6.44	8.01
4.	*Diplodia mutila*	_	_	_	29.35	17.68	11.88	_	_	_
5.	*Epicoccum nigrum*	19.02	16.80	10.93	_	_	_	_	_	_
6.	*Fusarium oxysporum*	19.21	18.43	12.33	12.00	15.62	18.20	6.32	1.55	4.21
7.	*Pestalotiopsis funereal*	22.54	26.73	18.22	_	_	_	_	_	_
8.	*Phoma crysanthemicola*	23.94	15.31	17.64	_	_	_	_	_	_
9.	*Pithomyces chartarum*	26.32	21.20	18.64	_	_	_	_	_	_
10.	*Stachybotrys chartarum*	_	_	_	_	_	_	_	_	_
11.	*Stachybotrys nilgirica*	14.56	17.20	15.92	_	_	_	_	_	_
12.	*Trimmatostroma hughesii*	_	_	_	18.65	14.32	12.80	_	_	_

TABLE 1.11
Colonization Frequency (%) of Endophytic Fungi Isolated from *Vitex negundo*

S. no.	Endophytes	Colonization frequency (%) of endophytic fungi isolated from *Vitex negundo* in three different seasons								
		Monsoon			Winter			Summer		
		Stem	Leaf	Petiole	Stem	Leaf	Petiole	Stem	Leaf	Petiole
1.	*Acremonium fusidioide*	35.61	21.90	26.38	–	–	–	–	–	–
2.	*Aspergillus nidulans*	–	–	–	–	–	–	5.71	9.05	7.55
3.	*Aspergillus stellatus*	–	–	–	–	–	–	4.60	9.88	8.44
4.	*Cladosporium cladosporioides*	30.92	23.92	27.81	10.48	21.50	12.87	2.55	1.78	6.11
5.	*Colletotrichum gloeosporiodes*	21.65	16.33	13.93	–	–	–	–	–	–
6.	*Curvularia lunata*	20.92	14.66	12.84	40.26	26.57	33.80	9.63	7.70	5.88
7.	*Fusarium oxysporum*	22.11	29.34	18.55	35.88	28.45	24.66	10.36	4.80	8.02
8.	*Nigrospora oryzae*	16.80	14.20	18.19	–	–	–	–	–	–
9.	*Penicillium chrysogenum*	29.36	21.92	19.64	19.36	22.92	29.64	8.67	5.32	6.24
10.	*Phoma crysanthemicola*	16.45	14.77	17.11	–	–	–	–	–	–
11.	*Pithomyces chartarum*	26.14	22.44	31.48	19.64	14.20	16.24	–	–	–
12.	*Xylaria* sp.	29.04	17.16	13.04	–	–	–	–	–	–

TABLE 1.12
Colonization Frequency (%) of Endophytic Fungi Isolated from *Withania somnifera*

S. no.	Endophytes	Colonization frequency (%) of endophytic fungi isolated from *Withania somnifera* in three different seasons								
		Monsoon			Winter			Summer		
		Stem	Leaf	Petiole	Stem	Leaf	Petiole	Stem	Leaf	Petiole
1.	*Acremonium fusidioide*	21.90	16.20	27.92	–	–	–	–	–	–
2.	*Alternaria alternata*	–	15.66	22.06	–	16.22	22.08	–	2.83	5.65
3.	*Aspergillus nidulans*	–	–	–	–	–	–	1.80	8.12	4.24
4.	*Aspergillus stellatus*	–	–	–	–	–	–	4.20	5.92	7.72
5.	*Colletotrichum gloeosporiodes*	–	–	–	22.04	16.31	25.06	–	–	–
6.	*Curvularia lunata*	–	–	–	16.86	21.52	42.42	4.33	2.06	3.15
7.	*Diplodia mutila*	22.52	19.33	16.22	–	–	–	12.64	9.21	8.30
8.	*Epicoccum nigrum*	30.69	18.56	14.23	–	–	–	–	–	–
9.	*Fusarium oxysporum*	–	17.58	21.54	–	16.55	33.83	–	7.20	6.33
10.	*Penicillium chrysogenum*	–	28.33	32.47	–	25.21	14.77	–	6.23	2.47
11.	*Stachybotrys chartarum*	–	–	–	22.30	28.77	25.34	–	–	–
12.	*Stachybotrys nilgirica*	–	–	–	18.02	14.66	21.06	–	–	–
13.	*Trimmatostroma hughesii*	–	–	–	14.26	18.24	21.94	–	–	–

1.4 DISCUSSION

The diversity of endophytic fungi was studied in 12 medicinal plants collected from different localities of Melghat forest in Maharashtra, India. Some endophytes prefer growing in winter, i.e., at low temperature and moderate to high humidity, while some relish growing in less humidity and higher temperature. This indicates that colonization of endophytic fungi in hosts depends on several environmental factors such as temperature, rainfall and humidity, which make them sensitive to the host response. Due to their sensitive nature, endophytic fungi adjust to the responses or reactions of hosts. This sensitive nature of fungi is a result of the survival urge. Such survival urges finally lead to diversified survival attitudes of fungi and thus fungi exhibit diversity of extreme propensity.

The species of Hyphomycetes like *Alternaria*, *Curvularia*, *Cladosporium*, *Fusarium* and *Penicillium* were commonly reported as endophytes in various previously studied plants (Kharwar et al., 2008; Huang et al., 2009; Suryanarayanan et al., 2002; Tejesvi et al., 2006). Also, species of Coelomycetes such as *Colletotrichum*, *Pestalotiopsis* and *Phoma* were the most commonly studied endophytes from a wide variety of host plants (Suryanarayanan & Vijaykrishna, 2001; Brown et al., 1998; Gond et al., 2007; Ragazzi et al., 2001). In the present study, *A. alternata*, *C. lunata*, *C. cladosporioides*, *F. oxysporum*, *P. crysoginum*, *P. funereal* and *P. crysanthimicola* were isolated from all the host plants. This result suggested that few fungal genera are highly modified to being endophytes.

A literature survey suggested that rainfall plays an important role in the dispersion of waterborne and endophytic fungal spores which become inoculi to colonize the host tissues (Suryanarayanan et al., 1998; Wang & Guo, 2007). Humidity and rainfall increase the penetration and germination of fungal spores in the host plants, which influence the diversity of endophytic fungi. In the present study, during the investigation of *H. isora*, it was observed that the fungal endophytes like *A. alternata*, *C. lunata*, *C. cladosporiodes*, *F. oxysporum* and *P. chrysogenum* were the most dominant fungi throughout the study (Table 1.7), which further showed a decline in their colonization frequency as rainfall decreased from monsoon to summer. Most of the fungi isolated from *H. indicus* dominated and exhibited a higher colonization frequency during monsoon and winter, while some isolates like *A. fusidioide*, *A. hydei*, *A. phaeospermum*, *C. gloeosporiodes*, *P. crysanthemicola* and *T. hughesii* were isolated only in the monsoon season (Table 1.8). The results showed similarities with those obtained by Suryanarayanan et al. (2002). The colonization frequency among endophytic fungi of *M. koenigii* was significantly diverse between seasons. This host was more vulnerable to endophytes during the rainy season as compared to other seasons, which may have been due to the higher fungal population density during this period (Table 1.9).

The tissue specificity of endophytes has been studied by earlier workers for grasses (Clay, 1990), orchids (Bayman et al., 1997) and tree species (Arnold et al., 2001). During the present study of *W. somnifera*, tissue specificity was observed. The fungal endophytes like *A. alternata*, *F. oxysporum* and *P. chrysogenum* were restricted to stem tissues, and they were not found in the leaf and petiole tissues throughout all the seasons of the study (Table 1.12).

Among various endophytic fungi isolated from *C. procera* throughout the study, fungal isolates like *C. lunata*, *F. oxysporum* and *S. nilgirica* showed tissue specificity in their colonization (Table 1.3). These fungi were observed in leaf and petiole tissues only. These observations are in accordance with the results of Aharwal et al. (2014).

Variations in colonization frequencies in plant tissues have been observed which may be due to utilization of metabolites by endophytic fungi (Suryanarayanan & Vijaykrishna, 2001). Petrini and Fisher (1988) investigated *Pinus sylvestris* and observed higher colonization rates in stem tissues than xylem. These observations showed that colonization of endophytic fungi varies in different tissues. The occurrence of endophytic fungi seems to be influenced by environment, types of host tissue and seasonal variation (Rodrigues, 1994; Halmschlager et al., 1993). In the host *Dioscorea bulbifera*, the leaf and petiole harbored the greatest number of endophytic fungi with a high colonization frequency, which may be due to the large surface area of leaves exposed to the environment and the presence of stomata which provide an available passage for the entry of the fungi (Khan et al., 2007; Rajagopal et al., 2010; Petrini, 1991). Among all the isolates, most of the fungi showed a higher colonization frequency in leaf and petiole throughout the investigations (Table 1.6).

During the study of *A. vasica* it was found that there were great variations in isolation and colonization of endophytic fungi. Among the isolates, two species of *Stachybotrys*, i.e., *S. chartarum* and

S. nilgirica, were observed in monsoon and winter seasons and disappeared in summer (Table 1.1). This indicates that humidity plays an important role in the colonization of endophytic fungi.

Humidity may play an important role in increasing the endophytic infection rate. Penetration and germination of adherent fungal spores into the host tissues may increase due to atmospheric humidity which ultimately increases the colonization rates of endophytes (Wilson, 2000; Tejesvi et al., 2005; Suryanarayanan et al., 1998). The plants collected during the monsoon season were observed to harbor more endophytic fungi than the samples collected during the dry season (Suryanarayanan et al., 2002; Rodrigues, 1994). During the present study, the colonization rate and diversity of endophytic fungi in *D. metal* were also affected by temperature. They revealed a higher colonization frequency and diversity in humid conditions which decreased when humidity decreased. From all the isolated fungi of *V. negundo*, fungi such as *A. phaeospermum*, *A. hydei*, *C. gloeosporiodes* and *Xylaria* sp. may require more humidity to colonize the host (Table 1.5).

1.5 CONCLUSION

The morphotaxonomical study of endophytic fungi is essential to establish their diversity. During the present investigation, anamorphic (Hyphomycetes and Coelomycetes) fungi were predominantly found. The most dominant species were *A. alternata*, *A. nidulance*, *A. stelatus*, *C. cladosporiodes*, *C. lunata*, *F. oxysporum* and *P. chrysogenum*.

Association of endophytic fungi with the host is affected by many environmental factors such as temperature, humidity, rainfall and forest type. In the present work, variations were observed in the colonization of endophytes during different seasons. Endophytes showed a higher colonization frequency in the monsoon season, as well as in winter, as compared with summer. Conclusively, it is suggested that environmental factors play an important role in the establishment of endophytic fungal diversity.

Seasonal variations may cause tissue specificity in the colonization of endophytic fungi. Some species such as *P. chartarum*, *S. nilgirica* and *T. hughesii* isolated from *C. procera* revealed tissue specificity in their colonization. These isolates colonized only the leaf and petiole tissues of their host plant.

REFERENCES

Aharwal RP, Kumar S, Sandhu SS. 2014. Isolation and antibacterial property of endophytic fungi isolated from indian medicinal plant calotropis procera (linn.) r.br. *World J. Pharm. Pharm. Sci.* 3(5):678–691.

Arnold AE, Maynard Z, Gilbert GS. 2001. Fungal endophytes in dicotyledonous neotropical trees: patterns of abundance and diversity. *Mycol. Res.* 105(12):1502–1507.

Barnett HL, Hunter BB. 1972. *Illustrated Genera of Imperfect Fungi*. Burgess Publishing Company. pp. 240.

Bayman P, Lebrón LL, Tremblay RL, Lodge DJ. 1997. Variation in endophytic fungi from roots and leaves of *Lepanthes* (Orchidaceae). *New Phytol.* 135(1):143–149.

Brown KB, Hyde KD, Guest DI. 1998. Preliminary studies on endophytic fungal communities of *Musa acuminata* species complex in Hong Kong and Australia. *Fungal Divers.* 1:27–51.

Cannon PF, Simmons CM. 2002. Diversity and host preference of leaf endophytic fungi in the Iwokrama Forest Reserve, Guyana. *Mycologia.* 94:210–220.

Clay K. 1990. Fungal Endophytes of grasses. *Annu. Rev. Ecol. Systemat.* 21:275–298.

Ellis MB. 1971. *Dematiaceous hyphomycetes*. Commonwealth Mycological Institute, Kew, pp. 595.

Ellis MB. 1976. *More dematiaceous hyphomycetes*. Commonwealth Mycological Institute, Kew, pp. 494.

Gond SK, Verma VC, Kumar A, Kharwar RN. 2007. Study of endophytic fungal community from different parts of *Aegle marmelos* Correae (Rutaceae) from Varanasi (India). *World J. Microb. Biotechnol.* 23(10):1371–1375.

Hawksworth DL. 2001. The magnitude of fungal diversity: the 1.5 million species estimate revisited. *Mycol. Res.* 105:1422–1432.

Halmschlager E, Butin H, Donaubauer E. 1993. Endophytische Pilze in Bl ttern und Zweigen von *Quercus petraea*. *Eur. J. Pathol.* 23:51–63.

Huang WY, Cai YZ, Surveswaran S, Hyde KD, Corke H, Sun M. 2009. Molecular phylogenetic identification of endophytic fungi isolated from three *Artemisia* species. *Fungal Divers.* 36:69–88.

Karthikeyan S. 2009. Flowering plants of India in 19th and 21th centuries- A comparison *In Plant and Fungal Biodiversity & Bioprospecting. Edited by* S. Krishanan and D. J. Bhat, Goa University. pp. 19–30.

Khan R, Saleem SM, Choudary I, Khan SA, Ahmad A. 2007. Biodiversity of the endophytic fungi isolated from *Calotropis procera* (Ait) R.Br. *Pak. J. Bot.* 39(6):2233–2239.

Kharwar RN, Verma VC, Strobel G, Ezra D. 2008. The endophytic fungal complex of *Catharanthus roseus* (L.) G. *Don. Curr. Sci.* 95:228–233.

Kumaresan V, Suryanarayanan TS. 2001. Occurrence and distribution of endophytic fungi in a mangrove community. *Mycol. Res.* 105(11):1188–1193.

Manoharacharya C, Sridhar K, Singh R, Adholeya A, Suryanarayanan TS, Rawat S, Johri BS. 2005. Fungal biodiversity: Distribution, conservation and prospecting of fungi from India. *Curr. Sci.* 89:58–71.

Petrini O. 1991. Fungal endophytes of tree leaves. *In Microbial Ecology of leaves. Edited by* J. H. Andrews, and S. S. Monano, Springer-Verlag, NewYork, pp. 179–197.

Petrini O, Fisher PJ. 1988. A comparative study of fungal endophytes in xylem and whole stem of *Pinus sylvestris* and *Fagus sylvatica*. *Trans. Br. Mycol. Soc.* 91:233–238.

Photita W, Lumyong S, Lumyong S, Hyde KD. 2001. Endophytic fungi of wild banana (*Musa acuminata*) at Doi Suthep Pui National Park, Thailand. *Mycol. Res.* 105:1508–1513.

Ragazzi A, Moricca S, Capretti P, Dellavalle I, Mancini F, Turco E. 2001. Endophytic fungi in *Quercus cerris*: Isolation frequency in relation to phonological phase, tree health and the organ affected. *Phytopathol. Mediterr.* 40:165–171.

Rajagopal K, Kalavathy S, Kokila S, Karthikeyan S, Kathiravan G, Prasad R, Balasubraminan P. 2010. Diversity of fungal endophytes in few medicinal herbs of South India. *Asian J. Exp. Biol. Sci.* 1(2):415–418.

Rodrigues KF. 1994. The foliar fungal endophytes of the Amazonian palm *Euterpe oleracea. Mycologia.* 86:376–85.

Rodrigues KF, Samueles G J. 1999. Fungal endophytes of *Spondias mombin* leaves in Brazil. *J. Basic Microbiol.* 39:131–135.

Singh SK. 2012. Endophytic Fungi: Diversity. Ex situ Conservation and Biotech Potential. National Workshop on Taxonomy, Ex Situ Conservation and Bioprospecting of Fungi, Agarkar Research Institute, Pune. pp.46–51.

Subramanian CV. 1971. *Hyphomycetes*, Indian Council of Agricultural Research, New Delhi. pp. 930.

Suryanarayanan TS, Kumarsan V, Johnson JA. 1998. Foliar fungal endophytes from two species of the mangrove *Rhizophora*. *Can. J. Microbiol.* 44:1003–1006.

Suryanrayanan TS, Kumaresan V, Johnson JA. 2001. Fungal endophytes: The tropical dimension. *In Trichomycetes and Other Fungal Groups. Edited by* J K Misra and B W Horn. Science Publishers, Inc, Enfield, N. H. pp. 197–207.

Suryanarayanan TS, Murali TS, Venkatesan G. 2002. Ouccurance and distribution of fungal endophytes in tropical forests across a rain fall gradient. *Can. J. Bot.* 80:818–826.

Suryanarayanan TS, Vijaykrishna D. 2001. Fungal endophytes of aerial roots of *Ficus benghalensis. Fungal Divers.* 8:155–161.

Suryanarayanan TS, Venkatesan G, Murali TS. 2003. Endophytic fungal communities in leaves of tropical forest trees: diversity and distribution patterns. *Curr. Sci.* 85:489–493.

Sutton BC. 1980. *The Coelomycetes, Fungi Imperfecti with pycnidia Acervuli and Stromata*. Robert Mac Lechose and Co. Ltd. University of Glasgow, England. pp. 696.

Tejesvi MV, Mahesh B, Nalini MS, Prakash HS, Kini KR, Subbiah V, Shetty HS. 2005. Endophytic fungal assemblages from inner bark and twig of *Terminalia arjuna* W. & A. (Combretaceae). *World J. Microbiol. Biotechnol.* 21:1535–1540.

Tejesvi MV, Mahesh B, Nalini MS, Prakash HS, Kini KR, Subbiah V, Shetty HS. 2006. Fungal endophyte assemblages from ethnopharmaceutically important medicinal trees. *Can. J. Microbiol.* 52:427–435.

Wang Y, Guo LD. 2007. A comparative study of endophytic fungi in needles, bark, and xylem of *Pinus tabulaeformis*. *Can. J. Bot.* 85:911–917.

Wilson D. 2000. Ecology of woody plant endophytes. *In Microbial Endophytes Edited by* W. Bacon and I. F. White, Jr. Marcel Dekker, New York. pp. 389–420.

2 Diversity and Conservation Implications of Tropical Endophyte Fungi in Ecologically Vital Tropical Forests

Sabiha Naz and Seema Verma*
*Corresponding author.

2.1 INTRODUCTION

Tropical forests cover a significant portion of India's landmass and are characterized by high levels of species diversity and endemism. Despite their importance, there is limited understanding of the diversity and ecological functions of endophyte fungi in the context of Indian tropical forests. Tropical forests, renowned for their unparalleled biodiversity and ecological complexity, are home to a myriad of intricate symbiotic relationships that shape the dynamics of these ecosystems. Among the fascinating components of this intricate web of interactions are tropical endophyte fungi, a class of fungi that are associated with the roots of numerous plant species in mutualistic interactions also known as arbuscular mycorrhizal fungi (AMF); these organisms play a crucial role in nutrient cycling, plant health, and overall ecosystem functioning within the tropical forest biome. They coexist in mutually beneficial interactions with plants, stimulating growth and providing protection against external stresses. These fungi are essential for biodiversity and contribute to the overall health of tropical ecosystems (Sharma et al., 2015).

Endophyte fungi form a symbiotic association with the roots of most plants in tropical forests, creating a mycorrhizal network that enhances nutrient uptake for both the fungi and the host plants. The development of complex structures termed arbuscules inside plants root, which facilitate nutrient exchange, is a defining feature of this mutually beneficial interaction. In return for carbohydrates supplied by the host plant through photosynthesis, the tropical endophyte fungi contribute to the plant's nutrient acquisition, particularly phosphorus and nitrogen. The tropical environment, with its high temperatures, abundant rainfall, and diverse array of plant species, provides a unique setting for the activities of tropical endophyte fungi. These fungi have adapted to thrive in the complex and competitive conditions of tropical soils, forming associations with a wide range of plant species, from towering canopy trees to understory herbs. As such, they contribute significantly to the overall health and resilience of tropical forest ecosystems (Rodriguez et al., 2009).

This chapter describes the tropical endophytic fungi in tropical forests, exploring their ecological significance, the mechanisms of their symbiotic relationships, and their function in regulating the balance and sustainability of one of the Earth's most diverse and vital ecosystems. As we delve deeper into the complexities of these fungal–plant partnerships, we gain valuable insights into the

DOI: 10.1201/9781003486275-3

functioning of tropical forests and the delicate balance that sustains life within this biologically rich environment (Alam et al., 2021).

2.2 TROPICAL FUNGI: KEYSTONE ROLES AND ECOLOGICAL SIGNIFICANCE

Tropical endophyte fungi in tropical forests pursue a multitude of objectives that are integral to the ecological equilibrium and functioning of these ecosystems. The key goals are described in the following subsections.

2.2.1 Nutrient Cycling

They facilitate nutrient cycling by forming symbiotic relationships with plants, particularly trees. The fungi assist in the absorption of essential nutrients such as phosphorus and nitrogen, enhancing the overall nutrient availability in the ecosystem like *Amanita*, *Boletus*, *Cortinarius* and *Russula* (Huey et al., 2020).

2.2.2 Enhanced Plant Growth

The promote plant growth by aiding nutrient uptake. Many tropical endophyte fungi form mycorrhizal associations that help plant roots absorb more water and minerals. This helps plants grow and develop more quickly, e.g. *Genus rhizophagus* (Begum et al., 2019).

2.2.3 Biodiversity Support

They contribute to the diversity of plant species by forming specific associations with a wide range of host plants. This diversity is essential for the overall health and resilience of tropical forest ecosystems, e.g. the relationship between orchid mycorrhizal fungi and orchid plants in tropical forest ecosystems (Brockerhoff et al., 2017).

2.2.4 Drought Tolerance

Tropical endophytic fungi enhance the ability of plants to tolerate drought stress. The extensive mycorrhizal network can explore a larger soil volume, helping plants access water in dry conditions. This is crucial in tropical forests, where seasonal variations in precipitation can occur, e.g. *Glomus* sp. (Li et al., 2019).

2.2.5 Stabilization of Soil Structure

They assist in maintaining soil structure and preventing erosion. Tropical endophyte fungi's hyphal network helps bind soil particles together, reducing the likelihood of nutrient runoff and soil erosion during heavy rainfall, e.g. *Genus rhizoctonia* (Bashagaluke et al., 2018).

2.2.6 Carbon Sequestration

They contribute to carbon sequestration by promoting the growth of vegetation. Tropical endophytic fungi increased plant biomass which promotes greater carbon capture and storage in the form of plant tissues, assisting in the prevention of climate change. Shi et al. (2023) underscored the pivotal role of tropical endophyte fungi in enhancing carbon sequestration through the promotion of vegetation growth. Their research illustrates that the symbiotic relationship between these fungi and plants contributes significantly to increased biomass. As tropical endophyte fungi facilitate nutrient absorption for plants, the latter exhibit heightened growth and vitality. This amplified plant biomass, in turn, serves as a potent mechanism for capturing and storing carbon dioxide from the atmosphere. *Piriformospora indica* is an endophytic fungus known for its beneficial interactions with a wide range of plant species, including those found in tropical regions.

Shi et al. (2023) showed an important way that tropical endophyte fungi work together to help plants grow strong. This symbiotic interaction not only bolsters the resilience of plant ecosystems but also augments their overall carbon capture capacity. The study emphasizes the potential of this ecological partnership as a sustainable strategy for mitigating climate change. By elucidating the intricate dynamics between tropical endophyte fungi and vegetation, the research provides valuable insights into harnessing natural processes to combat the escalating challenges posed by global warming. These revelations pave the way for innovative and eco-friendly approaches for carbon sequestration, fostering a more sustainable and resilient future.

2.2.7 Disease Resistance

Endophytic fungi enhance plant resistance to diseases by forming a protective barrier around plant roots. The symbiotic relationship between tropical endophyte fungi and plants can contribute to increased resilience against certain soilborne pathogens.

Chauhan et al. (2022) proposed a novel approach to enhance plant resistance to diseases by establishing a protective barrier around plant roots. They highlight the potential of fostering a symbiotic relationship between plants and tropical endophytic fungi, emphasizing the role of these fungi in bolstering resilience against soilborne pathogens. The study underscores the significance of this interaction as a promising strategy for mitigating the impact of diseases on plant health, providing valuable insights for agricultural practices aiming to improve crop protection and yield. *Trichoderma* species are well known as biocontrol agents and beneficial endophytes that can form symbiotic relationships with a wide range of plant species, including those found in tropical regions.

2.2.8 Sustainable Ecosystem Functioning

Endophytes support the overall sustainability and resilience of tropical forest ecosystems by contributing to the efficient cycling of nutrients, the maintenance of soil health, and the promotion of diverse plant communities including the genus *Fusarium.*

Tahat et al. (2020) emphasize the crucial role of promoting nutrient cycling, maintaining soil health, and fostering diverse plant communities to support the sustainability and resilience of tropical forest ecosystems. Efficient nutrient cycling contributes to the ecosystem's overall health; ensuring essential elements are recycled for plant growth. Soil health maintenance is pivotal for sustaining biodiversity and ecological balance. The promotion of diverse plant communities enhances ecosystem resilience, providing stability in the face of environmental challenges. Implementing these practices can collectively contribute to the long-term well-being of tropical forest ecosystems.

2.2.9 Water Relations

Begum et al. (2019) emphasize that mycorrhizal associations played a crucial role in facilitating water uptake by plants, particularly in periods of water stress. These symbiotic relationships enhance the plant's capacity to access water from the soil, thereby bolstering the resilience of vegetation in tropical environments susceptible to fluctuations in water availability. Their study of *Glomus* sp. underscores the significance of mycorrhizal interactions in mitigating the impact of water scarcity on plant survival and adaptation in challenging climates.

2.3 ECOLOGICAL IMPORTANCE AND COMPLEXITIES OF ENDOPHYTIC FUNGI IN TROPICAL FOREST ECOSYSTEMS

Tropical endophyte fungi are essential to the ecosystems of tropical forests. These fungi have symbiotic interactions with the majority of the forest's plant roots, and they are essential to the ecosystems of tropical forests, creating a mycorrhizal association that benefits both the fungi and the plants (Begum et al., 2019). In tropical forests, where biodiversity is exceptionally high and ecological interactions are intricate, tropical endophyte fungi contribute significantly to plant growth and development and their nutrient cycling. Plants can obtain vital nutrients, such as nitrogen and phosphorus, which may be scarce in tropical soils, due to the mycorrhizal relationship. This is particularly vital in nutrient-poor tropical soils, where the rapid decomposition of organic matter can lead to nutrient leaching (Wahab et al., 2023).

The structure of the mycorrhizal association involves the penetration of fungal hyphae into the root cells of the host plant. This intimate connection forms arbuscules and vesicles, which are specialized structures that enable the fungus and host plant to exchange nutrients. The mycorrhizal network extends beyond individual plants, connecting multiple plant species in the forest ecosystem. This interconnected network, frequently called the "wood-wide web," allows for the sharing of nutrients and information among plants (Van Der Heijden et al., 2015).

The presence of endophyte fungi in tropical forests enhances plant resilience to environmental stressors, including drought and nutrient deficiency. The fungi improve water uptake by plants and aid in the absorption of essential minerals, contributing to the overall health and vitality of the forest ecosystem. Additionally, the mycorrhizal network can influence plant community dynamics and succession processes in tropical forests (Boczoń et al., 2021). Despite their ecological significance, the study of endophyte fungi in tropical forests is a complex task due to the diverse array of plant and fungal species involved. Researchers use molecular techniques, soil sampling and advanced imaging methods to unravel the intricacies of these symbiotic relationships and understand their implications for tropical forest ecology (Penton et al., 2016).

2.4 PLANT–FUNGAL INTERACTIONS

Adamo et al. (2021) thoroughly studied the complicated relationships between tropical endophyte fungi and their host plants. They explore the complex signaling pathways and molecular mechanisms that make these symbiotic relationships possible. The study highlighted the pivotal role of mycorrhizal interactions in facilitating nutrient exchange between plants and fungi, contributing improved nutrition absorption and overall host plant growth. They studied the specificity and selectivity of signaling events that control how these partnerships are made and maintained. This sheds light on how fungal and plant partners co-evolve in various fungi, e.g., *Glomus* sp., *Piriformospora indica*, *Serendipita* sp. and members of the Clavicipitaceae family.

2.5 TAXONOMIC DIVERSITY AND IDENTIFICATION OF TROPICAL ENDOPHYTE FUNGI IN TROPICAL FORESTS

Endophyte fungi exhibit a high degree of taxonomic diversity in tropical forests. This section reviews the major taxa found in these ecosystems, highlighting their distribution and host preferences. A review of current knowledge on the taxonomy and identification of endophytic fungi in tropical forests highlights recent advancements in molecular techniques for accurate species identification (Hariharan & Prasannath, 2021).

The fascinating relationship between ecological function and biodiversity can be seen in the study of taxonomic diversity and the identification of endophyte fungi in tropical forests. The lush expanses of tropical forests harbor a myriad of tropical endophyte fungi, forming symbiotic relationships with plant roots and significantly influencing nutrient cycling. Understanding the taxonomic diversity of these fungi is paramount for unraveling the complex web of ecological interactions within these ecosystems. Researchers delve into advanced molecular techniques, such as DNA bar-coding, to accurately identify and classify these elusive fungi, shedding light on their ecological roles and contributions to forest dynamics. The study of taxonomic diversity in endophyte fungi not only helps us understand tropical forest ecosystems better, but also has implications for conservation strategies and long-term management methods in a world where the environment is changing quickly (Frąc et al., 2018).

2.5.1 Distribution Patterns

In the study conducted by Edy et al. (2022), an in-depth exploration of the geographical distribution of endophyte fungi in various types of tropical forests was meticulously undertaken. The researchers delved into the intricate interplay of factors such as soil types, climate nuances, and host plant associations, shedding light on the intricate ecological dynamics within these ecosystems. The study showed different patterns of tropical endophyte fungal prevalence in different parts of the world. This gave us important information about how fungi and their host plants work together. Soil composition emerged as a pivotal determinant, with fungal communities exhibiting noteworthy variations based on soil characteristics. Additionally, the study underscored the influential role of climate, unraveling how temperature and precipitation gradients shape the mycorrhizal networks within tropical forests.

2.5.2 Spatial and Temporal Dynamics

Factors like soil type, climate, and plant community composition have an impact on the spatial and temporal dynamics of endophyte fungi in tropical forests. Understanding these dynamics is crucial for predicting how these symbiotic relationships may respond to environmental changes.

Wen et al. (2022) comprehensively investigated the spatial and temporal dynamics of endophyte fungi within tropical forests. The authors highlighted the multifaceted influences shaping these dynamics, emphasizing the pivotal roles played by soil type, climate, and plant community composition. Through meticulous analysis, they underscore the intricate interplay of these factors in shaping the symbiotic relationships between endophyte fungi and plants. Their research underscores the importance of comprehending these dynamics, as they serve as key predictors for anticipating how such symbiotic associations may respond to environmental changes.

2.5.3 Nutrient Cycling

Tropical endophyte fungi facilitate nutrient cycling in tropical forests by enhancing the uptake of essential elements, such as phosphorus and nitrogen, by host plants. This section examines the mechanisms through which these fungi contribute to nutrient cycling and overall ecosystem productivity. Wahab et al. (2023) studied the pivotal role of endophyte fungi in tropical forest ecosystems, specifically focusing on their ability to bolster nutrient cycling. The research underscores the mechanisms by which these fungi significantly enhance the uptake of crucial elements, including phosphorus and nitrogen, by host plants. Through intricate symbiotic relationships, endophyte fungi form mycorrhizal associations with plant roots, fostering a mutually beneficial exchange of nutrients. The findings reveal the fungi's adeptness in acquiring nutrients from the soil and delivering them to plants, thus promoting overall ecosystem productivity. The molecular and biochemical processes that underpin this nutrient transfer, elucidating the intricate pathways involved, are critical for understanding how nutrients are absorbed, transported and utilized within living organisms. The study not only advances our understanding of the ecological dynamics within tropical forests but also emphasizes the importance of endophyte fungi as key players in sustaining nutrient cycles critical for the vitality of these ecosystems.

2.5.4 Plant Health and Fitness

The mutualistic association between endophyte fungi and plants growing in tropical areas showed that it improved plant health and fitness. This included increased resistance to pathogens, improved stress tolerance, and enhanced growth. The symbiotic relationship between endophyte fungi and plants has been demonstrated to positively impact plant well-being and overall fitness. This association results in heightened resistance to pathogens, enhanced stress tolerance, and improved growth. Rudgers et al. (2005) extensively discussed the underlying mechanisms of this mutualistic interaction and explored its implications for plant community dynamics. These fungi contribute to the health and resilience of plants, shedding light on their potential applications in the management of ecosystems and agriculture.

2.6 DIVERSITY OF ENDOPHYTES IN VARIOUS TROPICAL FORESTS OF INDIA

Dreyfuss and Chapela (1994) found that there are approximately 1.3 million endophytic fungi species known to exist, most of which are found in tropical habitats. These findings are supported by various studies aimed at characterizing fungal communities connected to tropical plants. Fungal endophytes can be categorized into two main groups: generalists, which are common in many plant species, and singletons, which are uncommon and unique to a single plant host. While it is anticipated that tropical plants harbor a diverse array of endophytic fungi, only a limited number of tropical plant species have undergone screening for their presence. Research indicates that a wide variety of singleton species are found in tropical plants.

Hawksworth (2004) highlights the unknown extent of fungal diversity found in tropical forests, where many novel species remain unidentified. The highest fungal diversity is believed to occur in tropical forests due to the presence of a wide variety of angiosperms (Arnold et al., 2000). Supporting this notion, several endophytic fungal species have been found in association with plants across Asia, Africa, Central and South America, Australia, and Mexico, as well as some Atlantic and Pacific islands. However, within tropical forests, the richness of endophytic fungi may vary depending on the biome. For instance, Suryanarayanam et al. (2002) observed that Clavicipitaceae family endophytes (Hypocreales; Ascomycota) were found in a lower number in a dry tropical forest. Bioactive compounds have been reported to be produced by numerous species, particularly within genera such as *Balansia*, *Cordyceps*, *Claviceps*, *Myriogenospora* and *Epichloë/Neotyphodium.* In contrast, nonclavicipitaceous endophytes constitute a larger, less taxonomically defined group, primarily comprising species from the phyla Ascomycota and Basidiomycota. These include genera like *Aspergillus*, *Cladosporium*, *Alternaria*, *Arthrobotrys*, *Colletotrichum*, *Fusarium*, *Paecilomyces*, *Coprinellus*, *Curvularia*, *Phanerochaete*, *Penicillium* and *Phoma*, among others. Many species belonging to these two endophytic categories have been studied for their ability to create different kinds of bioactive compounds; tropical plant-associated species contribute significantly to this pool of metabolites (Table 2.1).

A variety of endophytes have been isolated, including *Phoma*, *Trichoderma*, *Chaetomium*, *Botryodiplodia* and *Acremonium*. *Penicillium islandicum*, *Pestalotiopsis*, *Cladosporium herbarum*, *F. graminearum*, *Alternaria alternata*, *Sphaeronema* and *Phomopsis* are some of the fungal species that are more rarely isolated. The most common endophytic fungi are *Colletotrichum* sp. (Photita, 2005), while *Pestalotiopsis* sp. are well-known endophytes found in numerous tropical tree species, including *Azadirachta indica* (Mahesh et al., 2005), *Terminalia arjuna* (Tejesvi et al., 2005) and many herbs and shrubs (Krishamurthy et al., 2008), as well as various species of tropical rainforest plants (Strobel, 2002). Since Hawksworth and Rossman (1997) predicted that a million different

TABLE 2.1
Endophytic Fungal Diversity in Medicinal Plants of Western Ghats, India

Medicinal plant	Plant parts	Season	Endophyte fungi
Tylophora asthmatica	–	Monsoon	*Acremonium strictum*, *Aspergillus terreus*, *Red pycnidia*, *Botryodiplodia theobromae*, *Chaetomium globosum*, *F. oxysporum*, *F. solani*, *F. verticilloides*
			Gelasinospora sp., *Heinesia rubi*, *Memnoniella* sp., *Myrothecium verrucaria*, *Nigrospora oryzae*, *Pestalotiopsis guepinii*, *Phoma* sp., *Penicillium* sp.
			Sphaeronema sp., sterile mycelia
	–	Winter	*Acremonium strictum*, *Colletotrichum dematium*, *C. lindemuthianum*, *Chaetomium globosum*, *Pestalotiopsis* sp., *Phoma* spp., sterile mycelia
Rubia cordifolia	Stem	Monsoon	*Acremonium strictum*, *Botryodiplodia theobromae*, *F. oxysporum*, *F. solani*, *Memnoniella* sp., *Periconia* sp., *Phoma* sp., *Penicillium* sp., pycnidial forms
		Winter	*Acremonium strictum*, *Alternaria alternate*, *Botryodiplodia theobromae*, *C. lindemuthianum*
			Curvularia lunata, *F. oxysporum*
Plumbago zeylanica		Monsoon	*Aspergillus terreus*, *F. oxysporum*, *F. verticilloides*, *Myrothecium verrucaria*, *Trichoderma harzianum*, *Verticillium albo-atrum*, pycnidial forms, sterile mycelia
		Winter	*Acremonium strictum*, *Chaetomium globosum*, *F. verticilloides*, *Penicillium* sp.
Phyllanthus amarus	Stem and root	Monsoon	*Acremonium strictum*, *P. islandicum*, *Botryodiplodia theobromae* (S), *Cladosporium herbarum* (S), *Curvularia lunata* (S), *Fusarium* sp. (R), *F. solani* (R), *F. verticilloides* (S), *Gelasinospora* sp. (S), *Pestalotiopsis* sp. (S), *P. islandicum* (S), *Penicillium* sp. (R), *T. viride* (R), *Sphaeronema* sp. (S), sterile mycelia (R)
	Stem and root	Winter	*Acremonium strictum* (R), *Aspergillus terreus* (R), *Botryodiplodia theobromae* (S), *Curvularia lunata* (S), *Fusarium* sp. (S), *F. solani* (S), *Myrothecium verrucaria* (S), *Penicillium* sp., *Trichoderma harzianum*, *T. viride* (R), sterile mycelia (R)
Eryngium foetidum	Stem and root	Monsoon	*Acremonium strictum*, *F. solani* (R), *F. pseudonygamai* (S), *F. verticilloides* (S), *Penicillium* sp. (R), *Trichoderma harzianum*, sterile mycelia (R)
	Stem and root	Winter	*Acremonium strictum*, *Alternaria alternate* (S), *Botryodiplodia theobromae* (R), *Chaetomium globosum* (R), *Chaetomium globosum* (S), *F. oxysporum* (R)
Centella asiatica	Stem and root	Monsoon	*Acremonium strictum* (St, R), *C. lindemuthianum* (R), *Curvularia lunata*, *F. graminearum* (S), *F. oxysporum* (R), *Phomopsis* sp. (S), *Penicillium* sp. (R), *Trichoderma harzianum* (R)
	Stem, stolen, and root	Winter	*Acremonium strictum* (St, R), *F. graminearum* (St), *Curvularia lunata* (R), *F. oxysporum* (R), *Penicillium* sp. (R), *Trichoderma harzianum* (R)
Zingiber sp.	Stem, root, stolen, and rhizome	Monsoon	*F. verticilloides* (S), *Verticillium albo-atrum* (S), *Cladosporium herbarum* (Rh), *Chaetomium globosum* (Rh), *Trichoderma harzianum* (Rh), *F. oxysporum* (R), *F. solani* (R), *F. pseudonygamai* (R)
	Stem, root and rhizome	Winter	*F. oxysporum* (Rh), *Trichoderma harzianum* (Rh)

S: stem, R: root, Rh: rhizome.

fungi species remain to be identified, it is crucial to screen more recent plant species for the isolation of fungal endophytes. There were variations in the rates at which endophytes colonized over two distinct seasons, and during the monsoon season, they showed greater isolation because rain impacts were more effective in dispersing the slimy conidia of spores from fungi, and climate conditions affected conidia germination (Wilson & Carroll, 1994).

Flora are numerous, and Melghat Forest is one of the regions with the greatest diversity of flora; however, research on the richness of endophytic fungi in this area is woefully lacking. With the goal to identify the endophytic fungi, the commonly used medicinal herb *Cissus quadrangularis* was examined. Using accepted techniques, a total of eight endophytes were found from various host parts (leaf, petiole, and stem). The endophytes *Arthrinium hydei*, *Epicoccum nigrum*, *Nigrospora oryzae*, *Stachybotrys nilgirica*, *S. chlorohalonata*, *S. chartarum*, *T. scutellare* and *Trimmatostroma hughesii* colonize all over the world depending upon a variety of soil conditions and host plants (Suradkar & Hande, 2017).

The endophytic fungi found within the Melghat Forest exhibit a fascinating seasonal pattern of colonization within *Adathoda vasica* plants. Among the 13 fungi isolated from plant samples collected between 2012 and 2013, *Alternaria alternata* predominated with the highest colonization frequency in the petiole region across all seasons. During the monsoon, *Arthrinium hydei* and *Arthrinium phaeospermum* were notably present, particularly thriving in the leaf area. However, during the summer months, *Stachybotrysi* species, including *S. chartarum* and *S. nilgirica*, disappeared despite their presence in the winter and monsoon seasons. Intriguingly, only during winter were *Virgaria nigra*, *Xylaria* sp., *Pestalotiopsis funerea*, *Nigrospora oryzae* and *Epicoccum nigrum* detected within the samples (Suradkar & Hande, 2018).

In all, 15 fungal endophytes were isolated in the 2013–14 season. The occurrence of each isolate varied with the season. During the monsoon, *Arthrinium phaeospermum* demonstrated a greater colonization rate in leaf tissue, while during the winter it demonstrated a larger colonization frequency in stem tissue. *Aspergillus stellatus* was found throughout the research, however, compared to previous seasons, it demonstrated a higher colonization frequency during winter. *Aspergillus nidulans*, *Curvularia lunata*, *Penicillium chrysogenum* and *Alternaria alternata* were among the very few fungi that were recovered throughout the summer with decreased colonizing frequency. During the summer and monsoon seasons, *Fusarium oxysporum* exhibited a higher colonization frequency in the petiole compared to the stem and leaf tissues. While *Phoma crysanthemicola* and *Colletotrichum gloeosporiodes* were discovered solely during the winter, *Phoma crysanthemicola*, *Arthrinium hydei*, *Stachybotrys nilgirica* and *Stachybotrys chartarum* were only found during the rainy season (Suradkar & Hande, 2018).

2.7 THREATS TO ENDOPHYTIC FUNGI IN TROPICAL FORESTS

Human actions such as alterations in land use and deforestation, and climate change, are significant threats to tropical forest ecosystems and the endophyte fungi that inhabit them. The process of removing substantial tracts of forest for infrastructure development, logging, agriculture or other uses is known as deforestation. This directly reduces the available habitat for fungi and disrupts the intricate ecological balance within the forest. Land-use changes, often associated with urbanization and agriculture expansion, further exacerbate the challenges faced by endophyte fungi. Altered land use can lead to fragmentation of natural habitats, isolating fungal populations, and hindering their ability to disperse and establish symbiotic relationships with plants. The loss of connectivity between different patches of forests can negatively impact the diversity and distribution of these fungi, affecting their vital roles in nutrient cycling and ecosystem resilience (Morris, 2010).

Climate change adds an additional layer of complexity to the threats faced by tropical forest ecosystems and their fungal inhabitants. The physiology and ecology of fungi can be directly impacted by temperature rises, changes in patterns of precipitation and extreme weather conditions.

Many fungi have specific temperature and moisture requirements, and changes in these environmental conditions may disrupt their life cycles and interactions with host plants. Additionally, climate-induced shifts in vegetation composition may influence the various species of fungi which can grow in a specific location.

The potential impacts of these threats on fungal diversity and function are profound. Fungal diversity is intricately linked to ecosystem health, as different species contribute to various ecological processes, including nutrient cycling, decomposition and mycorrhizal symbiosis. Changes in fungal diversity can disrupt these processes, leading to imbalances in nutrient availability and overall ecosystem functioning (Lawrence et al., 2022).

Furthermore, alterations in fungal communities can have cascading effects on plant communities, as many plants depend on mycorrhizal associations for nutrient uptake. The health and resilience of tropical forest ecosystems, therefore, hinge on the preservation of fungal diversity and function. Mitigating the impacts of deforestation, land-use changes and climate change is crucial for the long-term sustainability of these ecosystems and the vital roles played by the endophyte fungi within them. Conservation efforts should prioritize strategies that address these anthropogenic threats and promote the coexistence of human activities and healthy tropical forest ecosystems.

2.8 CONSERVATION STRATEGIES

Due to the ecological importance of tropical endophyte fungi in tropical forests, effective conservation strategies are crucial. This section discusses potential conservation measures, including habitat preservation, restoration efforts and sustainable land-use practices (Morri & Santolini, 2021). Conservation efforts should focus on preserving intact ecosystems where these fungi thrive, as intact habitats provide essential resources and suitable conditions for their survival and proliferation. Habitat preservation initiatives should prioritize protecting key areas of biodiversity, such as old-growth forests and untouched natural habitats, which serve as hotspots for endophyte fungal diversity (Camenzind, 2016). Furthermore, restoration efforts can play a significant role in recovering degraded ecosystems and creating suitable habitats for these fungi to colonize. Adopting appropriate land utilization techniques, such as reduced-impact logging and agroforestry, can support local livelihoods while reducing habitat damage and preserving ecological integrity (Morri & Santolini, 2021). By emphasizing the importance of preserving intact ecosystems and promoting sustainable land-use practices, conservation strategies can mitigate the impact of threats to tropical endophyte fungi and contribute to the overall health and resilience of tropical forests.

2.9 CONCLUSION

A thorough understanding of the diversity and ecological functions of endophytic fungi within tropical forest ecosystems illuminates their essential role in maintaining biodiversity and ecological harmony. These fungi are integral to the intricate web of life within these biodiverse environments, contributing significantly to the delicate balance that sustains them. By appreciating the significance of these symbiotic relationships, we gain deeper insights into the complexities of tropical forests and can develop well-informed conservation strategies to safeguard these vital ecosystems. Tropical forests are renowned for their astonishing array of plant and animal species, and amidst this richness, endophytic fungi emerge as key contributors to ecological stability. They play a crucial role in nutrient cycling, aiding in the decomposition of organic matter and enriching the soil. Additionally, their partnerships with plants enhance nutrient absorption, fostering the growth and resilience of various plant species. Through mycorrhizal associations, these fungi extend the reach of plant roots, enabling them to access essential nutrients that would otherwise be inaccessible. The synergistic relationship between endophytic fungi and plants underscores the intricate interdependence that sustains the vitality of these ecosystems. Recognizing and preserving the role of endophytic fungi is paramount in nurturing the health and resilience of tropical forests.

REFERENCES

Adamo, I., Piñuela, Y., Bonet, J. A., Castano, C., de Aragón, J. M., Parladé, J. & Alday, J. G. (2021). Sampling forest soils to describe fungal diversity and composition. Which is the optimal sampling size in mediterranean pure and mixed pine oak forests? *Fungal Biology*, *125*(6), 469–476.

Alam, B., Lǐ, J., Gě, Q., Khan, M. A., Gōng, J., Mehmood, S. & Gŏng, W. (2021). Endophytic fungi: From symbiosis to secondary metabolite communications or vice versa? *Frontiers in Plant Science*, *12*, 3060

Arnold, A. E., Maynard, Z., Gilbert, G. S., Coley, P. D. & Kursar, T. A. (2000). Are tropical fungal endophytes hyperdiverse? *Ecology Letters*, *3*(4), 267–274.

Bashagaluke, J. B., Logah, V., Opoku, A., Sarkodie-Addo, J. & Quansah, C. (2018). Soil nutrient loss through erosion: Impact of different cropping systems and soil amendments in Ghana. *PloS One*, *13*(12), e0208250.

Begum, N., Qin, C., Ahanger, M. A., Raza, S., Khan, M. I., Ashraf, M. & Zhang, L. (2019). Role of arbuscular mycorrhizal fungi in plant growth regulation: implications in abiotic stress tolerance. *Frontiers in Plant Science*, *10*, 1068.

Boczoń, A., Hilszczańska, D., Wrzosek, M., Szczepkowski, A. & Sierota, Z. (2021). Drought in the forest breaks plant–fungi interactions. *European Journal of Forest Research*, *140*(6), 1301–1321.

Brockerhoff, E. G., Barbaro, L., Castagneyrol, B., Forrester, D. I., Gardiner, B., González-Olabarria, J. R. & Jactel, H. (2017). Forest biodiversity, ecosystem functioning and the provision of ecosystem services. *Biodiversity and Conservation*, *26*, 3005–3035.

Camenzind, T. (2016). *Responses of arbuscular mycorrhizal fungi to nutrient additions in a tropical montane forest of Southern Ecuador* (Doctoral dissertation).

Chauhan, S., Mahawar, S., Jain, D., Udpadhay, S. K., Mohanty, S. R., Singh, A. & Maharjan, E. (2022). Boosting sustainable agriculture by arbuscular mycorrhiza under stress condition: Mechanism and future prospective. *BioMed Research International*, *2022*, 5275449.

Dreyfuss, M. & Chapela, I. (1994). Potential of fungi in the discovery of novel, low-molecular weight pharmaceuticals. *The Discovery of Natural Products with Therapeutic Potential (Gullo VP, ed)*, *26* (5), 49–80.

Edy, N., Barus, H. N., Finkeldey, R. & Polle, A. (2022). Host plant richness and environment in tropical forest transformation systems shape arbuscular mycorrhizal fungal richness. *Frontiers in Plant Science*, *13*, 1004097.

Frąc, M., Hannula, S. E., Bełka, M. & Jędryczka, M. (2018). Fungal biodiversity and their role in soil health. *Frontiers in Microbiology*, *9*, 707.

Hariharan, G. & Prasannath, K. (2021). Recent advances in molecular diagnostics of fungal plant pathogens: A mini review. *Frontiers in Cellular and Infection Microbiology*, *10*, 600234

Hawksworth, D. (2004). Fungal diversity and its implications for genetic resource collections. *Studies in Mycology*, *50*, 9–18.

Hawksworth, D. L., & Rossman, A. Y. (1997). Where are all the undescribed fungi? *Phytopathology*, *87*, 888–891.

Huey, C. J., Gopinath, S. C., Uda, M. N. A., Zulhaimi, H. I., Jaafar, M. N., Kasim, F. H. & Yaakub, A. R. W. (2020). Mycorrhiza: a natural resource assists plant growth under varied soil conditions. *3 Biotech*, *10*, 1–9.

Krishnamurthy, Y. L., Naik, S. & Shashikala, J. (2008). Fungal communities in herbaceous medicinal plants from the Malnad Region, Southern India. *Microbes and Environment*, 23, 24–28. 10.1264/jsme2.23.24

Lawrence, D., Coe, M., Walker, W., Verchot, L. & Vandecar, K. (2022). The unseen effects of deforestation: biophysical effects on climate. *Frontiers in Forests and Global Change*, *5*, 49.

Li, J., Meng, B., Chai, H., Yang, X., Song, W., Li, S. & Sun, W. (2019). Arbuscular mycorrhizal fungi alleviate drought stress in C3 (*Leymus chinensis*) and C4 (*Hemarthria altissima*) grasses via altering antioxidant enzyme activities and photosynthesis. *Frontiers in Plant Science*, *10*, 499.

Mahesh, B., Mysore, T., Nalini, M., Prakash, H. S. & Kini, R., Subbiah, V. & Shetty, S. (2005). Endophytic mycoflora of inner bark of Azadirachta indica A. Juss. *Current Science*, 88, 218–219.

Morri, E. & Santolini, R. (2021). Ecosystem services valuation for the sustainable land use management by nature-based solution (NbS) in the common agricultural policy actions: a case study on the foglia river basin (marche region, Italy). *Land, 11*(1), 57.

Morris, R. J. (2010). Anthropogenic impacts on tropical forest biodiversity: a network structure and ecosystem functioning perspective. *Philosophical Transactions of the Royal Society B: Biological Sciences, 365*(1558), 3709–3718.

Penton, C. R., Gupta, V. V., Yu, J. & Tiedje, J. M. (2016). Size matters: assessing optimum soil sample size for fungal and bacterial community structure analyses using high throughput sequencing of rRNA gene amplicons. *Frontiers in Microbiology, 7*, 824.

Photita, W., Taylor, P. W. J., Ford, R., Hyde, K. D. & Lumyong, S. (2005). Morphological and molecular characterization of *Colletotrichum* species from herbaceous plants in Thailand. *Fungal Diversity, 18*, 117–133.

Rodriguez, R. J., White Jr, J. F., Arnold, A. E. & Redman, A. R. A. (2009). Fungal endophytes: diversity and functional roles. *New Phytologist, 182*(2), 314–330.

Rudgers, J. A., Mattingly, W. B. & Koslow, J. M. (2005). Mutualistic fungus promotes plant invasion into diverse communities. *Oecologia, 144,* 463–71.

Sharma, A., Shukla, R. V., Chaubey, A. & Mahish, M. (2015). Diversity of endophytic fungi in tropical semi-evergreen forests trees of Chhattisgarh. *Proceedings of the National Academy of Sciences, India Section B: Biological Sciences, 85*, 253–261.

Shi, J., Wang, X. & Wang, E. (2023). Mycorrhizal symbiosis in plant growth and stress adaptation: From genes to ecosystems. *Annual Review of Plant Biology, 74*, 569–607.

Strobel, G. A. (2002). Microbial gifts from rain forests. *Canadian Journal of Plant Pathology, 249*(1), 14–20.

Suradkar, K. P. & Hande, D.V. (2017). Morphotaxonomy of endophytic fungi on *Cissus quadrangularis* from Amravati (MS) India. *Journal of Bacteriology Mycology Open Access, 5*(2), 253–258.

Suradkar, K. P. & Hande, D.V. (2018). Diversity of Fungal Endophytes Isolated from *Adathoda vasica* from Melghat Forest. *Aayushi International Interdisciplinary Research Journal, V*(III), 206–210.

Suryanarayanan, S., Murali, T. & Venkatesan, G. (2002). Ocurrence and distribution of fungal endophytes in tropical across a rainfall gradient. *Canadian Journal of Botany, 80*(8), 818–826.

Tahat, M. M., Alananbeh, M. K., Othman, A. Y. & Leskovar, I. D. (2020). Soil health and sustainable agriculture. *Sustainability, 12*, 4859.

Tejesvi, M. V., Mahesh, B., Nalini M. S. (2005). Endophytic fungal assemblages from inner bark and twig of Terminalia arjuna W. & A. (Combretaceae). *World Journal of Microbiology and Biotechnology, 21*(8–9), 1535–1540.

Van Der Heijden, M. G., Martin, F. M., Selosse, M. A. & Sanders, I. R. (2015). Mycorrhizal ecology and evolution: the past, the present, and the future. *New Phytologist, 205*(4), 1406–1423.

Wahab, A., Muhammad, M., Munir, A., Abdi, G., Zaman, W., Ayaz, A. & Reddy, S. P. P. (2023). Role of arbuscular mycorrhizal fungi in regulating growth, enhancing productivity, and potentially influencing ecosystems under abiotic and biotic stresses. *Plants, 12*(17), 3102.

Wen, J., Okyere, S. K., Wang, S., Wang, J., Xie, L., Ran, Y. & Hu, Y. (2022, Feb). Endophytic fungi: An effective alternative source of plant-derived bioactive compounds for pharmacological studies. *Journal of Fungi (Basel)*, 8(2), 205.

Wilson, D. & Carroll, G. C. (1994). Infection studies of Discula quercina, an endophyte of Quercus garryana. *Mycologia*, 86, 635–647.

Xavier Martins, W. F. & Rodrigues, B. F. (2020). Identification of dominant arbuscular mycorrhizal fungi in different rice ecosystems. *Agricultural Research, 9*, 46–55.

3 Omics' Approach to Discern the Functional Diversity of Fungi in Plants

Arkadeep Mukherjee, Giridhar Babu Anam and Young-Ho Ahn*
*Corresponding author.

3.1 INTRODUCTION

Phytomicrobiomes are diverse microbial populations that interact with different plant components. Terrestrial plants support a variety of microbial communities, including bacteria, fungi, nematodes, protists, and viruses (Lyu et al., 2021; Quiza, St-Arnaud, & Yergeau, 2015). Fungal communities are diverse in shape, ecology, and nutrition with a diverse aspect across numerous ecosystems. They live in soil, water, air, animals, plants, and extreme conditions including low or high temperatures, and high metal and salt concentrations (Coleine, Stajich, & Selbmann, 2022; Manoharachary et al., 2005). Next to insects, they are the largest group of organisms. About 150,000 of the 1.5–5 million fungal species in diverse habitats have been described (Bhunjun et al., 2022). However, ecology and the environment greatly affect fungal prevalence. Fungi are diverse and difficult to estimate, complicating their investigation (Hawksworth David & Lücking, 2017).

Fungal species, which are essential to plant health, may be beneficial or detrimental. Some fungi help the nutrient cycle by channeling C, N, P, and S via trophic levels, but others may cause many plant diseases. They may be symbiotic, pathogenic, or decompose with plants (Field & Pressel, 2018; Zeilinger et al., 2016). Saprophytic fungi shape nutrient recycling and biodiversity, while symbiotic fungal species limit nutrient availability to promote ecological diversity and protect plants from plantborne pathogens. Microorganism–plant interactions in nitrogen fixation, drought stress relief, and stress mitigation against environmental contaminants, notably heavy metals, safeguard and improve plant health (Hildebrandt, Regvar, & Bothe, 2007; Luo et al., 2014; Zou, Wu, & Kuča, 2021). Additionally, plant–microbe interactions promote cell division, seed sprouting, seedling growth, and phytohormone synthesis. Recognizing these symbiont-mediated adaptations emphasizes that plants are not isolated. Instead, the host microorganism and its microorganisms should be considered a "holobiont," making plant–fungi connections dynamic and multidimensional. Fungi help plants absorb nutrients and fight infections, whereas fugal colonies depend on plants for nutrient supplies, which can affect agricultural and natural ecosystems. Sustainable practices and ecological equilibrium require understanding and managing these connections. Scientists use molecular and the latest omics methods to investigate these complicated connections. These methods allow for in-depth investigation of plant–fungus biological molecules. The holistic method to understanding biological information is called "omics" and includes genomes, transcriptomics, proteomics, and metabolomics (Fadiji & Babalola, 2020).

This chapter examines how omics approaches have transformed our understanding of plant–fungus interactions. In addition to enhancing scientific understanding, omics research may lead to more effective and sustainable agricultural approaches.

DOI: 10.1201/9781003486275-4

3.2 PLANT-ASSOCIATED FUNGI: CATEGORIZATION AND FUNCTIONAL ROLES

Fungi that associate with plants are usually categorized by function. Due to their importance in plant nutrition, mycorrhizal fungi are among the most studied mutualistic fungus partners. However, non-mycorrhizal root-associated fungi also help plants get more nutrients. Endophytic fungi are another varied plant-associated microbe. Their functions vary and are often unknown. The well-known grass–systemic endophyte symbiosis affects host plants in many ways. Some taxonomic groups of forest tree leaf endophytes, such as the Xylariales, produce many antibiotic chemicals to fight infections and illnesses (Helaly, Thongbai, & Stadler, 2018). Some leaf endophytes have been suggested to be involved in litter decomposition, a vital process in nutrient cycling. Mycorrhizal fungi can also play a direct role in this process. Consequently, the fungal species responsible for decomposition may be considered as plant-related in a wider sense, particularly as saprotrophic fungi which may be associated and connected with higher plants. Therefore, it is important to take into account non-mycorrhizal soil fungi in the context of plant-associated fungal communities, since they play a role in the breakdown of root litter and the release of nutrients from minerals (Herman et al., 2012). Fungi suppress root diseases and fight plant infections with fungal enzymes, improving plant health. Fungi secrete antibodies that prevent other bacteria from developing, reducing competition. They produce many vitamins that help plants grow. Beneficial fungi compete for resources with health-threatening species to prevent their spread. Fungi also protect the host plant by producing webs and nets around the roots and leaves (Bastías, Johnson, & Card, 2020). *Trichoderma* is one such fungus that shields plant roots from pathogenic microbes (Adnan et al., 2019). In addition, fungi provide a protective covering that allows plant roots to receive phosphate and water during dry spells. Figure 3.1 illustrates various sources of fungal microbiota.

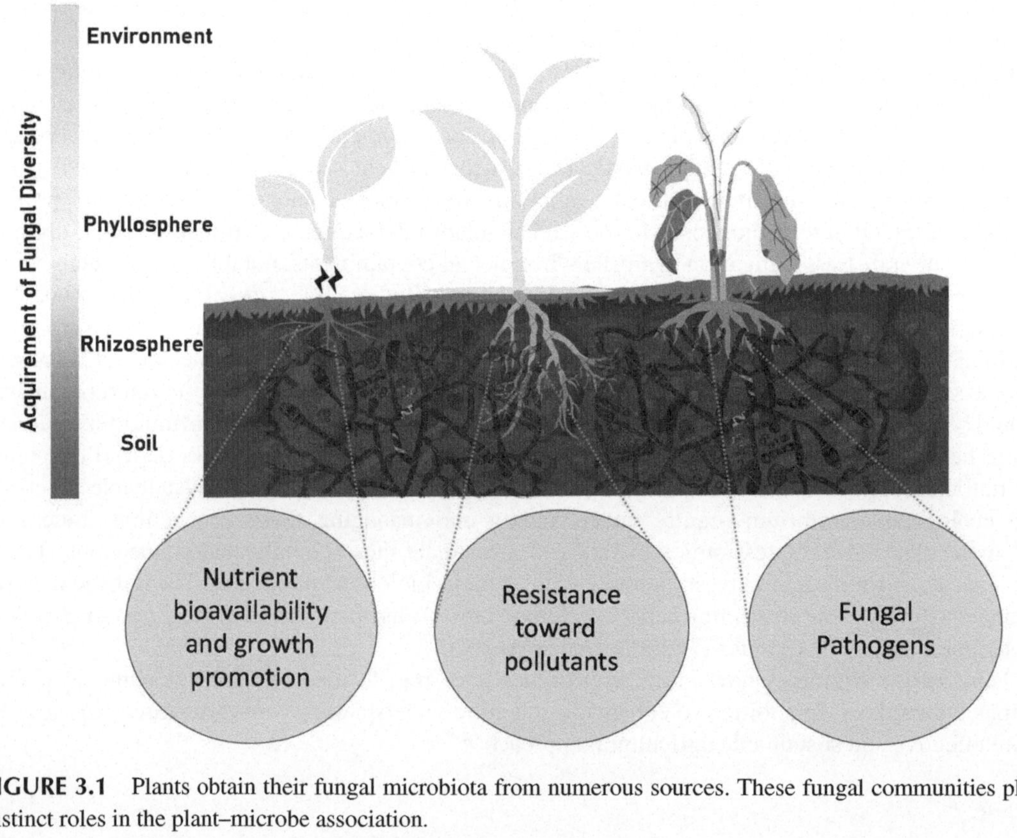

FIGURE 3.1 Plants obtain their fungal microbiota from numerous sources. These fungal communities play distinct roles in the plant–microbe association.

Fungal colonies induce wood deterioration, another important nutrient cycling process. Many forest fungi decompose wood, cycling nutrients and creating new habitats for other wood-dwelling organisms. Understanding the complicated interactions between lignocellulose-degrading fungal communities is essential for implementation (Arantes et al., 2011). This understanding has the potential to open up important supplies of biocatalysts that are essential for turning agricultural waste into commodity chemicals and biofuels (Charpentier-Alfaro et al., 2023). Oxidative and hydrolytic enzymes drive lignocellulose decomposition in compost and soil. The complicated connections between cellulose, hemicellulose and lignin – the main components of plant cell walls – are broken by these enzymes. This procedure requires mutually beneficial connections between fungal community groupings. Environmental factors affect the fungal community structure in several ways. Oxygen levels, plant material origin, temperature, pH, soil composition, lignocellulose breakdown stage, and biomass chemical makeup influence these communities. Each component affects microbial interactions differently. Enzymes release sugars and other resources from lignocellulose breakdown, which encourages microbial competition and generates a complex metabolic network (Aguilar-Paredes et al., 2023; Costa et al., 2023).

The plantborne and plant material-degrading communities contain hundreds to thousands of fungal species. These may differ from plant-related ones. Traditional methods of fungal identification involve close study of fungi in their natural environment or on growth media. Classical methods are still useful for studying fungal variations since some fungi have distinguishing traits. Since some fungi are difficult to grow, molecular approaches are important for taxonomic identification (Götz et al., 2006). Mycologists have used molecular methods to analyze conserved and newly discovered fungus samples in addition to morphological methods. Due to this, fungal taxonomists have proposed or established several new taxa.

3.3 META-OMIC INSIGHTS INTO FUNGAL DIVERSITY

The suffix -omics is used to designate large-scale data in the study of life science and to understand the functioning of various life processes. Metagenomics, metaproteomics and metatranscriptomics are among the meta-omics techniques that shed light on the complex interactions that occur between entophytic bacteria, fungi, and their host plants. These cutting-edge technologies have helped researchers study plant-associated microorganisms that support plant development by modulating phytohormones and increasing host defenses against environmental challenges (Bhat et al., 2022). Further, meta-omics reveals the innate capabilities of these fungi and bacteria, suggesting that they may be able to support plant development in a variety of settings (Künzler, 2018). The qualitative research of complicated functional properties of dwelling microorganisms requires extensive collection, DNA extraction, sequencing and analysis. Figure 3.2 represents different approaches to study fungal diversity.

To study the multi-dimensional interactions between plant and fungal communities, the multi-omics integrated approach makes it possible to identify important metabolic pathways, regulatory networks and molecular markers to understand the interactions with the host microorganism.

Traditional microbial identification methods relied on culturing, thus we only knew about lab-grown species. Culture-dependent approaches for microbial separation have limitations, including recovering just <1% of microorganisms, limiting their utility in agriculture. However, metagenomics – powered by NGS – has transformed this setting (Youseif Sameh et al., 2021). This novel method sequences environmental genetic material without isolating and cultivating bacteria. By identifying the genetic makeup of microbial species in a given environment, NGS-based metagenomics provides a complete picture. This technology shows the genetic blueprints of cultivated and uncultured bacteria, revealing their variety, functional potential and environmental roles. It extracts and sequences DNA or RNA from complex microorganism cultures. NGS's ability to quickly and cheaply sequence massive amounts of data is its genius. This deluge of data handled

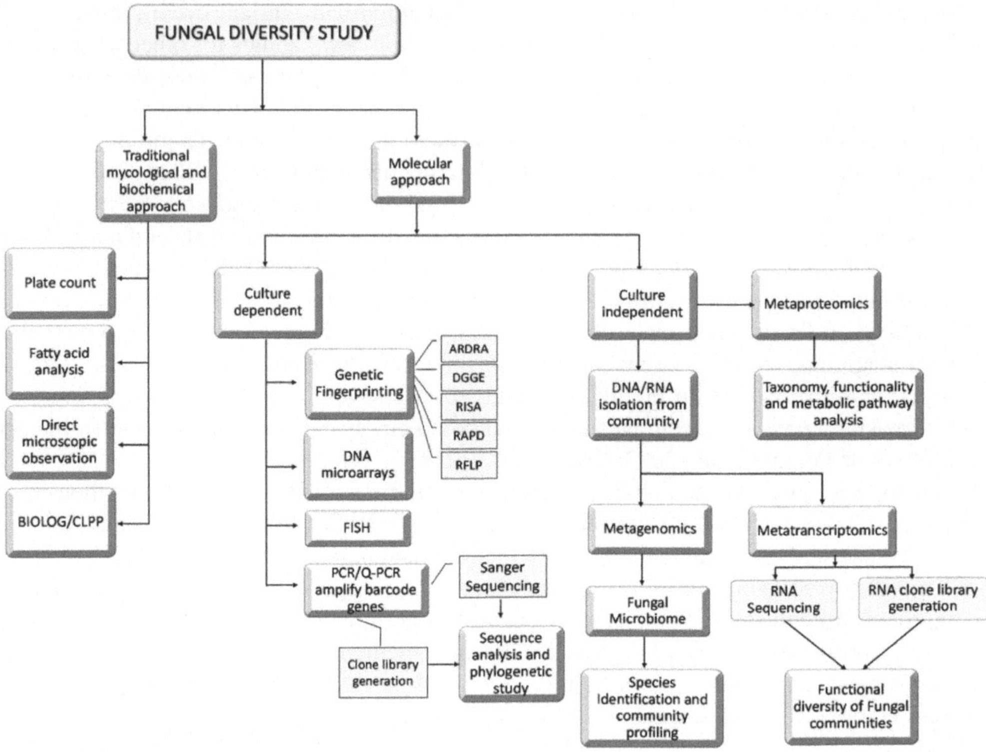

FIGURE 3.2 Different approaches to study fungal diversity. Traditional mycological studies and the molecular approach differ in a variety of ways. Culture-dependent molecular approaches to study sequence variations include genetic fingerprinting techniques such as amplified ribosomal DNA restriction analysis (ARDRA), denaturing gradient gel electrophoresis (DGGE), ribosomal intergenic spacer analysis (RISA), randomly amplified polymorphic DNA (RAPD), restriction fragment length polymorphism (RFLP), are a few examples. Fluorescence in situ hybridization (FISH) detects species by using specific nucleotide probes.

using advanced bioinformatics approaches can be used to build and analyze genomes and genetic fragments from a variety of bacteria in a sample (Satam et al., 2023).

DNA sequencing and bioinformatics have revolutionized our understanding of plant-ecosystem fungi. Large genomic datasets submitted to public repositories make endophytic genome comparisons easier. This has uncovered genes required for phytohormone synthesis, secretion systems, siderophore generation, phosphate metabolism, surface attachment and other activities (Adeleke, Ayangbenro, & Babalola, 2021; Kaul, Sharma, & Dhar, 2016). The complex interactions between microbes and plants are mediated by these genes, which impact the dynamics of plant development (Adeleke, Muller, & Babalola, 2023). Recent metagenomics research has provided insights into the dynamics and roles of bacteria, fungi, and archaea in the root endosphere (Kalu et al., 2021).

Metagenomics may improve our understanding of plant microbes and their connections. Many published metagenomic datasets for plant-associated microbial communities reveal their evolution over time and within changing environmental conditions. DNA extraction from plant tissues like leaves, roots and stems is essential for the plant metagenome. Scientists have studied plant endophytic communities' diversity and functionality (Bollmann-Giolai, Malone, & Arora, 2022). In most of these studies, a significant challenge is the difficulty of separating the DNA of plants from that of microorganisms. Separating plant DNA from microbes is difficult in most research. Plant DNA usually outnumbers endophytic microbe DNA, making metagenome sequencing challenging. Fungal community studies using metagenomics involve DNA isolation from plant sources, preprocessing, quality control and diversity analysis, which are discussed further.

3.3.1 DNA Isolation and Estimation for Plant-Associated Fungal Metagenome

DNA isolation is the first step toward metagenomic and computational investigations. Multiple sample processing methods affect DNA quality. These procedures are dependent on plant components, and the sampling strategy should be reproducible to provide an effective experiment. To eliminate contaminated species, plant pieces must be surface-sterilized before DNA isolation to recover plant-associated microbial communities. Initial metagenomics investigation and processing of plant-associated microbial community materials uses several standard chemical kits and extraction procedures. Each kit is designed to isolate genomic DNA from cells, tissues, seeds and other plants materials. The MoBio PowerPlant ProDNA extraction kit extracts entire cellular DNA from plant cells, tissues and seeds. Several extraction procedures are utilized to recover high-quality DNA from difficult material types such as strawberry leaf, cotton leaf, cotton seeds and pine needles (Yang et al., 2023). On the other hand, the Qiagen DNeasy Mini Plant DNA kit enables the isolation of pure total DNA, including mitochondrial, genomic and chloroplast DNA, from plant and fungal tissues (Neves & Forrest, 2011). The GeneJet Plant Genomic DNA Purification Kit from Thermofisher Scientific offers rapid and efficient purification specifically suited for various plant species and tissue types, utilizing silica-based membrane technology (Maropola, 2015). Meanwhile, the CTAB method, employing cetyltrimethyl ammonium bromide as a detergent, stands as a widely adopted technique for extracting DNA from plant tissues, with additives aiding in the removal of interfering compounds such as polyphenols (Porebski, Bailey, & Baum, 1997). These diverse kits and methods efficiently recover high-quality DNA from plant materials, enabling molecular investigation of plant-associated microbial populations. Evaluation of the quality of the extracted DNA is essential by quantifying its concentration and purity utilizing the absorbance of the DNA at wavelengths of 260 nm and 280 nm using specialized equipment such as a UV spectrophotometer or the NanoDrop 1000 (Thermo Fisher Scientific, Waltham, MA) (Bareither et al., 2013). The ultimate pure DNA samples may be kept at temperatures of –20°C or –80°C to facilitate further processing.

3.3.2 Metagenomic Community Analysis by DGGE and t-RFLP

Researchers have used metagenomic methods to study endophytic-rhizospheric microbiomes' rich makeup and various spectra. Denaturing gradient gel electrophoresis (DGGE) replicates phylogenetic marker sequences from metagenomic DNA from environmental materials using PCR. Target sequences are amplified by GC clamped primers. The amplicons are then electrophoresed on a denaturing acrylamide gel. Denaturants form a horizontal gradient in the gel along the electric field during electrophoresis. DNA molecules encounter this denaturant gradient and get trapped as they migrate toward the cathode leading to size separation (Dimitrakopoulou, Tzimotoudis, & Vantarakis, 2023). In theory, each species of fungi is expected to produce a distinct band on the gel due to the unique sequences of their phylogenetic markers. The intensity of these bands correlates with the abundance or richness of that specific species within the microbial community present in the sample. Terminal restriction fragment length polymorphism (t-RFLP) serves as a powerful tool to discern and analyze the diversity, structural differences and shifts in endophytic fungal communities (Dickie & FitzJohn, 2007). Its use helps researchers understand the complicated relationship between host plants and their bacterial residents, revealing the dynamics of these microbial ecosystems across plant species and environments. Restrictive enzymes cleave DNA at certain places in purified amplicons to create labeled terminal restriction fragments (t-RFs) with different sizes. These unique-sized pieces are separated using capillary electrophoresis. Each separated amplicon is an operational taxonomic unit (OTU), allowing plant-associated fungal species to be distinguished by their terminal restriction fragment lengths. Despite its limitations, t-RFLP gives valuable insights into endophytic microbial communities. This approach cannot accurately predict microbial community composition at the species level. It often ignores distinctive or rare elements in these groups (Nikolcheva & Bärlocher, 2020). In order to overcome these constraints and deepen our comprehension of endophytic microbial communities, it is imperative to include high-resolution models and innovative methodologies. The use of advanced high-resolution methods would enable researchers to do comprehensive structural and functional analyses of these microbial communities at a more intricate level.

3.3.3 Metagenomic Studies using High-Throughput Sequencing (HTS)

The advancements in sequencing technological are thought to be about the length of the data that can be read. This study presents a comprehensive examination of the aforementioned technologies, categorizing them into two distinct paradigms: short-read, also referred to as "second-generation" technologies, and long-read, alternatively known as "third-generation" technologies (Meslier et al., 2022). In this context, the dominant technologies for short-read sequencing are Illumina and Ion Torrent, whereas Pacific Biosciences and Oxford Nanopore technologies represent the long-read sequencing techniques(Cuber et al., 2023).

Second- and third-generation high-throughput sequencing (HTS) is used to study fungal populations. The variety of HT technologies increases the variety of laboratory procedures and analytical processes. Multiple DNA sequencing systems have shaped genetic research. Roche's 454 pyrosequencing, discontinued in 2016, was essential for fungal community investigations despite the expense and DNA region restrictions (Trofymow et al., 2023). Thermo Fisher Scientific's Ion Torrent PGM and GeneStudio provided cost-effective options but grappled with limitations in homopolymer-rich areas, straddling the space between pyrosequencing and Illumina technologies. Illumina, represented by the MiSeq, HiSeq and NovaSeq platforms, emerged as the dominant force in second-generation sequencing, boasting shorter read lengths yet ensuring high quality and substantial throughput (Benz & Mitra, 2023; Li et al., 2022; Tedersoo et al., 2022). The MiSeq Illumina, in particular, found extensive use in metabarcoding, metagenomic and metatranscriptomic investigations (Rui et al., 2019). Meanwhile, Pacific Biosciences' PacBio RSII and Sequel platforms specialized in generating lengthy reads, exceling in producing high-quality sequences crucial for specific amplicons, thereby enabling precise phylogenetic analysis. On another front, Oxford Nanopore's MinION, GridION and PrometION brought affordability and on-site sequencing capabilities but faced limitations due to a relatively higher error rate, restricting their application in microbial identification (Tedersoo, Tooming-Klunderud, & Anslan, 2018). However, ongoing efforts persist in enhancing nanopore sequencing's viability for ecological studies beyond genome sequencing. In HTS experiments, choosing the right primers, amplification conditions and genetic markers is essential. The principal fungal DNA barcode is the internal transcribed spacer (ITS) region of the ribosomal RNA (rRNA) operon, and sequencing it is a basis of HTS-based distinction (Baldrian et al., 2022). Figure 3.3 depicts the schematic representation of fungal rDNA.

ITS is not sufficiently accurate to distinguish fungal species. Alternative markers for HTS-based distinction of these lineages have not been used. SSU (18S) and LSU (28S) nuclear ribosomal RNA (rRNA) genes are commonly used in aquatic and arbuscular mycorrhizal fungus research.

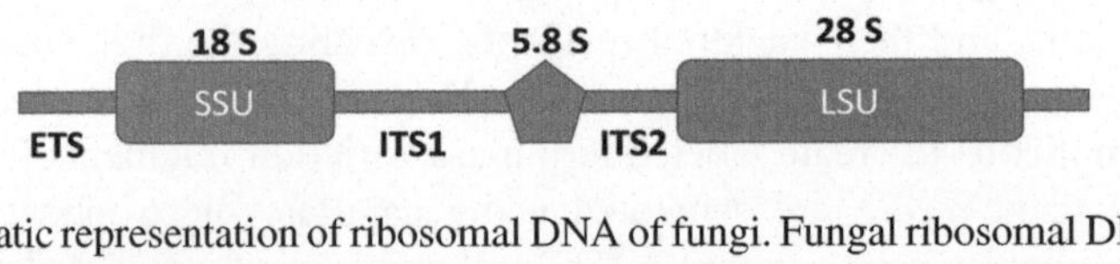

FIGURE 3.3 Schematic representation of ribosomal DNA of fungi. Fungal ribosomal DNA genes showing the position of ETS (external transcribed spacer) 18S SSU (small sub-unit), ITS1, 5.8S, 28S LSU (large sub-unit).

For ascomycetes and basidiomycetes, these markers are usually only relevant for taxonomic classifications beyond the species level, occasionally even the genus level (Vu et al., 2019). This is due to the little variation seen in SSU and LSU sequences across species within these fungal groupings, which hinders accurate and reliable determinations. High-throughput sequencing (HTS) studies primarily concentrate on the ITS1 or ITS2 subregion, which generally spans 250–400 bases (Blumenstein et al., 2021). The ITS2 subregion offers several advantages such as less length variation and a greater number of universal primer sites, which therefore leads to a decreased taxonomic bias compared to ITS1. It has been shown that the ITS1 and ITS2 subregions are suitable for second-generation HTS methods. However, the use of third-generation techniques, such as the Pacific Biosciences (PacBio) and Oxford Nanopore platforms, allows for the sequencing of the complete ITS region, as well as portions or the entirety of the adjacent rRNA genes. Targeting the entire ITS region, as opposed to its individual subregions, offers several advantages in terms of taxonomic resolution and minimizing the amplification of non-viable organisms (Nilsson, Anslan, et al., 2019). However, a drawback of this approach is its limited effectiveness when dealing with low-quality samples, such as deteriorated DNA where complete sequencing of the ITS region is no longer feasible (Tedersoo et al., 2017).

Since most primers have amplification limitations it is more relevant to use the ITS2 subsection region or the whole ITS region by considering a combination of degenerate primers to have extensive coverage of the fungal kingdom (Tedersoo et al., 2015). During HTS, primers inculcate unique identification indices, called tags/molecular barcodes, to identify samples during bioinformatics analysis. To prevent cross-contamination, barcodes are usually six bases long with three to four nucleotides variation. Additionally, sequencing adaptors for certain platforms can simplify processing according on the platform employed (Lindahl et al., 2013). Illumina sequencing uses barcodes during PCR amplification before HTS, which makes generating huge libraries time-consuming and challenging. NGS studies have used Ion Torrent, Illumina and Pacific Biosciences sequencing technology. Illumina sequencing is the primary sequencing method. Several businesses use sequencing by synthesis. This method uses fluorescently tagged reversible-terminator nucleotides (Rodriguez & Krishnan, 2023). In 2011, Illumina introduced the MiSeq, a machine capable of generating results at a faster pace. Hence it is crucial for endophytic research (Quail et al., 2012).

3.3.4 Data Preprocessing, Quality Analysis and Computational Tools for Fungal Taxonomic Annotations

Following sequencing, demultiplexing is a crucial processing step in high-throughput sequencing data analysis, enabling the assignment of sequences to their respective samples based on barcode information (Ståhlberg et al., 2017). Barcodes represent unique sequences integrated into individual genetic samples prior to their pooling. The sequencing facility may provide either separate FASTQ files for each sample or a single file containing all sequences with attached barcodes, necessitating subsequent splitting (Aragona et al., 2022). When sequences are merged, a "mapping" file is needed to link barcodes to samples. Illumina paired sequencing reads are suitable for reading short regions like ITS1 or ITS2. Illumina single-end reads can be used to sequence the whole ITS region. To improve taxonomic resolution, large parts of the genes encoding SSU, 5.8S and LSU that border the ITS1 and ITS2 subregions must be removed. USEARCH and VSEARCH filter chimaeras effectively too (Rognes et al., 2016).

The most common method for microbial high-throughput sequencing is grouping sequences into OTUs that approximate species categorization. It is customary to utilize ITS sequence similarity levels from 97.0% to 98% to balance taxonomic resolution at the species level with error mitigation during sequencing (Baldrian et al., 2021). To conserve taxon-specific markers, significant taxonomic groupings may have barcoding criteria. In fungal ITS data sets, single-linkage clustering algorithms perform well. These methods add a new sequence to an operational taxonomic unit (OTU) if it matches at least one sequence in that OTU (Edgar, 2010; Lindahl et al., 2013). Commonly used approaches for inferring OTUs include algorithms found in widely used software tools such as MOTHUR, USEARCH and VSEARCH (Op De Beeck et al., 2015; Schloss Patrick et al., 2009).

OTU clustering can be conducted as closed reference, de novo or open reference. Closed-reference clustering uses a reference sequence database to organize input sequences. De novo clustering classifies input sequences by pairwise similarity to all dataset sequences. The program constructs OTUs using a two-step process. The first step is to create a starting group of OTUs by gradually combining comparable amplicons. The abundance of amplicons is then used to uncover the internal structures of the OTUs and, if needed, divide them into smaller sub-OTUs. The MOTHUR software is a bundle of programs which clusters sequences according to genomic distance using three de novo clustering algorithms: average neighbor, furthest neighbor and closest neighbor. QIIME (v1.8.0) supports UCLUST and USEARCH (versions 5.2 and 6.1). Both programs have the capability to do de novo, closed-reference and open-reference clustering, with the exception of USEARCH 5.2 (Kopylova et al., 2016).

Upon picking and grouping OTUs, taxonomic classification is crucial. OTU taxonomy relies on reference database sequence similarity searches. The SILVA SSU Ref NR database for SSU sequences and the RDP database for LSU sequences are utilized for fungal kingdom reference databases. Also, Maarj-AM has been used to identify arbuscular mycorrhizal fungi SSU sequencing data (Öpik et al., 2010). There are three primary databases that cover the whole fungal kingdom: INSDC, UNITE and Warcup ITS (Deshpande et al., 2016; Nilsson, Larsson, et al., 2019; Nilsson et al., 2015). The UNITE database contains the latest sequencing and taxonomy data. Several unnamed fungal species-level lineages are also present. An approximate similarity of 97.0–98.5% in a pairwise alignment with query coverage of more than 90% confirms species-level taxonomic identification. For taxonomic annotation, a similarity of 90–97% implies genus, 80–85% family, 75–80% order, 70–75% class and ≥70% phylum. Some observations similar to fungal reference sequences may indicate OTUs that are fungi but may not be. Instead, they may be eukaryotic Stramenopiles (heterokonts), Alveolates and Rhizaria (SAR) supergroup members (Fadiji & Babalola, 2020). Table 3.1 presents a compilation of frequently used computational platforms and tools in the analysis of endophytic microbiomes in metagenomics research. Figure 3.4. depicts a flow diagram of important steps in metabarcode sequencing and data analysis.

TABLE 3.1
Different Computational Tools and Software to Study Endophytic Microbiomes

Software/platform	Module/program	Function	References
Trimmomatic Software	–	Read trimming	(Elmagzob, Ibrahim, & Zhang, 2019)
HMMER	Hmm-Search Program	Read trimming and alignment	(Chauhan, Gutama, & Aysa, 2019)
ACE Pyrotag pipeline	–	Read trimming	(Elmagzob, Ibrahim, & Zhang, 2019)
MUSCLE	Multiple sequence alignment	Alignment of sequence reads	(Abdelfattah et al., 2016)
MOTHUR	–	Determination of the OTUs or taxa, which were differentially represented between the samples	(Schloss Patrick et al., 2009; Tian, Cao, & Zhang, 2015)

TABLE 3.1 (Continued)
Different Computational Tools and Software to Study Endophytic Microbiomes

Software/platform	Module/program	Function	References
MG-RAST	–	Quality control	(Tian, Cao, & Zhang, 2015)
	KEGG	Determination of the metabolic pathway	(Tian, Cao, & Zhang, 2015)
	SEED	Annotation of Genomes	(Tian, Cao, & Zhang, 2015)
	PCoA	Principal coordinate analysis	(Tian, Cao, & Zhang, 2015)
	CCA	Canonical correspondence analysis	(Furtado Bliss et al., 2019)
CLC Genomics workbench	De novo assembly	Assembly of Genome	(Tian, Cao, & Zhang, 2015)
	RDP Classifier	Quality control	(Elmagzob, Ibrahim, & Zhang, 2019)
	Shannon, ACE, chao1 and Simpson index	Abundance base coverage, richness and diversity estimation	(Abdelfattah et al., 2016)
	UniFrac	Principal coordinate analysis	(Lozupone et al., 2011)
	Against SILVA database	Sequences quality check and alignment	(Akinsanya et al., 2015)
FLASH	–	Merged paired-end reads	(Elmagzob, Ibrahim, & Zhang, 2019)
SILVA NGS Server	–	Taxonomic classification	(Tian, Cao, & Zhang, 2015)
MEGA/MEGA5	Neighbor-Joining methods and Jukes-Canter Model	Constructions neighbor-joining trees and phylogenetic trees	(Tamura et al., 2013)
MEGAN	Lower common ancestor (LCA) algorithm	Taxonomic assignment of genes, species richness analysis	(Zimmerman & Vitousek, 2012)
UCHIME	–	Removal and detection of chimeras	(Akinsanya et al., 2015; Furtado Bliss et al., 2019)
UPARSE	–	Clustering of OTU's	(Akinsanya et al., 2015)
UNITE	–	Taxonomical classification	(Furtado Bliss et al., 2019)
STAMP	Two-sided Welch's t-test	Pairwise statistical comparison of taxonomy between samples	(Furtado Bliss et al., 2019)
QIIME	–	Denoising of sequences	(Elmagzob, Ibrahim, & Zhang, 2019)
	Blast Taxon Assigner	Assigning taxonomy to the sequences	(Caporaso et al., 2010)
USEARCH	Fast-filter command	Trimming of demultiplexed reads	(Wemheuer et al., 2019)
	Cluster-OTUs command	Clustering of dereplicated reads	(Wemheuer et al., 2019)
	Usearch-globus command	Assigning of OTUs to reads	(Wemheuer et al., 2019)
Blast against RDP database	–	Phylogenetic classification	(Abdelfattah et al., 2016)
Blast against COG database		Functional characterization	(Galperin et al., 2015)
Blast 2× against KEGG database		Functional annotation of reads	(Kanehisa et al., 2014)
PyNAST	Against Greengenes core reference set	Alignment of representative OTUs sequences	(Adeleke et al., 2018)

(continued)

TABLE 3.1 (Continued)
Different Computational Tools and Software to Study Endophytic Microbiomes

Software/platform	Module/program	Function	References
R software Package	Shanon index, Pielou's index	Richness, diversity and evenness estimation	(Cui, Vijayakumar, & Zhang, 2018, Wemheuer et al., 2019)
PAST	NMDS	Visualization of beta diversity metrics	(Chauhan, Gutama, & Aysa, 2019, Elmagzob, Ibrahim, & Zhang, 2019)
	ANOSIM	Analysis of the strength and statistical significance of sample groupings	(Carrell & Frank, 2015)
	PERMANOVA	Analysis of the strength and statistical significance of sample groupings	(Carrell & Frank, 2015)
UNITE	–	Taxonomical classification	(Abdelfattah et al., 2016)
FAST UniFrac	PCoA	Evaluation of differences in community structure and parameters	(Lozupone et al., 2011)
DEseq	–	Differential test among OTUs	(Gdanetz & Trail, 2017; Taudière et al., 2018)
RDP (RDP database Project)	Tree builder, RDP classifier and aligner	Sorting of sequences, trimming and alignment, biodiversity analysis and taxonomic classification of sequences	(Dong et al., 2018; Elmagzob, Ibrahim, & Zhang, 2019)
Infernal aligner	–	Aligning of multiple clean sequences	(Chimwamurombe, Grönemeyer, & Reinhold-Hurek, 2016)
Chimera slayer	–	Removal of chimera from sequences	(Carrell & Frank, 2015)
XLSTAT	Canonical correspondence analysis CCA	Determination of correlations between the microbial community	(Johnston-Monje & Raizada, 2011)
CANOCO	Redundancy analysis (RDA)	Visualized correlations between community structure and parameter	(Geisen et al., 2017)
PICRUSt	–	Functional prediction	(Huang, 2019)
LEfSe	–	Determination of the OTUs between two or more groups using relative abundances	(Sun et al., 2019)
DADA2	–	Determination of the OTUs or taxa, which were differentially represented between the samples	(Callahan et al., 2016; Kuźniar et al., 2020; Wagner, Busby, & Balint-Kurti, 2020)

Source: Data from Fadiji & Babalola (2020).

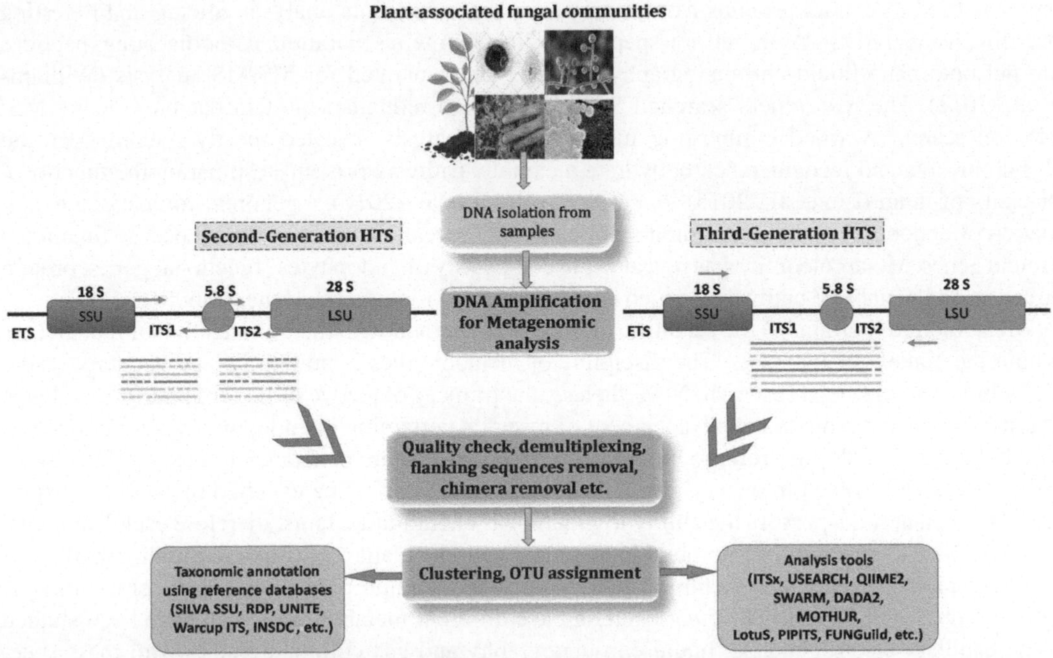

FIGURE 3.4 Prime stages in metabarcode analysis of plant-associated fungal communities. Diversity is studied by sequence analysis of markers such as ITS, SSU, LSU, etc. by different HTS sequencing approaches.

3.4 EXPLORING THE RELATIONSHIP BETWEEN FUNGAL DIVERSITY AND FUNCTIONAL ROLES

3.4.1 Metatranscriptomics

A useful tool for studying plant species' microbial populations is transcriptomics (Rastogi & Sani, 2011). The molecular approach compares transcriptomes of interacting species groups according to environmental changes using the ribosomal RNA operon or enzyme-coding genes (Franzosa et al., 2015). Reverse transcribing enzyme-encoding genes to identify different transcripts in fungal communities provides insights about their ecological functioning, however primer selectivity limits community research. A comprehensive study of transcribed enzyme-coding genes in an environmental sample will help us understand fungal metabolic diversity, functional activities, community dynamics and plant relationships. Functional roles of fungi-related grapevine plant woods carbohydrate metabolism (CAZymes) genes were more prevalent, while most expressed genes included peroxidases and cytochrome P450 (Nerva et al., 2022). These metabolites may be utilized by trees in exchange for the carbon sources that are derived from photosynthesis (Larsen et al., 2011). The metatranscriptomic research conducted on soybean plants has yielded findings indicating the existence of many short RNA sequences that are not associated with the soybean genome. The existence of several pathogenic, symbiotic and free-living microorganisms in different samples of soybean plants was demonstrated by a comparative analysis of the acquired sequences (Molina et al., 2012).

3.4.2 Metaproteomics and Metabolomics

This method investigates microorganism-synthesized proteins during sampling, reconstructing ecosystem-functioning microbial activity and metabolic pathways. Researchers have reported metaproteomic data from soil, activated sludge and leaf litter (Benndorf et al., 2007). SDS-phenol buffer is used to extract proteins from samples for metaproteomic analysis. Slicing and digesting one-dimensional SDS-PAGE gels into peptides follows protein separation. After digesting, peptides are put onto nano-liquid chromatography columns and separated for MS/MS analysis (Williams et al., 2012). The researchers searched SwissProt, a non-redundant protein database, using MS/MS ion search. A wood composting metaproteome analysis revealed mostly Ascomycota and Basidiomycota and recognized carbohydrate metabolic pathway proteins, estimating the functional diversity of fungi (Liu et al., 2015). According to Kaul et al. (2016), a genomic comparison examination of endophytes in water-limited soil plants may reveal bacteria- or fungi-specific functional protein genes. Metaproteomics has revealed the complexity of endophytes' functional genes, protein structure and metabolic pathways, which could be utilized to understand their functions.

Metabolomics pertains to the scientific discipline concerned with the investigation of metabolites within the context of live cells. The discipline of metabolomics is now in its nascent stage, especially in its use in fungal research. Nevertheless, the primary objective of disseminating knowledge via literature is to facilitate the advancement of research pertaining to endophytic bacteria by using metabolomics. Metabolites refer to biological compounds that are generated as a result of the metabolic processes involved in energy generation inside living cells. The excreted metabolites originating from plant roots possess the ability to function as chemo-attractants, therefore establishing the root zone as a focal point for microbial interactions with the plant root (Shah & Smith, 2020).

Using metabolomics in agricultural production can examine plant metabolite composition at different phases of growth. Different bacteria have different metabolites, which have been studied using capillary electrophoresis, liquid chromatography and gas chromatography with mass spectrometry. NMR is also utilized to determine microorganism metabolic pathways and metabolites. Microbial metabolites help plants grow and survive in harsh environments. The research results revealed a wide range of functional secondary metabolites that induce changes in microbial adaptability and help us understand plant–microbe interactions.

3.5 CONCLUSION

It is clear that microbe–plant interactions shape fungal populations in the phyllosphere, rhizosphere and pedosphere. Thus, ecological science has moved from studying individual plant and fungal species to holobionts. Identifying holobiont systems in natural ecosystems is difficult because plant community structure affects microbial communities associated with each host. Microbial community composition and function change over time. High-throughput sequencing and meta-omics are revealing functional variations between ecological fungal populations. This chapter discusses metagenomics and computational methods for analyzing sequenced data from environmental samples of plant-associated fungal populations. Several metagenomic methodologies and computational tools were used to study fungal communities' evolutionary relationship, taxonomic makeup and functions with plants. Taxonomic identification requires amplicon sequencing, which can be complicated by sequencing errors and PCR bias. Chimaera removal and de-noising can help. GenBank contains many HTS-derived ITS sequences from environmental materials, including uncultivable and dark taxa. This shows how modern methods were used to characterize important fungi. The advancements in DNA and RNA sequencing are helping researchers investigate fungi and comprehend their biology, ecology and classification. Over a billion HTS-derived ITS values are available in databases for mycological study. Using these data is essential for finding previously overlooked facts and creating new ideas, research questions and theories. Depositing all fungal cultures in culture collections and making them available to academics may improve fundamental taxonomy study. Focusing on active taxa and temporal variability may improve microbial community composition forecasting. Thus, meta-omics approaches give the best foundation for understanding plant-associated fungal communities' potentialities. These cutting-edge methods reveal the complex interactions between plants and microorganisms and unlock their sustainable agriculture potential.

REFERENCES

Abdelfattah, Ahmed, Michael Wisniewski, Maria Giulia Li Destri Nicosia, Santa Olga Cacciola, and Leonardo Schena. 2016. "Metagenomic analysis of fungal diversity on strawberry plants and the effect of management practices on the fungal community structure of aerial organs." *PLOS ONE* 11 (8):e0160470. doi: 10.1371/journal.pone.0160470.

Adeleke, Bartholomew Saanu, Ayansina Segun Ayangbenro, and Olubukola Oluranti Babalola. 2021. "Genomic assessment of Stenotrophomonas indicatrix for improved sunflower plant." *Current Genetics* 67 (6):891–907. doi: 10.1007/s00294-021-01199-8.

Adeleke, Bartholomew Saanu, Daniel Muller, and Olubukola Oluranti Babalola. 2023. "A metagenomic lens into endosphere microbial communities, promises, and discoveries." *Letters in Applied Microbiology* 76 (2):ovac030. doi: 10.1093/lambio/ovac030.

Adeleke, Rasheed A., Asnath R. Mashiane, Carlos C. Bezuidenhout, and George Chirima. 2018. "Community composition and functions of endophytic bacteria of Bt maize." *South African Journal of Science* 114 (7/8). doi: 10.17159/sajs.2018/20170018.

Adnan, Muhammad, Waqar Islam, Asad Shabbir, Khalid Ali Khan, Hamed A. Ghramh, Zhiqun Huang, Han Y. H. Chen, and Guo-dong Lu. 2019. "Plant defense against fungal pathogens by antagonistic fungi with Trichoderma in focus." *Microbial Pathogenesis* 129:7–18. doi: https://doi.org/10.1016/j.micpath.2019.01.042.

Aguilar-Paredes, Ana, Gabriela Valdés, Nicole Araneda, Emky Valdebenito, Felipe Hansen, and Marco Nuti. 2023. "Microbial Community in the Composting Process and Its Positive Impact on the Soil Biota in Sustainable Agriculture." *Agronomy* 13 (2). doi:10.3390/agronomy13020542.

Akinsanya, Mushafau Adewale, Joo Kheng Goh, Siew Ping Lim, and Adeline Su Yien Ting. 2015. "Metagenomics study of endophytic bacteria in Aloe vera using next-generation technology." *Genomics Data* 6:159–163. doi: https://doi.org/10.1016/j.gdata.2015.09.004.

Aragona, Maria, Anita Haegi, Maria T. Valente, Luca Riccioni, Laura Orzali, Salvatore Vitale, Laura Luongo, and Alessandro Infantino. 2022. "New-Generation Sequencing Technology in Diagnosis of Fungal Plant Pathogens: A Dream Comes True?" *Journal of Fungi* 8 (7). doi:10.3390/jof8070737.

Arantes, Valdeir, Adriane M. F. Milagres, Timothy R. Filley, and Barry Goodell. 2011. "Lignocellulosic polysaccharides and lignin degradation by wood decay fungi: the relevance of nonenzymatic Fenton-based reactions." *Journal of Industrial Microbiology and Biotechnology* 38 (4):541–555. doi: 10.1007/s10295-010-0798-2.

Baldrian, Petr, Tomáš Větrovský, Clémentine Lepinay, and Petr Kohout. 2021. "High-throughput sequencing view on the magnitude of global fungal diversity." *Fungal Diversity* 114:539–547.

Baldrian, Petr, Tomáš Větrovský, Clémentine Lepinay, and Petr Kohout. 2022. "High-throughput sequencing view on the magnitude of global fungal diversity." *Fungal Diversity* 114 (1):539–547. doi: 10.1007/s13225-021-00472-y.

Bastías, Daniel A., Linda J. Johnson, and Stuart D. Card. 2020. "Symbiotic bacteria of plant-associated fungi: friends or foes?" *Current Opinion in Plant Biology* 56:1–8. doi: https://doi.org/10.1016/j.pbi.2019.10.010.

Benndorf, Dirk, Gerd U. Balcke, Hauke Harms, and Martin von Bergen. 2007. "Functional metaproteome analysis of protein extracts from contaminated soil and groundwater." *The ISME Journal* 1 (3):224–234. doi: 10.1038/ismej.2007.39.

Benz, Saskia, and Suparna Mitra. 2023. "From Genomics to Metagenomics in the Era of Recent Sequencing Technologies." In *Metagenomic Data Analysis*, edited by Suparna Mitra, 1–20. New York, NY: Springer US.

Bhat, Basharat Ahmad, Lubna Tariq, Showkat Nissar, Sheikh Tajamul Islam, Shahid Ul Islam, Zahid Mangral, Noshin Ilyas, Riyaz Z. Sayyed, Govarthanan Muthusamy, Woong Kim, and Tanvir Ul Hasan Dar. 2022. "The role of plant-associated rhizobacteria in plant growth, biocontrol and abiotic stress management." *Journal of Applied Microbiology* 133 (5):2717–2741. doi: 10.1111/jam.15796.

Bhunjun, Chitrabhanu S., Tuula Niskanen, Nakarin Suwannarach, Nopparat Wannathes, Yi-Jyun Chen, Eric H. C. McKenzie, Sajeewa S. N. Maharachchikumbura, Bart Buyck, Chang-Lin Zhao, Yu-Guang Fan, Jing-Yi Zhang, Asha J. Dissanayake, Diana S. Marasinghe, Ruvishika S. Jayawardena, Jaturong Kumla, Mahajabeen Padamsee, Ya-Ya Chen, Kare Liimatainen, Joseph F. Ammirati, Chayanard Phukhamsakda, Jian-Kui Liu, Wiphawanee Phonrob, Émile Randrianjohany, Sinang Hongsanan, Ratchadawan

Cheewangkoon, Digvijayini Bundhun, Surapong Khuna, Wen-Jie Yu, Lun-Sha Deng, Yong-Zhong Lu, Kevin D. Hyde, and Saisamorn Lumyong. 2022. "The numbers of fungi: are the most speciose genera truly diverse?" *Fungal Diversity* 114 (1):387–462. doi: 10.1007/s13225-022-00501-4.

Blumenstein, Kathrin, Eeva Terhonen, Hui Sun, and Fred O. Asiegbu. 2021. "Chapter 3 – Methods for studying the forest tree microbiome." In *Forest Microbiology*, edited by Fred O. Asiegbu and Andriy Kovalchuk, 35–58. Academic Press.

Bollmann-Giolai, Anita, Jacob G. Malone, and Sanu Arora. 2022. "Diversity, detection and exploitation: linking soil fungi and plant disease." *Current Opinion in Microbiology* 70:102199. doi: https://doi.org/10.1016/j.mib.2022.102199.

Bareither, Christopher A., Georgia L. Wolfe, Katherine D. McMahon, and Craig H. Benson. 2013. "Microbial diversity and dynamics during methane production from municipal solid waste." *Waste Management* 33 (10):1982–1992. doi: 10.1016/j.wasman.2012.12.013.

Callahan, Benjamin J., Paul J. McMurdie, Michael J. Rosen, Andrew W. Han, Amy Jo A. Johnson, and Susan P. Holmes. 2016. "DADA2: High-resolution sample inference from Illumina amplicon data." *Nature Methods* 13 (7):581–583. doi: 10.1038/nmeth.3869.

Caporaso, J. Gregory, Justin Kuczynski, Jesse Stombaugh, Kyle Bittinger, Frederic D. Bushman, Elizabeth K. Costello, Noah Fierer, Antonio Gonzalez Peña, Julia K. Goodrich, Jeffrey I. Gordon, Gavin A. Huttley, Scott T. Kelley, Dan Knights, Jeremy E. Koenig, Ruth E. Ley, Catherine A. Lozupone, Daniel McDonald, Brian D. Muegge, Meg Pirrung, Jens Reeder, Joel R. Sevinsky, Peter J. Turnbaugh, William A. Walters, Jeremy Widmann, Tanya Yatsunenko, Jesse Zaneveld, and Rob Knight. 2010. "QIIME allows analysis of high-throughput community sequencing data." *Nature Methods* 7 (5):335–336. doi: 10.1038/nmeth.f.303.

Carrell, Alyssa, and Carolin Frank. 2015. "Bacterial endophyte communities in the foliage of coast redwood and giant sequoia." *Frontiers in Microbiology* 6:1008. doi: 10.3389/fmicb.2015.01008.

Charpentier-Alfaro, Camila, Jorge Benavides-Hernández, Marco Poggerini, Alfonso Crisci, Giacomo Mele, Gianni Della Rocca, Giovanni Emiliani, Angela Frascella, Tommaso Torrigiani, and Sabrina Palanti. 2023. Wood-Decaying Fungi: From Timber Degradation to Sustainable Insulating Biomaterials Production. *Materials* 16 (9):3547. doi:10.3390/ma16093547.

Chauhan, Nitin M., Abdissa D. Gutama, and Afras Aysa. 2019. "Endophytic fungal diversity isolated from different agro-ecosystem of Enset (Ensete ventericosum) in Gedeo zone, SNNPRS, Ethiopia." *BMC Microbiology* 19 (1):172. doi: 10.1186/s12866-019-1547-y.

Chimwamurombe, Percy Maruwa, Jann Lasse Grönemeyer, and Barbara Reinhold-Hurek. 2016. "Isolation and characterization of culturable seed-associated bacterial endophytes from gnotobiotically grown Marama bean seedlings." *FEMS Microbiology Ecology* 92 (6):fiw083. doi: 10.1093/femsec/fiw083.

Coleine, Claudia, Jason E. Stajich, and Laura Selbmann. 2022. "Fungi are key players in extreme ecosystems." *Trends in Ecology & Evolution* 37 (6):517–528. doi: https://doi.org/10.1016/j.tree.2022.02.002.

Costa, Diandra A. L., Thomas C. R. Williams, Luis Henrique Ferreira do Vale, and Edivaldo X. F. Filho. 2023. "Characterization of mannanases from Clonostachys byssicola involved in the breakdown of lignocellulosic substrates." *Biocatalysis and Agricultural Biotechnology* 50:102680. doi: https://doi.org/10.1016/j.bcab.2023.102680.

Cuber, Piotr, Darren Chooneea, Clementine Geeves, Silvia Salatino, Thomas J. Creedy, Claire Griffin, Laura Sivess, Ian Barnes, Ben Price, and Raju Misra. 2023. "Comparing the accuracy and efficiency of third generation sequencing technologies, Oxford Nanopore Technologies, and Pacific Biosciences, for DNA barcode sequencing applications." *Ecological Genetics and Genomics* 28:100181. doi: https://doi.org/10.1016/j.egg.2023.100181.

Cui, Jin-Long, Vinod Vijayakumar, and Gang Zhang. 2018. "Partitioning of fungal endophyte assemblages in root-parasitic plant cynomorium songaricum and its host nitraria tangutorum." *Frontiers in Microbiology* 9:666. doi: 10.3389/fmicb.2018.00666.

Deshpande, Vinita, Qiong Wang, Paul Greenfield, Michael Charleston, Andrea Porras-Alfaro, Cheryl R. Kuske, James R. Cole, David J. Midgley, and Nai Tran-Dinh. 2016. "Fungal identification using a Bayesian classifier and the Warcup training set of internal transcribed spacer sequences." *Mycologia* 108 (1):1–5. doi: 10.3852/14-293.

Dickie, I. A., and R. G. FitzJohn. 2007. "Using terminal restriction fragment length polymorphism (T-RFLP) to identify mycorrhizal fungi: a methods review." *Mycorrhiza* 17 (4):259–270. doi: 10.1007/s00572-007-0129-2.

Dimitrakopoulou, Maria-Eleni, Dimosthenis Tzimotoudis, and Apostolos Vantarakis. 2023. "Emulsion Polymerase Chain Reaction Coupled with Denaturing Gradient Gel Electrophoresis for Microbial Diversity Studies." In *PCR: Methods and Protocols*, edited by Lucília Domingues, 31–39. New York, NY: Springer US.

Dong, Linlin, Ruiyang Cheng, Lina Xiao, Fugang Wei, Guangfei Wei, Jiang Xu, Yong Wang, Xiaotong Guo, Zhongjian Chen, and Shilin Chen. 2018. "Diversity and composition of bacterial endophytes among plant parts of Panax notoginseng." *Chinese Medicine* 13 (1):41. doi: 10.1186/s13020-018-0198-5.

Edgar, Robert C. 2010. "Search and clustering orders of magnitude faster than BLAST." *Bioinformatics* 26 (19):2460–2461. doi: 10.1093/bioinformatics/btq461.

Elmagzob, Abubakr A., Muhammed M. Ibrahim, and Guo-Fang Zhang. 2019. Seasonal Diversity of Endophytic Bacteria Associated with Cinnamomum camphora (L.) Presl. *Diversity* 11 (7). doi:10.3390/d11070112.

Fadiji, Ayomide Emmanuel, and Olubukola Oluranti Babalola. 2020. "Metagenomics methods for the study of plant-associated microbial communities: A review." *Journal of Microbiological Methods* 170:105860. doi: https://doi.org/10.1016/j.mimet.2020.105860.

Field, Katie J., and Silvia Pressel. 2018. "Unity in diversity: structural and functional insights into the ancient partnerships between plants and fungi." *New Phytologist* 220 (4):996–1011. doi: https://doi.org/10.1111/nph.15158.

Franzosa, Eric A., Tiffany Hsu, Alexandra Sirota-Madi, Afrah Shafquat, Galeb Abu-Ali, Xochitl C. Morgan, and Curtis Huttenhower. 2015. "Sequencing and beyond: integrating molecular 'omics' for microbial community profiling." *Nature Reviews Microbiology* 13 (6):360–372. doi: 10.1038/nrmicro3451.

Furtado Bliss, Ursula, Marcin Gołębiewski, Monika Skorupa, Piotr Hulisz, and Katarzyna Hrynkiewicz. 2019. "Bacterial and Fungal Endophytic Microbiomes of Salicornia europaea." *Applied and Environmental Microbiology* 85 (13):e00305–19. doi: 10.1128/AEM.00305-19.

Galperin, Michael Y., Kira S. Makarova, Yuri I. Wolf, and Eugene V. Koonin. 2015. "Expanded microbial genome coverage and improved protein family annotation in the COG database." *Nucleic Acids Research* 43 (D1):D261–D269. doi: 10.1093/nar/gku1223.

Gdanetz, Kristi, and Frances Trail. 2017. "The wheat microbiome uder four management strategies, and potential for endophytes in disease protection." *Phytobiomes Journal* 1 (3):158–168. doi: 10.1094/PBIOMES-05-17-0023-R.

Geisen, Stefan, Olga Kostenko, Mark C. Cnossen, Freddy C. ten Hooven, Branko Vreš, and Wim H. van der Putten. 2017. "Seed and Root Endophytic Fungi in a Range Expanding and a Related Plant Species." *Frontiers in Microbiology* 8:1645. doi: 10.3389/fmicb.2017.01645.

Götz, Monika, Helgard Nirenberg, Sibylle Krause, Heike Wolters, Siegfried Draeger, Arno Buchner, Jana Lottmann, Gabriele Berg, and Kornelia Smalla. 2006. "Fungal endophytes in potato roots studied by traditional isolation and cultivation-independent DNA-based methods." *FEMS Microbiology Ecology* 58 (3):404–413. doi: 10.1111/j.1574-6941.2006.00169.x.

Hawksworth David, L., and Robert Lücking. 2017. "Fungal Diversity Revisited: 2.2 to 3.8 Million Species." *Microbiology Spectrum* 5 (4). doi: 10.1128/microbiolspec.funk-0052-2016.

Helaly, Soleiman E., Benjarong Thongbai, and Marc Stadler. 2018. "Diversity of biologically active secondary metabolites from endophytic and saprotrophic fungi of the ascomycete order Xylariales." *Natural Product Reports* 35 (9):992–1014. doi: 10.1039/C8NP00010G.

Herman, Donald J., Mary K. Firestone, Erin Nuccio, and Angela Hodge. 2012. "Interactions between an arbuscular mycorrhizal fungus and a soil microbial community mediating litter decomposition." *FEMS Microbiology Ecology* 80 (1):236–247. doi: 10.1111/j.1574-6941.2011.01292.x.

Hildebrandt, Ulrich, Marjana Regvar, and Hermann Bothe. 2007. "Arbuscular mycorrhiza and heavy metal tolerance." *Phytochemistry* 68 (1):139–146. doi: https://doi.org/10.1016/j.phytochem.2006.09.023.

Huang, Yonghong. 2019. "Illumina-based Analysis of Endophytic Bacterial Diversity of four Allium species." *Scientific Reports* 9 (1):15271. doi: 10.1038/s41598-019-51707-7.

Johnston-Monje, David, and Manish N. Raizada. 2011. "Conservation and Diversity of Seed Associated Endophytes in Zea across Boundaries of Evolution, Ethnography and Ecology." *PLOS ONE* 6 (6):e20396. doi: 10.1371/journal.pone.0020396.

Kalu, Chimdi Mang, Henry Joseph Oduor Ogola, Ramganesh Selvarajan, Memory Tekere, and Khayalethu Ntushelo. 2021. "Fungal and metabolome diversity of the rhizosphere and endosphere of Phragmites australis in an AMD-polluted environment." *Heliyon* 7 (3):e06399. doi: https://doi.org/10.1016/j.heliyon.2021.e06399.

Kanehisa, Minoru, Susumu Goto, Yoko Sato, Masayuki Kawashima, Miho Furumichi, and Mao Tanabe. 2014. "Data, information, knowledge and principle: back to metabolism in KEGG." *Nucleic Acids Research* 42 (D1):D199–D205. doi: 10.1093/nar/gkt1076.

Kaul, Sanjana, Tanwi Sharma, and Manoj K. Dhar. 2016. ""Omics" tools for better understanding the plant–endophyte interactions." *Frontiers in Plant Science* 7:955. doi: 10.3389/fpls.2016.00955.

Kopylova, Evguenia, A. Navas-Molina Jose, Céline Mercier, Zech Xu Zhenjiang, Frédéric Mahé, Yan He, Hong-Wei Zhou, Torbjørn Rognes, J. Gregory Caporaso, and Rob Knight. 2016. "Open-Source Sequence Clustering Methods Improve the State Of the Art." *mSystems* 1 (1). doi: 10.1128/msystems.00003-15.

Künzler, Markus. 2018. "How fungi defend themselves against microbial competitors and animal predators." PLOS Pathogens 14 (9):e1007184. doi: 10.1371/journal.ppat.1007184.

Kuźniar, Agnieszka, Kinga Włodarczyk, Jarosław Grządziel, Weronika Goraj, Anna Gałązka, and Agnieszka Wolińska. 2020. "Culture-independent analysis of an endophytic core microbiome in two species of wheat: Triticum aestivum L. (cv. 'Hondia') and the first report of microbiota in Triticum spelta L. (cv. 'Rokosz')." *Systematic and Applied Microbiology* 43 (1):126025. doi: https://doi.org/10.1016/j.syapm.2019.126025.

Larsen, Peter E., Avinash Sreedasyam, Geetika Trivedi, Gopi K. Podila, Leland J. Cseke, and Frank R. Collart. 2011. "Using next generation transcriptome sequencing to predict an ectomycorrhizal metabolome." *BMC Systems Biology* 5 (1):70. doi: 10.1186/1752-0509-5-70.

Li, Puyu, Ning Ju, Shengzhuo Zhang, Yuanyuan Wang, and Yulong Luo. 2022. "Evaluation of microbial diversity of Jiangshui from the Ningxia Hui autonomous region in China." *Food Biotechnology* 36 (2):173–190. doi: 10.1080/08905436.2022.2054818.

Lindahl, Björn D., R. Henrik Nilsson, Leho Tedersoo, Kessy Abarenkov, Tor Carlsen, Rasmus Kjøller, Urmas Kõljalg, Taina Pennanen, Søren Rosendahl, Jan Stenlid, and Håvard Kauserud. 2013. "Fungal community analysis by high-throughput sequencing of amplified markers – a user's guide." *New Phytologist* 199 (1):288–299. doi: https://doi.org/10.1111/nph.12243.

Liu, Dongming, Mingxiao Li, Beidou Xi, Yue Zhao, Zimin Wei, Caihong Song, and Chaowei Zhu. 2015. "Metaproteomics reveals major microbial players and their biodegradation functions in a large-scale aerobic composting plant." *Microbial Biotechnology* 8 (6):950–60. doi: 10.1111/1751-7915.12290.

Lozupone, Catherine, Manuel E. Lladser, Dan Knights, Jesse Stombaugh, and Rob Knight. 2011. "UniFrac: an effective distance metric for microbial community comparison." *The ISME Journal* 5 (2):169–172. doi: 10.1038/ismej.2010.133.

Luo, Zhi-Bin, Chenhan Wu, Chao Zhang, Hong Li, Ulrike Lipka, and Andrea Polle. 2014. "The role of ectomycorrhizas in heavy metal stress tolerance of host plants." *Environmental and Experimental Botany* 108:47–62. doi: https://doi.org/10.1016/j.envexpbot.2013.10.018.

Lyu, Dongmei, Jonathan Zajonc, Antoine Pagé, Cailun A. S. Tanney, Ateeq Shah, Nadia Monjezi, Levini A. Msimbira, Mohammed Antar, Mahtab Nazari, Rachel Backer, and Donald L. Smith. 2021. "Plant Holobiont theory: The phytomicrobiome plays a Central Role in Evolution and Success." *Microorganisms* 9 (4). doi:10.3390/microorganisms9040675.

Manoharachary, C., K. Sridhar, Reena Singh, Alok Adholeya, T. S. Suryanarayanan, Seema Rawat, and B. N. Johri. 2005. "Fungal biodiversity: Distribution, conservation and prospecting of fungi from India." *Current Science* 89 (1):58–71.

Maropola, Mapula Kgomotso Annah, Jean-Baptiste Ramond, and Marla Trindade. 2015. "Impact of metagenomic DNA extraction procedures on the identifiable endophytic bacterial diversity in Sorghum bicolor (L. Moench)." *Journal of Microbiological Methods* 112:104–117. doi: https://doi.org/10.1016/j.mimet.2015.03.012

Meslier, Victoria, Benoit Quinquis, Kévin Da Silva, Florian Plaza Oñate, Nicolas Pons, Hugo Roume, Mircea Podar, and Mathieu Almeida. 2022. "Benchmarking second and third-generation sequencing platforms for microbial metagenomics." *Scientific Data* 9 (1):694. doi: 10.1038/s41597-022-01762-z.

Molina, L. G., G. C. da Fonseca, G. L. de Morais, L. F. de Oliveira, J. B. de Carvalho, F. R. Kulcheski, and R. Margis. 2012. "Metatranscriptomic analysis of small RNAs present in soybean deep sequencing libraries." *Genetics Molecular Biology* 35 (1 (suppl)):292–303. doi: 10.1590/s1415-47572012000200010.

Nerva, Luca, Jadran F. Garcia, Francesco Favaretto, Gaetano Giudice, Loredana Moffa, Marco Sandrini, Dario Cantu, Alessandro Zanzotto, Massimo Gardiman, Riccardo Velasco, Giorgio Gambino, and Walter

Chitarra. 2022. "The hidden world within plants: metatranscriptomics unveils the complexity of wood microbiomes." *Journal of Experimental Botany* 73 (8):2682–2697. doi: 10.1093/jxb/erac032.

Neves, Susana S., and Laura L. Forrest. 2011. "Plant DNA Sequencing for Phylogenetic Analyses: From Plants to Sequences." In *Network Biology: Methods and Applications*, edited by Gerard Cagney and Andrew Emili, 183–235. Totowa, NJ: Humana Press.

Nikolcheva, Liliya G., and Felix Bärlocher. 2020. "Terminal Restriction Fragment Length Polymorphism (T-RFLP) to Estimate Fungal Diversity." In *Methods to Study Litter Decomposition: A Practical Guide*, edited by Felix Bärlocher, Mark O. Gessner and Manuel A. S. Graça, 311–318. Cham: Springer International Publishing.

Nilsson, R. Henrik, Sten Anslan, Mohammad Bahram, Christian Wurzbacher, Petr Baldrian, and Leho Tedersoo. 2019. "Mycobiome diversity: high-throughput sequencing and identification of fungi." *Nature Reviews Microbiology* 17 (2):95–109. doi: 10.1038/s41579-018-0116-y.

Nilsson, R. Henrik, Leho Tedersoo, Martin Ryberg, Erik Kristiansson, Martin Hartmann, Martin Unterseher, Teresita M. Porter, Johan Bengtsson-Palme, Donald M. Walker, Filipe de Sousa, Hannes Andres Gamper, Ellen Larsson, Karl-Henrik Larsson, Urmas Kõljalg, Robert C. Edgar, and Kessy Abarenkov. 2015. "A comprehensive, automatically updated fungal ITS sequence dataset for Reference-Based Chimera Control in environmental sequencing efforts." *Microbes and Environments* 30 (2):145–150. doi: 10.1264/jsme2.ME14121.

Nilsson, Rolf Henrik, Karl-Henrik Larsson, Andy F S Taylor, Johan Bengtsson-Palme, Thomas S. Jeppesen, Dmitry Schigel, Peter Kennedy, Kathryn Picard, Frank Oliver Glöckner, Leho Tedersoo, Irja Saar, Urmas Kõljalg, and Kessy Abarenkov. 2019. "The UNITE database for molecular identification of fungi: handling dark taxa and parallel taxonomic classifications." *Nucleic Acids Research* 47 (D1):D259–D264. doi: 10.1093/nar/gky1022.

Op De Beeck, Michiel, Bart Lievens, Pieter Busschaert, Francois Rineau, Mark Smits, Jaco Vangronsveld, and Jan V. Colpaert. 2015. "Impact of metal pollution on fungal diversity and community structures." *Environmental Microbiology* 17 (6):2035–2047. doi: https://doi.org/10.1111/1462-2920.12547.

Öpik, M., A. Vanatoa, E. Vanatoa, M. Moora, J. Davison, J. M. Kalwij, Ü Reier, and M. Zobel. 2010. "The online database MaarjAM reveals global and ecosystemic distribution patterns in arbuscular mycorrhizal fungi (Glomeromycota)." *New Phytologist* 188 (1):223–241. doi: https://doi.org/10.1111/j.1469-8137.2010.03334.x.

Porebski, Sue, L. Grant Bailey, and Bernard R. Baum. 1997. "Modification of a CTAB DNA extraction protocol for plants containing high polysaccharide and polyphenol components." *Plant Molecular Biology Reporter* 15 (1):8–15. doi: 10.1007/BF02772108.

Quail, Michael A., Miriam Smith, Paul Coupland, Thomas D. Otto, Simon R. Harris, Thomas R. Connor, Anna Bertoni, Harold P. Swerdlow, and Yong Gu. 2012. "A tale of three next generation sequencing platforms: comparison of Ion Torrent, Pacific Biosciences and Illumina MiSeq sequencers." *BMC Genomics* 13 (1):341. doi: 10.1186/1471-2164-13-341.

Quiza, Liliana, Marc St-Arnaud, and Etienne Yergeau. 2015. "Harnessing phytomicrobiome signaling for rhizosphere microbiome engineering." *Frontiers in Plant Science* 6:507. doi: 10.3389/fpls.2015.00507.

Rastogi, Gurdeep, and Rajesh K. Sani. 2011. "Molecular Techniques to Assess Microbial Community Structure, Function, and Dynamics in the Environment." In Microbes and Microbial Technology: Agricultural and Environmental Applications, edited by Iqbal Ahmad, Farah Ahmad and John Pichtel, 29–57. New York, NY: Springer New York. doi : 10.1007/978-1-4419-7931-5_2

Rodriguez, Raphaël, and Yamuna Krishnan. 2023. "The chemistry of next-generation sequencing." *Nature Biotechnology* 41 (12):1709–1715. doi: 10.1038/s41587-023-01986-3.

Rognes, Torbjørn, Tomáš Flouri, Ben Nichols, Christopher Quince, and Frédéric Mahé. 2016. "VSEARCH: a versatile open source tool for metagenomics." *PeerJ* 4:e2584. doi: 10.7717/peerj.2584.

Rui, Ying, Peng Wan, Guijie Chen, Minhao Xie, Yi Sun, Xiaoxiong Zeng, and Zhonghua Liu. 2019. "Analysis of bacterial and fungal communities by Illumina MiSeq platforms and characterization of Aspergillus cristatus in Fuzhuan brick tea." *LWT* 110:168–174. doi: https://doi.org/10.1016/j.lwt.2019.04.092.

Satam, Heena, Kandarp Joshi, Upasana Mangrolia, Sanober Waghoo, Gulnaz Zaidi, Shravani Rawool, Ritesh P. Thakare, Shahid Banday, Alok K. Mishra, Gautam Das, and Sunil K. Malonia. 2023. "Next-generation sequencing technology: Current trends and advancements." *Biology* 12 (7):997. doi:10.3390/biology12070997.

Schloss Patrick, D., L. Westcott Sarah, Thomas Ryabin, R. Hall Justine, Martin Hartmann, B. Hollister Emily, A. Lesniewski Ryan, B. Oakley Brian, H. Parks Donovan, J. Robinson Courtney, W. Sahl Jason, Blaz Stres, G. Thallinger Gerhard, J. Van Horn David, and F. Weber Carolyn. 2009. "Introducing mothur: Open-source, platform-independent, community-supported software for describing and comparing microbial communities." *Applied and Environmental Microbiology* 75 (23):7537–7541. doi: 10.1128/AEM.01541-09.

Shah, Ateeq, and Donald L. Smith. 2020. "Flavonoids in agriculture: Chemistry and roles in, biotic and abiotic stress responses, and microbial associations." *Agronomy* 10 (8):1209. doi:10.3390/agronomy10081209.

Ståhlberg, Anders, Paul M. Krzyzanowski, Matthew Egyud, Stefan Filges, Lincoln Stein, and Tony E. Godfrey. 2017. "Simple multiplexed PCR-based barcoding of DNA for ultrasensitive mutation detection by next-generation sequencing." *Nature Protocols* 12 (4):664–682. doi: 10.1038/nprot.2017.006.

Sun, Xiang Dong, Hong Shan, Lili Li, Ping Su, Jing Lan, Lin Zhao, and Huan Chun Yang. 2019. "Field fungal diversity in freshly harvested maize." *International Journal of Biochemistry Research & Review* 26 (1):1–13. doi: 10.9734/ijbcrr/2019/v26i130088.

Tamura, Koichiro, Glen Stecher, Daniel Peterson, Alan Filipski, and Sudhir Kumar. 2013. "MEGA6: Molecular Evolutionary Genetics Analysis Version 6.0." *Molecular Biology and Evolution* 30 (12):2725–2729. doi: 10.1093/molbev/mst197.

Taudière, Adrien, Jean-Michel Bellanger, Christopher Carcaillet, Laetitia Hugot, Finn Kjellberg, Aarón Lecanda, Annick Lesne, Pierre-Arthur Moreau, Karin Scharmann, Sebastian Leidel, and Franck Richard. 2018. "Diversity of foliar endophytic ascomycetes in the endemic Corsican pine forests." *Fungal Ecology* 36:128–140. doi: https://doi.org/10.1016/j.funeco.2018.07.008.

Tedersoo, Leho, Sten Anslan, Mohammad Bahram, Sergei Põlme, Taavi Riit, Ingrid Liiv, Urmas Kõljalg, Veljo Kisand, Henrik Nilsson, Falk Hildebrand, Peer Bork, and Kessy Abarenkov. 2015. "Shotgun metagenomes and multiple primer pair-barcode combinations of amplicons reveal biases in metabarcoding analyses of fungi." *MycoKeys* 10:1–43.

Tedersoo, Leho, Mohammad Bahram, Rasmus Puusepp, R. Henrik Nilsson, and Timothy Y. James. 2017. "Novel soil-inhabiting clades fill gaps in the fungal tree of life." *Microbiome* 5 (1):42. doi: 10.1186/s40168-017-0259-5.

Tedersoo, Leho, Mohammad Bahram, Lucie Zinger, R. Henrik Nilsson, Peter G. Kennedy, Teng Yang, Sten Anslan, and Vladimir Mikryukov. 2022. "Best practices in metabarcoding of fungi: From experimental design to results." *Molecular Ecology* 31 (10):2769–2795. doi: https://doi.org/10.1111/mec.16460.

Tedersoo, Leho, Ave Tooming-Klunderud, and Sten Anslan. 2018. "PacBio metabarcoding of Fungi and other eukaryotes: Errors, biases and perspectives." *New Phytologist* 217 (3):1370–1385. doi: https://doi.org/10.1111/nph.14776.

Tian, Bao-Yu, Yi Cao, and Ke-Qin Zhang. 2015. "Metagenomic insights into communities, functions of endophytes and their associates with infection by root-knot nematode, Meloidogyne incognita, in tomato roots." *Scientific Reports* 5 (1):17087. doi: 10.1038/srep17087.

Trofymow, John A., Philip-Edouard Shay, Bradley Tomm, Jean A. Bérubé, and Tod Ramsfield. 2023. Differences in Soil Fungal Communities between forested reclamation and forestry sites in the Alberta Oil Sands Region. *Journal of Fungi* 9 (11):1110. doi:10.3390/jof9111110.

Vu, D., M. Groenewald, M. de Vries, T. Gehrmann, B. Stielow, U. Eberhardt, A. Al-Hatmi, J. Z. Groenewald, G. Cardinali, J. Houbraken, T. Boekhout, P. W. Crous, V. Robert, and G. J. M. Verkley. 2019. "Large-scale generation and analysis of filamentous fungal DNA barcodes boosts coverage for kingdom fungi and reveals thresholds for fungal species and higher taxon delimitation." *Studies in Mycology* 92 (1):135–154. doi: 10.1016/j.simyco.2018.05.001.

Wagner, Maggie R., Posy E. Busby, and Peter Balint-Kurti. 2020. "Analysis of leaf microbiome composition of near-isogenic maize lines differing in broad-spectrum disease resistance." *New Phytologist* 225 (5):2152–2165. doi: https://doi.org/10.1111/nph.16284.

Wemheuer, Franziska, Bernd Wemheuer, Rolf Daniel, and Stefan Vidal. 2019. "Deciphering bacterial and fungal endophyte communities in leaves of two maple trees with green islands." *Scientific Reports* 9 (1):14183. doi: 10.1038/s41598-019-50540-2.

Williams, Timothy J., Emilie Long, Flavia Evans, Mathew Z. DeMaere, Federico M. Lauro, Mark J. Raftery, Hugh Ducklow, Joseph J. Grzymski, Alison E. Murray, and Ricardo Cavicchioli. 2012. "A metaproteomic assessment of winter and summer bacterioplankton from Antarctic Peninsula coastal surface waters." *The ISME Journal* 6 (10):1883–1900. doi: 10.1038/ismej.2012.28.

Yang, Hao, Jianting Zhang, Haiwen Zhang, Rongxiang Cao, Donglan Tang, and Liangju Wang. 2023. "5-Aminolevulinic acid against strawberry Fusarium wilt: bidirectional regulation of biocontrol agents and pathogens." *Horticultural Plant Journal* 10 (6): 1349–1361. doi: https://doi.org/10.1016/j.hpj.2023.02.014.

Youseif Sameh, H., H. Abd El-Megeed Fayrouz, A. Humm Ethan, Maskit Maymon, H. Mohamed Akram, A. Saleh Saleh, and M. Hirsch Ann. 2021. "Comparative analysis of the cultured and total bacterial community in the wheat rhizosphere microbiome using culture-dependent and culture-independent approaches." *Microbiology Spectrum* 9 (2):e00678–21. doi: 10.1128/Spectrum.00678-21.

Zeilinger, Susanne, Vijai K. Gupta, Tanya E. S. Dahms, Roberto N. Silva, Harikesh B. Singh, Ram S. Upadhyay, Eriston Vieira Gomes, Clement Kin-Ming Tsui, and Chandra Nayak S. 2016. "Friends or foes? Emerging insights from fungal interactions with plants." *FEMS Microbiology Reviews* 40 (2):182–207. doi: 10.1093/femsre/fuv045.

Zimmerman, Naupaka B., and Peter M. Vitousek. 2012. "Fungal endophyte communities reflect environmental structuring across a Hawaiian landscape." *Proceedings of the National Academy of Sciences* 109 (32):13022–13027. doi: 10.1073/pnas.1209872109.

Zou, Y. N., Q. S. Wu, and K. Kuča. 2021. "Unravelling the role of arbuscular mycorrhizal fungi in mitigating the oxidative burst of plants under drought stress." *Plant Biology* 23 (S1):50–57. doi: https://doi.org/10.1111/plb.13161.

Unit 2

Endophytic Fungi from Medicinal Plants as a Reservoir of Bioactive Natural Products

4 Advances and Opportunities in the Bio-Prospection of Bioactive Metabolites from Medicinal Plant-Associated Endophytes

Pragya Tiwari and Mangalam Bajpai*
*Corresponding author.

4.1 NATURAL COMPOUNDS AND TRENDS IN DRUG DISCOVERY

The ancient wisdom of natural products, including animals, plants and minerals, has been the basis of the treatment of human diseases and will remain an important source of future medicines and therapeutics (Atanasov et al., 2021; Tiwari et al., 2016; Tiwari et al., 2019). For centuries, bioactive natural products have been used, with references from the beginning of recorded history. Various civilizations have used natural products for a variety of purposes such as antibiotics from the moldy curd of soybean, and tetracycline antibiotics from beer produced by *Streptomyces* sp. Other examples include poppy juice (*Papaver somniferum*), cypress oil (*Cupressus sempevirens*), myrrh (*Commiphora* sp.), cedar oil (*Cedrus* sp.) and licorice oil (*Glycyrrhiza glabra*) (Cragg & Newman, 2001). The Greek natural scientist and philosopher, Theophrastus (~300 BC) described the medicinal properties of herbs and their cultivation in his book entitled *History of Plants* (Spainhour, 2005). Dioscorides (100 AD), wrote *De Materia Medica* in which he accounted for the use of herbs as medicine. Avicenna (a Persian philosopher) wrote *Canon Medicinae* for cataloging Greek and Roman medicines (Spainhour, 2005).

At the present time, natural-product-based new molecular entities (NMEs) have also been used to target many pathways and indications such as G-protein-coupled receptors (GPCRs), opioid receptors, or neurotransmitter receptors. Cyclooxygenases are targeted by salicylic acid-based medicines, isolated from willow trees (Mahdi et al., 2006). Statin-based drugs are used for the treatment of hypercholesterolemia. These drugs were derived from *Penicillium* spp. and target 3-hydroxy-3-methyl-glutaryl-coenzyme A reductase (HMGR) enzyme (Endo, 2017). An analysis of all US-FDA-approved NMEs suggested that natural products and their derivatives account for more than one-quarter of all NMEs. Nearly half of these are derived from mammals, one-fourth from plants, and one-fourth from microorganisms (Patridge et al., 2016). Recent research did not promote the direct usage of natural products but favors the high-throughput screening of these products for bioactive molecules.

Plant-based NMEs were more than one-fifth of the total NMEs present before 1950, however, following the discovery of penicillin, the golden era of antibiotics was heralded which shifted the paradigm towards microbe-based NMEs. For this reason, there has been a decrease in plant-based NMEs and an increase in microbe-based new molecules (Monciardini et al., 2014). Microbe-based

DOI: 10.1201/9781003486275-6

new molecules are dominated by molecules isolated from bacteria and fungi, and mainly comprise anticancer and antiinfective agents. More than 200 natural product-based antibiotics characterized were active against fungi, bacteria and human cancer cells. Examples include those derived from *Digitalis*, *Salix*, *Guaiacum*, *Penicillium*, *Cephalosporium* and *Streptomyces* sp. (Gerwick & Moore, 2012). At the present time, the emergence of the "omics era" and molecular biology tools has made it possible to develop more delicate biological and cell-based assays. These assays are harvested to discover and develop drugs from a series of natural sources.

Genome sequencing and elicitation approaches are applied to increase the production of bioactive compounds from plants, fungi and bacteria (Guo et al., 2023; Tiwari, 2022b). A combination of chemical profiles built by LC-HRMS or NMR with multivariate data analysis (MDA) enables scientists to compare and detect differential metabolites (Wu et al., 2015). LC-MS metabolome profiling of *Myxococcus* spp. made it possible to quickly identify known myxobacterial metabolites with subsequent identification of the putative novel compound determined by its exact mass and isotopic pattern (Krug et al., 2008). The potential to map the entire metabolome of a microorganism during symbiotic interactions is a pivotal approach for exploring the untapped pool of anti-infectives. Comparison of differential gene expression patterns with metabolite production allows quick recognition of the gene/gene clusters responsible for producing the biologically active molecules (Yang et al., 2022). Taxol, the billion-dollar anticancer drug, isolated from endophytes of *Taxomyces andreanae*, and subsequently from other endophytic strains, remains the best example to date, revolutionizing the pharmaceutical industry. Discussion and insights on the metabolic potential of endophytes from medicinal plants, case studies on key metabolites, knowledge gaps and achievements are discussed.

4.2 BIO-PROSPECTION OF ENDOPHYTES FOR NOVEL COMPOUNDS

Natural product-based drugs have regained interest due to key attributes, such as structural diversity, low cost and the presence of multiple active compounds. Plants are a major focus in this area since there exists a long-established use of plants as medicine sources (Tiwari, 2022b). Exploring unheard endophyte-based natural products increases the chances of the discovery of new compounds. Plants have been known to have in-house microorganisms, collectively known as endophytes (Tiwari, 2022a; Tiwari & Bae, 2020). The rapid increase in drug resistance demands novel or new drugs to be used as therapeutic agents. β-Lactam-based antibiotics (penicillin and cephalosporin) are broad-spectrum antibacterial drugs derived from microbes. The screening of endophyte-based novel compounds defines a prospective approach in this field. Although a wide range of bioactive compounds has been reported from endophytic species, they are still a largely untapped source of new natural products (Ryan et al., 2007; Tiwari & Bae, 2022). External stimuli like foreign infection or nutritional changes are responsible for the generation of secondary metabolites in plants (Baker et al., 2000). Only 6% of the total of investigated plants have been screened for bioactivity (Tiwari et al., 2023a; Verpoorte & Alfermann, 2000), while many species possessing pharmacological properties remains less explored.

Medicinal plant-associated endophytes include endophytic bacteria, endophytic actinomycetes and endophytic fungi, exhibiting enormous diversity (Choudhary et al., 2021; Rani et al., 2022; Wang et al., 2023). Endophytes show symbiotic interaction, colonize healthy plant tissues and do not adversely impact their plant hosts (Gouda et al., 2016; Tiwari et al., 2023b). The present trend shows that endophytic species have been isolated from many species of medicinal plants and beneficially impact plant growth and development by carrying out multiple processes, namely plant hormones, and regulating the biosynthesis of bioactive constituents including other functions (Hashem et al., 2023; Tshikhudo et al., 2023). The biotransformation potential of endophytes is quite remarkable, and the production of novel analogs illustrates great prospects in the pharmaceutical industry. Endophytes develop host-like features by producing similar secondary metabolites. Endophytic fungi produce taxol on a variety of *Taxus* spp. (Subban & Kempken, 2023). *Colletotrichum gloeosporioides*, which is an endophytic fungus of the taxol-producing plant *Justicia gendarussa*, is involved in the production of the potent anticancer compound taxol (Gangadevi & Muthumary, 2008). *Cladosporium*, *Alternaria*, *Aspergillus*, *Fusarium*, *Monochaetia*, *Pithomyces*, *Pestlotia*, *Pestalotiopsis*, *Penicillium* and *Xylaria* were some of the endophytic fungi involved in the production of taxol from *Taxus* and non-*Taxus* plants (Heinig et al., 2013). Endophytes associated with plants produce many structurally diverse compounds with less cell toxicity to eukaryotes as compared to epiphytes and soil microbes. Thus, these compounds are a better option for the search for future drugs. A variety of antitubercular compounds were identified from different microbial sources (Fernandes et al., 2017). Endophytes have the potential to biosynthesize a wide array of secondary metabolites, which helps them to compete with other microorganisms surrounding the plant (Nishad et al., 2019; Tiwari et al., 2023b). These include steroids, ergosterols, terpenoids, isocoumarins, xanthones, chinones, phenols, cytochalasines, benzopyranones, tetralones and enniatines (Hashem et al., 2023). Figure 4.1 provides a diagrammatic representation of key metabolites produced from medicinal plant-associated endophytes demonstrating pharmacological properties.

Isolation of endophytes from plant hosts was a successful approach for identifying (novel) species producing novel compounds with high antimicrobial activity (Prado et al., 2013). Also, endophytes are associated with producing insecticidal and herbicidal compounds (Kusari et al., 2013; Tiwari et al., 2023b). Leucinostatin A is produced from the endophytic fungus of *Taxus* spp. but it is toxic to other plant species (Strobel & Hess, 1997). From an unknown endophyte associated with *Abies balsam* (balsam fir) two novel diterpene compounds namely 9a-hydroxy-1,8(14),15-isopimaratrien-3,11-dione and 9a-hydroxy-1,8(14),15-isopimaratrien-3,7,11-trion were isolated. These compounds act as toxins to the cells and larvae of spruce budworm (*Chmistonerra fumifiwana*) (Findlay et al., 1995). Azadirachtin derivatives possessing insecticidal activities were reported in the endophytic fungus *Eupenicillium parvum* from the neem tree (*Azadirachta indica*) (Kusari et al., 2012). Sometimes the synthesis of compounds by the endophytes is dependent on their symbiotic relationships with the other species. *Rhizopus*, an endophytic fungus, synthesizes a compound Rhizoxin in the presence of a symbiotic relationship with the bacterium *Burkholderia* (Partida-Martinez & Hertweck, 2005). Different types of anticancer drugs were produced by the endophytes associated with different unrelated plant species. *Pestalotiopsis microspora*, an endophytic fungus, produces the anticancerous compound torreyanic acid (Lee et al., 1996); podophyllotoxin is produced from the fungus *Phialocephala fortinii* from *Podophyllum peltatum* (Eyberger et al., 2006). Rohitukine, an anticancer precursor, was isolated from *Fusarium* sp. from *Dysoxylum binectariferum* (Mohana Kumara et al., 2014). *Trametes hirsute*, a novel endophytic fungus, is also involved in the production of podophyllotoxin (Puri et al., 2006). Similarly, hypericin and emodin were reported to be produced by the endophytic fungus INFU/Hp/KF/34B from *Hypericum perforatum.* Similarly, endophytic bacteria also produce analogous compounds similar to their hosts. *Alternanthera brasiliana* extracts contain the antimicrobial compound oxylipins, which was also derived from its endophytic bacteria (Trapp et al., 2015).

Endophytic fungi also exhibited antifungal activity. A triterpene glycoside compound Enfumafungin was isolated from an endophytic fungus of genus *Hormonema* from *Juniperus communis* having potential antifungal activities against *Candida* and *Aspergillus* (Peláez et al., 2000). An endophytic fungus, *Aspergillus fumigates*, associated with *Juniperus communis*, was found to produce deoxypodophyllotoxin, an anticancer prodrug. Furthermore, Kusari et al. (2009) identified the broad-spectrum antimicrobial potential of deoxypodophyllotoxin. *Actinomycetes* spp. derived from a variety of sources produces various key drugs and biomedical agents. Therefore, actinomycetes existing as endophytes will serve as a potent source of bioactive compounds (Bose et al., 2022). Mangromicins were isolated from *Lechevalieria aerocolonigenes*, Lariatins from *Rhodococcus jostii*, and Actinohivin possessing anti-HIV activity from *Longispora albida*. Actinomycetes produce enediyne

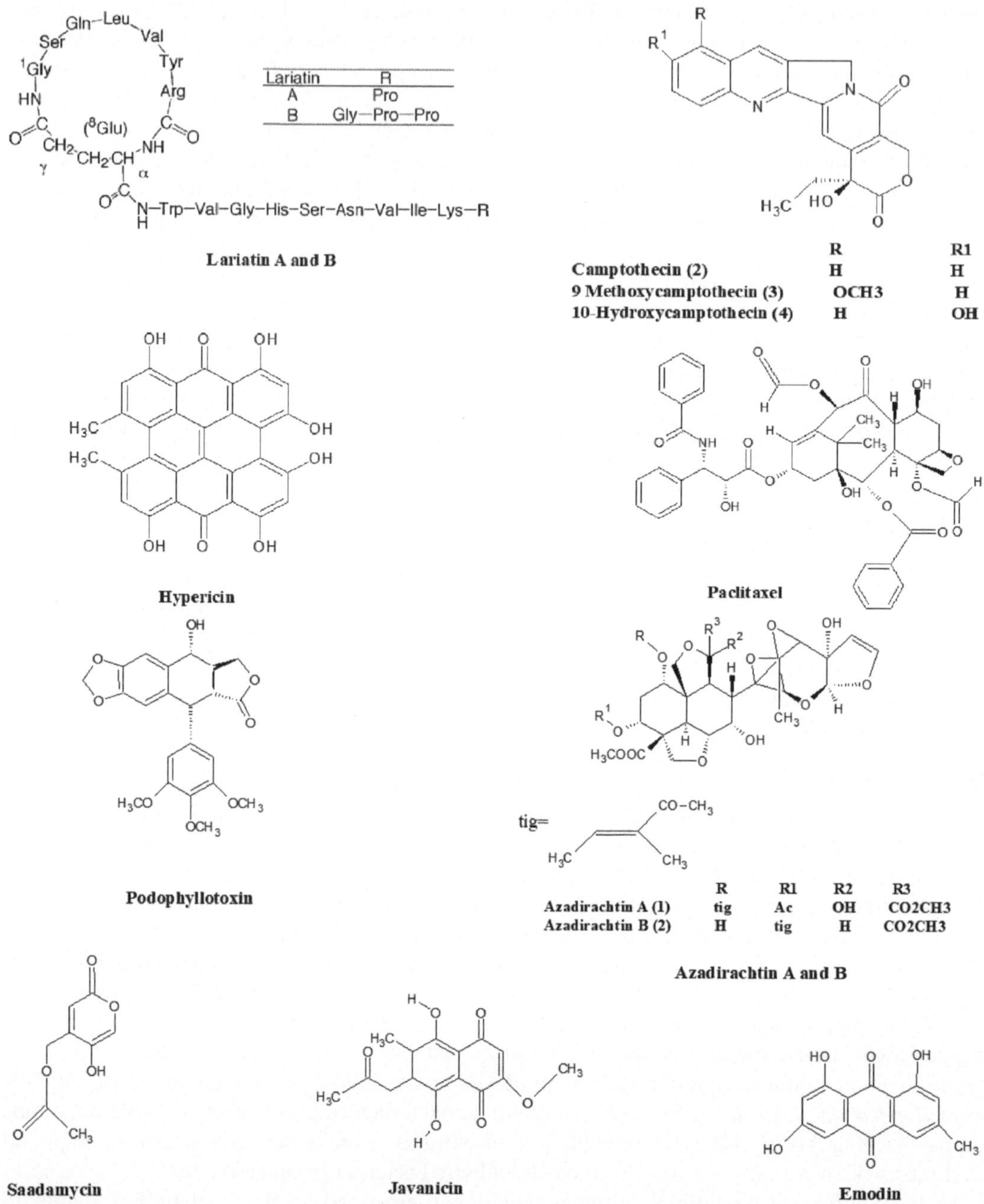

FIGURE 4.1 Key metabolites produced from medicinal plant-associated endophytes demonstrating pharmacological properties.

antibiotics, including *Micromonospora* strain which produces dynamicin, and *Actinomadura* strain which produces Maduropeptin (Matsumoto & Takahashi, 2017). Anthraquinone derivatives were isolated from *Phoma sorghina*, an endophyte of *Tithonia diversifolia* (Borges & Pupo, 2006). Novel pimarane cytochalasins and diterpenes are compounds that were isolated from *Eutypella scoparia* PSU-D44, an endophytic fungus from *Garcinia atroviridis* (Pongcharoen et al., 2006; Phongpaichit et al., 2006). Three novel metabolites including Phomonitroester, Deacetylphomoxanthone B and Phomoenamide were isolated from *Phomopsis* sp. PSU-D15 from *Garcinia dulcis* (Rukachaisirikul et al., 2008). An endophytic strain, *Streptomyces* sp. CS, isolated from the medicinal plant *Maytenus hookeri*, contains a novel compound Naphthomycin K, which possesses cytotoxic activity (Lu & Shen, 2007). Compounds showing antimicrobial, antitubercular and antimalarial properties were reported from endophytic fungus PSU-N24 from *Garcinia nigrolineata* (Sommart et al., 2008). Compounds isolated from *Muscodor crispans* isolated from *Ananas ananassoides* showed potential bioactivity against a drug-resistant strain of *M. tuberculosis* (Mitchell et al., 2010). Four diterpenes and a nor-diterpenoid (Elisabethin D, Elisapterosin A, 3-epielisabanolide, and Elisapterosin B, and Elisabethin D acetate) with strong antituberculosis activity were reported from *Pseudopterogorgia elisabethae* (Rodriguez et al., 2000). Endophytic actinomycetes from *Trewia nudiflora* (*Streptomyces* sp. M27m3 and *Streptomyces* sp. 5B) were involved in producing Ansamycins (Arachevaleta et al., 1989). *Mycobacterium* inhibitory compounds, Diaportheone A and B, were extracted from the fungus *Diaporthe* sp. P133 isolated from the leaves of *Pandanus amaryllifolius* (Bungihan et al., 2011). Huperzine A, a cholinesterase-inhibiting alkaloid, was synthesized from several endophytic fungi of the originally Huperzine A-producing plant *Huperzia serrata*, among which the fungus *Shiraia* sp. is the most prominent producer (Zhu et al., 2010). The endophytic fungus *Alternaria alternata* from *Maytenus hookeri* is a source of nine biologically novel bioactive molecules: ergosta-7,24(28)-dien-3-ol, alternariol monomethyl ether, 5-epialtenuene, ergosta-4,6,8,22-tetraen-3-one, altenuene, adenosine, uridine, ACTG toxin-E and alternariol (Ma et al., 2010) and structural elucidation of these compounds was performed.

Cinnamomum zeylanicum contains a novel endophytic fungus, *Muscodor albus*, which produces volatile compounds having inhibitory effects on other fungi and bacteria (Strobel et al., 2001). These compounds contain alcohols, esters, ketones, acids and lipids, having an inhibitory effect on test bacteria and fungi (Strobel et al., 2001). The insect-repellent naphthalene is produced by the novel fungi *Muscodor vitigenus* isolated from *Paullinia paullinioides.* Naphthalene isolated from this fungus contained a repellent effect on the wheat stem sawfly, *Cephus cinctus.* Thus, *M. vitigenus* may act as a source for mass production of insect repellents (Daisy et al., 2002). A quinoxaline antibiotic structured with a novel antibiotic Kakadumycin A was extracted from endophytic *Streptomyces* spp. from *Grevillea pteridifolia.* This novel antibiotic showed broad-spectrum antibiotic activity, especially against Gram-positive bacteria *Bacillus anthracis* and *Plasmodium falciparum* (Castillo et al., 2003). From the endophytic fungi colonizing mangroves, a novel 1-isoquinoline analog named Marinamide (A) and its methyl ester were isolated by a mixed fermentation technique (Zhu & Lin, 2006). Marine microorganisms have recently been proven to be a source of novel and diverse compounds. Therefore, endophytic microorganisms associated with marine plants constitute a target for modern research. An endophytic fungus, *Penicillium* sp. GD6, was isolated from a Chinese mangrove plant *Bruguiera gymnorrhiza.* This endophytic fungus contained a novel pyrrolizidine alkaloid named Penibruguieramine A (Zhou et al., 2014). A cytotoxic compound, Camptothecin, and its derivatives were extracted from endophytic fungi, namely *Nodulisiporium* sp., *Fusarium solani*, *Neurospora* sp. and *Entrophospora infrequens* from *Nethapodytes fortida* (Kaur et al., 2020). Some endophytic fungi protect their host plants from insects and grazing animals by secreting insect-feeding deterrents and toxic compounds. *Epichloë* and *Neotyphodium* sp. fungi produce Lolitrems, which cause neurotoxic effects in grazing vertebrates and also have potential insecticidal activity (Philippe, 2016). Similarly, the alkaloid Peramine acts as an insect-feeding deterrent (Saikkonen et al., 2016). Spiroquinazine alkaloids were reported from an endophytic fungus, *Eupenicillum* sp., isolated from the leaves of *Murraya paniculata*; structural elucidation was performed by MS-MS and 2D NMR experiments (Barros et al., 2005), and these metabolites were compared with those produced by the host plant. Endophytic fungi belonging to the genus *Pestalotiopsis* are widely known for the presence of unique and structurally diverse compounds (Wang et al., 2015). These endophytes produce secondary metabolites that are not present in normal conditions when induced by an environmental or nutritional factor. When chemicals including DNA methyltransferase inhibitor, histone deacetylase inhibitor, suberoylanilide hydroxamic acid and 5-azacytidine, were added to the culture medium of a plant endophytic fungus, production of some novel aromatic and polyketide compounds was achieved. These were Pestalotiollide B, Scirpyrone A, Endocrocin, Pestalotiopyrone, 4,6-dihydroxy-7-hydroxymethyl-3-methylcoumarin and 7-hydroxy-2-(2-hydroxypropyl)-5-methylchromone,4,6-dihydroxy-3,7-dimethylcoumarin G (Yang et al., 2013). The Penicillide derivatives and alpha-pyrone analogs isolated from the endophytic fungus *Pestalotiopsis sydowiana* demonstrated 20S proteasome inhibitory activities (Xia et al., 2016).

The ethyl acetate extract of endophytic *Phomopsis* sp. was isolated from the halotolerant plant *Adenocarpus foliolosus* and showed potent antifungal and antibacterial activity against *Septoria tritici* and *Pyricularia oryzae*. Also, novel compounds were isolated from the extract, namely 6-hydroxy-6-isopropylcyclohex-1-enecarboxylic acid, Phomosines D–G, 3,4a,8,8 tetramethyl-octahydro-1aH-naphtho[1-b]oxirene-4-carboxylic acid (Dai et al., 2005). Among the metabolites, Phomosine A demonstrated good antibacterial activities, while 5-methylmellein possessed potent fungicidal properties. The endophytic fungus *Pestalotiopsis mangiferae* associated with *Mangifera indica* contains a bioactive compound, 4-(2,4,7-trioxa-bicyclo[4.1.0]heptan-3-yl) phenol, which was active against fungal and bacterial pathogens such as *Candida albicans*, *Klebsiella pneumoniae*, *Bacillus subtilis*, *Micrococcus luteus*, *Escherichia coli* and *Pseudomonas aeruginosa* (Subban et al., 2013). Sequoiatones A and B were reported from a fungal endophyte, *Aspergillus parasiticus*, associated with the inner bark of the *Sequoia sempervirens*. These compounds showed antitumor activity (Stierle et al., 1999). Sequoiamonascins A–D and Sequoiatones C–F were the other anticancer compounds extracted from this endophytic fungus from *Sequoia sempervirens* (Stierle et al., 2003). A novel compound possessing antifungal activity against *Cladosporium phlei* was isolated from an endophytic fungus *Epichloe typhina* associated with the timothy plant (*Phleum pretense*). Stereochemical and mass spectroscopy analysis confirmed the peptide nature of the compound (Seto et al., 2005). Various endophytes were associated with common bacteria including *Burkholderia*, *E. coli*, *Bacillus* and *Pseudomonas*. These endophytes are a rich source of diverse secondary metabolites such as anticancer, antifungal, antibiotic, insecticidal and immunosuppressant compounds (Ryan et al., 2008). One member of the plant-associated fluorescent *Pseudomonas viridiflava* has been reported from within the tissues of many grass species (Miller et al., 1998). This microbe produces novel antimicrobial compounds designated as Ecomycins. Ecomycins constitute a family of novel lipopeptides made up of unusual amino acids like ß-hydroxy aspartic acid and homoserine. These compounds could inhibit human pathogens, including *Candida albicans* and *Cryptococcus neoformans* (Miller et al., 1998).

4.3 TRENDS IN THE DISCOVERY OF PHARMACOLOGICAL COMPOUNDS FROM ENDOPHYTES ASSOCIATED WITH MEDICINAL PLANTS

Medicinal plants have been explored for their potential health benefits and disease treatment since ancient times. Several bioactive constituents are isolated and formulated to treat several ailments, and include flavonoids, terpenoids, steroids, glycosides, alkaloids, saponins, etc. (Tiwari & Bae, 2020; Tiwari et al., 2015; Tiwari et al., 2014). The endophytic association with medicinal plants and their favorable implications in the biosynthesis of pharmacologically active metabolites have garnered scientific attention across the globe and rapidly expanded the horizons of endophytic biology and biotechnology. Studies into understanding medicinal plant–endophyte association utilization aim to improve plant quality and develop new/novel, high-value products from endophytes (Figure 4.1).

At the present time, endophytic species from medicinal plants are increasingly studied for the presence and isolation of bioactive metabolites. *Dendrobium* is a medicinal plant, classified in the family Orchidaceae, and includes bioactive constituents with gastrointestinal, anticancer and antifatigue properties (Wu et al., 2023). Endophytes from Orchidaceae have been isolated (Adil et al., 2023) from different plants; Wang et al. (2019) reported and isolated endophytic bacteria from *Dendrobium officinale* cultivars and classified them into 43 species and characterized their biocontrol functions. A Chinese medicinal plant, *Atractylodes macrocephala* Koidz., was investigated for the presence of endophytes. Wu et al. (2021) studied the presence of cultivable endophytic bacteria in different plant parts and its correlation with bioactive metabolite synthesis. Xu et al. (2019) isolated 608 endophytic bacteria from different cultivars of mulberry, an important edible and medicinal plant due to its rich content of bioactive constituents. Several decades of research on medicinal plants have focused on the screening of plant-associated endophytes: the key examples comprise endophytic bacteria from *Calendula officinalis* L., *Handroanthus impetiginosus*, *Endostemon obtusifolius*, *Prunus domestica*, *Centella asiatica* and *Archidendron pauciflorum*, among others. Similarly, representative species of endophytic fungi have been reported from medicinal plants with studies for their role in the production of pharmacologically important metabolites. A medicinal plant (rich in alkaloids), *Aconitum heterophyllum*, used in the Chinese system of medicine, was screened for the presence of endophytic fungi; 328 fungal isolates from different plant parts were isolated and 12 fungal species were characterized (Hafeez et al., 2023). Wang and coworkers isolated multiple endophytic fungi from *Salvia miltiorrhiza*, among which *Colletotrichum gloeosporioids* and *Epicoccus* sp. SX19 demonstrated potent biocontrol of pathogens. Another key example showed the presence and isolation of endophytic fungi from the roots of *Codonopsis pilosula* from different provinces in China. This is a medicinal plant popularly employed for control of diseases of the kidney, central nervous system, gastrointestinal, metabolism, etc. Besides, several other medicinal plant species including *Crocus sativus*, mint (*Endostemon obtusifolius*), *Codonopsis pilosula*, etc. showed the presence of diverse endophytic fungal species.

4.4 FAVORABLE IMPLICATIONS OF MEDICINAL PLANT-ASSOCIATED ENDOPHYTES

Studies have shown the beneficial impact of endophytes on medicinal plants, assisting and promoting plant growth, conferring stress tolerance, and plant disease management. In a key study, co-inoculation of endophytes from *Astragalus mongholicus* with *Trichoderma* strains improved drought tolerance, by improving root biomass, content of calycosin-7-O-β-D-glucoside, and root length, among others (Li et al., 2022). Endophytes promote plant growth through phytohormone secretion and modulate plant hosts' physiological and biochemical mechanisms. Similarly, plant adaptability and survival in fluctuating climates are improved by the associated endophytes through the enhanced expression of stress resistance genes, enzyme activity and antimicrobial compounds to confer disease resistance in plants (Godara & Ramakrishna, 2023; Wu et al., 2021). An integral aspect of plant–endophyte association is the beneficial role of endophytes in promoting plant growth in normal and stressful conditions. Mathur et al. (2022) isolated a gibberellin-producing endophyte, *Aspergillus niger*, from *Albizia lebbeck* (L.) Benth. which promoted seed germination in crop plants. In another example, Khan et al. isolated *Acremonium* sp. Ld-03 from the medicinal plant, *Lilium davidii*, which showed antibacterial activity. In addition to the above, several actinomycetes species are also reported from medicinal plants (Khan et al., 2021).

4.5 ACHIEVEMENTS AND TRANSLATIONAL SUCCESS IN MEDICINAL PLANT-ASSOCIATED ENDOPHYTES: A SPECIAL REFERENCE TO SECONDARY METABOLITE BIOSYNTHESIS

The recent era has witnessed increased exploration and studies on medicinal plants for screening endophytes and their metabolic correlations. The medicinal plant *Bupleurum scorzonerifolium* Willd. constitutes Saikosaponin d, showing medicinal properties including immunomodulatory, antitumor and antiinflammation. Two endophytic fungi producing Saikosaponin d were isolated from the plant employing UPLC/Q-TOF-MS (Cheng et al., 2022). Similarly, an endophytic fungus, *Penicillium* sp., isolated from *Baphicacanthus cusia* (Nees) bremek. produced 7-methoxy-13-dehydroxypaxilline, an anticancer diterpenoid (Yu et al., 2022). *Archidendron pauciflorum*, an endophytic bacterium, produced 1,4-dihydroxy-2-methyl-anthraquinone that showed promising antibacterial effects (Priyanto et al., 2023). Another key study on the endophytic fungus, *Xylaria feejeensis*, colonizing medicinal plant, *Geophila repens*, discussed the presence of integrated acids, an important source of antibiotics (Rajendran et al., 2023).

Several metabolites possessing pharmacological effects including alkaloids, coumarins, terpenoids, phenolics and saponins, among others have been isolated from plant-associated endophytes (Ahmed et al., 2023; Hashem et al., 2023; Tshikhudo et al., 2023; Wang et al., 2023). These naturally occurring bioactive substances demonstrate antibacterial, antioxidant, anticancer, antiinflammatory and other bioactivities (Tshikhudo et al., 2023; Hashem et al., 2023; Tiwari & Bae, 2022). The present trends in drug discovery are aimed at discovering novel metabolites from natural sources, and their identification via high-throughput technologies (Mursyidah et al., 2023). Advances and techniques in genome sequencing have been employed to decipher the structural features and diversity of plant-associated endophytes (Chen et al., 2019; Pan et al., 2023; Riva et al., 2022), to address knowledge gaps, and to explore the microbial strains associated with medicinal plants. Although rich microbial diversity is associated with medicinal plants, only very few have been isolated, and optimization of cultivation methods is essential to obtain key microbial resources. The screening and isolation of new drug molecules from endophytes show enormous prospects in drug discovery programs and in addressing the knowledge gaps in understanding plant–endophyte dynamics. An interesting example of a diterpene, taxol, was isolated from *Taxus chinensis*, and witnessed enormous translational success as an anticancer drug worldwide. However, the presence of taxol in low amounts led to investigations of multiple endophyte strains from different plant species (Zhou et al., 2010). In this direction, a new taxol production pathway has been developed (Cao et al., 2022; Gill & Vasundhara, 2019). The endophytes associated with *Taxus chinensis* are good producers of taxol, and the isolation/purification of bioactive constituents is carried out after the fermentation process (Liu & Liu, 2018). However, limitations with loss of metabolic potential of endophytes on repeated subculturing led to decreased biosynthesis of bioactive metabolites (Sabzehzari et al., 2020). These concerns can be addressed by co-culturing endophyte strains, optimizing culture conditions, activating taxol biosynthesis gene clusters, etc. (El-Sayed et al., 2020; Zhou et al., 2010). In addition to addressing the gaps, strategies for enhanced bioactive metabolite production and maintenance of consistent biological activities of endophytes are required to harness the maximum positive attributes of endophytes associated with medicinal plants.

4.6 CONCLUSIONS AND FUTURE DIRECTIONS

Endophytes associated with medicinal plants illustrate enormous prospects in the biosynthesis of metabolites with pharmacological properties and drug-like characteristics. Since ancient times, medicinal plants and their bioactive constituents have been integral to the traditional system of medicine and curing multiple disease ailments. Literature has documented the beneficial impacts of endophytes on their medicinal plant counterparts and this needs to be explored further to harness biosynthetic potential and bioactive entities from alternative microbial resources. Alternatively, the use of rare medicinal plant species can be replaced with endophytes capable of mimicking plant metabolism and exhibiting biotransformation potential, toward the generation of new/novel metabolic candidates. The bio-prospection and generation of novel candidates from endophytes offer great hope and potential in identifying new drug candidates and their utilization in drug discovery programs.

Acknowledgments

The authors express gratitude to their academic/research institutions for consistent encouragement and support. No conflict of interests was declared.

REFERENCES

Adil, M, P Tiwari, J Chen, and S Kanwal. 2023. "Major bioactive metabolites and antimicrobial potential of Orchidaceae Fungal endophytes." In *Advances in Orchid Biology, Biotechnology, and Omics,* edited by Pragya Tiwari, and Jentsung Chen. Springer Publishers. ISBN 13 978-9819910786.

Ahmed, AM, BK Mahmoud, N Millan-Aguinaga, UR Abdelmohsen, and MA Fouad. 2023. "The endophytic *Fusarium* strains: a treasure trove of natural products." *RSC Advances* 13, no. 2:1339–1369.

Akira, E. 2017. "Discovery and development of statins." *Natural Product Communications* 12, no. 8:1934578X1701200801.

Arachevaleta, M, CW Bacon, CS Hoveland, and DE Radclife. 1989. "Effect of tall fescue endophyte on plant response to environmental stress." *Agronomy Journal* 81, no. 1:83–90.

Atanasov, AG, SB Zotchev, and VM Dirsch. et al. 2021. "Natural products in drug discovery: advances and opportunities." *Nature Reviews Drug Discovery* 20, no. 3:200–216.

Baker, D, U Mocek, U, and C Garr. 2000. "Natural products vs. combinatorial: A case study." In *Biodiversity: New leads for pharmaceutical and agrochemical industries*, edited by SK Wrigley, MA Hayes, R Thomas, EJT Chrystal, and N Nicholson (pp. 66–72). The Royal Society of Chemistry

Barros, FAP and E Rodrigues-Filho. 2005. "Four spiroquinazoline alkaloids from *Eupenicillium* sp. isolated as an endophytic fungus from leaves of *Murraya paniculata* (Rutaceae)." *Biochemical Systematics and Ecology* 33, no. 3:257–268.

Borges, W, and M Pupo. 2006. "Novel anthraquinone derivatives produced by *Phoma sorghina*, an endophyte found in association with the medicinal plant *Tithonia diversifolia* (Asteraceae)." *Journal of the Brazilian Chemical Society* 17, no. 5:929–934.

Bose, SK, M Bajpai, J Das, A Gautam, and P Tiwari. 2022. "Actinomycetes endophytes: Overview and significance in the production of bioactive compounds." In *Endophytes: Types, Potential Uses, and Mechanisms of Action*, edited by P. Tiwari. Nova Publishers. ISBN: 979-8-88697-045-6.

Bungihan, ME, MA Tan, and M Kitajima, et al. 2011. "Bioactive metabolites of *Diaporthe* sp. P133, an endophytic fungus isolated from *Pandanus amaryllifolius*." *Journal of Natural Medicine* 65:606–609.

Cao, X, L Xu, L, J Wang, M Dong, C Xu, G Kai, G, W Wan, and J Jiang. 2022. "Endophytic fungus *Pseudodidymocyrtis lobariellae* KL27 promotes taxol biosynthesis and accumulation in *Taxus chinensis*." *BMC Plant Biology* 22:12.

Castillo, U, JK Harper, GA Strobel, JSears, K Alesi, E Ford, J Lin, M Hunter, M Maranta, H Ge, D Yaver, JB Jensen, H Porter, R Robison, D Millar, WM Hess, M Condron, and D Teplow. 2003. "Kakadumycins, novel antibiotics from *Streptomyces* sp. NRRL 30566, an endophyte of *Grevillea pteridifolia*." *FEMS Microbiology Letters* 224, no. 2:183–190.

Chen, P, CM Zhang, XY Ju, YW Xiong, K Xing, and S Qin. 2919. "Community composition and metabolic potential of endophytic Actinobacteria from coastal salt marsh plants in Jiangsu, China." *Frontiers in Microbiology* 10:1063.

Cheng, Y, G Liu, Z Li, Y Zhou, and Gao, N. 2022. "Screening Saikosaponin d (SSd)-producing endophytic fungi from *Bupleuru scorzonerifolium* Willd." *World Journal of Microbiology and Biotechnology* 38, no. 12:242.

Choudhary, M, S Gupta, MK Dhar, and S Kaul. 2021. "Endophytic fungi-mediated biocatalysis and biotransformations paving the way toward green chemistry." *Frontiers in Bioengineering and Biotechnology* 9:664705.

Cragg, GM, and DJ Newman. 2001. "Natural product drug discovery in the next millennium." *Pharmaceutical Biology* 39, no. 1:8–17.

Dai, J, K Krohn, U Flörke, D Gehle, HJ Aust, S Draeger, B Schulz, and J Rheinheimer. 2005. "Novel highly substituted biraryl ethers, Phomosines D–G, isolated from the endophytic fungus *Phomopsis* sp. from *Adenocarpus foliolosus*." European *Journal of Organic Chemistry* 2005, no. 23:5100–5105.

Daisy, BH, GA Strobel, U Castillo, D Ezra, J Sears, DK Weaver, and JB Runyon. 2002. "Naphthalene, an insect repellent, is produced by *Muscodor vitigenus*, a novel endophytic fungus." *Microbiology (Reading)* 148, no. 11:3737–3741.

El-Sayed, ASA, MT El-Sayed, MT, A Rady, N Zein, G Enan, A Shindia, S El-Hefnawy, M Sitohy, and B Sitohy. 2020. "Exploiting the biosynthetic potency of Taxol from Fungal endophytes of Conifers plants; genome mining and metabolic manipulation." *Molecules* 25, no. 13:3000.

Eyberger, AL, R Dondapati, and JR Porter. 2006. "Endophyte fungal isolates from *Podophyllum peltatum* produce podophyllotoxin." *Journal of Natural Products* 69, no. 8:1121–1124.

Fernandes, GFDS, C Man Chin, and JL Dos Santos. 2017. "Advances in drug discovery of new antitubercular multidrug-resistant compounds." *Pharmaceuticals (Basel)* 10, no. 2:51.

Findlay, JA, G Li, PE Penner, and JD Miller. 1995. "Novel diterpenoid insect toxins from a conifer endophyte." *Journal of Natural Product* 58, no. 2:197–200.

Gangadevi, V, and J Muthumary. 2008. "Isolation of *Colletotrichum gloeosporioides*, a novel endophytic taxol producing fungus from the leaves of a medicinal plant *Justicia gendarussa*." *Mycol. Balcanica* 5, no. 2008:1–4.

Gerwick, WH, and BS Moore. 2012. "Lessons from the past and charting the future of marine natural products drug discovery and chemical biology." *Chemical Biology* 19, no. 1:85–98.

Gill, H, and M Vasundhara. 2019. "Isolation of taxol producing endophytic fungus *Alternaria brassicicola* from non-Taxus medicinal plant *Terminalia arjuna*." *World Journal of Microbiology and Biotechnology* 35:74.

Godara, H, and W Ramakrishna. 2023. "Endophytes as nature's gift to plants to combat abiotic stresses." *Letters in Applied Microbiology* 76:ovac067.

Gouda, S, G Das, SK Sen, HS Shin, and JK Patra. 2016. "Endophytes: a treasure house of bioactive compounds of medicinal importance." *Frontiers in Microbiology* 7:1538.

Guo, M, H Lv, H Chen, S Dong, J Zhang, W Liu, L He, Y Ma, H Yu, S Chen, and H Luo. 2023. "Strategies on biosynthesis and production of bioactive compounds in medicinal plants." *Chinese Herbal Medicines* 16:13–26. https://doi.org/10.1016/j.chmed.2023.01.007.

Hafeez, S, S Yaqoob, AR Magray, AN Kamili, and BA Ganai. 2023. "Molecular characterization of fungal endophyte diversity isolated from *Aconitum heterophyllum*: a critically endangered medicinal plant of Kashmir Himalaya." *International Microbiology* 26, no. 3:651–662.

Hashem, AH, MS Attia, EK Kandil, MM Fawzi, AS Abdelrahman, MS Khader, MA Khodaira, AE Emam, MA Goma, and AM Abdelaziz. 2023. "Bioactive compounds and biomedical applications of endophytic fungi: a recent review." *Microbial Cell Factories* 22:107.

Heinig, U, S Scholz, and S Jennewein. 2013. "Getting to the bottom of Taxol biosynthesis by fungi." *Fungal Diversity* 60:161–170.

Kaur, P, V Kumar, R Singh, P Dwivedi, A Dey, and DK Pandey. 2020. "Biotechnological strategies for the production of camptothecin from fungal and bacterial endophytes." *South African Journal of Botany* 134:135–145,

Khan, MS, J Gao, I Munir, M Zhang, Y Liu, TS Moe, J Xue, and X Zhang. 2021. "Characterization of endophytic fungi, *Acremonium* sp., from *Lilium davidii* and analysis of its antifungal and plant growth-promoting effects." *BioMed Research International* 2021:9930210.

Krug, D, G Zurek, B Schneider, R Garcia, and R Müller. 2008. "Efficient mining of myxobacterial metabolite profiles enabled by liquid chromatography-electrospray ionization-time-of-flight mass spectrometry and compound-based principal component analysis." *Anal Chim Acta.* 624, no. 1:97–106

Kusari, S, M Lamshöft, and M Spiteller. 2009. "*Aspergillus fumigatus Fresenius*, an endophytic fungus from *Juniperus communis* L. Horstmann as a novel source of the anticancer pro-drug deoxypodophyllotoxin." *Journal of Applied Microbiology* 107, no. 3:1019–1030.

Kusari, S, SP Pandey, and M Spiteller. 2013. "Untapped mutualistic paradigms linking host plant and endophytic fungal production of similar bioactive secondary metabolites." *Phytochemistry* 91:81–87.

Kusari, S, VC Verma, M Lamshoeft, and M Spiteller. 2012. "An endophytic fungus from *Azadirachta indica A. Juss*. that produces azadirachtin." *World Journal of Microbiology and Biotechnology* 28:1287–1294.

Lee, JC, GA Strobel, E Lobkovsky, and J Clardy. 1996. "Torreyanic acid: a selectively cytotoxic quinone dimer from the endophytic fungus *Pestalotiopsis microspora*." *The Journal of Organic Chemistry* 61:3232–3233.

Li, M, Y Ren, C He, J Yao, M Wei, and X He. 2022. "Complementary effects of Dark septate endophytes and *Trichoderma* strains on growth and active ingredient accumulation of *Astragalus mongholicus* under drought stress." *Journal of Fungi* 8:920.

Liu, J, and G Liu. 2018. "Analysis of secondary metabolites from plant Endophytic fungi." *Methods in Molecular Biology* 1848:25–38.

Lu, C, and Y Shen. 2007. "A novel Ansamycin, Naphthomycin K from *Streptomyces* sp." *Journal of Antibiotics* 60, no. 10:649–653.

Ma, YT, LR Qiao, and WQ Shi. et al. 2010. "Metabolites produced by an endophyte *Alternaria alternata* isolated from *Maytenus hookeri*." *Chemistry of Natural Compounds* 46:504–506.

Mahdi, JG, AJ Mahdi, AJ Mahdi, and ID Bowen. 2006. "The historical analysis of aspirin discovery, its relation to the willow tree and antiproliferative and anticancer potential." *Cell Proliferation* 39, no. 2:147–155.

Mathur, P, P Chaturvedi, C Sharma, and P Bhatnagar. 2022. "Improved seed germination and plant growth mediated by compounds synthesized by endophytic *Aspergillus niger* (isolate 29) isolated from *Albizia lebbeck* (L.) Benth." *3 Biotech* 12, no. 10:271.

Matsumoto, A, and Y Takahashi. 2017. "Endophytic actinomycetes: a promising source of novel bioactive compounds." *Journal of Antibiotics* 70, no. 5:514–519.

Miller, CM, RV Miller, D Garton-Kenny, B Redgrave, J Sears, MM Condron, DB Teplow, and GA Strobel. 1998. "Ecomycins, unique antimycotics from *Pseudomonas viridiflava*." *Journal of Applied Microbiology* 84, no. 6:937–944.

Mitchell, AM, GA Strobel, E Moore, R Robison, and J Sears. 2010. "Volatile antimicrobials from *Muscodor crispans*, a novel endophytic fungus." *Microbiology (Reading)* 156, no. 1:270–277.

Mohana Kumara, P, KN Soujanya, G Ravikanth, R Vasudeva, KN Ganeshaiah, and R Uma Shaanker. 2014. "Rohitukine, a chromone alkaloid and a precursor of flavopiridol, is produced by endophytic fungi isolated from *Dysoxylum binectariferum* Hook.f and *Amoora rohituka* (Roxb). Wight & Arn." *Phytomedicine* 21, no. 4:541–546.

Monciardini, P, M Iorio, S Maffioli, M Sosio, and S Donadio. 2014. "Discovering new bioactive molecules from microbial sources." *Microbial Biotechnology* 7, no. 3:209–220.

Mursyidah, AK, M Hafizzudin-Fedeli, NAN Muhammad, A Latiff, M Firdaus-Raih, and KL Wan. 2023. "Dissecting the biology of *Rafflesia* species: current progress and future directions made possible with High-throughput sequencing data." *Plant Cell Physiology* 64, no. 4:368–377.

Nishad, JH, A Singh, VS Gautam, D Kumar, J Kumar, and RN Kharwar. 2019. "Endophytic fungi: A cryptic fountainhead for biodiversity, functional metabolites, host stress tolerance, and myco-mediated nanoparticles (Nps) Synthesis." In *Endophytes and Secondary Metabolites,* edited by S. Jha. Reference Series in Phytochemistry. Springer. https://doi.org/10.1007/978-3-319-76900-4_18-1.

Pan, WJ, LY Miao, SP Fan, PW Lv, AH Lin, H Geng, FJ Song, and P Zhang. 2023. "New insights into the composition and diversity of endophytic bacteria in cultivated *Huperzia serrata*." *Canadian Journal of Microbiology* 69, no. 8:321–327.

Partida-Martinez, LP, and C Hertweck. 2005. "Pathogenic fungus harbors endosymbiotic bacteria for toxin production." *Nature* 437, no. 7060:884–888.

Patridge, E, P Gareiss, MS Kinch, and D Hoyer. 2016. "An analysis of FDA-approved drugs: natural products and their derivatives." *Drug Discovery Today* 21, no. 2:204–207.

Peláez, F, A Cabello, G Platas, MT Díez, A González del Val, A Basilio, I Martán, F Vicente, GF Bills, RA Giacobbe, RE Schwartz, JC Onishi, MS Meinz, GK Abruzzo, AM Flattery, L Kong, MB Kurtz. 2000. "The discovery of Enfumafungin, a novel antifungal compound produced by an Endophytic *Hormonema* species biological activity and taxonomy of the producing organisms." *Systematic and Applied Microbiology* 23, no. 3:333–343.

Philippe G. 2016. "Lolitrem B and Indole diterpene alkaloids produced by Endophytic fungi of the Genus *Epichloë* and their toxic effects in Livestock." *Toxins (Basel)* 8, no. 2:47.

Phongpaichit, S, N Rungjindamai, V Rukachaisirikul, and J Sakayaroj. 2006. "Antimicrobial activity in cultures of endophytic fungi isolated from *Garcinia* species." *FEMS Immunol Med Microbiol.* 48, no. 3:367–372.

Pongcharoen, W, V Rukachaisirikul, S Phongpaichit, N Rungjindamai, and J Sakayaroj. 2006. "Pimarane diterpene and cytochalasin derivatives from the endophytic fungus *Eutypella scoparia* PSU-D44." *Journal of Natural Product* 69, no. 5:856–858.

Prado, S, D Buisson, I Ndoye, M Vallet, and B Nay. 2013. "One-step enantioselective synthesis of (4S)-isosclerone through biotranformation of juglone by an endophytic fungus." *Tetrahedron Letters*, 54, no. 10:1189–1191.

Priyanto, JA, ME Prastya, RI Astuti, and R Kristiana. 2023. "The antibacterial and antibiofilm Activities of the endophytic bacteria associated with *Archidendron pauciflorum* against multidrug-resistant strains." *Applied Biochemistry and Biotechnology* 2023:1–22.

Puri, S C, A Nazir, R Chawla, R Arora, S Riyaz-ul-Hasan, T Amna, B Ahmed, V Verma, S Singh, R Sagar, A Sharma, R Kumar, RK Sharma, and GN Qazi. 2006. "The endophytic fungus *Trametes hirsuta* as a novel alternative source of podophyllotoxin and related aryl tetralin lignans." *Journal of Biotechnology* 122, no. 4:494–510.

Rajendran, S, LP Robertson, L Kosgahakumbura, C Fernando, U Goransson, H Wang, C Hettiarachchi, and S Gunasekera. 2023. "Antibacterial eremophilane sesquiterpenoids from *Xylaria feejeensis*, an endophytic fungi of the medicinal plant *Geophila repens*." *Fitoterapia* 167:105496.

Rani, S, P Kumar, P Dahiya, R Maheshwari, AS Dang, and P Suneja. 2022. "Endophytism: a multidimensional approach to plant-prokaryotic microbe interaction." *Frontiers in Microbiology* 13:861235.

Riva, V, F Mapelli, A Bagnasco, A Mengoni, and S Borin. 2022. "A meta-analysis approach to defining the culturable core of plant endophytic bacterial communities." *Applied Environment Microbiology* 88, no. 6: e0253721.

Robert, V, and AW Alfermann. 2000. *Metabolic engineering of plant secondary metabolism*. Springer, ISBN: 978-90-481-5475-3.

Rodríguez, AD, C Ramírez, II Rodríguez, and CL Barnes. 2000. "Novel terpenoids from the West Indian sea whip *Pseudopterogorgia elisabethae* (Bayer). Elisapterosins A and B: rearranged diterpenes possessing an unprecedented cagelike framework." *Journal of Organic Chemistry* 65, no. 5:1390–1398.

Rukachaisirikul, V, U Sommart, S Phongpaichit, J Sakayaroj, and K Kirtikara. 2008. "Metabolites from the endophytic fungus *Phomopsis* sp. PSU-D15." *Phytochemistry* 69, no. 3:783–787.

Ryan, RP, K Germaine, A Franks, DJ Ryan, and DN Dowling. 2008. "Bacterial endophytes: recent developments and applications." *FEMS Microbiology Letters*, 278, no. 1:1–9.

Sabzehzari, M, M Zeinali, and MR Naghavi. 2020. "Alternative sources and metabolic engineering of Taxol: advances and future perspectives." *Biotechnology Advances* 43:107569.

Saikkonen, K, CA Young, and M Helander, et al. 2016. "Endophytic *Epichloë* species and their grass hosts: from evolution to applications." *Plant Molecular Biology* 90:665–675.

Seto, Y, Y Kogami, T Shimanuki, K Takahashi, H Matsuura, and T Yoshihara. 2005. "Production of Phleichrome by *Cladosporium phlei* as stimulated by diketopiperadines of *Epichloe typhina*." *Biosci Biotechnol Biochem* 69, no. 8:1515–1519.

Sommart, U, V Rukachaisirikul, Y Sukpondma, S Phongpaichit, J Sakayaroj, and K Kirtikara. 2008. "Hydronaphthalenones and a dihydroramulosin from the endophytic fungus PSU-N24." *Chem Pharm Bull* 56, no. 12:1687–1690.

Spainhour, CB. 2005. "Natural Products. Drug Discovery". Handbook, Editor(s): Shayne Cox Gad. https://doi.org/10.1002/0471728780.ch1

Stierle, DB, AA Stierle, and T Bugni. 2003. "Sequoiamonascins a-d: novel anticancer metabolites isolated from a redwood endophyte." *Journal of Organic Chemistry* 68, no. 12:4966–4969.

Stierle, AA, DB Stierle, and T Bugni. 1999, July. "Sequoiatones A and B: Novel antitumor metabolites isolated from a redwood endophyte." *Journal of Organic Chemistry*, 64, no. 15:5479–5484.

Strobel, GA, and WM Hess, 1997. "Glucosylation of the peptide leucinostatin A, produced by an endophytic fungus of European yew, may protect the host from leucinostatin toxicity." *Chemistry & Biology* 4, no. 7: 529–536.

Strobel, GA, E Dirkse, J Sears, and C Markworth. 2001. "Volatile antimicrobials from *Muscodor albus*, a novel endophytic fungus." *Microbiology (Reading)* 147, no. 11:2943–2950.

Strobel, GA, and WM Hess, 1997. "Glucosylation of the peptide leucinostatin A, produced by an endophytic fungus of European yew, may protect the host from leucinostatin toxicity." *Chemistry & Biology* no. 4(7): 529–536.

Subban, K, and F Kempken. 2023. "Insights into Taxol biosynthesis by endophytic fungi." *Applied Microbiology and Biotechnology* 107:6151–6162.

Subban, K, R Subramani, and M Johnpaul. 2013. "A novel antibacterial and antifungal phenolic compound from the endophytic fungus *Pestalotiopsis mangiferae*." *Natural Product Research* 27, no. 16:1445–1449.

Subban, K, and F Kempken. 2023. "Insights into Taxol biosynthesis by endophytic fungi." *Applied Microbiology and Biotechnology* no. 107:6151–6162.

Tiwari, P. 2022a. *Endophytes- Types, Potential Uses, and Mechanisms of Action*. Nova Publishers, USA, ISBN: 979-8-88697-045-6.

Tiwari, P. 2022b. "Triterpenoids from *Gymnema sylvestre* R. Br: biological significance in the treatment of Diabetes." In *Phytochemicals and Medicinal Plants in Food Design: Strategies and Technologies for Improved Healthcare*, edited by Megh R. Goyal, Preeti Birwal, Santosh K Mishra. Innovations in Agricultural and Biological Engineering, Apple Academic Press, Book volume 31, ISBN 9781774639450.

Tiwari, P, and H Bae. 2020. "Horizontal gene transfer and Endophytes: an implication for the acquisition of novel traits." *Plants* 9:305.

Tiwari, P, and H Bae. 2022. "Endophytic fungi: insights, prospects, and challenges in natural product drug discovery." *Microorganisms* 10:360.

Tiwari, P, K Ahmed, and H Baig. 2016. "*Gymnema sylvestre* for diabetes: from traditional herb to future therapeutic." *Current Pharmaceutical Design* 23:1–10.

Tiwari, P, M Bajpai, and A Sharma. 2023a. "Antimicrobials from medicinal plants: key examples, success stories and prospects in tackling antibiotic resistance." *Letters in Drug Design and Discovery* 20:420–438.

Tiwari, P, S Kang, and H Bae. 2023b. "Plant-endophyte associations, rich yet under-explored sources for novel bioactive compounds and applications." *Microbiological Research* 266:127241.

Tiwari, P, A Katyal, MF Khan, GM Ashraf, and K Ahmed. 2019. "Lead optimization resources in drug discovery studies." *Endocrine, Metabolic, and Immuno Disorders-Drug Targets* 19, no. 6:754–774.

Tiwari, P, RS Sangwan, Asha, BN Mishra, F Sabir, and NS Sangwan. 2014. "Molecular cloning and biochemical characterization of a recombinant sterol 3-O-glucosyltransferase from *Gymnema sylvestre* R.Br. catalyzing biosynthesis of sterol glucosides." *BioMed Research International* 2014:1–14.

Tiwari, P, P Sharma, F Khan, NS Sangwan, BN Mishra, and RS Sangwan. 2015. "Structure-Activity Relationship studies of Gymnemic acid analogs for antidiabetic activity targeting PPARγ." *Current Computer-Aided Drug Design* 11, no. 1:57–71

Trapp, MA, M Kai, A Mithöfer, and E Rodrigues-Filho. 2015. "Antibiotic oxylipins from *Alternanthera brasiliana* and its endophytic bacteria." *Phytochemistry* 110:72–82.

Tshikhudo, PP, K Ntushelo, and FN Mudau. 2023. "Sustainable applications of endophytic bacteria and their physiological/biochemical roles on medicinal and herbal plants: review." *Microorganisms* 11:453.

Wang, SS, JM Liu, J Sun, YF Sun, JN Liu, N Jia, B Fan, and XF Dai. 2019. "Diversity of culture-independent bacteria and antimicrobial activity of culturable endophytic bacteria isolated from different *Dendrobium* stems." *Scientific Reports* 9:10389.

Wang, X, X Zhang, and L Liu, *et al.* 2015. "Genomic and transcriptomic analysis of the endophytic fungus *Pestalotiopsis fici* reveals its lifestyle and high potential for synthesis of natural products." *BMC Genomics*, 16, no. 28. https://doi.org/10.1186/s12864-014-1190-9

Wang, ZC, L Wang, YP Pan, XX Zheng, XN Liang, LL Sheng, D Zhang, Q Sun, and Q Wang. 2023. "Research advances on endophytic fungi and their bioactive metabolites." *Bioprocess Biosystem Engineering* 46:165–170.

Wu, C, HK Kim, GP van Wezel, and YH Choi. 2015. "Metabolomics in the natural products field – a gateway to novel antibiotics." *Drug Discovery Today: Technologies* 13:11–17.

Wu, W, Y Lin, MA, Farag, Z Li, and P Shao. 2023. "*Dendrobium* as a new natural source of bioactive for the prevention and treatment of digestive tract diseases: a comprehensive review with future perspectives." *Phytomedicine* 114:154784.

Wu, W, S Wang, J Wu, B He, B Zhu, and L Qin. 2021. "Influence of tissue and geographic locality on culturable endophytic bacteria of *Atractylodes macrocephala*." *Microbiology* 167:001109.

Wu, W, WH Chen, SY Liu, JJ Wu, YT Zhu, LP Qin, and B Zhu. 2021. "Beneficial relationships between endophytic bacteria and medicinal plants." *Frontiers in Plant Science* 12:646146.

Xia, X, S Kim, C Liu, and SH Shim. 2016. "Secondary metabolites produced by an endophytic fungus *Pestalotiopsis sydowiana* and their 20S proteasome inhibitory activities." *Molecules (Basel, Switzerland),* 7:944.

Xu, W, F Wang, M Zhang, T Ou, R Wang, G Strobel, Z Xiang, Z Zhou, and J Xie. 2019. "Diversity of cultivable endophytic bacteria in mulberry and their potential for antimicrobial and plant growth-promoting activities." *Microbiological Research* 229:126328.

Yang, H, H Li, and Q Li. 2022. "Biosynthetic regulatory network of flavonoid metabolites in stems and leaves of *Salvia miltiorrhiza*." *Scientific Reports* 12:18212.

Yang, XL, T Awakawa, T Wakimoto, and I Abe. 2013. "Induced production of novel prenyldepside and coumarins in endophytic fungi *Pestalotiopsis acaciae*." *Tetrahedron Letters* 43, no. 54:5814–5817.

Yu, J, JP Wang, SF Liu, CY, Yin, DY, Tang, YH Li, and LX Zhang. 2022. "7-Methoxy-13-dehydroxypaxilline: new indole diterpenoid from an endophytic fungus *Penicillium* sp. Nb 19." *Natural Product Research* 38, no. 1–9:103–111.

Zhou, X, H Zhu, L Liu, J Lin, and K Tang. 2010. "A review: recent advances and future prospects of taxol-producing endophytic fungi." *Applied Microbiology and Biotechnology* 86:1707–1717.

Zhou, ZF, T Kurtán, XH Yang, A Mándi, MY Geng, BP Ye, O Taglialatela-Scafati, and YW Guo. 2014. "Penibruguieramine A, a novel pyrrolizidine alkaloid from the endophytic fungus *Penicillium* sp. GD6 associated with Chinese mangrove *Bruguiera gymnorrhiza*." *Organic Letters* 16, no. 5:1390–1393.

Zhu, F, and Y Lin. 2006. "Marinamide, a novel alkaloid and its methyl ester produced by the application of mixed fermentation technique to two mangrove endophytic fungi from the South China Sea." *Chinese Science Bulletin* 51:1426–1430.

Zhu, D, J Wang, Q Zeng, Z Zhang, and R Yan. 2010. "A novel endophytic Huperzine A-producing fungus, *Shiraia* sp. Slf14, isolated from *Huperzia serrata*." *Journal of Applied Microbiology* 109, no. 4:1469–1478.

5 Pharmaceutical-Based Applications of Plant-Associated Fungi

*Meghna Dhar, Deepansh Sharma and Sunny Dholpuria**
*Corresponding author.

5.1 INTRODUCTION

Global warming and climate change have a significant impact on agriculture, including changing rainfall patterns, high temperatures, and increased CO_2 levels, salinity and drought. These changes create abiotic stress conditions for plants, making it difficult for them to thrive leading to a reduction in food productivity. While environmental issues cannot be completely regulated, there are ways to help plants endure these abnormal abiotic conditions and one such approach is incorporating the plant microbiome (Verma et al., 2022). The plant microbiome refers to the diverse community of microorganisms that live in and around plants. These microorganisms include bacteria, fungi, archaea and viruses, which have positive or negative interactions with particular plant species. The plant microbiome is divided into three main compartments: the rhizosphere, which is the soil surrounding the plant roots; the phyllosphere, which is the aerial part of the plant, including the leaves, stem, flowers; and the endosphere, which is the internal tissues of the plant, such as the stems and roots. Each of these compartments has its own unique microbial community, and these microorganisms play important roles in plant health and growth (Pandey et al., 2022). Endophytes, which are associated with plants, represent a great example of the symbiotic relationships found in nature. These microorganisms are renowned for colonizing the tissues of plants without causing any harmful symptoms or adverse changes, thus establishing themselves as a beneficial and integral part of the plant's biology. The term "endophyte" was coined by Anton de Bary in 1866 and refers to any organism that grows within the tissues of a plant. The word "endo" means "within" and "phyto" means "plant," so "endophyte" literally translates to "within plant" (Collinge et al., 2022). Furthermore, endophytes are recognized as valuable bio-reservoirs of plant-derived natural products, highlighting their potential in the fields of agriculture and pharmacology. Their ability to produce bioactive compounds that can benefit their host plants makes them a promising area of study for scientists seeking natural solutions for plant health and productivity (Ahmed et al., 2023). The discovery of taxol, however, elevated this recognition to an entirely new level. Harvested from the endophytic fungus *Taxomyces andreanae*, taxol showcased the ability of these microorganisms to produce therapeutically significant compounds (El-Sayed et al., 2020).

5.2 PLANT–ENDOPHYTE RELATIONSHIP

The interaction between plants and fungi has evolved over the course of time with changing environmental conditions. It has been seen that in many cases specific types of fungi harbor specific plant species, and mostly a number of different fungi are found in many plant species. Endophytic fungi and their host plants have developed various types of relationships over time, including mutualistic,

DOI: 10.1201/9781003486275-7

antagonistic and neutral associations (Baron & Rigobelo, 2022). Fungal endophytes typically remain dormant within a specific plant part, either throughout the host's lifespan (referred to as neutralism) or until favorable conditions make them active again (referred to as mutualism or antagonism). Mostly, three different types of transmission mechanisms are found to be associated with endophytic fungi. First, vertical transmission in which the transmission occurs through seeds or hyphae of infected plant species, as seen in the case of *Neotyphodium* sp. Second, horizontal transmission in which infected sexual spores are formed by the fungi harboring the infected plant, as observed in the case of *Epichloe* spp. The last one is mixed transmission, showing both vertical and horizontal transmission, as seen in *Epichloe* spp. (Omomowo et al., 2023).

One of the major benefits provided by the endophytes is the enhancement of the defense mechanism of plants via both direct and indirect mechanisms of action. Direct mechanisms involve the endophytes physically interacting with the pathogens, which leads to the production of antimicrobial compounds like antibiotics, thereby inhibiting or killing the pathogens (Tsalgatidou et al., 2023). Some endophytes are also capable of outcompeting the pathogen for nutrients and space, essentially starving the pathogen and preventing it from establishing itself within the plant. In some cases, endophytes can degrade the toxins produced by the pathogens, neutralizing their harmful effects (Tiwari & Bae, 2022). Indirect mechanisms, on the other hand, are more about bolstering the plant's innate defenses rather than directly confronting the pathogens. Endophytes can induce systemic resistance within the plant, preparing it to resist future pathogen attacks. They can modulate the plant's hormonal pathways, enhancing its immune response. Some endophytes promote plant growth and increase the plant's overall health, making it more resistant to diseases. By employing these mechanisms, endophytic bacteria and fungi serve as an integral part of a plant's defense system, safeguarding it from harmful pathogens and ensuring its survival (Sharma et al., 2022). A number of bioactive compounds are released by the endophytic fungi and these harbor a great deal of industrial and biomedical applications (Figure 5.1). For instance, alkaloids, which are nitrogenous organic compounds, possess a wide range of biological activities. Some alkaloids of endophytic origin have shown significant potential as anticancer and antitumor agents due to their ability to interfere with cell division and growth in abnormally dividing cells. Flavonoids, on the other hand, are polyphenolic compounds that are best known for their antioxidant activity, which can help to neutralize the harmful effects of free radicals in the body. Some flavonoids have also been found to possess antimicrobial activity, making them effective against various bacterial and fungal infections. For instance, quercetin flavonoid found in fruits, vegetables and grains exhibits activity against various bacterial and fungal strains (Thebti et al., 2023). Similarly, a number of other bioactive compounds are being produced as a result of plant–fungi association.

5.3 PHARMACOLOGICAL RELEVANCE OF BIOACTIVE COMPOUNDS

Bioactive compounds obtained from endophytic fungi are creating waves in the pharmaceutical world, potentially serving as novel drug candidates. They offer unique biochemical pathways that could provide solutions to a range of health issues. As the pharmaceutical landscape evolves, these compounds are paving the way for cutting-edge drug development. The applications of these compounds are diverse, from the creation of antimicrobial agents that fight off resistant pathogens, to antiinflammatory drugs that regulate immune responses. The multifaceted nature of numerous diseases calls for comprehensive treatment strategies, and the use of bioactive compounds from endophytic fungi can be a game-changer. These compounds can be instrumental in multimodal treatment approaches as they possess a variety of health-promoting properties that can be harnessed to design pharmaceutical drugs targeting multiple facets of a disease at once (Tiwari et al., 2023). This concept goes beyond simply addressing symptoms because it seeks to tackle the root causes and interconnected factors of diseases, leading to a more holistic treatment. The synergistic approach of

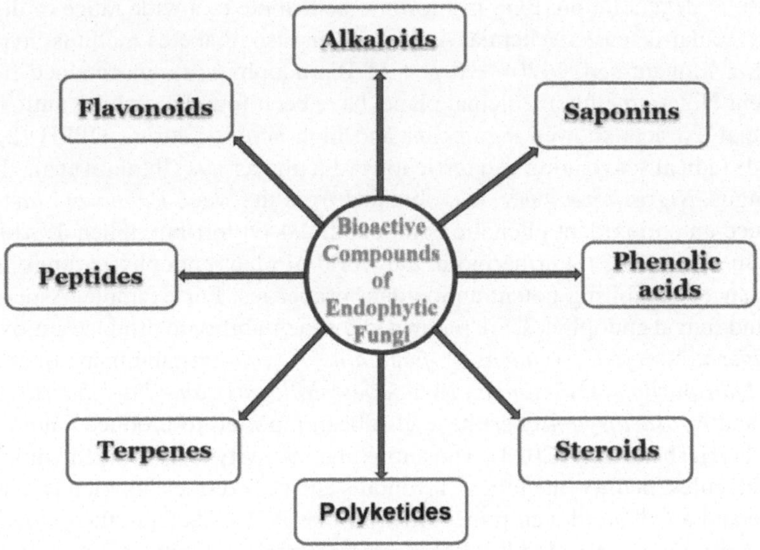

FIGURE 5.1 Broad-spectrum bioactive compounds of endophytic fungi.

TABLE 5.1
Endophytic Fungal Compounds Hving Pharmacological Relevance

Class of Bioactive Compounds	Biological activity	References
Depsipeptides	Cardiovascular protective	Wijesekara & Xu, 2023
Isoprenoids	Antibacterial	Savidov et al., 2018
Alkaloids	Antidiabetic	Khader et al., 2024
Steroids	Anticancer	Chen et al., 2016
Phenols and shikimates	Antiviral	Selim et al., 2018
Xanthones	Antiarthritis	Tawfeeq et al., 2023
Terpenoids	Antioxidant	Kumar et al., 2024
Quinones	Immunomodulatory	El-Hawary et al., 2021
Isocoumarines	Antiinflammatory	Koopklang et al., 2024.
Polyketide and glycosides	Antifungal	Gao et al., 2020; Wang et al., 2022.

using bioactive compounds from endophytic fungi not only has the potential to improve treatment outcomes but also to reduce possible side effects. By targeting multiple disease aspects simultaneously, it minimizes the need for multiple drugs, thereby reducing the risk of adverse reactions and side effects (Gupta et al., 2020; Wijesekara & Xu, 2023). A number of secondary metabolites with pharmacological relevance have been derived from fungal endophytes and these compounds can be used in treating various human diseases (Table 5.1).

5.3.1 Antioxidant Activity

The efficacy of a compound that has antioxidant activity lies in its ability to protect cells from the damage caused by the formation of reactive oxygen species (ROS) and free radicals. These damaging substances can lead to various negative effects, including DNA damage, the development of cancer and cellular degeneration. ROS is known to contribute to a wide range of diseases such as cancer, cardiovascular disease, ischemia, Alzheimer disease, diabetes mellitus, hypertension and aging (Ahmed & Mohammed, 2020). Extracts of 18 endophytic fungi obtained from the leaves of eight different Nigerian ethnomedicinal plants have been found to exhibit antioxidant activity. The crude fungal extracts showed significant and high activity during DPPH (2,2-diphenyl-1-picryl-hydrazyl) radical scavenging and ferric ion reduction assays (Ibrahim et al., 2021). Another endophytic fungus, *Nigrospora sphaerica*, obtained from the weed *Euphorbia hirta* L., has been found to produce an antioxidant phenolic compound, 2,4-Di-tert-butylphenol, which is a potent antioxidant (Sun et al., 2023). Furthermore, the world of plant endophytic fungi harbors diverse bioactive substances exhibiting potent antioxidant properties. For example, Aspergilli, the most commonly found fungal endophytes, are renowned for their ability to produce antioxidants such as *A. minisclerotigens*, *A. oryzae*, *A. flavus*, *A. fumigatus*, *A. nidulans* and many more. Other fungal species, like *Cladosporium tenuissimum*, *Pseudopestalotiopsis camelliae-sinensis*, *Lasiodiplodia venezuelensis* and *Neosartorya fischeri* have also been reported to produce a number of antioxidant compounds (Hashem et al., 2023). The antioxidant activity is associated with different types of bioactive molecules, mainly phenolics, flavonoids and polyketides. Ophicirsins, a heterozygote of a polyketide and an alkaloid, can protect neuron-like PC1 cells from the oxidative damage of hydrogen peroxide. This compound has been obtained from endophytic fungi, *Ophiobolus cirsii* LZU-1509, and it also possesses more free radical scavenging activity than resveratrol, which has been confirmed by DPPH radical scavenging assay (Guo et al., 2023). A similar kind of antioxidant effect has been observed in case of endophytic fungus, *Chaetomium globosum*, crude extract which is found in association with the ethnomedicinal plant *Dillenia indica* Linn. The composition analysis of extract showed the presence of a number of phenolics which are potent antioxidants and the same has been confirmed via DPPH and hydrogen peroxide assays (Kumar & Prasher, 2024). Endophytic fungi, constituting three *Penicillium* and two *Aspergillus* species, isolated from the roots of *Taxus wallichiana* exhibit significant antioxidant activity as confirmed by DPPH assay. The composition of crude extract shows the presence of phenolics and alkaloids as the main antioxidant agents along with others like taxol, gallic acid and fatty acids (Adhikari & Pandey, 2024).

5.3.2 Antiinflammatory Effect

Inflammation is a natural defense mechanism of the human body but sometimes it can create certain biological complications. It has the potential to damage our tissues and contribute to the development of chronic inflammation which poses a significant risk factor for heart disease, autoimmune disorders, as well as certain cancers. Consequently, mitigating the adverse effects of inflammation becomes vital in maintaining our overall health (Brandum et al., 2021). The antiinflammatory compounds produced by endophytic fungi work in a number of ways to counter balance the inflammatory reactions. The fine-tuned regulatory effects produced by these change the overall immune response and prevent an excessive inflammatory reaction that can damage healthy tissues (Mazumder et al., 2021). Furthermore, the antiinflammatory effects of endophytic compounds have been observed in modulating the activity of immune cells, such as macrophages, T lymphocytes and B lymphocytes, contributing to the resolution of chronic inflammation (Deshmukh et al., 2009). It has been observed that compounds produced by *P. brefeldianum*, an endophytic fungus, derived from *Acalypha hispida* leaves, effectively exhibited antiinflammatory properties in the carrageenan-induced paw edema model (Saleh et al., 2023). Another endophytic fungus, *Colletotrichum gloeosprioides* JS0419, obtained from the leaves of *Suaeda japonica*, can produce four β-resorcylic acid derivatives and seven β-resorcylic acid lactones. These compounds exhibit significant antiinflammatory activity along with a reduction in nitric oxide production, inhibition of interlukin-6 and tumor necrosis factor-α secretion (Kim et al., 2023). Lipopolysaccharide (LPS) present on the outer membrane of Gram-negative bacteria is a potent inflammatory agent. The compounds obtained from the mangrove endophytic fungus *Amorosia* sp. SCSIO 41026 directly suppress LPS-induced nitric oxide (NO) accumulation. Apart from this, some of the compounds also inhibit the release of pro-inflammatory factors like IL-6 (interleukin-6) and TNF-α (tumor necrosis factor-α) (Ren et al., 2022). Another mangrove endophytic fungus, *Daldinia eschscholtzii* KBJYZ-1, has been a source of many polyketide compounds and of these compounds eschscholin B shows significant suppression of iNOS (inducible nitric oxide synthase) and COX-2 (cyclooxygenase) in LPS-induced RAW264.7 macrophages, thereby providing antiinflammatory activity (Wang et al., 2022). A large number of other compounds have been retrieved from endophytic fungi possessing antiinflammatory activities (Pal et al., 2020).

5.3.3 Antidiabetic Compounds from Endophytes

Currently, diabetes is one of the major problems which not only the adults or older people face, but which also has become a problem in children due to a lack of awareness, life style and genetic factors. The potential of endophytic fungi as sources for antidiabetic agents has been increasingly recognized. The compounds released by these fungi help in overcoming the diabetic condition in different ways. They may inhibit a particular enzyme whose inactivity can reverse the diabetic condition or they may directly mimic the insulin molecule (Malik et al., 2020). An exopolysaccharide, galactomannan, isolated from mangrove endophytic fungus of *Penicillium janthinellum* N29 proved to be an inhibitor of α-glucosidase. This compound reduces carbohydrate absorption from the gut and hence may help in overcoming type 2 diabetic conditions (Shao et al., 2023). Another compound, 3-hydroxybenzoic acid, isolated from the endophytic fungus *Diaporthe* sp. of the medicinal plant *Simarouba glauca*, shows similar inhibitory properties against α-glucosidase (Mugaranja & Kulal, 2022). A variety of other substances such as 8-hydroxy-6,7-dimethoxy-3-methylisocoumarine, anthraquinonoid [(*S*)-1,3,6-trihydroxy-7-(1-hydroxyethyl) anthracene-9,10-dione], polymethoxylated flavones, all of which are of endophytic fungal origin and show similar inhibitory effects on α-glucosidase (Indrianingsih & Tachibana,, 2017; Fan et al., 2022; Ruzieva et al., 2022). Apart from this inhibitory activity, some compounds have been found to overcome the insulin hyposecretion problem. Two compounds (2,3-dihydrosorbicillin and chrysophanol) of an endophytic fungus, *Penicillium* sp. SSP-1CLG, have been observed to stimulate insulin secretion from pancreatic β-cells in rats (Lee et al., 2023). Another bioactive compound, 2,4,6-triphenylaniline, of the mangrove endophytic fungi *Alternaria longipes* strain VITN14G has been identified to upregulate the activity of AMP-activated protein kinase and thereby increase insulin sensitivity (Ranganathan & Mahalingam, 2020). Apart from these compounds, a huge number of other bioactive substances have been obtained from endophytic fungi which have great potential in addressing the diabetes problem (Khader et al., 2024).

5.3.4 Antimicrobial Activity

Antimicrobial resistance has become one of the major issues the world is facing today. The disorganized and uncontrolled use of antibiotics has created this major issue which demands innovative solutions to combat the growing number of microbial pathogens. Traditional antibiotics are gradually losing their effectiveness, necessitating a change in the approach to treating microbial infections. As such, it is crucial to explore new avenues and develop alternative strategies to address this pressing global health concern (Tang et al., 2023). Bioactive compounds of plant-associated fungi are showing positive effects and have been found to act differently depending upon the nature of the compound. They may disrupt microbial cell membranes, inactivate crucial biomolecules or interfere with the cellular pathway during their mechanism of action. The leaf tissue of the medicinal plant *Callistemon subulatus* contains *Aspergillus* sp. ASCLA and the extract of this plant has been found to possess a compound called isoshamixanthone, which exhibits antimicrobial properties against pathogenic microorganisms. Furthermore, the isolation of *A. flavus* from *Ocimum basilicum* has been found to produce secondary metabolites with potential antimicrobial activity (Kamel et al., 2020; Hashem et al, 2023). It has also been observed that one of the ethyl acetate extracts of endophytic fungi obtained from *Hyssopus officinalis* possesses potential antimicrobial activity. This extract contains the endophytic fungi *Chaetomium elatum* which produces penicillin acid with immense antimicrobial potential against *Bacillus subtilis* and *Staphylococcus aureus* (Eshboev et al., 2023). Another compound, 4a-methyl-dodecahydro-1H-pyrrolo [3,4-b] quinoline-6-one, a bioactive substance, has been isolated from the fungal strain *Aspergillus niger* E12 which is obtained from leaf samples of the plant *Dodonaea viscosa*. This compound shows antibacterial activity against a number of species but mostly against the Gram-negative bacteria *Pseudomonas aeruginosa*, which is a multidrug-resistant pathogen (Kumar & Prasher, 2022). A similar phenomenon has been observed in the case of crude extracts of endophytic fungi obtained from the tropical medicinal plant *Agave americana*. A total of 32 fungal types (15 from stem and 17 from leaf) have been obtained from this plant and of those more than 80% exhibit potential antimicrobial activity against *E. coli* ATTC 25902, *S. aureus* ATTC 14775 and *B. subtilis* NRRL 5109 (Khadka et al., 2024). This endophytic fungal strain not only shows antimicrobial activity against common bacterial species but it has also been observed that a number of endophytic fungi possess metabolites that can act against multidrug-resistant bacterial strains. Extracts of the endophytic fungi *Aspergillus niger* (OL519514) obtained from *O. ficusindica* fruit peels have been found to contain three major phytochemicals with antibacterial properties: dihydroauroglaucin, isotetrahydroauroglaucin and cristatumin B. When these compounds were tested separately, they exhibited antibacterial activity against carbapenem-resistant *Klebsiella pneumoniae*, multidrug-resistant *Pseudomonas aeruginosa*, methicillin-resistant *Staphylococcus aureus* and vancomycin-resistant *Enterococcus faecalis* (Elkady et al., 2022). There are many other compounds of endophytic origin which act strongly against pathogenic bacteria and exhibit a great deal of possible applications for the pharmaceutical industry in the context of antimicrobial agents (Caruso et al., 2022).

5.3.5 Anticancer Activity

Despite so much advancement in the pharmaceutical industry we are still unable to win the battle against cancer and every year millions of people are affected by this deadly disease. With advancements in research and technology, we are constantly seeking new solutions to improve cancer treatment and endophytic fungi-based bioactive compounds have emerged as novel substances to combat this. The identification of taxol-producing endophytic fungi and their application in cancer treatment have encouraged the search for more potential anticancerous agents (Zhang et al., 2009; El-Sayed et al., 2020). Ethyl acetate extract of endophytic fungal strain, *Fusarium perseae* WDF12, obtained from the plant *Cephalotaxus mannii* Hook f., has been found to exhibit cytotoxic activity against A549 (human lung cancer) and MCF7 (human breast cancer) cell lines (Vu et al., 2023). Another fungal metabolite, isoshamixanthone, obtained from the endophytic fungal strain *Aspergillus* sp. ASCLA, showed anticancer activity against a human cervix carcinoma cell line (KB-3-1). This fungal stain has been isolated from the leaf extract of the medicinal plant *Callistemon subulatus* (Kamel et al., 2020). Polyketides also have been found in several different types of endophytes and have a broad spectrum of biological activities, including antibacterial, antifungal, anticancer and immunosuppressive effects. A marine endophytic fungus, *Alternaria* sp., when grown on rice solid media produced many bioactive polyketides and of those alternariol-9-methyl ether showed positive impacts against the A549 lung cancer cell line and PC3 prostate cancer cell line (Mahmoud et al., 2022). The aforementioned endophytes are particularly interesting due to their ability to produce a biomolecule called podophyllotoxin. This is a non-alkaloid toxin lignan which is found in the roots and rhizomes of *Podophyllum* species. It has attracted substantial attention from the pharmaceutical industry for its significant antineoplastic activity. Podophyllotoxin works by inhibiting the assembly of microtubules during cell division. By preventing the normal formation of the mitotic spindle, it effectively arrests the dividing cells in the metaphase, leading to cell death. This unique mechanism of action has made podophyllotoxin a valuable resource in cancer treatment (Wijesekara & Xu, 2023). The exact mechanism of action of many metabolites remains unknown but some of them have been characterized, such as in the case of the endophytic strain *Aspergillus TRL1*, isolated from *Tabebuia rosea*, which produces a compound called pulchranin which is effective against human liver, breast and colorectal cancer cell lines. This metabolite functions by inhibiting cyclin-dependent kinases (Moussa et al., 2020). Another fungal metabolite, 8-n-butyrylneosolaniol, functions by inducing cell apoptosis in human hepatoma cells. This compound belongs to the category of trichothecenes and has been isolated from *Fusarium sporotrichioides*, which is an endophyte to the plant *Rauvolfia yunnanensis* (Wang et al., 2022).

5.3.6 Antiviral Substances of Endophytic Origin

Viral infection is one of the common problems associated not only with humans but also with plants and animals. Many antiviral agents have been made but these generally possess many side effects. In this context, antiviral agents of biological origin show positive implications and the compounds of endophytic origin are one of these. These fungal metabolites act in a number of ways to combat the viral pathogenesis such as by inhibiting viral entry, reverse transcriptase activity, protease activity and integrase activity. In case of human immunodeficiency virus-1 (HIV-1) strains, two compounds of endophytic fungus, *Chaetomium globosum Kunze* 1705, act as inhibitor of chemokine receptors CCR5 and thereby inhibit viral entry in $CD4^+$ cells, whereas another compound alters toxins obtained from another plant-associated fungi, *Alternaria tenuissima*, inhibiting the activity of HIV-1 reverse transcriptase enzyme (Nzimande et al., 2023). Similarly, extract of endophytes obtained from the stem bark sample of *Morus alba* exhibits anti-HIV activity by inhibiting the protease enzyme (Vora et al., 2021). Apart from these anti-HIV agents, an endophytic fungal compound, 14-nordrimane-type sesquiterpenoid, of *Phoma* sp. has been isolated from the roots of *Aconitum vilmorinianum* and has been proved to be effective against influenza A virus (Liu et al., 2019). The endophytic fugal metabolites have also shown positive aspects against the coronavirus that resulted in COVID-19 and in silico analyses have revealed them to be potent inhibitors of COVID-19 protease (El-Hawary et al., 2021).

5.4 CONCLUSION

There are a large number of plant-associated fungi which have tremendous biological applications (Table 5.2). The positive implications of endophytic fungal compounds appear to be the future of the pharmaceutical industry. Many bioactive compounds of endophytic fungal origin still awaits study for the betterment of human lives. The fungal extracts are generally a mixture of many bioactive compounds and sometimes a single compound is not as efficient in providing a particular biological activity. Therefore, identifying the exact compound and its production on a large scale remain a challenging task.

TABLE 5.2
Plant-Associated Endophytic Fungi and Their Biological Properties

Biological Property	Endophytic fungi	Host plant	References
Antioxidant activity	*Aspergillus flavus*	*Ocimum basilicum*	Sharaf et al., 2022
	Aspergillus minisclerotigens	*Mangifera casturi*	Nuraini et al., 2019
	Aspergillus nidulans	*Passiflora incarnate*	da Silva et al., 2020
	Aspergillus oryzae	*Kosterm*	Nuraini et al, 2019.
	Aspergillus terreus	*Artemisia annua*	Sayed et al., 2022
	Penicillium roqueforti	*Solanum surattense*	Ikram et al., 2019
	Penicillium pinophilum	*Alloteropsis cimicina*	Nischitha et al., 2021
	Alternaria alternata	*Lithospermum officinale*	Mollaei et al., 2019
	Alternaria alternata	*Picrorhiza kurroa*	Chandra et al, 2021
	Fusarium oxysporum	*Otoba gracilipes*	Caicedo et al., 2019
	Chaetomium cruentum	*Conyza blinii* H. Lév	Zhao et al., 2021
	Nigrospora sphaerica	*Euphorbia hirta* (dudhi) L	Gautam et al., 2022
	Nigrospora oryzae	*Tinospora cordifolia*	Vig et al., 2022
Antimicrobial activity	*Aspergillus* spp.	*Anthopleura xanthogrammica*	Liu et al., 2020
	Aspergillus fumigatus	*Albizia lucidior,*	Hussein et al., 2022
	Aspergillus fumigatus	*Ocimum basilicum*	Sharaf et al., 2022
	Aspergillus neoniger	*Ficus carica*	Abdou et al., 2021
	Aspergillus nidulans	*Ocimum basilicum*	Sharaf et al., 2022
	Aspergillus niger	*Sonneratia apetala*	Nurunnabi et al., 2020
	Aspergillus versicolor	*Pulicaria crispa* Forssk	Mohamed et al., 2020
	Penicillium sp.	*Stephania dielsiana*	Wu et al., 2020
	Penicillium citrinum	*Azadirachta indica*	Kumari et al., 2021
	Penicillium funiculosum	*Ficus elastic*	Ding et al., 2019
	Alternaria sp. SZMC 23772	*Hypericum perforatum*	Vigneshwari et al., 2019
	Alternaria alternata	*Ziziphus spina-christi*	Elghaffar et al., 2022
	Alternaria tenuissima PE2	*Psidium guajava* L.	Chatterjee et al., 2022
	Fusarium equiseti	*Sonneratia apetala*	Nurunnabi et al., 2020
	Fusarium proliferatum	*Cissus quadrangularis* L.	Singh et al., 2021
	Fusarium oxysporum	*Sceletium tortuosum* L.	Manganyi et al., 2019
	Chaetomium sp. HQ-1	*Astragalus chinensis*	Liu et al., 2019
	Chaetomium globosum	*Litsea cubeba*	Kaur et al., 2020
	Trichoderma harzianum	*Rosmarinus officinalis*	Abdulhadi et al., 2023
	Trichoderma harzianum	*Ficus elastica*	Ding et al., 2019
	Trichoderma virens QA-8	*Artemisia argyi*	Shi et al., 2019
	Nigrospora sphaerica	*Dillenia indica* L.	Kumar et al., 2021
	Nigrospora sphaerica	*Adiantum philippense* L.	Ramesha et al., 2020
	Epicoccum nigrum	*Hypericum perforatum*	Vigneshwari et al., 2019
	Colletotrichum gloeosporioides	*Sonneratia apetala*	Nurunnabi et al., 2020
	Colletotrichum coccodes	*Houttuynia cordata*	Talukdar et al., 2020
Anticancer activity	*Aspergillus* sp. TRL1	*Tabebuia rosea*	Moussa et al., 2020
	Aspergillus sp. GZWMJZ-258	*Garcinia multiflora*	He et al., 2019
	Aspergillus candidus LDJ-5	*Rhizophora apiculata*	Zhou et al., 2020
	Aspergillus flavus	*Tylophora ovata*	Liu et al., 2019
	Aspergillus micronesiensis	*Kappaphycus alvarezii*	Luyen et al., 2019
	Aspergillus terreus	*Ficus elastica*	El-Sayed et al., 2021
	Penicillium vsp. ct-28	*Corydlis tomentella*	Ming et al., 2022
	Penicillium ochrochloronthe	*Taxus media*	Zhao et al., 2019
	Penicillium simplicissimum	*Loranthus micranthus*	Govindappa et al., 2023
	Alternaria sp. MG1	*Vitis quinquangularis*	Wu et al., 2019
	Alternaria sp. sb23	*Schisandra sphenanthera*	Gao et al., 2020

TABLE 5.2 (Continued)
Plant-Associated Endophytic Fungi and Their Biological Properties

Biological Property	Endophytic fungi	Host plant	References
	Alternaria alternate	*Paeonia lactiflora*	Wang et al., 2021
	Alternaria brassicicola	*Terminalia arjuna*	Gill et al., 2019
	Chaetomium sp. SYP-F7950	*Panax notoginseng*	Peng et al., 2019
	Trichoderma harzianum	*Zingiber officinale*	Harwoko et al., 2021
	Nigrospora sphaerica	*Helianthus annuus*	Supaphon et al., 2019
	Epicoccum nigrum	*Salix* sp.	Harwoko et al., 2021
	Epicoccum nigrum	*Terminalia arjuna*	El-Sayed et al., 2020
Antidiabetic activity	*Aspergillus* sp. 16-5c	Mangrove	Ye et al., 2021
	Alternaria destruens	*Calotropis gigantean*	Kaur et al., 2020
	Curvularia lunata	*Ficus religiosa L*	Ye et al., 2021
Antiviral activity	*Trichoderma harzianum*	*Kadsura angustifolia*	Han et al., 2019
	Phoma sp.	*Aconitum vilmorinianum*	Liu et al, 2019
	Curvularia papendorfii	*Vernonia amygdalina*	Khiralla et al., 2020

REFERENCES

Abdulhadi, S.Y., Hasan, G.Q. and Gergees, R.N. 2023. Molecular detection and antimicrobial activity of endophytic fungi isolated from a medical plant Rosmarinus officinalis. *arXiv preprint arXiv:2303.05242*.

Abdou, R., Alqahtani, A.M. and Attia, G.H. 2021. Bioactive metabolites of aspergillus neoniger, an endophyte of the medicinal plant Ficus carica. *Indian Journal of Pharmaceutical Sciences*, 83.

Adhikari, P. and Pandey, A. 2024. Phytochemicals, and antioxidants and antibacterial activities of Taxus wallichiana Zucc. root associated endophytic fungi. *Biologia*:1–13.

Ahmed, A.M., Mahmoud, B.K., Millán-Aguiñaga, N., Abdelmohsen, U.R. and Fouad, M.A. 2023. The endophytic Fusarium strains: a treasure trove of natural products. *RSC Advances* 13:1339–1369.

Ahmed, O.M. and Mohammed, M.T. 2020. Oxidative stress: the role of reactive oxygen species (ROS) and antioxidants in human diseases. *Plant Archives* 20:4089–4095.

Baron, N.C. and Rigobelo, E.C. 2022. Endophytic fungi: a tool for plant growth promotion and sustainable agriculture. *Mycology* 13:39–55.

Brandum, E.P., Jørgensen, A.S., Rosenkilde, M.M. and Hjortø, G.M. 2021. Dendritic cells and CCR7 expression: an important factor for autoimmune diseases, chronic inflammation, and cancer. *International Journal of Molecular Sciences* 22:8340.

Caicedo, N.H., Davalos, A.F., Puente, P.A., Rodríguez, A.Y. and Caicedo, P.A. 2019. Antioxidant activity of exo-metabolites produced by Fusarium oxysporum: An endophytic fungus isolated from leaves of Otoba gracilipes. *MicrobiologyOpen*, 8:903.

Caruso, D.J., Palombo, E.A., Moulton, S.E. and Zaferanloo, B. 2022. Exploring the promise of endophytic fungi: a review of novel antimicrobial compounds. *Microorganisms* 10:1990.

Chandra, H., Kumari, P., Prasad, R., Gupta, S.C. and Yadav, S. 2021. Antioxidant and antimicrobial activity displayed by a fungal endophyte Alternaria alternata isolated from Picrorhiza kurroa from Garhwal Himalayas, India. *Biocatalysis and Agricultural Biotechnology*, 33:101955.

Chatterjee, S., Ghosh, S. and Mandal, N.C. 2022. Potential of an endophytic fungus Alternaria tenuissima PE2 isolated from Psidium guajava L. for the production of bioactive compounds. *South African Journal of Botany*, 150:658–670.

Chen, L., Zhang, Q.Y., Jia, M., Ming, Q.L., Yue, W., Rahman, K., Qin, L.P. and Han, T., 2016. Endophytic fungi with antitumor activities: Their occurrence and anticancer compounds. *Critical Reviews in Microbiology*, 42:454–473.

Collinge, D.B., Jensen, B. and Jørgensen, H.J. 2022. Fungal endophytes in plants and their relationship to plant disease. *Current Opinion in Microbiology* 69:102177.

da Silva, M.H.R., Cueva-Yesquén, L.G., Júnior, S.B., Garcia, V.L., Sartoratto, A., de Angelis, D.D.F. and de Angelis, D.A. 2020. Endophytic fungi from Passiflora incarnata: an antioxidant compound source. *Archives of Microbiology*, 202:2779–2789.

Deshmukh, S.K., Mishra, P.D., Kulkarni-Almeida, A., Verekar, S., Sahoo, M.R., Periyasamy, G., Goswami, H., Khanna, A., Balakrishnan, A. and Vishwakarma, R. 2009. Anti-inflammatory and anticancer activity of ergoflavin isolated from an endophytic fungus. *Chemistry & Biodiversity* 6:784–789.

Ding, Z., Tao, T., Wang, L., Zhao, Y., Huang, H., Zhang, D., Liu, M., Wang, Z. and Han, J. 2019. Bioprospecting of novel and bioactive metabolites from endophytic fungi isolated from rubber tree Ficus elastica leaves. *Journal of Microbiology and Biotechnology*, 29(5):731–738.

Elghaffar, R.Y.A., Amin, B.H., Hashem, A.H. and Sehim, A.E. 2022. Promising endophytic Alternaria alternata from leaves of Ziziphus spina-christi: phytochemical analyses, antimicrobial and antioxidant activities. *Applied Biochemistry and Biotechnology*, 194:3984–4001.

El-Hawary, S.S., Mohammed, R., Bahr, H.S., Attia, E.Z., El-Katatny, M.M.H., Abelyan, N., Al-Sanea, M.M., Moawad, A.S. and Abdelmohsen, U.R. 2021. Soybean-associated endophytic fungi as potential source for anti-COVID-19 metabolites supported by docking analysis. *Journal of Applied Microbiology* 131:1193–1211.

Elkady, W.M., Raafat, M.M., Abdel-Aziz, M.M., Al-Huqail, A.A., Ashour, M.L. and Fathallah, N. 2022. Endophytic fungus from Opuntia ficus-indica: a source of potential bioactive antimicrobial compounds against multidrug-resistant bacteria. *Plants* 11:1070.

El-Sayed, A.S., El-Sayed, M.T., Rady, A.M., Zein, N., Enan, G., Shindia, A., El-Hefnawy, S., Sitohy, M. and Sitohy, B. 2020. Exploiting the biosynthetic potency of taxol from fungal endophytes of conifers plants; genome mining and metabolic manipulation. *Molecules* 25:3000.

El-Sayed, A.S., Khalaf, S.A., Azez, H.A., Hussein, H.A., EL-Moslamy, S.H., Sitohy, B. and El-Baz, A.F. 2021. Production, bioprocess optimization and anticancer activity of Camptothecin from Aspergillus terreus and Aspergillus flavus, endophytes of Ficus elastica. *Process Biochemistry*, 107:59–73.

El-Sayed, E.R., Ahmed, A.S. and Abdelhakim, H.K. 2020. A novel source of the cardiac glycoside digoxin from the endophytic fungus Epicoccum nigrum: isolation, characterization, production enhancement by gamma irradiation mutagenesis and anticancer activity evaluation. *Journal of applied microbiology*, 128:747–762.

Eshboev, F., Karakozova, M., Abdurakhmanov, J., Bobakulov, K., Dolimov, K., Abdurashidov, A., Baymirzaev, A., Makhnyov, A., Terenteva, E., Sasmakov, S. and Piyakina, G. 2023. Antimicrobial and cytotoxic activities of the secondary metabolites of endophytic fungi isolated from the medicinal plant Hyssopus officinalis. *Antibiotics* 12:1201.

Fan, J., Lv, C., Li, Z., Guo, M., Yin, Y., Wang, H., Wang, W. and Sun, S. 2022. α-Glucosidase inhibitory effect of an anthraquinonoid produced by Fusarium incarnatum GDZZ-G2. *Journal of Basic Microbiology* 62:1360–1370.

Gao, Y., Wang, L., Kalscheuer, R., Liu, Z. and Proksch, P. 2020. Antifungal polyketide derivatives from the endophytic fungus Aplosporella javeedii. *Bioorganic & Medicinal Chemistry*, 28:115456.

Gao, Y., Zhou, J. and Ruan, H. 2020. Trichothecenes from an endophytic fungus Alternaria sp. sb23. *Planta Medica*, 86:976–982.

Gautam, V.S., Singh, A., Kumari, P., Nishad, J.H., Kumar, J., Yadav, M., Bharti, R., Prajapati, P. and Kharwar, R.N. 2022. Phenolic and flavonoid contents and antioxidant activity of an endophytic fungus Nigrospora sphaerica (EHL2), inhabiting the medicinal plant Euphorbia hirta (dudhi) L. *Archives of Microbiology*, 204:140.

Gill, H. and Vasundhara, M. 2019. Isolation of taxol producing endophytic fungus Alternaria brassicicola from non-Taxus medicinal plant Terminalia arjuna. *World Journal of Microbiology and Biotechnology*, 35:1–8.

Govindappa, M., Vishaka, A., Akshatha, B.S., Popli, D., Sunayana, N., Srinivas, C., Pugazhendhi, A. and Raghavendra, V.B. 2023. An endophytic fungus, Penicillium simplicissimum conjugated with C60 fullerene for its potential antimitotic, anti-inflammatory, anticancer and photodegradation activities. *Environmental Technology*, 44:817–831.

Guo, X.W., Yu, Z.Q., Xi, J., Ren, H., Xiang, X.Y., Wu, J., Fang, J. and Wu, Q.X. 2023. Isolation and identification of novel antioxidant polyketides from an endophytic fungus Ophiobolus cirsii LZU-1509. *Journal of Agricultural and Food Chemistry* 71:1593–1606.

Gupta, S., Chaturvedi, P., Kulkarni, M.G. and Van Staden, J. 2020. A critical review on exploiting the pharmaceutical potential of plant endophytic fungi. *Biotechnology Advances* 39:107462.

Han, M., Qin, D., Ye, T., Yan, X., Wang, J., Duan, X. and Dong, J. 2019. An endophytic fungus from Trichoderma harzianum SWUKD3. 1610 that produces nigranoic acid and its analogues. *Natural Product Research*, 33:2079–2087.

Harwoko, H., Daletos, G., Stuhldreier, F., Lee, J., Wesselborg, S., Feldbrügge, M., Müller, W.E., Kalscheuer, R., Ancheeva, E. and Proksch, P. 2021. Dithiodiketopiperazine derivatives from endophytic fungi Trichoderma harzianum and Epicoccum nigrum. *Natural product research*, 35:257–265.

Hashem, A.H., Attia, M.S., Kandil, E.K., Fawzi, M.M., Abdelrahman, A.S., Khader, M.S., Khodaira, M.A., Emam, A.E., Goma, M.A. and Abdelaziz, A.M. 2023. Bioactive compounds and biomedical applications of endophytic fungi: a recent review. *Microbial Cell Factories* 22:1–23.

He, W., Xu, Y., Fu, P., Zuo, M., Liu, W., Jiang, Y., Wang, L. and Zhu, W. 2019. Cytotoxic indolyl diketopiperazines from the Aspergillus sp. GZWMJZ-258, endophytic with the medicinal and edible plant Garcinia multiflora. *Journal of agricultural and food chemistry*, 67:10660–10666.

Hussein, M.E., Mohamed, O.G., El-Fishawy, A.M., El-Askary, H.I., El-Senousy, A.S., El-Beih, A.A., Nossier, E.S., Naglah, A.M., Almehizia, A.A., Tripathi, A. and Hamed, A.A. 2022. Identification of antibacterial metabolites from endophytic fungus Aspergillus fumigatus, isolated from Albizia lucidior leaves (Fabaceae), utilizing metabolomic and molecular docking techniques. *Molecules*, 27:1117.

Ibrahim, M., Oyebanji, E., Fowora, M., Aiyeolemi, A., Orabuchi, C., Akinnawo, B. and Adekunle, A.A. 2021. Extracts of endophytic fungi from leaves of selected Nigerian ethnomedicinal plants exhibited antioxidant activity. *BMC Complementary Medicine and Therapies* 21:1–13.

Ikram, M., Ali, N., Jan, G., Hamayun, M., Jan, F.G. and Iqbal, A. 2019. Novel antimicrobial and antioxidative activity by endophytic Penicillium roqueforti and Trichoderma reesei isolated from Solanum surattense. *Acta Physiologiae Plantarum*, 41:1–11.

Indrianingsih, A.W. and Tachibana, S. 2017. α-Glucosidase inhibitor produced by an endophytic fungus, Xylariaceae sp. QGS 01 from Quercus gilva Blume. *Food Science and Human Wellness* 6:88–95.

Kamel, R.A., Abdel-Razek, A.S., Hamed, A., Ibrahim, R.R., Stammler, H.G., Frese, M., Sewald, N. and Shaaban, M. 2020. Isoshamixanthone: a new pyrano xanthone from endophytic Aspergillus sp. ASCLA and absolute configuration of epiisoshamixanthone. *Natural Product Research* 34:1080–1090.

Kaur, N. and Arora, D.S. 2020. Prospecting the antimicrobial and antibiofilm potential of Chaetomium globosum an endophytic fungus from Moringa oleifera. *AMB Express*, 10:206.

Khader, B.M., Abu-Elsaoud, A.M., Darwish, A.M. and Abdel-Azeem, A.M. 2024. Endophytic fungi hosted anti-diabetic medicinal plants as a source of α-Amylase and α-Glucosidase inhibitors. *Microbial Biosystems* 8:44–56.

Khadka, G., Shetty, K.G., Annamalai, T., Tse-Dinh, Y.C. and Jayachandran, K. 2024. Characterization and antimicrobial activity of endophytic fungi from medicinal plant Agave Americana. *Letters in Applied Microbiology* 77 :ovae025.

Khiralla, A., Spina, R., Varbanov, M., Philippot, S., Lemiere, P., Slezack-Deschaumes, S., André, P., Mohamed, I., Yagi, S.M. and Laurain-Mattar, D. 2020. Evaluation of antiviral, antibacterial and antiproliferative activities of the endophytic fungus Curvularia papendorfii, and isolation of a new polyhydroxyacid. *Microorganisms*, 8:1353.

Kim, J., Baek, J.Y., Bang, S., Kim, J.Y., Jin, Y., Lee, J.W., Jang, D.S., Kang, K.S. and Shim, S.H. 2023. New anti-inflammatory β-resorcylic acid lactones derived from an endophytic fungus, Colletotrichum sp. *ACS Omega* 8:3530–3538.

Koopklang, K., Choodej, S., Hantanong, S., Intayot, R., Jungsuttiwong, S., Insumran, Y., Ngamrojanavanich, N. and Pudhom, K. 2024. Anti-inflammatory properties of oxygenated isocoumarins and xanthone from thai mangrove-associated endophytic fungus Setosphaeria rostrata. *Molecules*, 29:603.

Kumar, V. and Prasher, I.B. 2021. Phytochemical analysis and antimicrobial potential of Nigrospora sphaerica (Berk. & Broome) Petch, a fungal endophyte isolated from Dillenia indica L. *Advances in Traditional Medicine*, 1–13.

Kumar, V. and Prasher, I.B. 2022. Antimicrobial potential of endophytic fungi isolated from Dillenia indica L. and identification of bioactive molecules produced by Fomitopsis meliae (Undrew.) Murril. *Natural Product Research* 36:6064–6068.

Kumar, V. and Prasher, I.B. 2024. Phytochemical analysis and antioxidant activity of endophytic fungi isolated from Dillenia indica Linn. *Applied Biochemistry and Biotechnology* 196:332–349.

Kumari, P., Singh, A., Singh, D.K., Sharma, V.K., Kumar, J., Gupta, V.K., Bhattacharya, S. and Kharwar, R.N. 2021. Isolation and purification of bioactive metabolites from an endophytic fungus Penicillium citrinum of Azadirachta indica. *South African Journal of Botany*, 139:449–457.

Lee, D., Kim, J., Choi, S., Choi, J., Lee, J.W., Kang, K.S. and Shim, S.H. 2023. 2, 3-Dihydrosorbicillin and chrysopanol stimulate insulin secretion in INS-1 cells. *Bioorganic & Medicinal Chemistry Letters* 83:129186.

Liu, S., Ahmed, S., Zhang, C., Liu, T., Shao, C. and Fang, Y. 2020. Diversity and antimicrobial activity of culturable fungi associated with sea anemone Anthopleura xanthogrammica. *Electronic Journal of Biotechnology*, 44:41–46.

Liu, S.S., Jiang, J.X., Huang, R., Wang, Y.T., Jiang, B.G., Zheng, K.X. and Wu, S.H. 2019. A new antiviral 14-nordrimane sesquiterpenoid from an endophytic fungus Phoma sp. *Phytochemistry Letters 29:*75–78.

Luyen, N.D., Huong, L.M., Thi Hong Ha, T., Cuong, L.H., Thi Hai Yen, D., Nhiem, N.X., Tai, B.H., Gardes, A., Kopprio, G. and Van Kiem, P. 2019. Aspermicrones AC, novel dibenzospiroketals from the seaweed-derived endophytic fungus Aspergillus micronesiensis. *Journal of Antibiotics*, 72:843–847.

Mahmoud, M.M., Abdel-Razek, A.S., Soliman, H.S., Ponomareva, L.V., Thorson, J.S., Shaaban, K.A. and Shaaban, M. 2022. Diverse polyketides from the marine endophytic Alternaria sp. LV52: structure determination and cytotoxic activities. *Biotechnology Reports* 33:00628.

Malik, A., Ardalani, H., Anam, S., McNair, L.M., Kromphardt, K.J., Frandsen, R.J.N., Franzyk, H., Staerk, D. and Kongstad, K.T. 2020. Antidiabetic xanthones with α-glucosidase inhibitory activities from an endophytic Penicillium canescens. *Fitoterapia* 142:104522.

Manganyi, M.C., Regnier, T., Tchatchouang, C.D.K., Bezuidenhout, C.C. and Ateba, C.N. 2019. Antibacterial activity of endophytic fungi isolated from Sceletium tortuosum L.(Kougoed). *Annals of Microbiology*, 69:659–663.

Mazumder, K., Ruma, Y.N., Akter, R., Aktar, A., Hossain, M.M., Shahina, Z., Mazumdar, S. and Kerr, P.G. 2021. Identification of bioactive metabolites and evaluation of in vitro anti-inflammatory and in vivo antinociceptive and antiarthritic activities of endophyte fungi isolated from Elaeocarpus floribundus blume. *Journal of Ethnopharmacology* 273:113975.

Ming, Q., Li, Y., Jiang, X., Huang, X., He, Y., Qin, L., Liu, Y., Tang, Y. and Gao, N. 2022. Xanthones and benzophenones isolated from the endophytic fungus Penicillium sp. ct-28 of Corydlis tomentella and their cytotoxic activity. *Fitoterapia*, 157:105127.

Mohamed, G.A., Ibrahim, S.R. and Asfour, H.Z. 2020. Antimicrobial metabolites from the endophytic fungus Aspergillus versicolor. *Phytochemistry Letters*, 35:152–155.

Mollaei, S., Khanehbarndaz, O., Gerami-Khashal, Z. and Ebadi, M. 2019. Molecular identification and phytochemical screening of endophytic fungi isolated from Lithospermum officinale L. roots: A new source of shikonin. *Phytochemistry*, 168:112116.

Moussa, A.Y., Mostafa, N.M. and Singab, A.N.B. 2020. Pulchranin A: first report of isolation from an endophytic fungus and its inhibitory activity on cyclin dependent kinases. *Natural Product Research* 34:2715–2722.

Mugaranja, K.P. and Kulal, A. 2022. Inhibition of alpha-glucosidase enzyme and antioxidant activity of the metabolites purified from an endophytic fungus of Simarouba glauca DC. *South African Journal of Botany* 149:134–144.

Nischitha, R. and Shivanna, M.B. 2021. Metabolite fingerprinting, in vitro antimicrobial and antioxidant activities and in-silico docking in Alloteropsis cimicina and its endophytic fungus Penicillium pinophilum. *Molecular Biology Reports*, 48:4021–4037.

Nuraini, F.R., Setyaningsih, R. and Susilowati, A. 2019, July. Antioxidant activity of bioactive compound produced by endophytic fungi isolated from endemic plant of South Kalimantan Mangifera casturi Kosterm. In *AIP conference proceedings* (Vol. 2120, No. 1). AIP Publishing.

Nurunnabi, T.R., Sabrin, F., Sharif, D.I., Nahar, L., Sohrab, M.H., Sarker, S.D., Rahman, S.M. and Billah, M.M. 2020. Antimicrobial activity of endophytic fungi isolated from the mangrove plant Sonneratia apetala (Buch.-Ham) from the Sundarbans mangrove forest. *Advances in Traditional Medicine*, 20:419–425.

Nzimande, B., Makhwitine, J.P., Mkhwanazi, N.P. and Ndlovu, S.I. 2023. Developments in exploring fungal secondary metabolites as antiviral compounds and advances in HIV-1 Inhibitor screening assays. *Viruses* 15:1039.

Omomowo, I.O., Amao, J.A., Abubakar, A., Ogundola, A.F., Ezediuno, L.O. and Bamigboye, C.O. 2023. A review on the trends of endophytic fungi bioactivities. *Scientific African* 20:e01594.

Pal, P.P., Shaik, A.B. and Begum, A.S. 2020. Prospective leads from endophytic fungi for anti-inflammatory drug discovery. *Planta Medica* 86:941–959.

Pandey, S.S., Jain, R., Bhardwaj, P., Thakur, A., Kumari, M., Bhushan, S. and Kumar, S. 2022. Plant Probiotics–Endophytes pivotal to plant health. *Microbiological Research* 263:127148.

Peng, F., Hou, S.Y., Zhang, T.Y., Wu, Y.Y., Zhang, M.Y., Yan, X.M., Xia, M.Y. and Zhang, Y.X. 2019. Cytotoxic and antimicrobial indole alkaloids from an endophytic fungus Chaetomium sp. SYP-F7950 of Panax notoginseng. *RSC Advances*, *9*:28754–28763.

Ramesha, K.P., Mohana, N.C., Nuthan, B.R., Rakshith, D. and Satish, S. 2020. Antimicrobial metabolite profiling of Nigrospora sphaerica from Adiantum philippense L. *Journal of Genetic Engineering and Biotechnology*, 18:66.

Ranganathan, N. and Mahalingam, G. 2020. Validation of the antidiabetic potential of isolated and nanoemulsified endophytic fungal metabolite 2, 4, 6-triphenylaniline through AMPK activation pathway. *Journal of Molecular Liquids* 316:113836.

Ren, X., Chen, C., Ye, Y., Xu, Z., Zhao, Q., Luo, X., Liu, Y. and Guo, P. 2022. Anti-inflammatory compounds from the mangrove endophytic fungus Amorosia sp. SCSIO 41026. *Frontiers in Microbiology* 13:976399.

Ruzieva, D., Gulyamova, T., Nasmetova, S., Mukhammedov, I. and Rasulova, G. 2022. Identification of bioactive compounds of the endophytic fungus Aspergillus egypticus-HT166S inhibiting the activity of pancreatic α-Amylase. *Turkish Journal of Pharmaceutical Sciences* 19:630.

Saleh, A., Negm, W.A., El-Masry, T.A., Eliwa, D., Alotaibi, B., Alosaimi, M.E., Alotaibi, K.N., Magdeldin, S., Mahgoub, S. and Elekhnawy, E. 2023. Anti-inflammatory potential of Penicillium brefeldianum endophytic fungus supported with phytochemical profiling. *Microbial Cell Factories* 22:83.

Savidov, N., Gloriozova, T.A., Poroikov, V.V. and Dembitsky, V.M. 2018. Highly oxygenated isoprenoid lipids derived from fungi and fungal endophytes: Origin and biological activities. *Steroids*, 140:114–124.

Sayed, A.M., Sherif, N.H., El-Gendy, A.O., Shamikh, Y.I., Ali, A.T., Attia, E.Z., El-Katatny, M.M.H., Khalifa, B.A., Hassan, H.M. and Abdelmohsen, U.R. 2022. Metabolomic profiling and antioxidant potential of three fungal endophytes derived from Artemisia annua and Medicago sativa. *Natural Product Research*, 36:2404–2408.

Selim, K.A., Elkhateeb, W.A., Tawila, A.M., El-Beih, A.A., Abdel-Rahman, T.M., El-Diwany, A.I. and Ahmed, E.F. 2018. Antiviral and antioxidant potential of fungal endophytes of Egyptian medicinal plants. *Fermentation*, 4:49.

Shao, Z., Tian, Y., Liu, S., Chu, X. and Mao, W. 2023. Anti-diabetic activity of a novel exopolysaccharide produced by the mangrove endophytic fungus Penicillium janthinellum N29. *Marine Drugs* 21:270.

Sharaf, M.H., Abdelaziz, A.M., Kalaba, M.H., Radwan, A.A. and Hashem, A.H. 2022. Antimicrobial, antioxidant, cytotoxic activities and phytochemical analysis of fungal endophytes isolated from ocimum basilicum. *Applied Biochemistry and Biotechnology*, 1–19.

Sharma, A., Kaushik, N., Sharma, A., Marzouk, T. and Djébali, N. 2022. Exploring the potential of endophytes and their metabolites for bio-control activity. *3 Biotech* 12:277.

Shi, X.S., Meng, L.H., Li, X.M., Li, X., Wang, D.J., Li, H.L., Zhou, X.W. and Wang, B.G. 2019. Trichocadinins B–G: antimicrobial cadinane sesquiterpenes from Trichoderma virens QA-8, an endophytic fungus obtained from the medicinal plant Artemisia argyi. *Journal of natural products*, 82:2470–2476.

Singh, A., Kumar, J., Sharma, V.K., Singh, D.K., Kumari, P., Nishad, J.H., Gautam, V.S. and Kharwar, R.N. 2021. Phytochemical analysis and antimicrobial activity of an endophytic Fusarium proliferatum (ACQR8), isolated from a folk medicinal plant Cissus quadrangularis L. *South African Journal of Botany*, 140:87–94.

Supaphon, P. and Preedanon, S. 2019. Evaluation of in vitro alpha-glucosidase inhibitory, antimicrobial, and cytotoxic activities of secondary metabolites from the endophytic fungus, Nigrospora sphaerica, isolated from Helianthus annuus. *Annals of Microbiology*, 69:1397–1406.

Sun, Y., Ren, G., Shi, Q., Zhu, H., Zhou, N., Kong, X., Jiang, D. and Liu, C. 2023. Identification of a novel Coumarins biosynthetic pathway in the endophytic fungus Fusarium oxysporum GU-7 with antioxidant activity. *Applied and Environmental Microbiology* 89:01601–22.

Talukdar, R., Wary, S., Mili, C., Roy, S. and Tayung, K. 2020. Antimicrobial secondary metabolites obtained from endophytic fungi inhabiting healthy leaf tissues of Houttuynia cordata Thunb., an ethnomedicinal plant of Northeast India. *Journal of Applied Pharmaceutical Science*, 10:099–106.

Tang, K.W.K., Millar, B.C. and Moore, J.E. 2023. Antimicrobial resistance (AMR). *British Journal of Biomedical Science* 80:11387.

Tawfeeq, M.H., Aldhaher, A.H., Issa, A.H. and Alrubayae, I.M. 2023. Dothideoxanthone, immunosuppressive and antioxidant agents from endophytic fungus Aspergillus Costaricensis. *History of Medicine* 9:2585–2599.

Thebti, A., Meddeb, A., Ben Salem, I., Bakary, C., Ayari, S., Rezgui, F., Essafi-Benkhadir, K., Boudabous, A. and Ouzari, H.I. 2023. Antimicrobial activities and mode of flavonoid actions. *Antibiotics* 12:225.

Tiwari, P. and Bae, H. 2022. Endophytic fungi: key insights, emerging prospects, and challenges in natural product drug discovery. *Microorganisms* 10:360.

Tiwari, P., Kang, S. and Bae, H. 2023. Plant-endophyte associations: rich yet under-explored sources of novel bioactive molecules and applications. *Microbiological Research* 266:127241.

Tsalgatidou, P.C., Thomloudi, E.E., Delis, C., Nifakos, K., Zambounis, A., Venieraki, A. and Katinakis, P. 2023. Compatible consortium of endophytic Bacillus halotolerans strains Cal. l. 30 and Cal. f. 4 promotes plant growth and induces systemic resistance against Botrytis cinerea. *Biology* 12:779.

Verma, A., Shameem, N., Jatav, H.S., Sathyanarayana, E., Parray, J.A., Poczai, P. and Sayyed, R.Z. 2022. Fungal endophytes to combat biotic and abiotic stresses for climate-smart and sustainable agriculture. *Frontiers in Plant Science* 13:953836.

Vig, R., Bhadra, F., Gupta, S.K., Sairam, K. and Vasundhara, M. 2022. Neuroprotective effects of quercetin produced by an endophytic fungus Nigrospora oryzae isolated from Tinospora cordifolia. *Journal of Applied Microbiology*, 132:365–380.

Vigneshwari, A., Rakk, D., Németh, A., Kocsubé, S., Kiss, N., Csupor, D., Papp, T., Škrbić, B., Vágvölgyi, C. and Szekeres, A. 2019. Host metabolite producing endophytic fungi isolated from Hypericum perforatum. *PLoS One*, 14:0217060.

Vora, J., Velhal, S., Sinha, S., Patel, V. and Shrivastava, N. 2021. Bioactive phytocompound mulberroside C and endophytes of Morus alba as potential inhibitors of HIV-1 replication: a mechanistic evaluation. *HIV Medicine* 22:690–704.

Vu, T.H.N., Quach, N.T., Le, P.C., Pham, Q.A., Do, T.T., Chu, H.H. and Phi, Q.T. 2023. Bioprospecting endophytic fungi isolated from Cephalotaxus mannii Hook f. as prolific sources of antibacterial, anticancer, and antioxidant agents. *Microbiology* 92:284–292.

Wang, G., Yin, Z., Wang, S., Yuan, Y., Chen, Y. and Kang, W. 2022. Diversified polyketides with anti-inflammatory activities from mangrove endophytic fungus Daldinia eschscholtzii KBJYZ-1. *Frontiers in Microbiology* 13:900227.

Wang, J.T., Ma, Z.H., Wang, G.K., Xu, F.Q., Yu, Y., Wang, G., Peng, D.Y. and Liu, J.S. 2021. Chemical constituents from plant endophytic fungus Alternaria alternata. *Natural Product Research*, 35:1199–1206.

Wang, Y.J., Ma, N., Lu, Y.F., Dai, S.Y., Song, X., Li, C., Sun, Y. and Pei, Y.H. 2022. Structure elucidation and anti-tumor activities of Trichothecenes from endophytic fungus fusarium sporotrichioides. *Biomolecules* 12:778.

Wijesekara, T. and Xu, B. 2023. Health-Promoting Effects of Bioactive Compounds from Plant Endophytic Fungi. *Journal of Fungi* 9:997.

Wu, H., Yan, Z., Deng, Y., Wu, Z., Xu, X., Li, X., Zhou, X. and Luo, H. 2020. Endophytic fungi from the root tubers of medicinal plant Stephania dielsiana and their antimicrobial activity. *Acta Ecologica Sinica*, 40:383–387.

Wu, J.C., Hou, Y., Xu, Q., Jin, X.J., Chen, Y., Fang, J., Hu, B. and Wu, Q.X. 2019. (±)-Alternamgin, a pair of enantiomeric polyketides, from the endophytic fungi Alternaria sp. MG1. *Organic letters*, 21:1551–1554.

Ye, G., Huang, C., Li, J., Chen, T., Tang, J., Liu, W. and Long, Y. 2021. Isolation, structural characterization and antidiabetic activity of new diketopiperazine alkaloids from mangrove endophytic fungus Aspergillus sp. 16-5c. *Marine Drugs*, 19:402.

Zhang, P., Zhou, P.P. and Yu, L.J. 2009. An endophytic taxol-producing fungus from Taxus media, Cladosporium cladosporioides MD2. *Current Microbiology* 59:227–232.

Zhao, S., Wu, X., Duan, X., Zhou, C., Zhao, Z., Chen, H., Tang, Z., Wan, Y., Xiao, Y. and Chen, H. 2021. Optimal extraction, purification and antioxidant activity of total flavonoids from endophytic fungi of Conyza blinii H. Lév. *PeerJ*, 9:11223.

Zhao, T., Xu, L.L., Zhang, Y., Lin, Z.H., Xia, T., Yang, D.F., Chen, Y.M. and Yang, X.L. 2019. Three new α-pyrone derivatives from the plant endophytic fungus Penicillium ochrochloronthe and their antibacterial, antifungal, and cytotoxic activities. *Journal of Asian natural products research*, 21:851-858.

Zhou, G., Chen, X., Zhang, X., Che, Q., Zhang, G., Zhu, T., Gu, Q. and Li, D. 2020. Prenylated p-terphenyls from a mangrove endophytic fungus, Aspergillus candidus LDJ-5. *Journal of natural products*, 83:8–13.

6 Endophytic Fungi from Indian Medicinal Plants

A Goldmine of Novel Biomolecules with Potential Biological Applications

Reenu Kumari and Surender Kumar*
*Corresponding author.

6.1 INTRODUCTION

Endophytes are a vital part of the plant micro-ecosystem having unusual associations that alter plant metabolite synthesis or production (Jia et al., 2016). Endophytes are ubiquitous endosymbiotic microorganisms that can be bacteria or fungi and which colonize living internal plant tissues. The term endophyte was first introduced by the German scientist De Bary in 1866 for organisms living inside plant tissue. In 1991, Petrini described endophytes as any organism which, at some part of its life cycle, is capable of colonizing healthy plant tissues without causing any damage to plants (De Bary, 1866; Petrini, 1991; Caruso et al., 2020). There was a revolutionary finding in 1993 when paclitaxel, a diterpenoid anticancerous compound, was isolated from an endophytic fungus, *Taxomyces andreanae*, inhabiting a medicinal plant, *Taxus brevifolia*. Subsequently, many endophytes were reported to produce several secondary metabolites like alkaloids, steroids, quinones, saponins, terpenoids, tannins and phenolic acids having bioactive properties (Gouda et al., 2016). These findings developed much interest in the crosstalk of the host–endophyte relationship.

Fungal endophytes form an integrated element of the plant micro-ecosystem where they may influence the growth and disease resistance of host plants. These endophytes also exhibit host-specific interactions. These interactions are recognized as mutualism, antagonism and neutralism. The ecological habitats and the genetic background of the medicinal plants determine the diversity of endophytic fungi in these host plants (Jia et al., 2016). The environmental conditions and age of the host plant determine the conversion of endophytic fungi from asymptomatic behavior to a symptom-producing pathogen (Schulz & Boyle, 2005). Fungal endophytes serve as a natural reservoir of potentially useful novel bioactive compounds with therapeutic properties. They contribute to nearly 70% of anticancer and antimicrobial natural bioactive compounds (Newman & Cragg, 2020; Rai et al., 2021). Fungal symbiotic associations with host plants promote their growth, enhance the synthesis of secondary metabolites, and stimulate the defense-signaling response against various stresses. These secondary metabolites are naturally derived compounds of pharmaceutical importance. Therefore, fungal endophytes are also utilized as elicitors for secondary metabolites as medicinal plants have smaller amounts of these commercially important secondary metabolites (Parniske, 2000; Satheesan & Sabu, 2020).

India is a recognized mega-biodiversity nation and has four biodiversity hotspots, namely, the Himalayas, Indo-Burma (Northeast India and the Andaman Islands), Western ghats (and Sri Lanka) and Sundalands (Nicobar Islands) among the 34 biodiversity hotspots known worldwide. Indian

DOI: 10.1201/9781003486275-8

biodiversity accounts for approximately 49,000 plant species, of which 4,900 species are endemic. Medicinal plants are thriving in 15 agroclimatic zones in India having approximately 6,000–7,000 species of medicinal importance (Nautiyal et al., 2020). About 2,500 medicinal species are used in the traditional Indian medicine system and of these only 880 species are traded. Now, the popularity of the herbal medicine industry has increased tremendously, and it is partly dependent upon the collection of rare wild flora. The collection is made from different plant parts such as leaves, roots, bark, fruits, seeds or the whole plant. However, their over-exploitation has led to habitat destruction and depletion of the plant's genetic resources, making them endangered or taking them to the verge of extinction (Chandra, 2016).

Considering the threat of the endangered/protected medicinal plants' extinction and the rising need for natural bioactive compounds procured from wild plant resources to meet the global market requirements, continuous efforts are ongoing to identify novel resources of these bioactive compounds. In this scenario, fungal endophytes associated with medicinal plants can serve as a good alternative for the isolation of natural bioactive compounds. There are well-established examples of fungal endophytes known to synthesize the same bioactive compounds as their host plants, e.g., paclitaxel from *Taxomyes andreanae* (host: *Taxus bravifolia*) (Strobel, 2003), myrtucommulones from *Neofusicoccum australe* (host: *Myrtus communis*) (Nicoletti et al., 2014), camptothecin from *Entrophospora infrequens* (host: *Nothapodytes foetida*) (Puri et al., 2005), podophyllotoxin from *Phialocephala fortinii* (host: *Podophyllum peltatum*) (Eyberger et al., 2006) and vinca alkaloids from *Fusarium oxysporum* (host: *Catharanthus roseus*) (Zhang et al., 2000).

6.2 TYPES OF FUNGAL ENDOPHYTES

Fungal endophytes, depending upon their plant hosts, taxonomy, phylogenetic relatedness, transmission pattern and ecological roles, are divided into two main groups: the clavicipitaceous and non-clavicipitaceous endophytes. Clavicipitaceous endophytes (also referred to as class I endophytes) are represented by the phylogenetically related small number of grass species. They usually develop a systemic infection within plant shoots. Class I endophyte species range from pathogenic or symptomatic (type I) character to mixed (type II) and asymptomatic character (type III). Primarily, class I endophytes exhibit vertical transmission and benefit host plants by enhancing plant biomass, providing drought tolerance and deterring herbivores (Rodriguez et al., 2009).

Non-clavicipitaceous endophytes are represented by a diverse group including mostly ascomycetes and a few basidiomycetes. Their diversity ranges from tropical to tundra biome. They are differentiated into three classes (classes 2–4) having a broad host range. Class 2 inhabits both aerial tissues (stem and leaves) and underground tissues, whereas class 3 are restricted to aerial parts and class 4 colonizes only root tissues. Class 2 endophytes are known to develop immense infections and show horizontal as well as vertical transmission. They confer stress-induced tolerance in plants by mediating plant defense responses (Rodriguez et al., 2009). Class 3 endophytes generally exhibit localized infection within tissues and display horizontal transmission. The ecological roles of this group are understudied, although a few studies support their role in disease resistance (Arnold et al., 2003). Class 4 endophytes include dark-pigmented septate fungi in root tissues which are distinct from mycorrhizal fungi, and have worldwide distribution. These fungi exhibit horizontal transmission and the exact role of these diverse dark septate endophytes remains unknown but is probably significant in plant ecophysiology (Rodriguez et al., 2009).

6.3 FUNGAL ENDOPHYTES AS A SOURCE OF NOVEL BIOMOLECULES

The bioactive secondary metabolites biosynthesized by fungal endophytes can be categorized into the following classes.

6.3.1 Alkaloids

These are isolated from clavicipitaceous endophytes providing them with protections from insects and herbivores. They include amines and amides (ergovaline and phomoenamide), indole derivatives (vinblastine and vincristine), pyrimidine and pyrrolizidine (asperfumoid and lolines), quinoline and isoquinoline (camptothecin and penicinoline derivatives) with bioactive properties such as being antibacterial, antifungal, cytotoxic and antitumorous (Owen & West, 1971; Keglevich et al., 2012; Huang et al., 2016; Li et al., 2015; Shao et al., 2010; Rukachaisirikul et al., 2008; Alam et al., 2021).

6.3.2 Terpenoids

Terpenoids primarily serve in plant defense, as pollination attractants and as allelochemicals. Sesquiterpenes (chokols, heptelidic acid and hydroheptelidic acid) having fungicidal and insecticidal properties, and diterpenes (paclitaxel and subglutinol A and B) with anticancer and immune-suppressive activity, have been isolated from fungal endophytes (Hiroyuki et al., 1989; Calhoun et al., 1992; Tan & Zou, 2001).

6.3.3 Lignans

These secondary metabolites protect the host plant against pathogenic attack under mechanical injury. These include coniferin, syringin, phillyrin, podophyllotoxin, syringaresinol, sesamin and 4-ketopinoresinol, which are known to have the following biological properties: antiinflammatory, antioxidant, antitumor, antihypertensive, antidiabetic, antiviral and they also can slow down cellular senescence via activating genes involved (Alam et al., 2021; Kim et al., 2016; Liang et al., 2021; Singh et al., 2021).

6.3.4 Flavonoids

Flavonoids are pigments found in edible plant parts and include various derivatives such as flavones, flavanols, isoflavones, flavanones, anthocyanins and catechins. Apigenin, cajanol, chalcone, curcumin, kaempferol, luteolin, quercetin, rotenone, rutin, silymarin and vitexin are examples of plant-derived flavonoids isolated from fungal endophytes with antioxidant, anticancer, anti-inflammatory, antibacterial, antifungal, antimicrobial, antiaging, antidiabetic, antispasmodic, cardioprotective, insecticidal and neuroprotective properties (Alam et al., 2021; Singh et al., 2021).

6.3.5 Quinones and Xanthones

These are oxygenated heterocyclics generated from single polyketide chain cyclization via the polyacetate or polymalonate pathway. This group includes 1,7-dihydroxyxanthone, anthraquinone, eugenitin, hypericin, pachybasin, phomarin, plumbagin, shikonin and tanshinone having anticancer, antioxidant, antibacterial, antidepressant, glucoamylase activation, antiviral and antiinflammatory properties (Borges & Pupo, 2006; Hawas et al., 2012; Ming et al., 2012; Mollaei et al., 2019; Venkateswarulu et al., 2018; Vigneshwari et al., 2019; Singh et al., 2021).

6.3.6 Phenols and Phenolic Acids

These are also synthesized by fungal endophytes in addition to plants. This group includes 2-hydroxy-6-methyl benzoic acid, tyrosol, *cis*- and *trans*-p-coumaric acids, and colletotric acid isolated from endophytes having antidiabetic, antifungal and antimicrobial properties (Rodriguez-Morato et al., 2015; Zou et al., 2000; Alam et al., 2021).

6.4 FUNGAL ENDOPHYTES ISOLATED FROM INDIAN MEDICINAL PLANTS

India has a rich repository of medicinal plants and, since ancient times, several herbs have been used as traditional medicines or in households for the treatment of various common diseases. The history of their usage and formulation is noted in ancient Indian literature. Today, India has its familiar traditional medicine system; AYUSH (including Ayurveda, yoga, Yunani, Siddha, and Homeopathy) (Adhikari & Paul, 2018; Ahmad et al., 2021). Tulsi, giloy, amla, black pepper, cinnamon, ginger, turmeric, garlic and ashwagandha are commonly used as herbal remedies in India (Ahmad et al., 2021). Besides these, there are many wild and indigenous plants from forest flora having bioactive compounds that are also used for therapeutic purposes. As bioactive compounds can be of endophyte origin, various studies have been carried out for the isolation of fungal endophytes from medicinal plants in diverse parts of India and some of these are listed in Table 6.1.

6.5 BIOACTIVE COMPOUNDS WITH POTENTIAL APPLICATIONS

6.5.1 Tyrosol

Tyrosol (2-[4-hydroxyphenyl ethanol]) is a phenolic secondary metabolite having antioxidant properties, produced by both plants and fungi (*Alternaria tagetica*, *Colletotrichum gleosporoides*, *C. crassipes*, *Diplodia seriata*, *Diaporthella cryptica*, *Lasiodiplodia* spp., *Neofusicoccum parvum* and *N. australe*). It was also reported with potential applications in stroke therapy and quorum-sensing inhibitory activities (Bu et al., 2007; Guimarães et al., 2009; Rajesh & Rai, 2013).

6.5.2 Resveratrol

Resveratrol is a well-known stilbene polyphenol that exists in two isomeric forms: *cis* and *trans*. It acts like phytoalexin in plants which functions against fungal pathogens and toxins. It has been reported with many biological properties including free radical scavenging, antitumor activities, antioxidant activity, antiinflammatory, cardioprotective, neuroprotective, antimicrobial and antidiabetic (Nawaz et al., 2017; Salehi et al., 2018; Saxena & Srivastava, 2014). This polyphenol came into the spotlight when it was found that resveratrol enhanced yeast cells' life span by Sirtuin 1 gene activation (Hubbard & Sinclair, 2014). A preliminary report for resveratrol extraction from endophytic fungi came from Powell and his associates' study on the *Poaceae* family (*Festuca versuta*, *Festuca arundinacea*, *Stipa robusta*, *Poa alsodes*, *Lolium* sp., *Hordeum* sp.) when they isolated resveratrol from *Acremonium* species (Powell et al., 1994).

6.5.3 Phomalactone

Phomalactone exhibited differential biological activities such as antifungal, insecticidal, phytotoxic, antitumor and antimicrobial (Khambay et al. 2000; Meepagala et al., 2015; Prathumpai & Kocharin, 2016). This metabolite is reported to be produced by many fungal species: *Nigrospora sphaerica* (Kim et al., 2001; Ramesha et al., 2020), *Hirsutella thompsonii* (Krasnoff & Gupta, 1994), *Paecilomyces canteniobliquus* (Wu et al., 2012), *Ophiocordyceps communis* (Prathumpai & Kocharin, 2016) and *Verticillium chlamydosporium* (Khambay et al., 2000).

6.6 ENDOPHYTIC FUNGI FOR ANTIMICROBIAL APPLICATION

Endophytic fungi produce a variety of bio-active compounds having pharmacological activities of host plants and serve as a source of novel biomolecules with antimicrobial properties. Singh et al. (2013) synthesized silver nanoparticles (size 25 nm) from an endophytic fungus *Penicillium* sp. isolated from *Curcuma longa* with antibacterial potential against *Escherichia coli*, *Pseudomonas*

TABLE 6.1
List of Some of the Fungal Endophytes Isolated from Different Plants in India

Host plant	Endophytic fungi	Parts used	References
Adiantum philippense	*Nigrospora sphaerica*	Leaf	Ramesha et al., 2020
Aegle marmelos	*Muscodor kashayum*	Leaf, twigs	Meshram et al., 2013
Azadirachta indica	*Fusarium avenaceum*	Leaf, stem, root, fruit	Rajagopal & Suryanarayanan, 2000;
Berberis aristata	*Alternaria alternate, Alternaria macrospora, Alternaria solani, Aspergillus flavus, Cercospora citrullina, Colletotrichum coccoides, C. gleosporioides, C. coffeanum, C. kahawae, Clonostachys rosea, Fusarium falciforme, F. oxysporum, F. solani, F. lateritium, F. nematophilum, Myrothecium inundatum, Phyllosticta capitalensis, Phomopsis tersa*		Sharma et al., 2018
Catharanthus roseus	*Talaromyces radicus*	Leaf, root, stem, bark	Palem et al., 2015
Centella asiatica	*Aureobasidium* sp., *Bipolaris bicolor, Colletotrichum gloeosporioides, Lasiodiplodia viticola, Fusarium merismoides, Preussia* spp., *Pestalotiopsis longiseta, Pithomyces chartarum*	Leaf, root, stem	Sharma et al., 2020
Curcuma longa	*Aspergillus* sp., *Penicillium* sp., *Cladosporium* sp., *Fusarium* sp.	Rhizome, leaf, stem	Deepak & Virk, 2022
Euphorbia hirta	*Nigrospora sphaerica*	Leaf	Gautam et al., 2022
Equisetum sp.	*Anguillospora longissimi*	Root	Singh & Sati, 2020
Ginkgo biloba	*Colletotrichum gloeosporoides, Pestalotiopsis* sp., *Phyllosticta* sp., *Nigrospora sp. and Xyalaria* sp.	–	Pawle & Singh, 2014
Moringa oleifera	*Aspergillus fumigatus*	Bark	Kaur et al., 2020
Nothapodytes foetida	*Entrophospora infrequens*	Bark	Puri et al., 2005
Ocimum sanctum	*Cladosporium cladosporioides*	Leaf, stem	Bhagat et al., 2012
Picrorhiza kurroa	*Alternaria alternata*		Chandra et al., 2021
Podophyllum hexandrum	*Trametes hirsuta, Fusarium solani*	Rhizome, root	Nadeem, 2012; Puri et al., 2006
Rauvolfia densiflora	*Aspergillus carneus, A. versicolor, Bionectria ochroleuca, Chaetomium globosum, Clonostachys rosea, Colletotrichum boninense, C. cliviicola, C. gloeosporioides, Colletotrichum* sp., *Fusarium fujikuroi, F. oxysporum, Fusarium* sp., *Neocosmospora haematococca, Neocosmospora solani, Neocosmospora* sp., *Paecilomyces penicillatus, Penicillium chrysogenum, Sarocladium kiliense*	Leaf, stem	Bhavana et al., 2019
Zingiber officinale	*Aspergillus niger, Ascomycota* sp., *Fusarium sp., Gliocladiopsis* sp., *Acremonium furcatum, Trichothecium* sp., *Colletotrichum crassipes, Fusarium oxysporum, Earliella scabrosa, Pseudolagarobasidium* sp.	Rhizome	Anisha & Radhakrishnan, 2017

aeruginosa and *Klebsiella pneumoniae*. Volatile organic compounds (VOCs) produced by *Muscodor kashayum*, an endophytic fungus isolated from the Bael tree (*Aegle marmelos*), showed complete growth inhibition of *E. coli*, *Staphylococcus aureus* and *P. aeruginosa* after 48–72 h of exposure. This fungus produces more than 23 volatile compounds, which are responsible for its pungent smell (Meshram et al., 2013). Pawle and Singh (2014) demonstrated the antimicrobial activity of the ethyl extract of *Nigrospora* sp. isolated from *Ginkgo biloba* against *E. coli*, *Klebsiella* sp., *S. aureus*, *Candida albicans* and *Geotrichum* sp. The MIC value against all tested organisms in broth dilution was 2.5 mg/mL. VOCs of *Muscodor tigerii* isolated from *Cinnamomum camphora* inhibit the growth of *S. aureus* (81%) and *P. aeruginosa* (51%) (Saxena et al., 2015). *Fusarium oxysporum* isolated from *Zingiber officinale* exhibited antimicrobial activity against *S. aureus* and *Bacillus subtilis* with a 14- and 18-mm zone of inhibition (Anisha & Radhakrishnan, 2017).

The ethyl extract of *Xylaria* sp. FPL-25(M), mycoendophyte of *Ficus pumila*, displayed broad-spectrum activity against various human and phytopathogenic bacterial/fungal strains with the following growth inhibition zones: *B. subtilis* (28.66 mm), *Listeria monocytogenes* (25.66 mm), *S. aureus* (30 mm), *E. coli* (26.33 mm), *P. aeruginosa* (20.33 mm), *Shigella flexineri* (30.33 mm), *S. typhi* (28.66), *Vibrio parahaemolyticus* (27.66), *Xanthomonas campestris* (20.33), *A. fumigatus* (28.66 mm), *C. albicans* (25.66 mm), *Fusarium moniliforme* (24.66 mm), *Microsporum canis* (30 mm) and *M. gypseum* (40 mm). The bioactive compound isolated from the fungal ethyl extract was xylobovide-9-methyl ester and the MIC value of the purified compound in broth dilution assay for Gram-positive and Gram-negative bacteria ranged from 3.12 to 12.5 μg/mL. The MIC value for fungal pathogens ranged from 0.78 to 1.56 μg/mL (Rakshith et al., 2020). Kaur et al. (2020) found antimicrobial activity in the chloroform extract of *Aspergillus fumigatus* isolated from *Moringa oleifera* against various clinical strains. The extract also exhibited antibiofilm activity against *S. aureus*, *K. pneumoniae* and *C. albicans*, which was evaluated through initial cell attachment assay, disruption of preformed biofilms and quantification of metabolic activity by XTT assay. A freshwater pteridophyte fungal endophyte, *Anguillospora longissima*, showed antagonistic activity (measured as percentage inhibition by dual culture method) against pathogenic fungi such as *Tilletia indica* (79.76%), *Botrytis cinerea* (55.55%), *Pyricularia oryzae* (58.21%) and *Rhizoctonia solani* (55.63%) (Singh & Sati, 2020). Chandra et al. (2021) found the antimicrobial activity in ethyl acetate extract of *A. alternata*, isolated from *Picrorhiza kurroa*, against *B. subtilis* and *S. aureus* with MIC values of 64.0 μg/mL and 128 μg/mL, respectively. Talukdar et al. (2021) found that an organic extract of *Colletotrichum coccodes* from *Houttuynia cordata*, an ethnomedicinal plant from North-East India, displayed antimicrobial activity against *C. albicans*, *S. aureus* and *E. coli* with a MIC value of 125 μg/mL; and for *P. aeruginosa* with a MIC value of 250 μg/mL. The purification of the organic extract showed the presence of tyrosol, which was confirmed by NMR and ESIMS spectra.

6.7 ENDOPHYTIC FUNGI AS AN INHIBITOR FOR PATHOGENIC FUNGI

Many endophytic fungi also act as inhibitors for other fungi. This inhibition of the growth of others is due to mycoparasitism, competition for space or nutrients, and the release of chemical compounds like volatile organic compounds (VOCs). *Trichoderma gamsii*, an endophytic fungus isolated from the lateral roots of lentils, showed an antagonistic effect against *Cladosporium oxysporum* (ITCC3714), *Alternaria alternata* (ITCC1920), *Fusarium pallidoroseum* (ITCC3886), *Fusarium oxysporum* (ITCC4219), *Fusarium solani* (ITCC5017) and *Phomopsis archeri* (ARIFCC1445) (ITCC: Indian Type Culture Collection, Indian Agricultural Research Institute, New Delhi, India) (Rinu et al., 2014).

Muscodor kashayum VOCs showed broad-spectrum antifungal activity, as it completely killed 15 tested pathogenic fungi (out of 20) with a variable inhibition in the growth rate of pathogenic *Aspergillus flavus and Alternaria alternata* (20–60%). Also, it completely inhibits the growth of *Candida albicans* strains (yeast) (Meshram et al., 2013). Saxena et al. (2015) found that the volatile compounds of *Muscodor tigerii* isolated from *Cinnamomum camphora* completely suppressed the growth of fungal pathogens like *A. alternata* and *Cercospora beticola*, whereas for *Rhizoctonia solani*, *Penicillium marneffei* and *Aspergillus flavus* the rate was up to 72%, 42% and 40.2%, respectively. Also, it suppressed the growth of *Candida* isolates by 54–63%. *Gliocladiopsis* sp. from zinger showed significant bioactivity against the pathogenic fungus *Pythium myriotylum* by dual culture assay (Anisha & Radhakrishnan, 2017). Endophytic *Trichoderma* isolates, named *T. longibrachiatum* strain-1 (MK425639) and *T. longibrachiatum* strain-2 (MK751759), isolated from *Juniperus lutchuensis* Koidz, *T. pleuroti* (MK751757) from *Leucas aspera* (Willd.) Link. and *T. harzianum* (MK751758) from *Pterocarpus santalinus* L. showed inhibition of soilborne pathogens by producing VOCs (Rajani et al., 2021). These studies showed that endophytic fungi are the best alternative for synthetic fungicides and that they do not have a negative impact on the environment.

6.8 ENDOPHYTIC FUNGI WITH ANTIOXIDANT ACTIVITY

Antioxidants help to counteract the free radical damage caused by certain chronic diseases by free radical scavenging. Besides plants, endophytic fungi can also serve as a natural alternative to secondary metabolites with antioxidant potential. Secondary metabolites such as carotenoids, phenolics and tocopherols are known to have antioxidant properties (Hamilton et al., 2012; Boutigny et al., 2008). The following findings describe the antioxidant potential of fungal endophytes of Indian medicinal plants. *Phyllosticta* sp., a fungal endophyte from *Guazuma tomentosa* found with potent antioxidant potential having an EC_{50} value of 580.02 ± 0.57 µg/mL (in ABTS method) and 2030.25±0.81 µg/mL (in DPPH assay). The ethanol extract of the fungus showed the presence of phenolics (18.33 ± 0.68 GAE/g) and flavonoid (6.44 ± 1.24 µg/mg of quercetin equivalent) (Srinivasan et al., 2010).

The ethyl extract of *Nigrospora* sp. from *Ginkgo biloba* showed the occurrence of phenolics and anthraquinones, and displayed antioxidant activity having an IC_{50} value of 9.28 µg/mL through DPPH radical scavenging assay (Pawle & Singh, 2014). Significant antioxidant activity was observed in fungal endophytic extracts (*Chaetomium* sp., *Aspergillus* sp., *A. peyronelii* and *A. niger*) of *Eugenia jambolana* through DPPH radical-scavenging activity, hydrogen peroxide-scavenging assay and reducing power assay, and these strains were reported with a higher phenolic content in the range of 55–60 mg of GAE/g of extract (Yadav et al., 2014). Kaur et al. (2020) found radical-scavenging activities in the chloroform extract of *Aspergillus fumigatus* (isolated from *Moringa oleifera*) through DPPH assay (IC_{50} value of 40.07 µg) and by ABTS method (IC_{50} value of 54.28 µg). The fungal chloroform extract exhibited total phenolic and flavonoid contents of 20.31±0.005 mg/100 mg GA and 1.437±0.0060 mg/100 mg, respectively. Gautam et al. (2022) found higher antioxidant activity in the ethyl acetate extract of *Nigrospora sphaerica* (EHL2) isolated from the leaves of *Euphorbia hirta* through DPPH (having an IC_{50} value of 148.5 µg) and ABTS method (having an IC_{50} value of 32.44 µg/ml). The higher level of total phenolic and flavonoid contents observed in the ethyl acetate fraction were 77.74±0.04 mgGAE/g and 230.59±2.0 mgRE/g, respectively.

6.9 ENDOPHYTIC FUNGI AS A SOURCE OF ENZYMES

Endophytic fungi are also a source of valuable enzymes with insecticidal, antidiabetic, antioxidant, fibrinolytic and laccase properties. Endophytic fungi can provide protection to their respective host plants against different kinds of insects or pests by producing toxins that can inhibit the latter's digestive enzymes, affecting their growth and development. Singh et al. (2015) observed alpha-glucosidase inhibitory activity in *Cladosporium* sp. (TN-9S) isolated from *Tinospora cordifolia*. The insecticidal property of this purified inhibitor was tested against *Spodoptera litura*. There was a severe effect on the development rate and survival of respective leafworm larvae that ultimately increased the mortality rate. Katoch et al. (2017) isolated seven endophytic fungi strains from *Viola odorata* with pancreatic lipase inhibitor activity (IC_{50} value < 10 µg/mL). *Aspergillus* sp. (VOLF4) was found to have the most potent lipase inhibitory activity, i.e., an IC_{50} value of 3.8 μg/mL. These isolates could serve as potential candidates for antiobesity drugs. *Alternaria alternata*, an endophytic fungus isolated from *Vinca rosea*, produces "altenuene" which inhibits cholinesterase, i.e., acetylcholinesterase (78%) and butyrylcholinesterase (73%). Altenuene also had good antioxidant and insecticidal activity (checked for *Spodoptera*) (Bhagat et al., 2016).

Xylaria curta, an endophytic fungus isolated from *Cathranthus roseus* twigs, showed the production of "xylarinase." Under in vitro conditions, it showed both plasmin and tissue plasminogen activator-like activity and showed fibrinolytic activity by hydrolyzing Aα and Bβ chains of the fibrinogen (Meshram et al., 2016). These results showed that it can be used for the breakdown of blood clots. Rao et al. (2019) isolated 22 endophytic fungi from *Euphorbia milii*, commonly known as the crown of thorns, and screened them for laccase enzyme production. In this study, *Irpex lacteus* was found to have the most potent laccase activity. Optimization of production conditions leads to an increase by a factor of 52.63 and exhibits 38.25 U/L of laccase activity. Jagannath et al. (2021) isolated 203 endophytic fungi from different parts (leaves, flower, stem and root) of *Baliospermum montanum*, belonging to 29 species of fungi. The isolated fungi showed positive results to produce amylase (83%), cellulase (79%), phosphatase (77%), protease (72%) and lipase (59%) enzymes. Most isolates produce amylase and cellulase, having a role in the degradation of complex carbohydrates. None of the isolates produced laccase enzyme. L-asparaginase is commonly used to treat acute lymphoblastic leukemia, a common form of pediatric cancer. Endophytic fungi, *Talaromyces pinophilus*, isolated from *Curcuma amada* rhizomes, were identified to produce L-asparaginase (Krishnapura & Belur, 2020). Using solid-state fermentation (SSF) and optimizing the reaction conditions, enzyme production was increased to 120.3 U/mL.

6.10 ENDOPHYTIC FUNGI WITH BIOMOLECULES WITH ANTICANCEROUS PROPERTIES

Cancer is a major risk to human health. An estimated 600,920 cancer-related deaths and 1,688,780 new cases were projected to occur in the US in 2017 (Siegel et al., 2017; Uzma et al., 2018). Several endophytic fungi isolated from medicinal plants are known to produce pharmacological important compounds. Rohitukine is a chromane alkaloid compound having antiinflammatory, anticancer and immuno-modulatory properties. *Fusarium proliferatum* (MTCC 9690) is an endophytic fungus isolated from the inner bark tissue *of Dysoxylum binectariferum* and is reported to produce rohitukine, representing an important alternative to rohitukine production (Kumara et al., 2012). Bhagat et al. (2012) observed that *Cladosporium cladosporioides* isolated from *Ocimum sanctum* at a concentration of 30 μg/mL showed above 50% growth inhibition in the following cancer cell lines: lung (A-549), colon (COLO-205), neuroblastoma (IMR-32) and prostrate (MCF-7 and DU-145).

Taxol is another major compound known for its antimitotic properties and is highly effective against lung, ovarian, breast, head-neck cancer and Kaposi's sarcoma. It binds to the β-subunit of the tubulin heterodimers, promotes its polymerization, and stops cell proliferation (Kovacs et al., 2007). However, for complete treatment, each patient may require 2 g of taxol administered many times over many months. Taxol is extracted from the bark of the pacific yew tree (*Taxus brevifolia*). Approximately 10,000 kg of bark from 3000 yew trees is required for the extraction of 1 kg of taxol, and the current demand is 250 kg per annum. This insufficiency of taxol can be overcome by an alternate source. *Fusarium redolens*, isolated from *Taxus baccata*, has been shown to produce taxol and successfully suppressed the tumor induced by *Agrobacterium tumefaciens* (MTCC No. 431) in potato disc tumor induction assay (Garyali et al., 2013).

Vincristine (VCR; Oncovin) and vinblastine (VBL; Velbe) are the two major anticancer alkaloids that inhibit cell proliferation by binding to microtubules, leading to blockage of mitotic division and finally cell death. They are present in very low concentrations in the leaves of *Catharanthus roseus* and approximately 300 tons of dried leaves are required for the extraction of 3 kg of alkaloids.

Talaromyces radicus-CrP20 endophyte isolated from the leaves of *C. roseus* was shown to produce vincristine (670 μg/mL) in modified M2 medium and vinblastine (70 μg/mL) in potato dextrose broth. The cytotoxic activity of partially purified fungal vincristine showed good results in different human cancer cell lines (Palem et al., 2015). These studies offer important alternate sources for the production of rohitukine, taxol, vincristine and vinblastine from an endophytic fungus.

6.11 ENDOPHYTIC FUNGI AS QUORUM-SENSING INHIBITORS

Rajesh and Rai (2013) found quorum-sensing inhibitor activity in *Fusarium graminearum* and *Lasidiplodia* sp., endophytic isolates of *Ventilago madraspatana*. They observed a reduction in the production of violacein in *Chromobacterium violaceum* CV026. *Kwoniella* sp. PY016 isolated from *Terminalia bellirica* produces sugar, sugar alcohol, carboxylic acid, lipid and phenolic classes of metabolites (predominantly xylitol) which exhibit inhibition of anti-biofilm and quorum sensing of *Chromobacterium violaceum* (Mookherjee et al., 2018). *Alternaria alternata* endophyte isolated from *Carica papaya* produces bioactive molecules, namely, sulfurous acid, 2-propyl tridecyl ester, 1,2-benzenedicarboxylic acid, and bis(2-methylpropyl) ester, which exhibit antiquorum-sensing activity against *P. aeruginosa* (Rashmi et al., 2018). Chloroform extracts of *Aspergillus nidulans* (an endophytic fungi isolated from *Acacia nilotica*) showed antifungal activity against biofilm-forming *Candida albicans* (74.86%) (Meenambiga & Rajagopal, 2018).

6.12 CONCLUSIONS

Endophytic fungi can serve as a good alternative for the extraction of novel bioactive compounds rather than entirely depending on medicinal plants. In the last few years, several endophytic fungi having bioactive compounds with antimicrobial, anticancerous, antioxidant, antiinflammatory, insecticidal, antidiabetic and plant growth-promoting properties have been reported. The varying bioactive compounds of endophytic fungi/type of endophytic fungi could be one of the factors for the region-specific medicinal properties of a particular medicinal plant. India is rich in medicinal flora and several efforts are on-going for the isolation of fungal endophytes with novel bioactive compounds having potential biological applications from ethnobotanical plants. There is currently a need to put more focus on the structural characterization of novel bioactive compounds and the elucidation of their inhibition mode. The latter work requires confirmation by in vitro and in vivo experiments/tests, and then assessment of novel bioactive compounds for their safety and efficacy.

REFERENCES

1. Adhikari, P. P., and S. B. Paul. 2018. History of Indian traditional medicine: A medical inheritance. *Asian Journal of Pharmaceutical and Clinical Research*, 11(1): 421–426. https://doi.org/10.22159/ajpcr.2018.v11i1.21893
2. Ahmad, S., S. Zahiruddin, B. Parveen et al. 2021. Indian medicinal plants and formulations and their potential against COVID-19–preclinical and clinical research. *Frontiers in Pharmacology*, 11: 578970. https://doi.org/10.3389/fphar.2020.578970
3. Alam, B., J. Li, Q. Ge et al. 2021. Endophytic fungi: From symbiosis to secondary metabolite communications or vice versa? *Frontiers in Plant Science*, 12: 3060. https://doi.org/10.3389/FPLS.2021.791033/XML/NLM
4. Anisha, C., and E. K. Radhakrishnan. 2017. Metabolite analysis of endophytic fungi from cultivars of *Zingiber officinale* Rosc. identifies myriad of bioactive compounds including tyrosol. *3 Biotech*, 7(2): 146. https://doi.org/10.1007/S13205-017-0768-8
5. Arnold, A. E., L. C. Mejía, D. Kyllo et al. 2003. Fungal endophytes limit pathogen damage in a tropical tree. *Proceedings of the National Academy of Sciences, USA*, 100: 15649–15654.

6. Bhagat, J., A. Kaur, R. Kaur et al. 2016. Cholinesterase inhibitor (Altenuene) from an endophytic fungus *Alternaria alternata*: Optimization, purification and characterization. *Journal of Applied Microbiology*, 121(4): 1015–1025.
7. Bhagat, J., A. Kaur, M. Sharma, A. K. Saxena, and B. S. Chadha. 2012. Molecular and functional characterization of endophytic fungi from traditional medicinal plants. *World Journal of Microbiology and Biotechnology*, 28(3): 963–971. https://doi.org/10.1007/s11274-011-0894-0
8. Bhavana, N. S., H. S. Prakash, and M. S. Nalini. 2019. Antioxidative and L-asparaginase potentials of fungal endophytes from *Rauvolfia densiflora* (*Apocynaceae*), an ethnomedicinal species of the Western Ghats. *Czech Mycology*, 71(2): 187–203. https://doi.org/10.33585/cmy.71205
9. Borges, W. D., and M. T. Pupo. 2006. Novel anthraquinone derivatives produced by *Phoma sorghina*, an endophyte found in association with the medicinal plant *Tithonia diversifolia* (*Asteraceae*). *Journal of Brazilian Chemical Society*, 17: 929–934. https://doi:10.1590/S0103-50532006000500017
10. Boutigny, A. L., F. Richard-Forget, and C. Barreau. 2008. Natural mechanisms for cereal resistance to the accumulation of *Fusarium trichothecenes*. *European Journal of Plant Patholo*gy, 121: 411–423. https://doi:10.1007/s10658-007-9266-x
11. Bu, Y., S. Rho, J. Kim et al. 2007. Neuroprotective effect of tyrosol on transient focal cerebral ischemia in rats. *Neuroscience Letters*, 414(3): 218–221. https://doi.org/10.1016/j.neulet.2006.08.094
12. Calhoun, L. A., J. A. Findlay, J. D. Miller, and N. J. Whitney. 1992. Metabolites toxic to spruce budworm from balsam fir needle endophytes. *Mycological Research*, 96: 281–286.
13. Caruso, G., M. T. Abdelhamid, A. Kalisz, and A. Sekara. 2020. Linking endophytic fungi to medicinal plants therapeutic activity. A case study on *Asteraceae*. *Agriculture*, 10(7): 1–23. https://doi.org/10.3390/agriculture10070286
14. Chandra, L. D. 2016. Bio-diversity and conservation of medicinal and aromatic plants. *Advances in Plants and Agriculture Research*, 5(4). https://doi.org/10.15406/APAR.2016.05.00186
15. Chandra, H., P. Kumari, R. Prasad, S. Chandra Gupta, and S. Yadav. 2021. Antioxidant and antimicrobial activity displayed by a fungal endophyte *Alternaria alternata* isolated from *Picrorhiza kurroa* from Garhwal Himalayas, India. *Biocatalysis and Agricultural Biotechnology*, 33: 101955. https://doi.org/10.1016/j.bcab.2021.101955
16. De Bary, A. 1866. *Morphologie und Physiologie der Pilze, Flechten und Myxomyceten Vol. 2*. Leipzig.
17. Deepak, H., and V. Virk. 2022. Optimization of surface sterilization method for the isolation of endophytic fungi associated with *Curcuma longa* and their antibacterial activity. *J Adv Biotechnol Exp Ther*, 5(2): 334–346. https://doi.org/10.5455/jabet.2022.d119
18. Eyberger, A. L., R. Dondapati, and J. R. Porter. 2006. Endophyte fungal isolates from *Podophyllum peltatum* produce podophyllotoxin. *Journal of Natural Products*, 69(8): 1121–1124. https://doi.org/10.1021/np060174f
19. Garyali, S., A. Kumar, and M. S. Reddy. 2013. Taxol production by an endophytic fungus, *Fusarium redolens*, isolated from Himalayan yew. *Journal of Microbiology and Biotechnology*, 23(10): 1372–1380.
20. Gautam, S., A. Singh, P. Kumari et al. 2022. Phenolic and flavonoid contents and antioxidant activity of an endophytic fungus *Nigrospora sphaerica* (EHL2), inhabiting the medicinal plant *Euphorbia hirta* (dudhi) L. *Archives of Microbiology*, 204: 140. https://doi.org/10.1007/s00203-021-02650-7
21. Gouda, S., G. Das, S. K. Sen, H. S. Shin, and J. K. Patra. 2016. Endophytes: A treasure house of bioactive compounds of medicinal importance. *Frontiers in Microbiology*, 7. https://doi.org/10.3389/FMICB.2016.01538
22. Guimaraes, D. O., K. B. Borges, P. S. Bonato, and M. T. Pupo. 2009. A simple method for the quantitative analysis of tyrosol by HPLC in liquid Czapek Cultures from endophytic fungi. *Journal of the Brazilian Chemical Society*, 20(1): 188–194. https://doi.org/10.1590/S0103-50532009000100028
23. Hamilton, C. E., P. E. Gundel, M. Helander, and K. Saikkonen. 2012. Endophytic mediation of reactive oxygen species and antioxidant activity in plants: A review. *Fungal Diversity*, 54:1–10. https://doi.org/10.1007/s13225-012-0158-9
24. Hawas, U. W., A. A. El-Beih, and A. M. El-Halawany. 2012. Bioactive anthraquinones from endophytic fungus *Aspergillus versicolor* isolated from red sea algae. *Archives of Pharmacal Research*, 35: 1749–1756. https://doi:10.1007/s12272-012-1006-x

25. Hiroyuki, K., T. Satoshi, T. Shun-ichi et al. 1989. New fungitoxic sesquiterpenoids, chokols AG, from stromata of *Epichloe typhina* and the absolute configuration of chokol E. *Agricultural Biology and Chemistry*, 53: 789–796.
26. Huang, S., H. Chen, W. Li, X. Zhu, W. Ding, and C. Li. 2016. Bioactive chaetoglobosins from the mangrove endophytic fungus *Penicillium chrysogenum*. *Mar. Drugs*, 14: 172.
27. Hubbard, B. P., and D. A. Sinclair. 2014. Small molecule SIRT1 activators for the treatment of aging and age-related diseases. *Trends in Pharmacological Sciences*, 35(3): 146–154. https://doi.org/10.1016/j.tips.2013.12.004
28. Jagannath, S., N. Konappa, A. Lokesh et al. 2021. Bioactive compounds guided diversity of endophytic fungi from *Baliospermum montanum* and their potential extracellular enzymes. *Analytical Biochemistry*, 614: 114024.
29. Jia, M., L. Chen, H. L. Xin et al. 2016. A friendly relationship between endophytic fungi and medicinal plants: A systematic review. *Frontiers in Microbiology*, 7: 906. https://doi.org/10.3389/FMICB.2016.00906/BIBTEX
30. Katoch, M., A. Paul, G. Singh, and S. N. C. Sridhar. 2017. Fungal endophytes associated with Viola odorata Linn. as bioresource for pancreatic lipase inhibitors. *BMC Complementary and Alternative Medicine*, 17: 385. https://doi.org/10.1186/s12906-017-1893-y
31. Kaur, N., D. S. Arora, N. Kalia, and M. Kaur. 2020. Antibiofilm, antiproliferative, antioxidant and antimutagenic activities of an endophytic fungus *Aspergillus fumigatus* from *Moringa oleifera*. *Molecular Biology Reports*, 47(4): 2901–2911. https://doi.org/10.1007/s11033-020-05394-7
32. Keglevich, P., L. Hazai, G. Kalaus, and C. Szantay. 2012. Modifications on the basic skeletons of vinblastine and vincristine. *Molecules*, 17: 5893–5914.
33. Khambay, B. P. S., J. M. Bourne, S. Cameron, B. R. Kerry, and M. J. Zaki. 2000. A nematicidal metabolite from *Verticillium chlamydosporium*. *Pest Management Science*, 56:1098–1099.
34. Kim, J. C, G. J. Choi, J. H. Park, H. T. Kim, and K. Y. Cho. 2001. Activity against plant pathogenic fungi of phomalactone isolated from *Nigrospora sphaerica*. *Pest Management Science*, 57: 554–559.
35. Kim, J., S. Young Cho, S. Hwan Kim et al. 2016. Ginseng berry and its biological effects as a natural phytochemical. *Natural Product Chemistry and Research*, 4(2): 209. https://doi.org/10.4172/2329-6836.1000209
36. Kovacs, P., G. Csaba, E. Pallinger and R. Czaker. 2007. Effects of taxol treatment on the microtubular system and mitochondria of *Tetra hymena*. *Cell Biology International*, 31: 724–32.
37. Krasnoff, S. B., and S. Gupta. 1994. Identification of the antibiotic phomalactone from the entomopathogenic fungus *Hirsutella thompsonii* var. *synnematosa*. *Journal of Chemical Ecology*, 20: 293–302.
38. Krishnapura, P. R., and P. D. Belur. 2020. L-asparaginase production using solid-state fermentation by an endophytic *Talaromyces pinophilus* isolated from rhizomes of *Curcuma amada*. *Journal of Pure and Applied Microbiology*, 14: 307–318.
39. Kumara, M. P., S. Zuehlke, V. Priti et al. 2012. *Fusarium proliferatum*, an endophytic fungus from *Dysoxylum binectariferum* Hook.f, produces rohitukine, a chromane alkaloid possessing anti-cancer activity. *Antonie van Leeuwenhoek*, 101(2): 323–329. https://doi.org/10.1007/s10482-011-9638-2
40. Li, S., Z. Li, L. Qian et al. 2015. Fumigaclavine I, a new alkaloid isolated from endophyte *Aspergillus terreus*. *Chinese Journal of Natural Medicine*, 13: 937–941.
41. Liang, Y., G. Wei, K. Ning et al. 2021. Contents of lobetyolin, syringin, and atractylolide III in Codonopsis pilosula are related to dynamic changes of endophytes under drought stress. *Chinese Medicine (United Kingdom)*, 16(1): 1–16. https://doi.org/10.1186/S13020-021-00533-Z/FIGURES/8
42. Meenambiga, S.S., and K. Rajagopal. 2018. Antibiofilm activity and molecular docking studies of bioactive secondary metabolites from endophytic fungus *Aspergillus nidulans* on oral *Candida albicans*. *Journal of Applied Pharmaceutical Science*, 8(3): 037–045.
43. Meepagala, K. M., R. D. Johnson, N. Techen, D. E. Wedge, and S. O. Duke. 2015. Phomalactone from a Phytopathogenic Fungus Infecting *Zinnia elegans* (ASTERACEAE) Leaves. *Journal of Chemical Ecology*, 41(7): 602–612. https://doi.org/10.1007/s10886-015-0602-x
44. Meshram, V., N. Kapoor, and S. Saxena. 2013. *Muscodor kashayum* sp. nov.–a new volatile antimicrobial producing endophytic fungus. *Mycology*, 4(4): 196–204. https://doi.org/10.1080/21501203.2013.877990

45. Meshram, V., S. Saxena, and K. Paul. 2016. Xylarinase: A novel clot-busting enzyme from an endophytic fungus *Xylaria curta*. *Journal of Enzyme Inhibition and Medicinal Chemistry*, 31(6): 1502–1511.
46. Ming, Q. L., T. Han, W. Li et al. 2012. Tanshinone IIA and tanshinone I production by *Trichoderma atroviride* D16, an endophytic fungus in *Salvia miltiorrhiza*. *Phytomedicine*, 19: 330–333. https://doi:10.1016/j.phymed.2011.09.076
47. Mollaei, S., O. Khanehbarndaz, Z. Gerami-Khashal, and M. Ebadi. 2019. Molecular identification and phytochemical screening of endophytic fungi isolated from Lithospermum officinale L. roots: A new source of shikonin. *Phytochemistry*, 168: 112–116. https://doi:10.1016/j.phytochem.2019.112116
48. Mookherjee, A., R. Dineshkumar, N. N. Kutty et al. 2018. Quorum sensing inhibitory activity of the metabolome from endophytic Kwoniella sp. PY016: Characterization and hybrid model-based optimization. *Applied Microbiology and Biotechnology*, 102(17): 7389–7406.
49. Nadeem, M. 2012. *Fusarium solani*, P1, a new endophytic podophyllotoxin-producing fungus from roots of *Podophyllum hexandrum*. *African Journal of Microbiology Research*, 6(10). https://doi.org/10.5897/ajmr11.1596
50. Nautiyal, S., K. C. Smitha, and H. Kaechele. 2020. Medicinal plant biodiversity in India: Harnessing opportunities for promoting livelihood and food security. In *Socio-economic and Eco-biological Dimensions in Resource Use and Conservation: Strategies for Sustainability*, ed. N. Roy, S. Roychoudhary, S. Nautiyal, S. K. Agarwal, and S. Baksi, 135–169. Switzerland: Nature Springer Publishers.
51. Nawaz, W., Zhou, Z., Deng, S. et al. 2017. Therapeutic versatility of resveratrol derivatives. *Nutrients*, 9(11): 1188. https://doi.org/10.3390/NU9111188
52. Newman, D. J., and G. M. Cragg. 2020. Natural products as sources of new drugs over the nearly four decades from 01/1981 to 09/2019. *Journal of Natural Products*, 83(3): 770–803. https://doi.org/10.1021/acs.jnatprod.9b01285
53. Nicoletti, R., P. Ferranti, S. Caira et al. 2014. Myrtucommulone production by a strain of Neofusicoccum australe endophytic in myrtle (Myrtus communis). *World Journal of Microbiology and Biotechnology*, 30(3): 1047–1052. https://doi.org/10.1007/s11274-013-1523-x
54. Owen, E. and D. West. 1971. Isolation and identification of 7, 8-dimethyl-10-(2'-hydroxyethyl) isoalloxazine from natural sources. In *Methods in Enzymology*, ed D. B. McCormick, and L. D. Wright, 574–579. Amsterdam: Elsevier.
55. Palem, P. P. C., G. C. Kuriakose, and C. Jayabaskaran. 2015. An endophytic fungus, *Talaromyces radicus*, isolated from *Catharanthus roseus*, produces vincristine and vinblastine, which induce apoptotic cell death. *PLoS ONE*, 10(12). https://doi.org/10.1371/journal.pone.0144476
56. Parniske, M. 2000. Intracellular accommodation of microbes by plants: A common developmental program for symbiosis and disease? *Current Opinion in Plant Biology*, 3(4): 320–328.
57. Pawle, G., and S. Singh. 2014. Antimicrobial, antioxidant activity and phytochemical analysis of an endophytic species of *Nigrospora* isolated from living fossil *Ginkgo biloba*. *Current Research in Environmental and Applied Mycology*, 4(1): 1–9. https://doi.org/10.5943/cream/4/1/1
58. Petrini, O. 1991. Fungal endophytes of tree leaves. In *Microbial Ecology of Leaves*, ed. J. H. Andrews, and S. S. Hirano, 179–197. New York, Springer-Verlag.
59. Powell, R. G., M. R. Tepaske, R. D. Plattner et al. 1994. Isolation of resveratrol from *Festuca versuta* and evidence for the widespread occurrence of this stilbene in the poaceae. *Phytochemistry*, 35(2): 335–338.
60. Prathumpai, W., and K. Kocharin. 2016. Phomalactone optimization and production of entomopathogenic fungi by *Ophiocordyceps communis* BCC 1842 and BCC 2763. *Preparative Biochemistry and Biotechnology*, 46(1): 44–48. https://doi.org/10.1080/10826068.2014.970691
61. Puri, S. C., A. Nazir, R. Chawla et al. 2006. The endophytic fungus *Trametes hirsuta* as a novel alternative source of podophyllotoxin and related aryl tetralin lignans. *Journal of Biotechnology*, 122(4): 494–510. https://doi.org/10.1016/j.jbiotec.2005.10.015
62. Puri, S. G., V. Verma, T. Amna, G. N. Qazi and M. Spiteller. 2005. An endophytic fungus from *Nothapodytes foetida* that produces camptothecin. *Journal of Natural Products*, 68(12): 1717–1719. https://doi.org/10.1021/np0502802
63. Rai, N., P. K. Keshri, A. Verma et al. 2021. Plant associated fungal endophytes as a source of natural bioactive compounds. *Mycology*, 12(3): 139–159. https://doi.org/10.1080/21501203.2020.1870579

64. Rajagopal, K., and T. S. Suryanarayanan. 2000. Isolation of endophytic fungi from leaves of neem (*Azadirachta indica* A. Juss.). *Current Science*, 78(11): 1375–1378. https://www.researchgate.net/publication/289435056
65. Rajani, P., C. Rajasekaran, M. M. Vasanthakumari, S. B. Olsson, G. Ravikanth, and R. U. Shaanker. 2021. Inhibition of plant pathogenic fungi by endophytic Trichoderma spp. through mycoparasitism and volatile organic compounds. *Microbiological Research*, 242: 126595.
66. Rajesh, P. S., and V. R. Rai. 2013. Hydrolytic enzymes and quorum sensing inhibitors from endophytic fungi of *Ventilago madraspatana* Gaertn. *Biocatalysis and Agricultural Biotechnology*, 2(2): 120–124. https://doi.org/10.1016/j.bcab.2013.01.002
67. Rakshith, D., D. M. Gurudatt, H. C. Y. Rao et al. 2020. Bioactivity-guided isolation of antimicrobial metabolite from *Xylaria* sp. *Process Biochemistry*, 92: 378–385. https://doi.org/10.1016/j.procbio.2020.01.028
68. Ramesha, K. P., N. C. Mohana, B. R. Nuthan, D. Rakshith, and S. Satish. 2020. Antimicrobial metabolite profiling of *Nigrospora sphaerica* from *Adiantum philippense* L. *Journal of Genetic Engineering and Biotechnology*, 18(1). https://doi.org/10.1186/s43141-020-00080-4
69. Rao, A., N. Ramakrishna, S. Arunachalam, and M. Sathiavelu. 2019. Isolation, screening and optimization of laccase-producing endophytic fungi from *Euphorbia milii*. *Arabian Journal for Science and Engineering*, 44(1): 51–64.
70. Rashmi, M., H. Meena, C. Meena et al. 2018. Anti-quorum sensing and antibiofilm potential of *Alternaria alternata*, a foliar endophyte of *Carica papaya*, evidenced by QS assays and in-silico analysis. *Fungal Biology*, 122(10): 998–1012.
71. Rinu, K., P. Sati, and A. Pandey. 2014. *Trichoderma gamsii* (NFCCI 2177): A newly isolated endophytic, psychrotolerant, plant growth promoting and antagonistic fungal strain. *Journal of Basic Microbiology*, 54(5): 408–417.
72. Rodriguez, R. J., J. F. White, A. E. Arnold, and R. S. Redman. 2009. Fungal endophytes: Diversity and functional roles. *New Phytologist*, 182(2): 314–330. https://doi.org/10.1111/J.1469-8137.2009.02773.X
73. Rodríguez-Morato, J., L. Xicota, M. Fito, M. Farre, M. Dierssen, and R. de la Torre. 2015. Potential role of olive oil phenolic compounds in the prevention of neurodegenerative diseases. *Molecules*, 20: 4655–4680.
74. Rukachaisirikul, V., U. Sommart, S. Phongpaichit, J. Sakayaroj, and K. Kirtikara. 2008. Metabolites from the endophytic fungus *Phomopsis* sp. PSU-D15. *Phytochemistry*, 69: 783–787.
75. Salehi, B., A. P. Mishra, M. Nigam et al. 2018. Resveratrol: A Double-Edged Sword in Health Benefits. *Biomedicines*, 6(3):91. https://doi.org/10.3390/biomedicines6030091
76. Satheesan, J., and K. K. Sabu. 2020. Endophytic fungi for a sustainable production of major plant bioactive compounds. In *Plant-derived Bioactives: Production, Properties and Therapeutic Applications*, ed. M. K. Swamy, 195–207. Singapore: Springer.
77. Saxena, S., and A. Srivastava. 2014. Resveratrol: Biological activities and therapeutic Potentia. *Journal of Pharmaceutical Technology, Research and Management*, 2(2): 145–157. https://doi.org/10.15415/jptrm.2014.22010
78. Saxena, S., V. Meshram, and N. Kapoor. 2015. *Muscodor tigerii* sp. nov.-Volatile antibiotic producing endophytic fungus from the Northeastern Himalayas. *Annals of Microbiology*, 65(1): 47–57. https://doi.org/10.1007/s13213-014-0834-y
79. Schulz, B., and C. Boyle. 2005. The endophytic continuum. *Mycological Research*, 109: 661–686. https://doi:10.1017/S095375620500273X
80. Shao, C. L., C. Y. Wang, Y. C. Gu et al. 2010. Penicinoline, a new pyrrolyl 4-quinolinone alkaloid with anunprecedented ring system from an endophytic fungus *Penicillium* sp. *Bioorg. Medicinal Chemistry Letters*, 20: 3284–3286.
81. Sharma, R., S. Tangjang, and A. C. Shukla. 2020. Molecular characterization of endophytic fungi associated with *Centella Asiatica* linn. inhabiting wildly in Arunachal Pradesh. *The Journal of Indian Botanical Society*, 100(3–4): 160–168. https://doi.org/10.5958/2455-7218.2020.00038.8
82. Sharma, S., S. Gupta, M. K. Dhar, and S. Kaul. 2018. Diversity and bioactive potential of culturable fungal endophytes of medicinal shrub *Berberis aristata* DC: A first report. *Mycobiology*, 46(4): 370–381. https://doi.org/10.1080/12298093.2018.1538068
83. Siegel, R. L., K. D. Miller, and A. Jemal. 2017. Cancer Statistics, 2017. CA *Cancer Journal for Clinicians*, 67: 7–30. https://doi:10.3322/caac.21387

84. Singh, A., D. K. Singh, R. N. Kharwar, J. F. White, and S, K. Gond. 2021. Fungal Endophytes as Efficient Sources of Plant-Derived Bioactive Compounds and Their Prospective Applications in Natural Product Drug Discovery: Insights, Avenues, and Challenges. *Microorganisms*, 9(1): 1–42. https://doi.org/10.3390/MICROORGANISMS9010197
85. Singh, B., T. Kaur, S. Kaur, R. K. Manhas, and A. Kaur. 2015. An alpha-glucosidase inhibitor from an endophytic Cladosporium sp. with potential as a biocontrol agent. *Applied Biochemistry and Biotechnology*, 175: 2020–34. doi: 10.1007/s12010-014-1325-0.
86. Singh, D., V. Rathod, S. Ninganagouda, J. Herimath and P. Kulkarni. 2013. Biosynthesis of silver nanoparticle by endophytic fungi *Pencillium* sp. isolated from *Curcuma longa* (turmeric) and its antibacterial activity against pathogenic gram-negative bacteria. *Journal of Pharmacy Research*, 7(5): 448–453. https://doi.org/10.1016/J.JOPR.2013.06.003
87. Singh, L., and S. C. Sati. 2020. Antagonistic activity of isolated root endophytic freshwater fungus *Anguillospora longissima* against pathogenic fungi. *National Academy Science Letters*, 43(2): 201–205. https://doi.org/10.1007/s40009-019-00818-w
88. Srinivasan, K., L. K. Jagadish, R. Shenbhagaraman, and J. Muthumary. 2010. Antioxidant activity of endophytic fungus *Phyllosticta* sp. isolated from *Guazuma tomentosa*. *Journal of Phytology*, 2(6): 37–41.
89. Strobel, G. A. 2003. Endophytes as sources of bioactive products. *Microbes and Infection*, 5: 535–544. www.elsevier.com/locate/micinf
90. Talukdar, R., S. Padhi, A. Rai et al. 2021. Isolation and characterization of an endophytic fungus *Colletotrichum coccodes* producing tyrosol from *Houttuynia cordata* Thunb. using ITS2 RNA secondary structure and molecular docking study. *Frontiers in Bioengineering and Biotechnology*, 9. https://doi.org/10.3389/fbioe.2021.650247
91. Tan, R. X., and W. X. Zou. 2001. Endophytes: A rich source of functional metabolites. *Natural Product Reports*, 18: 448–459. https://doi.org/10.1039/b100918o
92. Uzma, F., C. D. Mohan, A. Hashem et al. 2018. Endophytic fungi—alternative sources of cytotoxic compounds: A review. *Frontiers in Pharmacology*, 9: 309. https://doi:10.3389/fphar.2018.00309
93. Venkateswarulu, N., S. Shameer, P. V. Bramhachari, S. T. Basha, C. Nagaraju, and T. Vijaya. 2018. Isolation and characterization of plumbagin (5-hydroxyl-2-methylnaptalene-1, 4-dione) producing endophytic fungi *Cladosporium delicatulum* from endemic medicinal plants. *Biotechnology Reports*, 20: e00282. https://doi:10.1016/j.btre.2018.e00282
94. Vigneshwari, A., D. Rakk, A. Nemeth et al. 2019. Host metabolite producing endophytic fungi isolated from *Hypericum perforatum*. *PLoS ONE*, 14: e0217060. https://doi:10.1371/journal.pone.0217060.
95. Wu, H.Y., Y. L. Wang, J. L. Tan et al. 2012. Regulation of the growth of cotton bollworms by metabolites from the entomopathogenic fungus *Paecilomyces cateniobliquus*. *Journal of Agricultural Food Chemistry*, 60: 5604–5608.
96. Yadav, M., A. Yadav, and J. P. Yadav. 2014. In vitro antioxidant activity and total phenolic content of endophytic fungi isolated from Eugenia jambolana Lam. *Asian Pacific Journal of Tropical Medicine*, 7(S1): S256–S261. https://doi.org/10.1016/S1995-7645(14)60242-X
97. Zhang, L., B. Guo, H. Li et al. 2000. Preliminary study on the isolation of endophytic fungus of Catharanthus roseus and its fermentation to produce products of therapeutic value. *Chinese Traditional and Herbal Drugs*, 31: 805–807.
98. Zou, W., J. Meng, H. Lu et al. 2000. Metabolites of *Colletotrichum gloeosporioides*, an endophytic fungus in *Artemisia mongolica*. *Journal of Natural Products*, 63: 1529–1530.

7 Discovering the Chemotherapeutic Potential of Bioactive Compounds Synthesized by Endophytic Fungi from Medicinal Plants

*Sherin Varghese, K.C. Rajeshkumar and M.S. Jisha**
*Corresponding author.

7.1 INTRODUCTION

Cancer, a complex disease with around 300 types, has historically been linked to environmental and genetic factors (Nagai et al., 2017; Bray et al., 2018). It is the second leading cause of death worldwide, following cardiovascular disease (Bray et al., 2021), with approximately 9.6 million deaths since 2018 (Shah et al., 2019). Treatments include radiotherapy, chemotherapy, targeted surgery, immunotherapy and emerging gene therapies, tailored to specific conditions (Oh et al., 2020; Niculescu & Grumezescu, 2022). Conventional therapies target cancer cells or tissues, but many cytotoxic compounds harm healthy tissues due to a lack of specificity (Huang et al., 2017; Schirrmacher, 2019). Additionally, chemoresistance complicates treatment and reduces effectiveness (Falzone et al., 2018). This underscores the need for new or synergistic drugs (Varghese et al., 2020).

Cutting-edge research highlights the vital role of natural products in developing innovative pharmaceuticals, offering unique templates for future drugs (Calixto, 2019; Wilson et al. 2020). The growing understanding of microbial synthesis and interactions in natural product generation makes endophyte research a key driver in transforming drug discovery (Strobel, 2018; Abdul-Razek et al., 2020). Endophytes, microorganisms living asymptomatically within plant tissues, benefit their hosts and produce various bioactive metabolites (Uzma et al., 2018; Newman & Cragg, 2020). Despite their potential, only about 0.75–1.50% of plant species have been thoroughly studied for endophytes, indicating vast untapped potential for discovering new bioactive compounds (Christenhusz & Byng, 2016; Strobel, 2018).

This unique group of microorganisms has captured researchers' interest by revealing hundreds of diverse chemical metabolites through careful isolation, growth, purification and characterization (Aly et al., 2010). Endophytes are particularly promising as sources of cytotoxic compounds like paclitaxel (taxol) (Stierle et al., 1993), suggesting potential for anticancer drug development. Notably, fusarubins from *Cladosporium* and *Fusarium* species, such as anhydrofusarubin and fusarubin (FUS), show significant cytotoxicity against K-562 human leukemia and Vero cells (Khan et al., 2016, 2018). Initially known for its antibacterial properties, FUS has recently demonstrated

 DOI: 10.1201/9781003486275-9

cytotoxic activity. The molecular mechanism of fusarubin's cytotoxic effects from *Cladosporium* in *Rauwolfia serpentina* leaves has also been documented (Adorisio et al., 2019).

These findings have paved the way for commercializing medicinal compounds derived from endophytic fungi. Growing demand for natural products, coupled with challenges in direct extraction from plants, underscores endophytes as promising targets for isolating host-derived bioactive compounds (Staneik et al., 2014; Knox, 2019). While medicinal plants are valuable sources of therapeutic compounds, exploring their endophytes is essential for effective extraction of these substances. Endophytes are valued for their sustainability, widespread availability and eco-friendly nature, offering diverse and novel bioactive compounds due to their extensive diversity and adaptability (Omeje et al., 2017).

This chapter emphasizes fungal endophytes as valuable sources for developing advanced anticancer agents. It explores fungal metabolites from medicinal plants, highlighting their potential as potent anticancer compounds effective against various types of cancer. The chapter discusses chemotaxonomic profiling of these metabolites, along with computational and experimental studies aimed at developing effective cancer treatments. It also explores emerging trends in endophyte-based drug discovery, including artificial intelligence, machine learning, multi-omics approaches and high-throughput strategies. In conclusion, the chapter addresses challenges and future directions for leveraging these fungal metabolites in cancer therapy.

7.2 SUSTAINABLE SOLUTIONS TO MEDICINAL PLANT CHALLENGES: LEVERAGING ENDOPHYTIC FUNGI FOR BIOACTIVE COMPOUND SYNTHESIS

Diverse medicinal plants in natural ecosystems provide healing compounds, offering alternatives to commercial drugs (Michel et al., 2020). In many developing countries, 80% of the population relies on herbal remedies for primary health care. Research shows that 51% of drugs developed between 1981 and 2014 came from medicinal plants, mainly for treating infectious diseases and cancer (Lu et al., 2020). The global demand for herbal pharmaceuticals is rising. However, challenges like overharvesting, ecological disturbances, anthropogenic activities and pest-related habitat destruction threaten these plant populations (Palanichamy et al., 2018).

Harnessing bioactive products from plants can be challenging due to limited product accumulation, long maturation times and difficulties in metabolite recovery. To fully unlock the potential of plant secondary metabolites, alternative methods for scaling up cultivation are essential. These methods include tissue culture, heterologous production, total chemical synthesis and using endophytic microorganisms like fungi to synthesize bioactive compounds (Sandargo et al., 2019). About 18% of plant-derived metabolites can originate from associated fungi (Chowdhary & Kaushik, 2015). For example, the bioactive compound taxol, found in the medicinal plant *Taxus*, is also derived from its fungal endophyte *Taxomyces andreanae* (Prakash et al., 2016). Taxol from *T. andreanae* shows promise, potentially surpassing its host plant in production via fermentation (Naik, 2019). Studies have documented the production of bioactive metabolites by medicinal plants and their endophytic fungi (Gomez & Luiz, 2018; Caruso et al., 2020).

7.3 OVERVIEW OF THE CHEMOTHERAPEUTIC POTENTIAL OF BIOACTIVE METABOLITES SYNTHESIZED BY ENDOPHYTIC FUNGI

Endophytes are microorganisms living within plant tissues symbiotically without causing disease (Arora et al., 2016). They can produce secondary metabolites similar to those of their host plants (Jia et al., 2016). For instance, fungal endophytes from *Podophyllum hexandrum*, *Camptotheca acuminata*, *Taxus chinensis* var. *mairei* and *Catharanthus roseus* have synthesized anticancer drugs like podophyllotoxin, camptothecin, taxol and vincristine/vinblastine (Nadeem, 2012; Palem et al.,

2015; Qiao et al., 2017; Ran et al., 2017). Additionally, these endophytes can produce unique chemical entities with novel structures (Pang et al., 2018; Chen et al., 2017).

The discovery of innovative structures in endophytic fungal metabolites has led to novel pharmaceuticals. Most approved small-molecule drugs are either natural compounds, synthetic compounds mimicking natural functions or synthetic analogs of natural compounds (Patridge et al., 2016). These drugs are crucial in drug discovery due to their diverse structures and physiological activities (Xiao et al., 2016). Endophytes produce biomolecules that enhance ecological adaptability, evolving with their host plants to synthesize functional natural compounds. For example, the endophytic fungus *Neotyphodium coenophialum* in *Festuca arundinacea* produces toxic alkaloids that repel herbivores, causing "fescue toxicosis" (Lyons et al., 1986). These alkaloids have potential medicinal value and could be explored for treating various diseases.

Torreyanic acid, an anticancer compound, is produced by the fungal endophyte *Pestalotiopsis microspora* in *Taxus taxifolia*. It induces apoptosis in cancer cells, especially those sensitive to protein kinase C agonists (Lee et al., 1996). Cytochalasins, anticancer alkaloids, are synthesized by about 20 species of endophytic fungi, including *Chalara*, *Hypoxylon*, *Xylaria* and *Phoma* (Wagenaar et al., 2000). Camptothecin is found in endophytic fungi like *Entrophospora infrequens* from *Nothapodytes foetida* (Puri et al., 2005). Topotecan and irinotecan, similar to camptothecin, were discovered in *Fusarium solani* from *Camptotheca acuminate*. Endophytic fungi such as *Phomopsis vaccinia*, *Colletotrichum gloeosporioides*, *Alternaria alternate* and *Fusarium nematophilum* from *Cyanea acuminate* also produce camptothecin (Kang et al., 2014). These compounds show potential as effective anticancer drugs in clinical studies (Kusari et al., 2009b).

The endophytic fungus *Trametes hirsuta* produces podophyllotoxin, a non-alkaloid lignan with significant anticancer properties (Puri et al., 2006). Other fungi such as *Aspergillus fumigatus* from *Juniperus communis*, *Fusarium oxysporum* from *Juniperus recurva*, and *Phialocephala fortinii* from *Podophyllum peltatum* also produce podophyllotoxin and its analogs (Kusari et al., 2009a; Kour et al., 2008; Eyberger et al., 2006), indicating its synthesis is not exclusive to *Trametes hirsuta*. Endophytic *Podophyllum hexandrum* produces lignans with cathartic, emetic and cholagog properties, showing anticancer effects. Resins from *P. emodi*, including etoposide and teniposide, also demonstrate substantial anticancer potential (Gouda et al., 2016). *Mycelia sterilia*, associated with *Catharanthus roseus*, synthesizes vincristine, a potent bioactive compound used in chemotherapy for nephroblastoma and acute lymphoblastic leukemia (Yang et al., 2004). These findings highlight fungal endophytes as valuable sources of anticancer compounds, emphasizing the importance of exploring natural sources for drug discovery. In addition to these bioactive compounds, this chapter elucidates a list of anticancer compounds synthesized by endophytic fungi in Table 7.1.

TABLE 7.1
Anticancer Compounds Isolated from Endophytic Fungi Residing in Medicinal Plants

Bioactive compounds	Endophytic fungi	Medicinal plant	Anticancer activity on cancer cell lines with IC_{50} values	References
Fusarithioamide B	*Fusarium solani*	*Anvillea arcinia* (Burm.f.) DC.	BT-549; 0.09 μM	Ibrahim et al. (2018)
Xylariphthalide A	*Diaporthe* spp.	*Tylophora ouata*	BGC-823; 1.5 μmol·L −1	Liu et al. (2018b)
Cis-4-hydroxy-6-deoxytalone	*Diaporthe* spp.	*Tylophora ouata*	BGC-823; 8.6 μmol·L −1	Liu et al. (2018b)

TABLE 7.1 (Continued)
Anticancer Compounds Isolated from Endophytic Fungi Residing in Medicinal Plants

Bioactive compounds	Endophytic fungi	Medicinal plant	Anticancer activity on cancer cell lines with IC_{50} values	References
Xylarolide A	*Diaporthe* spp.	*Datura inoxia*	MIAPaCa-2, PC-3; 20 μM and 14 μM	Sharma et al. (2018)
Pyrrocidine A	*Cylindrocarpon* spp.	*Sapium ellipticum*	A2780; 1.7 μM	Kamdem et al. (2018)
Colletotricone A	*Colletotrichum gloeosporioides* A12	*Aquilaria sinensis*	NCI-H460, HepG-2, and SF-268; 15.7–46.8 μM	Liu et al. (2018a)
Flavipin	*Chaetomium globosum*	*Couroupita guianensis* Aubl. leaves	A549, HT-29, and MCF-7; 9.89 μg/mL, 18 μg/mL and 54 μg/mL	Senthil Kumar et al. (2019)
Bellidisin D	*Phoma betae*	*Tricyrtis maculate* leaves	HL-60, A549, SMMC-7721, MCF-7, and SW480; 3.40–15.25 μM	Wang et al. (2019b)
Epicorazine A	*Epicoccum* spp.	*Salix* sp.	L5178Y, Ramos, and Jurkat J16; 1.3–28 mM	Harwoko et al. (2021)
Asperchalasin A–F	*Aspergillus* spp.	*Pinellia ternata* tubers	A-549; 23.3–70.2 μM	Xin et al. (2019)
Aspergisocoumrins A & B	*Aspergillus* spp. HN15-5D	*Acanthus ilicifolius* fresh leaves	MDA-MB-435; 5.08 ± 0.88 and 4.98 ± 0.74 μM	Wu et al. (2019)
Macrophin	*Phoma betae*	*Glycyrrhiza glabra* Linn	(MDA-MB-231, T47D, MCF-7, and MIAPaCa-2; 14.8, 8.12, 13.0 and 0.9 μM	Nalli et al. (2019)
Giluterrin	*Aspergillus terreus* P63	*Axonopus leptostachyus* roots	786-0 and PC-3; 22.93 μM and 48.55 μM	Gubiani et al. (2019)
Cytochalasin D1 (1) and C1	*Xylaria cf. cubensis*	*Solanum tuberosum* stem tissues	HL-60; 12.7 and 22.3 μM	Wang et al. (2019a)
Botryosulfuranol A	*Botryosphaeria* sp. strain E224	*Bixa orellana* fresh leaves	HT-29, HepG2, Caco-2, HeLa, IEC6, and vero cells; 8–23.5 μM	Barakat et al. (2019)
Cytosporins W	*Pseudopestalotiopsis* sp.	*Rhizophora racemosa* mangrove plants	L5178Y; 3.0 μM	Yu et al. (2020)
Lithocarin B & C, Tenellone H	*Diaporthe terebinthifolii* A740	*Morinda officinalis*	MCF-7; 30 -100 μM	Liu et al. (2019)
Integracin A and B	*Cytospora* spp.	*Ceriops tagal* (Chinese mangrove)	HepG2; 5.98 ± 0.12 μM and 9.97 ± 0.06 μM	Wei et al. (2020)
Allantopyrone E	*Aspergillus v.*	*Avicennia marina* mangrove	HeLa cells; 50.97 μM	Elsbaey et al. (2020)

(continued)

TABLE 7.1 (Continued)
Anticancer Compounds Isolated from Endophytic Fungi Residing in Medicinal Plants

Bioactive compounds	Endophytic fungi	Medicinal plant	Anticancer activity on cancer cell lines with IC_{50} values	References
Pramanicin A	*Aplosporella j.*	*Orychophragmus violaceus* (L.) O. E. Schul	Human lymphoma (Ramos) and leukemia (Jurkat J16) cells 4.7 and 4.4 μM	Gao et al. (2020)
Myrothecines H and I	*Paramyrothecium r*	*Morinda officinalis*	SF-268, NCI-H460 and HepG-2 tumor cell lines; 0.0002–16.2 μM	Liu et al. (2020)
Pestalotether D	*Pestalotiopsis terminaliae* (N635)	*Camellia sinensis* (Theaceae)	MCF-7; 22.6 M	Guo et al. (2020)
Sterigmatocystin	*Paecilamyces* spp. TE-540	*Nicotiana tabacum* L	MCF-7; 5.6–14.2 μM	Li et al. (2020)
Methyl 3-chloroasterric acid	*Pleosporales* spp. SK7.	*Kandelia candel* leaves	MDA-MB-435; 25.96 ± 0.32 μM	Wen et al. (2020)
Colletotrichalactone A and colletotrichalactone	*Colletotrichum* spp. JS-0361	*Morus alba*	MCF-7; 35.06 and 25.20 μM	Bang et al. (2020)

7.4 CHEMOTAXONOMIC PROFILING OF ENDOPHYTIC FUNGI FOR EXPLORING CANCER THERAPEUTICS

The study of endophyte characterization shows promise as medicinal plants host diverse microorganisms that produce bioactive compounds within their tissues (Hassan & Ullah, 2019). Current research aims to understand the structural diversity and functional properties of these compounds for therapeutic applications (Nivetha & Kharwar, 2019). Metabolomics has significantly advanced bioprospecting by enabling the discovery of novel chemical entities and facilitating the synthesis of plant-derived metabolites by endophytes (Mishra et al., 2022). Challenges such as seasonal variations and environmental factors impact the extraction of bioactive metabolites from plants (Ekiz et al., 2018). Developing efficient methods for extracting these chemicals from natural sources is crucial. Axenic cultivation, conducted through submerged or solid-state fermentation over several weeks, is standard for uncovering bioactive compounds from microbial cultures. Bioassay-based screening, evaluating properties like anticancer or antimicrobial activity, guides subsequent purification steps (Mohamad et al., 2018). These processes involve optimizing fermentation conditions and introducing elicitors to enhance metabolite production. Metabolomics accelerates the screening and identification of metabolites and new compounds in endophyte extracts (Mishra et al., 2021).

Chemotaxonomic profiling combined with two-dimensional analytical techniques is a powerful approach for identifying and characterizing secondary metabolites from endophytic fungi. This method provides insights into their chemical diversity and aids in discovering potent anticancer agents against cancer. Chemotaxonomic profiling merges chemical and taxonomic data to find novel natural products. Bioactive metabolite analysis can be conducted at various levels, including target analysis, metabolite profiling, metabolic fingerprinting and metabolomics (Shulaev et al., 2008; Porzel et al., 2014).

Multiple analytical platforms are increasingly used for both untargeted and targeted analysis of secondary metabolites (Segers et al., 2019). Advanced methods like combining chromatography and mass spectrometry provide robust tools for analyzing secondary metabolites in endophytic fungi. TLC and HPLC are commonly used to separate complex metabolite mixtures based on their physicochemical properties. Mass spectrometry (MS) then provides valuable structural

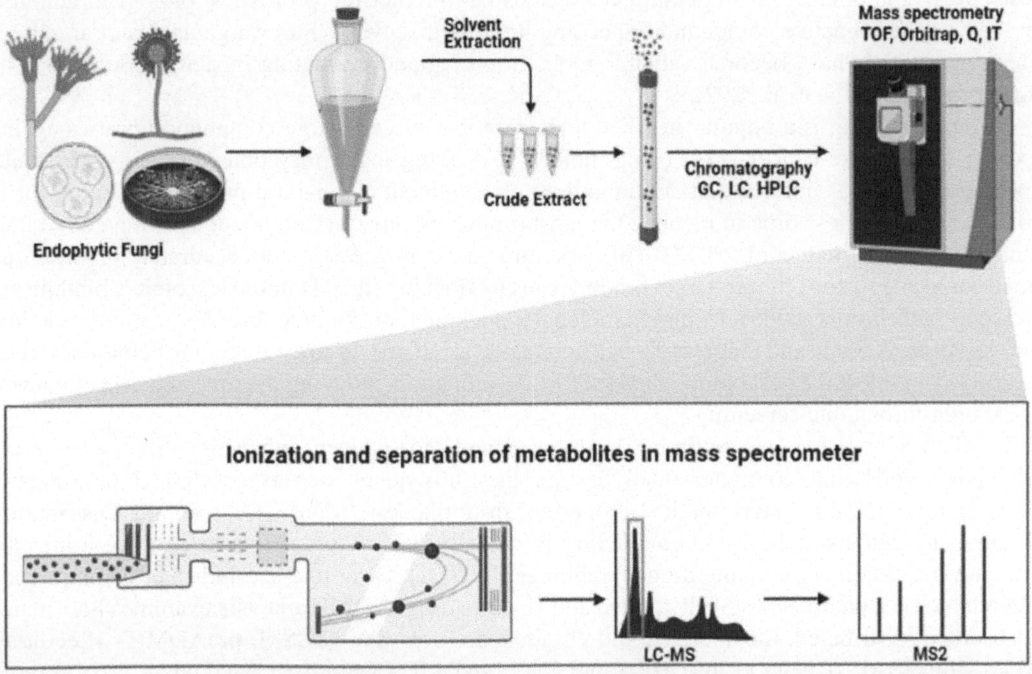

FIGURE 7.1 Schematic representation of secondary metabolite profiling of endophytic fungi coupled with solvent extraction and two-dimensional analytical techniques offering a powerful strategy to identify and characterize secondary metabolites from endophytic fungi, providing insights into the chemical diversity of these fungi and facilitating the discovery of potent anticancer agents.

information, such as molecular weight and fragmentation patterns, aiding in the identification and characterization of individual compounds. Separation methods like capillary electrophoresis (CE), liquid chromatography (LC) and gas chromatography (GC) are often used alongside MS. These complementary platforms can resolve numerous metabolites (Wei et al., 2020; Sagita et al., 2021). Figure 7.1 shows an overview of chemotaxonomic profiling of endophytic fungi using various 2D analytical techniques, which are frequently used to examine bioactive metabolites produced by endophytic fungi.

Analyzing secondary metabolites provides a comprehensive understanding of the chemical diversity in fungal endophytes and their structure–activity relationships, crucial for anticancer activity (Xiao et al., 2016). This method identifies specific chemical markers and correlates secondary metabolites with taxonomic classifications, aiding in targeted isolation and characterization (Patridge et al., 2016). These bioactive compounds can serve as lead structures for developing new pharmaceuticals or as templates for synthesizing analogs with improved efficacy, reduced toxicity or targeted delivery systems, paving the way for personalized cancer treatments. This integrated approach promises to advance cancer research and improve therapeutic outcomes (Bedi et al., 2018; Deshmukh et al., 2014).

7.5 CONTEMPORARY APPROACHES EMPLOYED IN THE DEVELOPMENT OF ENDOPHYTIC FUNGI-BASED CHEMOTHERAPEUTIC DRUGS

Traditionally, discovering new therapeutic compounds involves laborious and time-consuming experiments. Screening and characterizing endophytic fungal metabolites in vitro are particularly demanding and costly (Singh et al., 2017). In contrast, in silico studies provide a more efficient and cost-effective alternative. Computational methods can predict the properties, biological activities and toxicity of bioactive compounds, speeding up drug discovery (Purawargamatada et al., 2022). These approaches have become valuable tools in finding and evaluating bioactive compounds for cancer treatment (Jha et al., 2022).

Virtual screening is a popular in silico technique that screens large compound libraries against specific targets, like cancer cells, using molecular docking to identify potential leads. Molecular docking has gained significant attention in drug development and natural product screening for its efficiency. It takes less time to identify the most promising bioactive compounds compared to traditional methods (Singh et al., 2017). This structure-based drug design tool accurately predicts how small-molecule ligands fit into target binding sites by docking ligands into the protein's binding site and scoring their interactions. Common molecular docking tools include AutoDock, AutoDockVina, FlexX, SurflexX, Gold and Scigress-Dock (Lounnas et al., 2013). In silico screening helps researchers prioritize and select the best compounds for further validation, reducing the time and cost of conventional high-throughput screening.

In silico approaches can optimize lead compounds for cancer, enhancing their potency and selectivity. Following computer-aided design, the subsequent step is preclinical optimization focusing on refining physicochemical properties, drug likeness, ADME (Absorption, Distribution, Metabolism, and Excretion) and conducting in silico toxicity assessments to advance a hit molecule toward becoming a viable drug (Ibrahim et al., 2021). Drug likeness indicates a compound's potential to be a pharmacologically active and suitable drug. ADME analysis examines how a drug is absorbed, distributed, metabolized and eliminated. Tools like pkCSM, preADMET (Lee et al., 2004), admetSAR (Cheng et al., 2012) and SWISSADME (Ranjith & Ravikumar, 2019) predict ADME parameters from molecular structures. By integrating drug likeness prediction and ADME analysis, researchers can design and prioritize compounds with favorable properties, reducing the risk of expensive and time-consuming failures later in drug development.

Numerous in vitro assays and techniques have been developed to evaluate distinct characteristics of cancer (Hanahan & Weinberg, 2000, 2011). The selection of an assay is guided by the specific research questions. Today, a wide array of these assays supports the discovery of new cancer treatments (Bohlin & Bruhn, 2012). In cancer drug discovery, commonly used assays include antiproliferation assays, apoptosis assays, cell migration and invasion assessments, angiogenesis assays and reporter gene assays. These methods are pivotal in identifying new chemotherapeutic agents (Fox et al., 2006).

7.6 ADVANCEMENTS IN EXPANDING HORIZONS FOR ENDOPHYTIC FUNGI-BASED PHARMACEUTICALS

The field of endophytic fungi-based pharmaceuticals is advancing due to the discovery of bioactive metabolites, creating new opportunities for drug discovery from natural products. The pharmaceutical industry increasingly values plant–microbe associations, particularly endophytic fungi, attracting global research interest (Uzma et al., 2018; Varghese et al., 2020, 2021, 2024). With rising cancer cases and slowing drug development, natural sources for therapeutics are emphasized. High-throughput technologies and interdisciplinary approaches have advanced drug discovery programs. Computational techniques and multi-omics strategies are crucial in identifying bioactive compounds from plant–endophytic associations (Aghdam & Brown, 2021).

The "OSMAC" approach focuses on using a single strain to produce a variety of compounds by optimizing culture conditions and activating dormant gene clusters. Techniques like heterologous gene expression, ribosome engineering and studying promoters are employed to enhance gene expression diversity. Co-cultivating different endophytic fungal strains is effective in inducing gene expression and reactivating silent biosynthetic gene clusters (BGCs). Genetic methods, including ribosome engineering and using inducible/constitutive promoters, play a crucial role in activating these dormant gene clusters (Baral et al., 2018; Pan et al., 2019).

Artificial intelligence, including machine learning and deep learning, plays a vital role in analyzing extensive datasets to discover bioactive metabolites. These methods help predict plant microbiome distributions and synthesize bioactive compounds from endophytes. Advanced deep learning techniques like bidirectional long short-term memory (BiLSTM) neural networks and software such as DeepBGC are essential for enhancing drug discovery from microbial communities (Hannigan et al., 2019).

Incorporating models of metabolic genomics and advanced machine learning techniques aids in the bio-prospection of endophytes for bioactive metabolites. Chemi-informatics methods based on deep learning efficiently predict chemical entity diversity. Computational biology methods analyze chemical entities and utilize metabolomics strategies to predict plant microbiome and biochemical alterations. These innovative computational methods represent a significant breakthrough in drug discovery using endophytic fungal metabolites (Aghdam & Brown, 2021).

Technological advances in automation have revolutionized drug discovery, emphasizing high-throughput strategies. While combinatorial chemistry and cellular molecular biology identify molecular targets, natural products' diverse structures offer promising avenues for drug development. Understanding biochemical pathways enables precise control and manipulation of secondary metabolite production. Comparative genomics aids in predicting functional genes within conserved biosynthetic gene clusters, guiding the discovery of natural products through in silico predictions and experimental validation (Wright, 2019).

Epigenetic modifiers, chemical inducers and co-culture techniques activate dormant biosynthetic pathways in fungal endophytes, enhancing bioactive metabolite production. HDAC inhibitors, for instance, stimulate hidden metabolite synthesis in *Penicillium brevicompactum* (El-Hawary et al., 2018). Computational tools like ClusterMine 360, antiSMASH and RAVEN 2.0 streamline gene cluster discovery and annotation, aiding in genome-wide profiling of natural product pathways. These tools, coupled with network algorithms and metabolic modeling, enhance our understanding of microbial and endophytic metabolic interactions (Blin et al., 2019; Wang et al., 2018).

7.7 EXPLORING CHALLENGES AND FUTURE TRAJECTORIES IN ENDOPHYTIC FUNGI

Endophytic fungal research has gained considerable attention, yet significant challenges remain, particularly in the artificial cultivation of fungi that are difficult to culture. Overcoming these obstacles requires developing innovative bioengineering systems or adapting existing isolation methods (Zhang et al., 2006; Zhang et al., 2019). Successful identification and isolation of new fungal endophytes, and understanding their symbiotic signaling systems, hinge on carefully selecting suitable host plants and viable organs or tissues (Strobel, 2018).

Endophytic fungi present potential for producing unique bioactive compounds. Yet, challenges like low yield, incomplete understanding of complex fungal–plant interactions, scaling difficulties and lab constraints hinder progress. Replicating these natural metabolites in vitro faces significant obstacles, limiting commercial production success. Studying genetic pathways that activate specific genes for increased metabolite synthesis shows promising potential. Combinatorial chemical synthesis faces challenges in producing complex metabolites, but these hurdles can be overcome by using tailored inducers or elicitors to enhance desired metabolite yields. Understanding and mitigating negative feedback mechanisms caused by nonessential metabolites in growth media are critical for optimizing fermentation strategies (Zhao et al., 2010). Genetic engineering, elicitor addition and culture condition optimization are key research areas for enhancing the production of targeted secondary metabolites (Venugopalan & Srivastava, 2015).

Structural biology has advanced significantly in identifying new cancer drug targets and understanding chemotherapy resistance mechanisms. This progress prompts researchers to rethink approaches for discovering compounds to treat different cancer types. Exploring metabolites from endophytic fungi reveals broad cytotoxic effects against specific cancer cell lines in a dose-dependent manner, with some compounds showing potency comparable to or exceeding conventional chemotherapy drugs. However, the molecular mechanisms underlying their cytotoxicity remain poorly understood. Computational tools are increasingly crucial in drug discovery, helping to elucidate how these compounds interact with specific protein targets linked to cancer, thereby facilitating drug development processes. Further investigations in this field hold potential to revolutionize traditional pharmaceutical development (Parmar et al., 2020; Prajapati et al., 2021).

To harness endophytic metabolites for therapeutic use, several key steps are essential. This includes establishing regional and international repositories to preserve endophytic cultures and creating databases containing diverse bioactive compounds with varied chemical structures. These repositories safeguard cultures, including those from endangered species or remote locations, making them accessible for research. A comprehensive computational database detailing the chemical composition and biological activity of these metabolites is crucial. It validates compound authenticity and aids in understanding their mechanisms of action. Such efforts are vital for developing a repertoire of therapeutic compounds that can effectively target cancers with precision.

7.8 CONCLUSION

This chapter highlights the crucial role of endophytic fungi in producing bioactive compounds that could revolutionize cancer therapy. The discovery of taxol in *T. andreanae* has underscored the potential of these fungi as sources of chemotherapeutic drugs. Many metabolites from these fungi show promise for cancer treatment, and researchers are investigating their molecular mechanisms to better understand their biological activity. Exploring the interactions between these fungal metabolites and specific cellular targets is key to realizing their full potential. Advances in technology and a deeper understanding of their anticancer properties position endophytic fungi as a valuable source of economical, specific and minimally toxic chemotherapeutic agents. Using multi-omics technologies, advanced pharmacological tools and bioinformatics can optimize the synthesis of bioactive metabolites, making endophytic fungi a promising source of anticancer drugs.

ACKNOWLEDGMENTS

The authors acknowledge DST-PURSE (no. SR/PURSE/Phase 2/26 (C)) and DST-FIST (no. SR/FST/ LSI-660/2016 (C)), and National Fungal Culture Collection of India (NFCCI), Biodiversity and Palaeobiology (Fungi) Group, Agharkar Research Institute, Pune, for providing the necessary facilities. The first author would like to thank Mahatma Gandhi University for a Junior Research Fellowship (Section order no. 4446/AC A6/2022/MGU).

REFERENCES

Abdul-Razek, A.S., El-Nagar, M.E., Allam, A., Morsy, O.M., and Othman, S.I. 2020. Microbial natural products in drug discovery. *Processes* 8, 470.

Adorisio, S., et al. 2019. Fusarubin and Anhydrofusarubin isolated from a *Cladosporium* species inhibit cell growth in human cancer cell lines. *Toxins* 11, 503.

Aghdam, S.A., and Brown, A.M.V. 2021. Deep learning approaches for natural product discovery from plant endophytic microbiomes. *Environ. Microbiome*. 16, 6.

Aly, A.H., Debbab, A., Kjer, J., and Proksch, P. 2010. Fungal endophytes from higher plants: A prolific source of phytochemicals and other bioactive natural products. *Fungal Divers*. 41, 1–16.

Arora, D., et al. 2016. Isolation and characterization of bioactive metabolites from Xylaria psidii, an endophytic fungus of the medicinal plant Aegle marmelos and their role in mitochondrial dependent apoptosis against pancreatic cancer cells. *Phytomedicine* 23, 1312–1320. https://doi.org/10.1016/j.phymed.2016.07.004

Bang, S., Kwon, H.E., Baek, J.Y., Jang, D.S., Kim, S., Nam, S.J., Lee, D., Kang, K.S., and Shim, S.H. 2020. Colletotrichalactones A-Ca, Unusual 5/6/10-Fused Tricyclic Polyketides Produced by an Endophytic Fungus, Colletotrichum sp. JS-0361. *Bioorganic Chem.* 105, 104449.

Barakat, F., Vansteelandt, M., Triastuti, A., Jargeat, P., Jacquemin, D., Graton, J., Mejia, K., Cabanillas, B., Vendier, L., Stigliani, J.L. et al. 2019. Thiodiketopiperazines with Two Spirocyclic Centers Extracted from Botryosphaeria Mamane, an Endophytic Fungus Isolated from Bixa Orellana L. *Phytochemistry* 158, 142–148.

Baral, B., Akhgari, A., and Metsä-Ketelä, M. 2018. Activation of microbial secondary metabolic pathways: Avenues and challenges. *Synth. Syst. Biotechnol.* 3, 163–178.

Bedi, A., Adholeya, A., and Deshmukh, S.K., 2018. Novel anticancer compounds from endophytic fungi. *Curr. Biotechnol.* 7, 168–184.

Blin, K., et al. 2019. antiSMASH 5.0: Updates to the secondary metabolite genome mining pipeline. *Nucleic Acids Res.* 47, W81–W87.

Bohlin, L., and Bruhn, J. G. (Eds.) 2012. *Bioassay methods in natural product research and drug development* (Vol. 43). Springer Science and Business Media. doi: 10.1007/978-94-011-4810-8

Bray, F., Ferlay, J., Soerjomataram, I., Siegel, R.L., Torre, L.A., Jemal, A., and Global Cancer Statistics. 2018. GLOBOCAN estimates of incidence and mortality worldwide for 36 cancers in 185 countries. *CA Cancer J. Clin.* 68, 394–424.

Bray, F., Laversanne, M., Cao, B., Varghese, C., Mikkelsen, B., Weiderpass, E., and Soerjomataram, I. 2021. Comparing cancer and cardiovascular disease trends in 20 middle-or high-income countries 2000–19: A pointer to national trajectories towards achieving sustainable development goal target 3.4. *Cancer Treat. Rev.* 100, 102290.

Calixto, J.B. 2019. The role of natural products in modern drug discovery. *An. Acad. Bras. Cienc.* 91, e20190105.

Caruso, G., Abdelhamid, M.T., Kalisz, A., and Sekara, A. 2020. Linking endophytic fungi to medicinal plants therapeutic activity. A case study on asteraceae. *Agriculture* 10, 286.

Chen, L., et al. 2017. Trichoderpyrone, a unique polyketide hybrid with a cyclopentenone-pyrone skeleton from the plant endophytic fungus Trichoderma gamsii. *J. Nat. Prod.* 80, 1944–1947.

Cheng, F., Li, W., Zhou, Y., Shen, J., Wu, Z., and Liu, G. 2012. admetSAR: A comprehensive source and free tool for assessment of chemical ADMET properties. *J. Chem. Inf. Model.* 52, 3099–3105.

Chowdhary, K., and Kaushik, N. 2015. Fungal endophyte diversity and bioactivity in the Indian medicinal plant Ocimum sanctum Linn. *PLoS ONE* 10, e0141444.

Christenhusz, M.J.M., and Byng, J.W. 2016. The number of known plants species in the world and its annual increase. *Phytotaxa* 261, 201–217.

Deshmukh, S.K., Verekar, S.A., and Bhave, S.V. 2014. Endophytic fungi: A reservoir of antibacterials. *Front. Microbiol.* 5, 1–43.

Ekiz, G., Duman, S., and Bedir, E. 2018. Biotransformation of cyclocanthogenol by the endophytic fungus Alternaria eureka 1E1BL1. *Phytochem.* 151, 91–98. https://doi. org/10.1016/j.phytochem.2018.04.006

El-Hawary, S.S., Sayed, A.M., Mohammed, R., Hassan, H.M., Zaki, M.A., Rateb, M.E. … Abdelmohsen, U.R. 2018. Epigenetic modifers induce bioactive phenolic metabolites in the marinederived fungus Penicillium brevicompactum. *Marine Drugs* 16, 253.

Elsbaey, M., Tanaka, C., and Miyamoto, T. 2020. Allantopyrone E, a rare α-Pyrone metabolite from the mangrove derived fungus Aspergillus Versicolor. *Nat. Prod. Res.* 36(3) 1–5.

Eyberger, A.L., Dondapati, R., and Porter, J.R. 2006. Endophyte fungal isolates from Podophyllum peltatum produce podophyllotoxin. *J Nat Prod.* 69, 1121–1124.

Falzone, L., Salomone, S., and Libra, M. 2018. Evolution of cancer pharmacological treatments at the turn of the third millennium. *Front. Pharmacol.* 9, 1300.

Fox, S., Farr-Jones, S., Sopchak, L., Boggs, A., Nicely, H. W., Khoury, R., and Biros, M. 2006. High-throughput screening: Update on practices and success. *J. Biomol. Screen.* 11, 864–869. https://doi. org/10.1177/ 1087057106292473

Gao, Y., Stuhldreier, F., Schmitt, L., Wesselborg, S., Wang, L., Müller, W.E.G., Kalscheuer, R., Guo, Z., Zou, K., Liu, Z., et al. 2020. Sesterterpenes and Macrolide derivatives from the endophytic fungus Aplosporella Javeedii. *Fitoterapia* 146, 104652.

Gómez, O.C., Luiz, J.H.H. 2018. Endophytic fungi isolated from medicinal plants: Future prospects of bioactive natural products from Tabebuia/Handroanthus endophytes. *Appl. Microbiol. Biotechnol.* 102, 9105–9119.

Gouda, S., Das, G., Sen, S. K., Shin, H.-S., and Patra, J. K. 2016. Endophytes: A treasure house of bioactive compounds of medicinal importance. *Front. Microbiol.* 7, 1538.

Gubiani, J.R., Oliveira, M.C.S., Neponuceno, R.A.R., Camargo, M.J., Garcez, W.S., Biz, A.R., Soares, M.A., Araujo, A.R., da Bolzani, V.S., Lisboa, H.C.F., et al. 2019. Cytotoxic Prenylated Indole Alkaloid produced by the endophytic fungus Aspergillus Terreus P63. *Phytochem. Lett.* 32, 162–167.

Guo, L. Lin, J., Niu, S., Liu, S., and Liu, L. 2020. Pestalotiones A–D: Four new secondary metabolites from the plant endophytic fungus Pestalotiopsis Theae. *Molecules* 25, 470.

Hanahan, D., and Weinberg, R.A. 2000. The hallmarks of cancer. *Cell* 100, 57–70. doi: 10.1016/S0092-8674(00)81683-9

Hanahan, D., and Weinberg, R.A. 2011. Hallmarks of cancer: The next generation. *Cell* 144, 646–674. doi:10.1016/j. cell.2011.02.013

Hannigan, G.D., et al. 2019. A deep learning genome-mining strategy for biosynthetic gene cluster prediction. *Nucleic Acids Res.* 47, e110.

Harwoko, H., Daletos, G., Stuhldreier, F., Lee, J., Wesselborg, S., Feldbrügge, M., Müller, W.E.G., Kalscheuer, R., Ancheeva, E., and Proksch, P. 2021. Dithiodiketopiperazine derivatives from endophytic fungi Trichoderma Harzianum and Epicoccum Nigrum. *Nat. Prod. Res.* 35, 257–265.

Hassan, A., and Ullah, H. 2019. Antibacterial and antifungal activities of the medicinal plant Veronica biloba. *J. Chem.* 2019, 5264943.

Huang, C.-Y., Ju, D.-T., Chang, C.-F., Reddy, P.M., and Velmurugan, B.K. 2017. A review on the effects of current chemotherapy drugs and natural agents in treating non–small cell lung cancer. *Biomedicine* 7, 23.

Ibrahim, S.R.M., Mohamed, G.A., Al Haidari, R.A., Zayed, M.F., El-Kholy, A.A., Elkhayat, E.S., and Ross, S.A. 2018. Fusarithioamide B, a new Benzamide derivative from the endophytic Fungus fusarium Chlamydosporium with potent cytotoxic and antimicrobial activities. *Bioorg. Med. Chem.* 26, 786–790.

Ibrahim, Z.Y.U., Uzairu, A., Shallangwa, G.A., and Abechi, S.E. 2021. Pharmacokinetic predictions and docking studies of substituted aryl amine-based triazolopyrimidine designed inhibitors of Plasmodium falciparum dihydroorotate dehydrogenase (PfDHODH). *Future J. Pharm. Sci.* 7(1), 1–10.

Jha, V., Devkar, S., Gharat, K., Kasbe, S., Matharoo, D.K., Pendse, S., Bhosale, A., and Bhargava, A. 2022. Screening of phytochemicals as potential inhibitors of breast cancer using structure based multitargeted molecular docking analysis. *Phytomed. Plus.* 2(2), 100227.

Jia, M., et al. 2016. A friendly relationship between endophytic fungi and medicinal plants: A systematic review. *Front. Microbiol.* 7, 1–14.

Kamdem, R.S.T., Pascal, W., Rehberg, N., van Geelen, L., Höfert, S.P., Knedel, T.O., Janiak, C., Sureechatchaiyan, P., Kassack, M.U., Lin, W. et al. 2018. Metabolites from the endophytic Fungus Cylindrocarpon sp. isolated from tropical plant Sapium Ellipticum. *Fitoterapia* 128, 175–179.

Kang, J.C., et al. 2014. Medicinal plant endophytes produce analogous bioactive compounds morphological and molecular characterization of pathogenic Colletotrichum and Fusarium species and in vitro evaluation of antifungal activity of local plants extracts for potential biocontrol. *Chiang Mai J. Sci.* 41(1), 1–13.

Khan, M.I.H., et al. 2016. Cytotoxic and antibacterial Naphthoquinones from an endophytic fungus, Cladosporium sp. *Toxicol. Rep.* 3, 861–865.

Khan, N., et al. 2018. Endophytic *Fusarium solani*: A Rich Source of Cytotoxic and Antimicrobial Napthaquinone and Aza-Anthraquinone Derivatives. *Toxicol. Rep.* 5, 970–976.

Knox, R. 2019. Introduction to medicinal plants. In *Medicinal Plant Biotechnology*; Knox, R., Ed.; Scientific e-Resources; ED-Tech Press: Essex, UK, 32–60.

Kour, A., et al. 2008. Isolation and identification of an endophytic strain of Fusarium oxysporum producing podophyllotoxin from Juniperus recurva. *World J. Microbiol. Biotechnol.* 24,1115–1121.

Kusari, S., Lamshöft, M., and Spiteller, M. 2009a. Aspergillus fumigatus Fresenius, an endophytic fungus from Juniperus communis L. Horstmann as a novel source of the anticancer pro-drug deoxypodophyllotoxin. *J. Appl. Microbiol.* 107, 1019–1030.

Kusari, S., Zühlke, S., and Spiteller, M. 2009b. An endophytic fungus from Camptotheca acuminata that produces camptothecin and analogues. *J. Nat. Prod.* 72, 2–7.

Lee, J.C., Strobel, G.A., Lobkovsky, E., and Clardy, J. 1996. Torreyanic acid: A selectively cytotoxic quinone dimer from the endophytic fungus Pestalotiopsis microspora. *J. Org. Chem.* 61, 3232–3233.

Lee, S.K., Chang, G.S., Lee, I.H., Chung, J.E., Sung, K.Y., and No, K.T. 2004. The PreADME: Pc-based program for batch prediction of adme properties. *EuroQSAR* 9, 5–10.

Li, X.Q., Dong, Q.J., Xu, K., Yuan, X.L., Liu, X.M., and Zhang, P. 2020. Cytotoxic Xanthones from the plant endophytic fungus Paecilamyces sp. TE-540. *Nat. Prod. Res.* 35, 6134–6140.

Liu, Z., Zhao, J.Y., Li, Y., Lyu, X.X., and Liu, Y.B. 2018b. Investigations on secondary metabolites of endophyte Diaporthe sp. hosted in Tylophora ovata. *Zhongguo Zhong Yao Za Zhi Zhongguo Zhongyao Zazhi China J. Chin. Mater. Medica* 43, 2944–2949.

Liu, H.X., Tan, H.B., Chen, Y.C., Li, S.N., Li, H.H., and Zhang, W.M. 2018a. Secondary metabolites from the Colletotrichum Gloeosporioides A12, an endophytic fungus derived from Aquilaria Sinensis. *Nat. Prod. Res.* 32, 2360–2365.

Liu, H., Chen, Y., Li, S., Zhang, W., Liu, Z., Tan, H., and Zhang, W. 2020. Trichothecene Macrolides from the endophytic fungus Paramyrothecium Roridum and their cytotoxic activity. *Fitoterapia* 147, 104768.

Liu, H., Chen, Y., Li, H., Li, S., Tan, H., Liu, Z., Li, D., Liu, H., and Zhang, W. 2019. Four new metabolites from the endophytic fungus Diaporthe Lithocarpus A740. *Fitoterapia* 137, 104260.

Lounnas, V., Ritschel, T., Kelder, J., McGuire, R., Bywater, R.P., and Foloppe, N., 2013. Current progress in structure-based rational drug design marks a new mindset in drug discovery. *Comput. Struct. Biotechnol. J.* 5, e201302011. doi: 10.5936/csbj.201302011

Lu, S., Qiu, Y., Ni, D., He, X., Pu, J., and Zhang, J. 2020. Emergence of allosteric drug-resistance mutations: New challenges for allosteric drug discovery. *Drug Discov. Today* 25, 177–184.

Lyons, P.C., Plattner, R.D., and Bacon, C.W. 1986. Occurrence of peptide and clavine ergot alkaloids in tall fescue grass. *Science* 232, 487–489.

Michel, J., Abd Rani, N.Z., and Husain, K. 2020. A review on the potential use of medicinal plants from Asteraceae and Lamiaceae plant family in cardiovascular diseases. *Front. Pharmacol.* 11, 852.

Mishra, S., Bhardwaj, P., and Sharma, S. 2022. Metabolomic insights into endophyte-derived bioactive compounds. *Front. Microbiol.* 416, 1–11.

Mishra, S., Goyal, D., and Phurailatpam, L. 2021. Targeted 16S rRNA gene and ITS2 amplicon sequencing of leaf and spike tissues of Piper longum identifies new candidates for bioprospecting of bioactive compounds. *Arch. Microbiol.* 203, 3851–3867. doi: 10.1007/s00203-021-02356-w

Mohamad, O. A. A., Li, L., Ma, J. B., Hatab, S., Xu, L., Gua, J. W., et al. 2018. Evaluation of the antimicrobial activity of endophytic bacterial populations from Chinese traditional medicinal plant licorice and characterization of the bioactive secondary metabolites produced by Bacillus atrophaeus against Verticillium dahlia. *Front. Microbiol.* 9, 924. doi: 10.3389/fmicb.2018.00924

Nadeem, M. et al. 2012. Fusarium Solani, P1, a new endophytic Podophyllotoxin-producing fungus from roots of Podophyllum hexandrum. *Afr. J. Microbiol. Res.* 6, 2493–2499.

Nagai, H., and Kim, Y.H. Cancer prevention from the perspective of global cancer burden patterns. *J. Thorac. Dis.* 2017, 9, 448.

Naik, B.S. Developments in taxol production through endophytic fungal biotechnology: A review. *Orient. Pharm. Exp. Med.* 2019, 19, 1–13.

Nalli, Y., Arora, P., Khan, S., Malik, F., Riyaz-Ul-Hassan, S., Gupta, V., and Ali, A. 2019. Isolation, structural modification of macrophin from endophytic fungus Phoma Macrostoma and their cytotoxic potential. *Med. Chem. Res.* 28, 260–266.

Newman, D.J., and Cragg, G.M. 2020. Plant endophytes and epiphytes: Burgeoning sources of known and "unknown" cytotoxic and antibiotic agents? *Planta Med.* 86, 1095–1111.

Niculescu, A.G., and Grumezescu, A.M. 2022. Novel tumour-targeting nanoparticles for cancer treatment—A review. *IJMS*, 23, 5253.

Nivetha, L., and Kharwar, R. 2019. Antimicrobial and antioxidant activities of endophytes from Citrus Limon. *J. Med. Plant Res.* 7(6), 10–13.

Oh, C.M., Lee, D., Kong, H.J., Lee, S., Won, Y.J., Jung, K.W., and Cho, H. 2020. Causes of death among cancer patients in the era of cancer survivorship in Korea: Attention to the suicide and cardiovascular mortality. *Cancer Med.* 9, 1741–1752.

Omeje, E.O., et al. 2017. Endophytic fungi as alternative and reliable sources for potent anticancer agents. *Nat Prod Cancer Drug Discov.* 7, 141.

Palanichamy, P., Krishnamoorthy, G., Kannan, S., and Marudhamuthu, M. 2018. Bioactive potential of secondary metabolites derived from medicinal plant endophytes. *Egypt. J. Basic Appl. Sci.* 5, 303–312.

Palem, P.P.C., Kuriakose, G.C., and Jayabaskaran, C. 2015. An endophytic fungus, talaromyces radicus, isolated from catharanthus roseus, produces vincristine and vinblastine, which induce apoptotic cell death. *PloS One* 10, e0144476.

Pan, R., Bai, X., Chen, J., Zhang, H., and Wang, H. 2019. Exploring structural diversity of microbe secondary metabolites using OSMAC strategy: A literature review. *Front. Microbiol.* 10, 294.

Pang, X.J., et al. 2018. Two sesquiterpenoids with an unprecedented tricyclo [4,4,2,1]hendecane scaffold from the liquid cultures of endophytic fungus Emericella sp. XL 029. *Fitoterapia* 131, 55–58.

Parmar, P., et al. 2020. The rise of gingerol as anti-QS molecule: Darkest episode in the LuxR-mediated bioluminescence saga. *Bioorg. Chem.* 99, 103823.

Patridge, E., Gareiss, P., Kinch, M.S., and Hoyer, D. 2016. An analysis of FDA-approved drugs: Natural products and their derivatives. *Drug Discov. Today* 21, 204.

Porzel, A., Farag, M. A., Mülbradt, J., and Wessjohann, L. A. 2014. Metabolite profiling and fingerprinting of Hypericum species: A comparison of MS and NMR metabolomics. *Metabolomics* 10, 574–588. doi: 10.1007/s11306-013-0609-7

Prajapati, J., Patel, R., Goswami, D., Saraf, M., and Rawal, R.M. 2021. Sterenin M as a potential inhibitor of SARS-CoV-2 main protease identified from MeFSAT database using molecular docking, molecular dynamics simulation and binding free energy calculation. *Comput. Biol. Med.* 135, 104568.

Prakash, V., Rana, S., and Sagar, A. 2016. Taxomyces andreanae: A source of anticancer drug. *Int. J. Bot. Stud.* 1, 43–46.

Purawarga Matada, G.S., Dhiwar, P.S., Abbas, N., Singh, E., Ghara, A., Das, A., and Bhargava, S.V. 2022. Molecular docking and molecular dynamic studies: Screening of phytochemicals against EGFR, HER2, estrogen and NF-KB receptors for their potential use in breast cancer. *J. Biomol. Struct. Dyn.* 40(13), 6183–6192.

Puri, S.G., Verma V., Amna T.Q.G.N., and Spiteller, M. 2005. An endophytic fungus from Nothapodytes foetida that produces camptothecin. *J. Nat. Prod.* 68, 1717–1719.

Puri, S.C., et al. 2006. The endophytic fungus Trametes hirsuta as a novel alternative source of podophyllotoxin and related aryl tetralin lignans. *J. Biotechnol.* 122, 494–510.

Qiao, W., Ling, F., Yu, L., Huang, Y., and Wang, T., 2017. Enhancing taxol production in a novel endophytic fungus, Aspergillus aculeatinus Tax-6, isolated from Taxus chinensis var. mairei. *Fungal. Biol.* 121, 1037–1044.

Ran, X., Zhang, G., Li, S., and Wang, J., 2017. Characterization and antitumor activity of camptothecin from endophytic fungus *Fusarium solani* isolated from *Camptotheca acuminate. Afr. Health Sci.* 17, 566–574.

Ranjith, D., and Ravikumar, C., 2019. SwissADME predictions of pharmacokinetics and drug-likeness properties of small molecules present in Ipomoea mauritiana Jacq. *J. Pharmacogn. Phytochem.* 8(5), 2063–2073.

Sagita, R., Quax, W.J., and Haslinger, K. 2021. Current state and future directions of genetics and genomics of endophytic fungi for bioprospecting eforts. *Front. Bioeng. Biotechnol.* 15(9), 649906.

Sandargo, B., Chepkirui, C., Cheng, T., Chaverra-Muñoz, L., Thongbai, B., Stadler, M., and Hüttel, S. 2019. Biological and chemical diversity go hand in hand: Basidiomycota as source of new pharmaceuticals and agrochemicals. *Biotechnol. Adv.* 37, 107344. [PubMed]

Schirrmacher, V. 2019. From chemotherapy to biological therapy: A review of novel concepts to reduce the side effects of systemic cancer treatment. *Inter. J. Oncol.* 54, 407–419.

Segers, K., Declerck, S., Mangelings, D., Heyden, Y.V., and Eeckhaut, A.V. 2019. Analytical techniques for metabolomic studies: A review. *Bioanalysis* 11(24), 2297–2318.

Senthil Kumar, V., Kumaresan, S., Tamizh, M.M., Hairul Islam, M.I., and Thirugnanasambantham, K. 2019. Anticancer potential of NFKB targeting apoptotic molecule "Flavipin" isolated from endophytic Chaetomium Globosum. *Phytomedicine* 61, 152830.

Shah, S.C., Kayamba, V., Peek, R.M., and Heimburger, D. 2019. Cancer control in low-and middle-income countries: Is it time to consider screening? *J. Glob. Oncol.* 5, 1–8.

Sharma, V., Singamaneni, V., Sharma, N., Kumar, A., Arora, D., Kushwaha, M., Bhushan, S., Jaglan, S., and Gupta, P. 2018. Valproic acid induces three novel cytotoxic secondary metabolites in Diaporthe Sp., an endophytic fungus from Datura Inoxia Mill. *Bioorg. Med. Chem. Lett*. 28, 2217–2221.

Shulaev, V., Cortes, D., Miller, G., and Mittler, R. 2008. Metabolomics for plant stress response. *Physiol. Plant*. 132, 199–208. doi: 10.1111/j.1399-3054.2007.01025.x

Singh, A.N., Baruah, M.M., and Sharma, N., 2017. Structure Based docking studies towards exploring potential anti-androgen activity of selected phytochemicals against Prostate Cancer. *Sci. Rep*. 7(1), 1–8.

Staniek, A., et al. 2014. Natural products–learning chemistry from plants. *Biotechnol. J*. 9, 326–336.

Stierle, A., Strobel, G., and Stierle, D. 1993. Taxol and Taxane Production by Taxomyces Andreanae, an endophytic fungus of Pacific Yew. *Sci.-N. Y. Then Wash*. 260, 214.

Strobel, G. 2018. The emergence of endophytic microbes and their biological promise. *J. Fungi* 4, 57.

Uzma, F., et al. 2018. Endophytic fungi—Alternative sources of cytotoxic compounds: A review. *Front. Pharmacol*. 9, 309.

Varghese, S., Akshaya, C.S., and Jisha, M.S., 2021. Unravelling the bioprospects of mycoendophytes residing in withania somnifera for productive pharmaceutical applications. *Biocatal. Agric. Biotechnol*. 37, 102172.

Varghese, S., Jisha, M.S., Rajeshkumar, K.C., Gajbhiye, V., Haldar, N., and Shaikh, A., 2024. Molecular authentication, metabolite profiling and *in silico–in vitro* cytotoxicity screening of endophytic *Penicillium ramusculum* from *Withania somnifera* for breast cancer therapeutics. *3 Biotech* 14(3), 64.

Varghese, E., Samuel, S.M., Líšková, A., Samec, M., Kubatka, P., and Büsselberg, D. 2020. Targeting glucose metabolism to overcome resistance to anticancer chemotherapy in breast cancer. *Cancers* 12, 2252.

Venugopalan, A., and Srivastava, S. 2015. Endophytes as in vitro production platforms of high value plant secondary metabolites. *Biotechnol. Adv*. 33(Pt. 1), 873–887. doi: 10.1016/j.biotechadv.2015.07.004

Wagenaar, M.M., Corwin, J., Strobel, G., and Clardy, J. 2000. Three new cytochalasins produced by an endophytic fungus in the genus Rhinocladiella. *J Nat Prod*. 63, 1692–1695.

Wang, H., et al. 2018. RAVEN 2.0: A versatile toolbox for metabolic network reconstruction and a case study on Streptomyces coelicolor. *PLoS Comput. Biol*. 14, 1–17.

Wang, W.X., Li, Z.H., Ai, H.L., Li, J., He, J., Zheng, Y.S., Feng, T., and Liu, J.K. 2019a. Cytotoxic 19,20-Epoxycytochalasans from Endophytic Fungus Xylaria Cf. Curta. *Fitoterapia* 137, 104253.

Wang, W.X., Zheng, M.J., Li, J., Feng, T., Li, Z.H., Huang, R., Zheng, Y.S., Sun, H., Ai, H.L., and Liu, J.K. 2019b. Cytotoxic polyketides from endophytic fungus Phoma Bellidis Harbored in Ttricyrtis Maculate. *Phytochem. Lett*. 29, 41–46.

Wei, J., Chen, F., Liu, Y., Abudoukerimu, A., Zheng, Q., Zhang, X., Sun, Y., and Yimiti, D. 2020. Comparative metabolomics revealed the potential antitumor characteristics of four endophytic fungi of Brassica rapa L. *ACS Omega* 5(11), 5939–5950.

Wen, S., Fan, W., Guo, H., Huang, C., Yan, Z., and Long, Y. 2020. Two New Secondary Metabolites from the Mangrove Endophytic Fungus Pleosporales sp. SK7. *Nat. Prod. Res*. 34, 2919–2925.

Wilson, B.A., Thornburg, C.C., Henrich, C.J., Grkovic, T., and O'Keefe, B.R. 2020. Creating and screening natural product libraries. *Nat. Prod. Rep*. 37, 893–918.

Wright, G.D. 2019. Unlocking the potential of natural products in drug discovery. *Microb. Biotechnol*. 12, 55–57.

Wu, Y., Chen, S., Liu, H., Huang, X., Liu, Y., Tao, Y., and She, Z. 2019. Cytotoxic Isocoumarin Derivatives from the Mangrove Endophytic Fungus Aspergillus sp. HN15-5D. *Arch. Pharm. Res*. 42, 326–331.

Xiao, Z., Morris-Natschke, S.L., and Lee, K.H. 2016. Strategies for the optimization of natural leads to anticancer drugs or drug candidates. *Med Res Rev*. 36, 32–91.

Xin, X.Q., Chen, Y., Zhang, H., Li, Y., Yang, M.H., and Kong, L.Y. 2019. Cytotoxic Seco-Cytochalasins from an Endophytic Aspergillus sp. Harbored in Pinellia Ternata Tubers. *Fitoterapia* 132, 53–59.

Yang, X., Zhang, L., Guo, B., and Guo, S. 2004. Preliminary study of a vincristine-producing endophytic fungus isolated from leaves of Catharanthus roseus. *Zhongcaoyao* 35, 79–81.

Yu, X., Müller, W.E.G., Meier, D., Kalscheuer, R., Guo, Z., Zou, K., Umeokoli, B.O., Liu, Z., and Proksch, P. 2020. Polyketide derivatives from mangrove derived endophytic fungus Pseudopestalotiopsis Theae. *Mar. Drugs* 18, 129.

Zhang, H. W., Song, Y. C., and Tan, R. X. 2006. Biology and chemistry of endophytes. *Nat. Prod. Rep*. 23, 753–771.

Zhang, P., et al. 2019. Angularly prenylated indole alkaloids with antimicrobial and insecticidal activities from an endophytic fungus Fusarium sambucinum TE-6L. *J. Agric. Food Chem.* 67, 11994–12001.

Zhao, J., Mou, Y., Shan, T., Li, Y., Zhou, L., Wang, M., et al. 2010. Antimicrobial metabolites from the endophytic fungus Pichia guilliermondii isolated from *Paris polyphylla* var. *yunnanensis*. *Molecules* 15, 7961–7970. doi: 10.3390/molecules15117961

8 Harnessing the Therapeutic Potential of Endophytic Microbes in Medicinal Plants

Bioactives and Beyond

*Smita Rakshit, Sachin Awachar, Aniket Gade and Kapil Kamble**
*Corresponding author.

8.1 INTRODUCTION

Endophytes are microbes that live asymptomatically inside plant tissues without causing any negative effects. A complete definition of endophytes doesn't specify their exact role. They can range from friendly symbionts to potential pathogens, saprotrophs, or mutualistic partners. Predominant endophytes include bacteria, archaea, and fungi. The term endophyte was given by De Bary way back in 1866, but the first report about them appeared in 1904 (Tidke et al., 2017). The earliest mention of them was found in darnel, *Lolium temulentum.* Freeman (1904) made the discovery of endophytes when he found an endophytic fungus in this annual grass (Kandasamy & Kathirvel, 2023). The expanse of endophytic study is extensive; however, this chapter concentrates on elucidating the beneficial roles of endophytes and the bioactive compounds concomitant with plant interactions.

In general, endophytes influence the host plant through two basic mechanisms (Wani et al., 2015). First, endophytes exert their impact through the synthesis of a diverse array of chemicals. Among these compounds, phytohormones are recognized for their ability to promote plant growth. At the same time, endophytes' generation of oxygen radicals is known to provide a defense against oxidative stress (Lucas et al., 2014). Additionally, the synthesis of various secondary metabolites, including toxicants and antibiotics, serves to fortify plants against potential pathogens. Second, endophytes stimulate the production of genes that are essential for plant growth. Genes linked to nitrogen and carbon metabolism, defense systems, and general plant growth are reported to be upregulated when endophytes colonize plants. The particular genes that endophytes activate depend on their kinds and the plant species that they infect (Li et al., 2023) (Figure 8.1).

Endophytic bacteria produce a variety of bioactive compounds, some of which may be crucial in the creation of novel medications, in addition to providing the host plant with substantial support. These substances include those with antibacterial, antioxidant, and anticancer properties. There is always more to learn about the intricate relationships that exist between medicinal plants and endophytic bacteria, and new therapeutic prospects are frequently found. Every symbiotic interaction that exists between microorganisms and plants presents an opportunity to find new bioactive chemicals that have particular health benefits and offer innovative solutions to pressing medical problems. To ascertain their pharmacological potential, the bioactives mentioned below may be studied in a variety of clinical trials. (Kusari et al., 2014; Giridharan et al., 2012; Wijeratne et al., 2008).

DOI: 10.1201/9781003486275-10

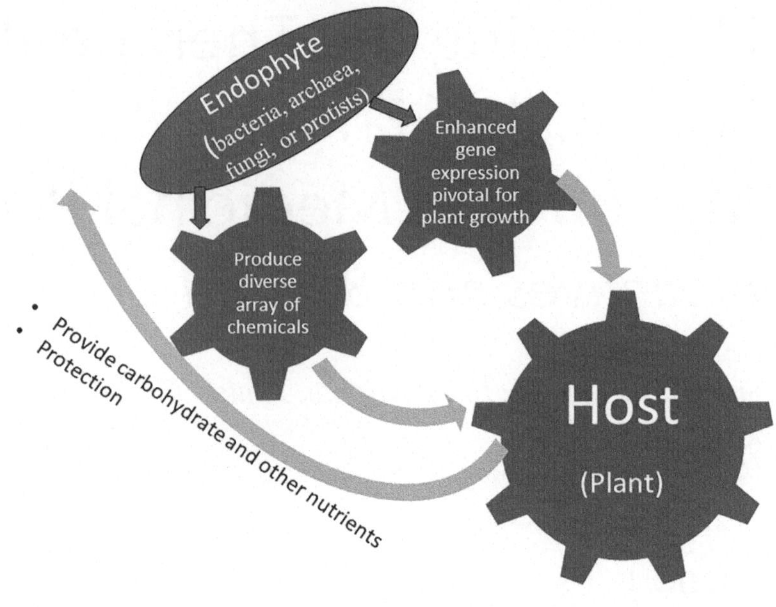

FIGURE 8.1 Mutualistic relationship between host (plant) and endophyte.

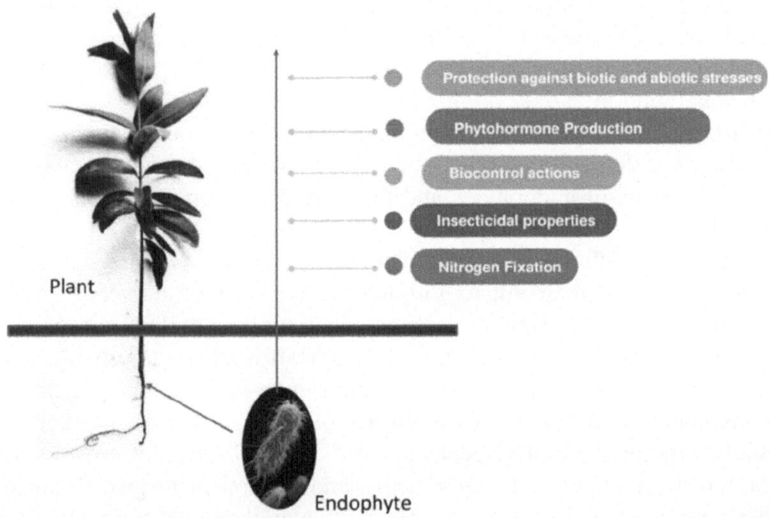

FIGURE 8.2 Role of endophytes in crop plants.

8.2 NONPATHOGENIC ROLES OF ENDOPHYTES

In nature, the reciprocity phenomenon is observed between plants and microbes (Figure 8.2). This symbiotic relationship is as ancient as the colonization of plants on Earth. In both agricultural and natural ecosystems, plants derive benefits from their association with microbes (Dolatabadian, 2020). The intricate symbiotic associations between diverse endophytic microorganisms and their respective host plants, as detailed in Table 8.1, unveil a spectrum of roles within the framework of plant–microbe association. The recognition of symbiotic nitrogen-fixing bacteria and their profound impact on plant nutrition dates back several decades. Pioneering studies in the early stages set the

TABLE 8.1
Benefits of Endophytes to the Host Plant

S. No.	Endophyte	Plant	Role	References
1	*Acetobacter diazotrophicus*	Sugarcane (*Saccharum officinarum* L)	Nitrogen fixation	Dong et al. (1994)
2	*Bradyrhizobium japonicum*	Soybean (*Glycine max*)	Nitrogen fixation	Bai et al. (2002)
3	*Bacillus cereus*, *Bacillus subtilis*	*Teucrium polium* L *Zea mays* L	Synthesis of IAA, ammonia, and the solubilization of phosphate	Hassan (2017)
4	*Brevibacillus agri*	*Phaseolus vulgaris* L.	Production of IAA and a number of hydrolytic enzymes, including as pectinase, cellulase, amylase and protease	Ismail et al. (2021)
5	*Penicillium* sp., *Aspergillus* sp.	*Taxus wallichiana*	Phosphate solubilization and phytase enzyme production	Adhikari & Pandey (2019)
4	*Enterobacter* sp.	*Capsicum chinense*	Phosphate solubilization	Mendoza-Arroyo et al. (2020)
5	*Trichoderma* sp. *Pichia guilliermondi*	Corn, tomato, pepper and watermelon	Enhanced shoot growth and biomass of tomato plants, resulting in increased fruit yields	Xia et al. (2019)
6	*Sphingomonas* sp. and *Serratia marcescens*	Soybean (*Glycine max*)	Modulate the levels of endogenous hormones and stimulate growth in crop plants	Asaf et al. (2017)
7	*Bacillus subtilis*	Lettuce plants	Synthesis of cytokinins with the aim of modulating both growth and endogenous hormone levels	Arkhipova et al. (2005)
8	*Bacillus amyloliquefaciens*	Rice (*Oryza sativa* L.)	Production of gibberellins and improving host-plant physiology	Shahzad et al. (2016)
9	*Penicillium* sp. *Aspergillus* sp.	*Monochoria vaginalis*	Synthesis of gibberellins	Ahmad et al. (2010)
10	*Trichoderma*	*Theobroma cacao*	Biocontrol of *M. roreri* causing frosty pod rot disease	Mejía et al. (2008)
11	*T. harzianum*, *P. aphanidermatum*	*Curcuma longa* L.	Demonstrating antagonistic efficacy against the pathogens that cause rhizome rot and leaf blight such as *Rhizoctonia solani* Kuhn and *Pythium aphanidermatum* (Edson) Fitzp.	Vinayarani & Prakash (2018)
12	*Beauveria bassiana*	*Phaseolus vulgaris*	Protection of plants from herbivory and diseases	Parsa et al. (2013)
13	*Bacillus subtilis*, *Pseudomonas fluorescens*	Cotton	Induction of systemic resistance without causing any impact to the environment	Rajendran et al. (2011)
14	*Rhizobacteria* (PGPR)	Rice	Enhancing resilience to both biotic and abiotic stressors through the induction of systemic resistance.	Lucas et al. (2014)
15	*Streptomyces albidoflavus*	*Azadirachta indica*	Exhibits the insecticidal activity	Chen et al. (2018)
16	*Sarocladium strictum*	*Cynanchum acutum*	Demonstrating effective insecticidal activity in opposition to *S. littoralis*	El-Sayed et al. (2020)

groundwork for comprehending the molecular interaction that underlie this symbiosis (Dolatabadian, 2020). The mutualistic association between certain bacteria, predominantly belonging to genera such as *Rhizobium*, *Bradyrhizobium* and *Azorhizobium*, and leguminous plants has long been acknowledged. This symbiotic relationship is characterized by the establishment of nodules on the plant roots, wherein the bacteria actively engage in biological nitrogen fixation.

Through the process of biological nitrogen fixation, atmospheric nitrogen is enzymatically transformed into ammonia, a form of nitrogen that plants can readily absorb. Within the specialized nodules, the symbiotic bacteria harbor nitrogenase enzymes, which catalyze the complex series of reactions leading to the conversion of atmospheric nitrogen into a biologically accessible form. In return for this essential nitrogen supply, the host plants provide the symbiotic bacteria with carbohydrates and other nutrients, fostering a mutually beneficial exchange. Noteworthy nitrogen-fixing endophytes include *Acetobacter diazotrophicus* inhabiting sugarcane (*Saccharum officinarum* L) and *Bradyrhizobium japonicum* in association with soybean (*Glycine max*). Dong et al. (1994) and Bai et al. (2002) contribute foundational insights into the nitrogen fixation studies of these endophytes, elucidating their imperative role in augmenting plant nitrogen availability and stimulating overall growth.

One important phytohormone that is involved in controlling many aspects of plant growth and development is indole acetic acid (IAA). The production of IAA by endophytes plays a significant role in regulating auxin levels in plants, which in turn affects various processes including root development, cell elongation, and overall plant architecture. *Bradyrhizobium japonicum* and *Acetobacter diazotrophicus* are two examples of endophytes that are known to synthesize ammonia, which may provide the host plant with additional nitrogen (Bai et al., 2002; Dong et al., 1994). Phosphate solubilization is another critical function performed by select endophytes. The ability to convert insoluble forms of phosphate into bioavailable forms directly influences the plant's phosphorus uptake. Phosphate is an essential nutrient for various cellular processes, including energy transfer and nucleic acid synthesis. Endophytes proficient in phosphate solubilization contribute to improving the phosphorus status of their host plants, thereby influencing their growth and metabolic activities. *Bacillus cereus* and *B. subtilis*, operational in felty germander and maize (Hassan, 2017), showcased versatility by manifesting these functionalities. *Brevibacillus agri*, residing within common bean, not only generates IAA but also synthesizes various hydrolytic enzymes (Ismail et al., 2021). *Penicillium* sp. and *Aspergillus* sp. are involved in phosphate solubilization and phytase enzyme production in *Taxus wallichiana* (Adhikari & Pandey, 2019). *Enterobacter* sp. are reported for phosphate solubilization within *Capsicum chinense* (Mendoza-Arroyo et al., 2020).

One important aspect of plant–microbe interactions is the generation of cytokinins by endophytic microbes, which reveals a dynamic biochemical connection within the host plant. As phytohormones, cytokinins are known for playing a key role in controlling a number of physiological processes that are essential to the growth and development of plants. The complexity of these symbiotic connections and their potential implications for plant health are highlighted by the emerging research that shows endophytes may produce and control cytokinin levels (Arkhipova et al., 2005).

One of the prominent positive impacts of cytokinins lies in their ability to stimulate cell division. Cytokinins act as mitogens, promoting the cell cycle and encouraging the generation of new cells. This pro-mitotic effect is particularly pronounced in meristematic tissues, such as shoot and root apical meristems, where cytokinins facilitate the formation of new organs, including leaves and lateral shoots. *Bacillus subtilis* in lettuce plants is reported to enhance plant growth through cytokinin production (Arkhipova et al., 2005).

Another important phytohormone produced by endophytes is gibberellin. The production of gibberellins by endophytes is hypothesized to modulate various aspects of plant growth. Gibberellins are recognized for their involvement in promoting stem elongation, seed germination and flowering, among other developmental processes. *Bacillus amyloliquefaciens* in rice and *Penicillium* sp. and

Aspergillus sp. in *Monochoria vaginalis* exert a positive influence through gibberellin production (Shahzad et al., 2016; Ahmad et al., 2010).

Through a variety of intricate interactions, endophytes restrict the growth, colonization and pathogenicity of phytopathogens. Numerous processes, including the generation of secondary metabolites that are antimicrobial, competition for nutrients and space, and modification of host plant immunological responses, contribute to the inhibition. The array of secondary metabolites that endophytes employ is noteworthy. These chemicals include terpenoids, phenolics and alkaloids, all of which have antibacterial qualities. These bioactive compounds provide a protective barrier for the host plant, reducing the growth of phytopathogens and lessening the harmful effects of several illnesses. Additionally, the plant's resistance is strengthened by endophytic competition for scarce resources and spatial occupation inside plant tissues, which suppresses phytopathogenic pathogens (Mohamad et al., 2018; Mamarasulov et al., 2023).

In diverse plant–pathogen interactions, various biocontrol agents showcase their efficacy in disease management. *Trichoderma*, when applied to *Theobroma cacao*, demonstrates biocontrol capabilities against *Moniliophthora roreri*, the causal agent of frosty pod rot disease (Mejía et al., 2008). Additionally, the synergistic action of *T. harzianum* and *P. aphanidermatum* proves effective in mitigating diseases caused by *Pythium aphanidermatum* and *Rhizoctonia solani* in *Curcuma longa* (Vinayarani & Prakash, 2018). *Beauveria bassiana* provides protection to *Phaseolus vulgaris* by acting against both herbivory and diseases (Parsa et al., 2013). Another notable example is the collaborative approach involving *Bacillus subtilis* and *Pseudomonas fluorescens*, which induces systemic resistance in cotton (Rajendran et al., 2011). These instances underscore the diverse strategies employed by biocontrol agents, ranging from fungal biopesticides to bacterial consortia, in safeguarding plants against various biotic stresses.

Moreover, endophytes extend their protective influence to abiotic stresses, including factors like drought, salinity and nutrient deficiencies. By fostering enhanced nutrient uptake, promoting water retention, and modulating stress-related signaling pathways, endophytes assist the host plant in coping with adverse environmental conditions. In rice, plant growth-promoting rhizobacteria (PGPR) confer protection against both biotic and abiotic stress, inducing systemic resistance (Lucas et al., 2014).

The insecticidal activity exhibited by endophytes is often attributed to the production of bioactive compounds with pesticidal properties. These secondary metabolites, synthesized by endophytes, can act as deterrents, feeding inhibitors or outright toxins against a spectrum of insect pests. Examples include alkaloids, terpenoids and phenolic compounds, which have been identified as key players in mediating the insecticidal effects of endophytes. Noteworthy instances of endophyte-induced insecticidal activity include *Streptomyces albidoflavus*, exhibiting potent insecticidal effects in *Azadirachta indica* (Chen et al., 2018). This highlights the capacity of endophytes to enhance the plant's defense mechanisms against herbivores, consequently impacting pest population dynamics. Similarly, *Sarocladium strictum* has been identified for its efficient insecticidal activity against *Spodoptera littoralis* in *Cynanchum acutum*, demonstrating the diverse repertoire of endophytic contributions to insect pest control (El-Sayed et al., 2020). In the context of sustainable environment and agriculture, endophytes have garnered attention for their potential to improve plant growth while reducing the reliance on synthetic inputs. Additionally, endophytes can contribute to the suppression of plant pathogens through the production of antimicrobial compounds, offering a natural and sustainable alternative to traditional disease management strategies.

8.3 BIOACTIVES OF ENDOPHYTES FOR MEDICINAL PURPOSES

It has been shown that endophytes are significant producers of a variety of secondary metabolites. The intended goal is not achieved when these metabolites are extracted using only isolated endophytes. Numerous external and internal factors, such as the host plant's properties, geographic location, climate and sample collection schedule, affect the synthesis of bioactives. The extensive regulatory network governing endophytic secondary metabolite production is complex and nuanced, as highlighted by the multiple interplay of various factors. A comprehensive understanding of this network is necessary (Morales-Sánchez et al., 2020).

8.4 BIOACTIVES OF ENDOPHYTES POSSESSING THERAPEUTIC PROPERTIES

It was thought that the valuable chemotherapeutic drug paclitaxel could only be produced by the *Taxus brevifolia* plant. However, it was later discovered that the fungal endophyte called *Taxomyces andreanae*, living inside the same plant, is responsible for synthesizing paclitaxel (Prajapati et al., 2021). This has accelerated the study of endophytic fungal bioactives against cancer. Paclitaxel is later studied against various types of cancers like lung, ovarian, head and neck carcinoma. This emphasizes the pharmaceutical significance of endophytes in providing natural sources for widely used therapeutic compounds (Kusari et al., 2014).

The endophyte *Cephalotheca faveolata* in *Eugenia jambolana* produces sclerotiorin, demonstrating activity in inducing apoptosis in colon cancer cells. This finding underscores the potential role of endophytes in cancer research and therapeutic applications (Giridharan et al., 2012). Zhang et al. (2010) identified anthracenedione from *Halorosellinia* sp. and *Guignardia* sp. in mangroves, showcasing anticancer activity. Tauranin from *Phyllosticta spinarum* in *Platycladus orientalis* inhibited cell proliferation in cancer cell lines (Wijeratne et al., 2008). Zhou et al. (2017) identified various bioactives, including 3-amino-5-hydroxy-5-vinyl-2-cyclopenten-1-one dimer, atrichodermone A, atrichodermone B and atrichodermone C, from *Trichoderma atroviride* in *Lycoris radiate*. These compounds demonstrated cytotoxicity against specific cell lines and an antiinflammatory effect, showcasing the versatility of endophytes. Hu et al. (2016) reported the presence of homoharringtonine (alkaloid) from *Alternaria tenuissima* in *Cephalotaxus hainanensis*, exhibiting antitumor activity. Budhiraja et al. (2013) identified 6-methyl-1,2,3-trihydroxy-7,8-cyclohepta-9,12-diene-11-one-5,6,7,8-tetralene-7-acetamide from *Aspergillus* sp. in *Gloriosa superba*, displaying antitumor activity. Eremophilanolides from *Xylaria* sp. in *Licuala spinosa* exhibited strong antioxidant and anticancer activities (Isaka et al., 2010). Nongkhlaw and Joshi (2015) identified gallic acid and quercetin from *Bacillus mycoides* and *Citrobacter youngae* in *Rubia cordifolia*. These compounds demonstrated antioxidant activity and antagonistic effects against various pathogens, highlighting the multifaceted roles of endophytes. Anwar et al. (2020) reported carbohydrates, alkaloids, flavonoids, glycosides, saponins, steroids, tannins and triterpenoids from *Pseudomonas* sp. in *Cissus quadrangularis*. These compounds showcased antioxidant and anti-osteosarcoma potential. Endophytic fungi hold great promise for discovering new bioactive molecules with potential uses in health care. To fully unlock their value, it is crucial to understand how these fungal substances interact with specific cell targets to produce their biological effects (Table 8.2).

8.5 BIOACTIVES OF ENDOPHYTES AS ANTIMICROBIAL AGENTS

Endophytes have gained attention for their bioactive compounds with remarkable antimicrobial actions (Table 8.3). These endophyte-produced bioactives have the capacity to stop the spread of a wide range of harmful microbes, such as viruses, fungi and bacteria. Endophytic bioactives are thought to possess antibacterial properties because of the variety of secondary metabolites they produce during their interactions with host plants (Semenzato et al., 2023). Examples include *Rhizoctonia solani* associated with *Cyperus rotundus*, producing solanioic acid with bactericidal activity against methicillin-resistant *Staphylococcus aureus* (MRSA), and *Chaetomium globosum* in *Nymphaea nouchali* producing chaetoglobosin A and C with antibacterial activities. Multiple endophytes collectively exert cumulative antimicrobial actions, exemplified by *Bacillus gibsonii*, *Methylobacterium radiotolerans* and others in *Centella asiatica*. Additionally, various endophytes

TABLE 8.2
Therapeutic Potential of Bioactives of Endophytes

S. No.	Endophyte	Medicinal plant	Bioactives	Activity	References
1	*Cephalotheca faveolata*	*Eugenia jambolana*	Sclerotiorin	Triggering apoptosis in colon cancer cells	Giridharan et al. (2012)
2	*Halorosellinia* sp. *Guignardia* sp.	*Mangrove*	Anthracenedione	Anticancer activity	Zhang et al. (2010)
3	*Taxomyces andreanae* and other several species	*Taxus brevifolia*	Taxol (paclitaxel)	Therapeutic intervention against various cancer types, including lung cancer, ovarian cancer, head and neck carcinoma, and other malignancies	Kusari et al. (2014)
4	*Phyllosticta spinarum*	*Platycladus orientalis*	Tauranin (C22H30O4)	Hindrance of cell proliferation in various cancer cell lines	Wijeratne et al. (2008)
5	*Trichoderma atroviride*	*Lycoris radiate*	3-Amino-5-hydroxy-5-vinyl-2-cyclopenten-1-one dimer, atrichodermone A, atrichodermone B and atrichodermone C	Exerting cytotoxic effects on HL60 and U937 cell lines	Zhou et al. (2017)
6	*Alternaria tenuissima*	*Cephalotaxus hainanensis*	Homoharringtonine (alkaloid)	Demonstrating antitumor activity and effectiveness in the treatment of human chronic myeloid leukemia	Hu et al. (2016)
7	*Aspergillus* sp.	*Gloriosa superba*	6-Methyl-1,2,3-trihydroxy-7, 8-cyclohepta-9,12-diene-11-one-5,6,7,8-tetralene-7-acetamide	Antitumor activity	Budhiraja et al. (2013)
8	*Xylaria* sp.	*Licuala spinosa*	Eremophilanolides	Strong antioxidant activity, anticancer activity	Isaka et al. (2010)
9	*Bacillus mycoides*, *Citrobacter youngae*	*Rubia cordifolia*	Gallic acid, quercetin Antioxidant and antiosteosarcoma potential	Exhibiting antioxidant activity and antagonistic effects against *T. rubrum*, *C. albicans*, *E. coli* and *S. aureus*	Nongkhlaw & Joshi (2015)
10	*Pseudomonas* sp.	*Cissus quadrangularis*	Carbohydrates, alkaloids, flavonoids, glycosides, saponins, steroids, tannins, triterpenoids	Antioxidant and antiosteosarcoma potential	Anwar et al. (2020)

TABLE 8.3
Antimicrobial Potential of Bioactives of Endophytes

S. No.	Endophyte	Medicinal plant	Bioactives	Activity	References
1	*Rhizoctonia solani*	*Cyperus rotundus*	Solanioic acid	Inhibitory activity against methicillin-resistant *Staphylococcus aureus* (MRSA)	Ratnaweera et al. (2015)
2	*Chaetomium globosum*	*Nymphaea nouchali*	Chaetoglobosin A and C	The compounds shows antibacterial efficacy against *Staphylococcus aureus* and *Bacillus cereus.*	Dissanayake et al. (2016)
3	*B. gibsonii, M. radiotolerans, P. agglomerans, P. agglomerans, E. tasmaniensis, E. soli, P. vermicola, P. fulva, Xanthomonas axonopodis*	*Centella asiatica* (L)	Cumulative action	Antibacterial against *E. coli, B. cereus, S. aureus* and *P. aeruginosa*	Rafat et al. (2012)
4	*A. alternata* AE1	*Azadirachta indica* A. Juss	Phenolics and favonoids	Bactericidal and antioxidant potential	Chatterjee et al. (2019)
5	*Paenibacillus, Bacillus, Brevundimonas, Kytococcus* and *Microbacterium*	*Ephedra foliata*	Alkaloid, aminoglycoside, flavonoid and terpenoid compounds	Antibacterial efficacy against *S. aureus, B. subtilis, P. aeruginosa* and *E. coli.* Exhibited activity against the cancerous cell line	Ghiasvand et al. (2020)
6	*Micrococcus, Bacillus, Pseudomonas, Stenotrophomonas, Pantoea* genus	*Vernonia anthelmintica, Saurauia scaberrinae* and *Chromolaena odorata*	β-Amylase, α-amylase, cellulase, chitinase and protease	Effective against *S. aureus, P. ultimum, S. sclerotiorum* and *R. solani*	Mamangkey et al. (2022)
7	*Arthrobacter* sp. OVS8	*Origanum vulgare*	β-Caryophyllene	Antagonistic activity against multidrug-resistant phenotype of *K. pneumoniae* strains	Semenzato et al. (2023)
8	*Aspergillus terreus*	*Artemisia annua*	Acetyl choline esterase	The compounds inhibited acetylcholinesterase and showed cytotoxicity against KB and HSC-T6 cell lines	Ge et al. (2010)

TABLE 8.3 (Continued)
Antimicrobial Potential of Bioactives of Endophytes

S. No.	Endophyte	Medicinal plant	Bioactives	Activity	References
9	*Firmicutes, Proteobacteria, Actinobacteria, Bacillus amyloliquefaciens*	*Salicornia brachiata*	Lipopeptide class based on ^{1}H-, ^{13}C	Exhibited versatile antibacterial and antifungal properties against various human, plant and aquatic infectious agents	Singh et al. (2021)
10	*Bacillus atrophaeus, Bacillus mojavensis*	*Glycyrrhiza uralensis*	1,2-Benzenedicarboxylic acid, methyl-ethyl esters	Control plant pathogens causing verticillium wilt disease in cotton and other crops	Mohamad et al. (2018)
11	*Bacillus pumilus, B. safensis*	*Thymus vulgaris, Mentha spicata*	Butanal, 3-methyl-, propene, butene, heptanone, 6-methyl-5-methylene- and 6-oxabicyclo hexane	Biocontrol of *Penicillium citrinum, Aspergillus fumigatus, Aspergillus flavus, Fusarium oxysporum* and *Rhizopus stolonifer* by antifungal activity	Erjaee et al. (2019)
12	*B. subtilis* ssp. (M10), *B. mojavensis* (M11), *B. amyloliquefaciens* (M13)	*Ajuga turkestanica* (Lamiaceae)	Mucolytic enzymes such as chitinase, protease and 1,3-glucanase	Antimicrobial activity against pathogenic test strains and phytopathogenic plant fungi	Mamarasulov et al. (2023)
13	*Bacillus, Aneurinibacillus, Pseudomonas* spp.	*Tinospora cordifolia*	Amylase, cellulase, gelatinase and catalase	Antiviral, anticancerous, antiinflammatory, antidiabetic and other activities	Duhan et al. (2020)
14	*Penicillium* sp.	*Cerberamanghas*	Pinazaphilones A and B	α-Glucosidase inhibitory activity	Liu et al. (2015)
15	*P. archeri*	*V. albindia*	Phomoarcherins A–C	Antimalarial	Hemtasin et al. (2011)
16	*Tolypocladium niveum, Cylindrocarpon lucidum, Enterobacter cloacae, Alternaria alternata, Pseudomonas, Bacillus* sp.	*Ocimum sanctum, Cinnamomum camphora, Arnebia euchroma, Taxus wallichiana, Picrorhiza kurroa* (Kutki) and *Pinus roxburghii*	Conjugative	For diverse conditions such as respiratory, gastrointestinal and dermatological ailments, as well as the restoration of oncogenic cells	Singh et al. (2023)
17	*Aspergillus flavus*	*Corchorus olitorius* (jute)	Nitrobenzoic acid, chlorophenyl ester and salsolidine	Antibiotic drug, antioxidants, anticancerous	Wulandari & Suryantini (2018)
18	*B. cereus, B. mycoides*	*Urtica dioica* L	Surfactants, polyphenols, and certain hydrolytic enzymes	Hydrolytic enzymes	Marchut et al. (2023)

associated with medicinal plants like *Azadirachta indica*, *Ephedra foliata*, *Vernonia anthelmintica* and *Origanum vulgare* showcase bioactives with diverse antimicrobial activities. *Rhizoctonia solani*, associated with *Cyperus rotundus*, produces solanioic acid, exhibiting in vitro antibacterial activity against MRSA (Ratnaweera et al., 2015). *Chaetomium globosum* in *Nymphaea nouchali* produces chaetoglobosin A and C, demonstrating antibacterial activities against *Staphylococcus aureus* and *Bacillus cereus*, with two strains exhibiting activity against *Pseudomonas aeruginosa* and *Escherichia coli* (Dissanayake et al., 2016). Multiple endophytes (*Bacillus gibsonii*, *Methylobacterium radiotolerans*, etc.) in *Centella asiatica* collectively exert a cumulative action against *E. coli*, *B. cereus*, *S. aureus* and *P. aeruginosa* (Rafat et al., 2012).

Alternaria alternata AE1 associated with *Azadirachta indica* A. Juss produces phenolics and flavonoids, demonstrating both bactericidal and antioxidant potential (Chatterjee et al., 2019). *Paenibacillus, Bacillus, Brevundimonas, Kytococcus* and *Microbacterium* within *Ephedra foliata* contribute alkaloids, aminoglycosides, flavonoids and terpenoids, exhibiting antibacterial activity against *S. aureus*, *B. subtilis*, *P. aeruginosa* and *E. coli*, along with cytotoxic activity against HeLa cancerous cells (Ghiasvand et al., 2020). *Micrococcus*, *Bacillus*, *Pseudomonas*, *Stenotrophomonas* and *Pantoea* genera associated with *Vernonia anthelmintic*, *Saurauia scaberrinae* and *Chromolaena odorata* produce β-amylase, α-amylase, cellulase, chitinase and protease, showcasing activity against *S. aureus*, *P. ultimum*, *S. sclerotiorum* and *R. solani* (Mamangkey et al., 2022). *Arthrobacter* sp. OVS8 from *Origanum vulgare* produces β-caryophyllene, exhibiting antagonistic activity against multidrug-resistant *K. pneumoniae* strains (Semenzato et al., 2023). *Aspergillus terreus* associated with *Artemisia annua* demonstrates an inhibitory effect against acetylcholinesterase and cytotoxic activity against KB and HSC-T6 cell lines (Ge et al., 2010). *Bacillus amyloliquefaciens* 5NPA-1, belonging to the Firmicutes, isolated from *Salicornia brachiata* was reported to be a potent producer of bioactives. The bioactives include lipopeptides identified through 1H- and 13C-based techniques, demonstrating broad-spectrum antibacterial and antifungal activities against human, plant and aquatic infectious agents (Singh et al., 2021). *Bacillus atrophaeus* and *Bacillus mojavensis* from *Glycyrrhiza uralensis* produced compounds like 1,2-benzenedicarboxylic acid and 9,12-octadecadienoic acid, among others. These bioactives exhibit efficacy against verticillium wilt disease and other phytopathogens in cotton and various crops (Mohamad et al., 2018). *Bacillus pumilus* and *B. safensis* from *Thymus vulgaris* and *Mentha spicata* produce compounds like butanal and 2-heptanone, showcasing antifungal activities against specific fungal strains, including *Aspergillus flavus* and *Fusarium oxysporum* (Erjaee et al., 2019).

Bacillus subtilis sp. (M10), *B. mojavensis* (M11) and *B. amyloliquefaciens* (M13) isolated from *Ajuga turkestanica* produce mucolytic enzymes such as chitinase, protease and 1,3-glucanase, showcasing antimicrobial activity against pathogenic test strains and phytopathogenic plant fungi (Mamarasulov et al., 2023). Various species of *Bacillus*, *Aneurinibacillus* and *Pseudomonas* found in *Tinospora cordifolia* produce enzymes like amylase, cellulase, gelatinase and catalase, demonstrating a wide range of activities including antiviral, anticancerous, antiinflammatory and antidiabetic effects (Duhan et al., 2020). *Penicillium* sp. from *Cerberamanghas* produces pinazaphilones A and B with α-glucosidase inhibitory activity (Liu et al., 2015), while *P. archeri* from *Vanilla albindia* produces phomoarcherins A–C with antimalarial activity (Hemtasin et al., 2011). Multiple microorganisms, including *Tolypocladium niveum*, *Cylindrocarpon lucidum*, *Enterobacter cloacae*, *Alternaria alternata*, *Pseudomonas*, *Bacillus* sp., etc. from various medicinal plants exhibit therapeutic potential against respiratory, gastrointestinal and skin diseases, as well as oncogenic cell repair (Singh et al., 2023). Additionally, *Aspergillus flavus* from *Corchorus olitorius* produces compounds with antibiotic, antioxidant and anticancerous properties (Wulandari & Suryantini, 2018). The bioactives produced by *B. cereus* and *B. mycoides* are polyphenols and certain hydrolytic enzymes (Marchut et al., 2023). The exploration of endophytic bioactives for their antimicrobial properties is significant in the context of addressing antibiotic resistance and discovering novel drugs against infectious diseases.

8.6 CONCLUSION

The insight into endophytic microbes within plants highlights their crucial contribution to plant well-being and growth. The wide array of bioactive compounds they produce, including those with anticancer and antioxidant properties, underscores their promising role in drug development. The ongoing exploration of the intricate connections between endophytic microbes and medicinal plants unveils new therapeutic avenues. Each plant–microbe partnership provides an opportunity to discover distinctive bioactive compounds with targeted health benefits, presenting innovative solutions to modern health issues. The pharmacological potential of the bioactives mentioned herein can be tested for other therapeutic studies.

Acknowledgment

Honourable Vice-Chancellor Sant Gadge Baba Amravati University, Amravati is gratefully acknowledged for providing infrastructural facilities for research dealing with the medicinal plants of Melghat forest.

REFERENCES

Adhikari, P. and Pandey, A. 2019. Phosphate solubilization potential of endophytic fungi isolated from *Taxus wallichiana* Zucc. roots. *Rhizosphere*, *9*, 2–9.

Ahmad, N., Hamayun, M., Khan, S. A., Khan, A. L., Lee, I. J. and Shin, D. H. 2010. Gibberellin-producing endophytic fungi isolated from *Monochoria vaginalis*. *Journal of Microbiology and Biotechnology*, *20*(12), 1744–1749.

Anwar, P. Z., Sezhian, U. G. and Narasingam, A. 2020. Endophytic bacteria from *Cissus Quadrangularis*, a promising source of bioactive compounds. *Biosciences Biotechnology Research Asia*, *17*(1), 115–126.

Arkhipova, T. N., Veselov, S. U., Melentiev, A. I., Martynenko, E. V. and Kudoyarova, G. R. 2005. Ability of bacterium Bacillus subtilis to produce cytokinins and to influence the growth and endogenous hormone content of lettuce plants. *Plant and Soil*, 272: 201–20.

Asaf, S., Khan, M. A. and Khan, A. L., et al. 2017. Bacterial endophytes from arid land plants regulate endogenous hormone content and promote growth in crop plants: an example of Sphingomonas sp. and *Serratia marcescens*. *Journal of Plant Interactions*, *12*(1), 31–38.

Bai, Y., D'Aoust, F., Smith, D. L., and Driscoll, B. T. 2002. Isolation of plant-growth-promoting Bacillus strains from soybean root nodules. *Canadian Journal of Microbiology*, *48*(3), 230–238.

Budhiraja, A., Nepali, K., Sapra, S., Gupta, S., Kumar, S. and Dhar, K. L. 2013. Bioactive metabolites from an endophytic fungus of Aspergillus species isolated from seeds of *Gloriosa superba* Linn. *Medicinal Chemistry Research*, *22*, 323–329.

Chatterjee, S., Ghosh, R. and Mandal, N. C. 2019. Production of bioactive compounds with bactericidal and antioxidant potential by endophytic fungus *Alternaria alternata* AE1 isolated from *Azadirachta indica* A. Juss. *PLoS One*, *14*(4), e0214744.

Chen, Y., Shafi, J., Li, M., Fu, D. and Ji, M. 2018. Insecticidal activity of endophytic actinomycetes isolated from *Azadirachta indica* against *Myzus persicae*. *Archives of Biological Sciences*, *70*(2), 349–357.

Dissanayake, R. K., Ratnaweera, P. B., Williams, D. E. et al. 2016. Antimicrobial activities of endophytic fungi of the Sri Lankan aquatic plant *Nymphaea nouchali* and chaetoglobosin A and C, produced by the endophytic fungus *Chaetomium globosum*. *Mycology*, *7*(1), pp.1–8.

Dolatabadian, A. 2020. Plant–microbe interaction. *Biology*, *10*(1), 15.

Dong, Z., Canny, M. J., McCully, M. E., et al. 1994. A nitrogen-fixing endophyte of sugarcane stems (a new role for the apoplast). *Plant Physiology*, *105*(4), 1139–1147.

Duhan, P., Bansal, P. and Rani, S. 2020. Isolation, identification and characterization of endophytic bacteria from medicinal plant *Tinospora cordifolia*. *South African Journal of Botany*, *134*, 43–49.

El-Sayed, A. S., Moustafa, A. H., Hussein, H. A., et al. 2020. Potential insecticidal activity of *Sarocladium strictum*, an endophyte of *Cynanchum acutum*, against *Spodoptera littoralis*, a polyphagous insect pest. *Biocatalysis and Agricultural Biotechnology*, *24*, 101524.

Erjaee, Z., Shekarforoush, S. S. and Hosseinzadeh, S. 2019. Identification of endophytic bacteria in medicinal plants and their antifungal activities against food spoilage fungi. *Journal of Food Science and Technology*, *56*, 5262–5270.

Freeman, E. M. 1904. Th e seed fungus of *Lolium temulentum* L. *Philosophical Transactions of the Royal Society of London B*, 196, 1–27.

Ge, H. M., Peng, H., Guo, Z. K., Cui, J. T., Song, Y. C. and Tan, R. X. 2010. Bioactive alkaloids from the plant endophytic fungus *Aspergillus terreus*. *Planta medica*, *76*, 822–824.

Ghiasvand, M., Makhdoumi, A., Matin, M. M. and Vaezi, J. 2020. Exploring the bioactive compounds from endophytic bacteria of a medicinal plant: *Ephedra foliata* (Ephedrales: Ephedraceae). *Advances in Traditional Medicine*, *20*, 61–70.

Giridharan, P., Verekar, S. A., Khanna, A., Mishra, P. D. and Deshmukh, S. K. 2012. Anticancer activity of sclerotiorin, isolated from an endophytic fungus *Cephalotheca faveolata* Yaguchi, Nishim & Udagawa.

Hassan, S. E. D. 2017. Plant growth-promoting activities for bacterial and fungal endophytes isolated from medicinal plant of *Teucrium polium* L. *Journal of Advanced Research*, *8*(6), s687–695.

Hemtasin, C., Kanokmedhakul, S., Kanokmedhakul, K., et al. 2011. Cytotoxic pentacyclic and tetracyclic aromatic sesquiterpenes from *Phomopsis archeri*. *Journal of Natural Products*, *74*(4), 609–613.

Hu, X., Li, W., Yuan, M., et al. 2016. Homoharringtonine production by endophytic fungus isolated from *Cephalotaxus hainanensis* Li. *World Journal of Microbiology and Biotechnology*, *32*, 1–9.

Isaka, M., Chinthanom, P., Boonruangprapa, T., Rungjindamai, N. and Pinruan, U. 2010. Eremophilane-type sesquiterpenes from the fungus Xylaria sp. BCC 21097. *Journal of Natural Products*, *73*(4), 683–687.

Ismail, M. A., Amin, M. A., Eid, A. M., et al. 2021. Comparative Study between exogenously applied plant growth hormones versus metabolites of microbial endophytes as plant growth-promoting for *Phaseolus vulgaris* L. *Cells*, *10*(5), 1059.

Kandasamy, G. D. and Kathirvel, P. 2023. Insights into bacterial endophytic diversity and isolation with a focus on their potential applications–A review. *Microbiological Research*, *266*, 127256.

Kusari, S., Singh, S. and Jayabaskaran, C. 2014. Rethinking production of Taxol®(paclitaxel) using endophyte biotechnology. *Trends in Biotechnology*, *32*(6), 304–311.

Li, Z., Xiong, K., Wen, W., Li, L. and Xu, D. 2023. Functional endophytes regulating plant secondary metabolism: current status, prospects and applications. *International Journal of Molecular Sciences*, *24*(2), p.1153.

Liu, Y., Yang, Q., Xia, G., et al. 2015. Polyketides with α-glucosidase inhibitory activity from a mangrove endophytic fungus, Penicillium sp. HN29-3B1. *Journal of Natural Products*, *78*(8), 1816–1822.

Lucas, J. A., García-Cristobal, J., Bonilla, A., Ramos, B. and Gutierrez-Manero, J. 2014. Beneficial rhizobacteria from rice rhizosphere confers high protection against biotic and abiotic stress inducing systemic resistance in rice seedlings. *Plant Physiology and Biochemistry*, *82*, 44–53.

Mamangkey, J., Mendes, L. W., Harahap, A., Briggs, D., and Kayacilar, C. 2022. Endophytic bacteria and fungi from Indonesian medicinal plants with antibacterial, pathogenic antifungal and extracellular enzymes activities: a review. *International Journal of Science, Technology & Management*, *3*(1), 245–255.

Mamarasulov, B., Davranov, K., Umruzaqov, A., et al. 2023. Evaluation of the antimicrobial and antifungal activity of endophytic bacterial crude extracts from medicinal plant *Ajuga turkestanica* (Rgl.) *Brig* (Lamiaceae). *Journal of King Saud University-Science*, *35*(4), 102644.

Marchut-Mikołajczyk, O., Chlebicz, M., Kawecka, M., et al. 2023. Endophytic bacteria isolated from Urtica dioica L.-preliminary screening for enzyme and polyphenols production. *Microbial Cell Factories*, *22*(1), 169.

Mejía, L. C. Rojas, E. I., Maynard, Z., et al. 2008. Endophytic fungi as biocontrol agents of *Theobroma cacao* pathogens. *Biological Control*, *46*(1), 4–14.

Mendoza-Arroyo, G. E., Chan-Bacab, M. J., Aguila-Ramírez, R. N., et al. 2020. Inorganic phosphate solubilization by a novel isolated bacterial strain Enterobacter sp. ITCB-09 and its application potential as biofertilizer. *Agriculture*, *10*(9), 383.

Mohamad, O. A. Li, L., Ma, J. B., et al. 2018. Evaluation of the antimicrobial activity of endophytic bacterial populations from Chinese traditional medicinal plant licorice and characterization of the bioactive secondary metabolites produced by *Bacillus atrophaeus* against *Verticillium dahliae*. *Frontiers in Microbiology*, *9*, 924.

Morales-Sánchez, V., Fe Andrés, M., Díaz, C.E. and González-Coloma, A. 2020. Factors affecting the metabolite productions in endophytes: biotechnological approaches for production of metabolites. *Current Medicinal Chemistry*, *27*(11), 1855–1873.

Nongkhlaw, F. M. and Joshi, S. R. 2015. Investigation on the bioactivity of culturable endophytic and epiphytic bacteria associated with ethnomedicinal plants. *Journal of Infection in Developing Countries*, *9*(09), 954–961.

Parsa, S., Ortiz, V. and Vega, F. E. 2013. Establishing fungal entomopathogens as endophytes: towards endophytic biological control. *JoVE (Journal of Visualized Experiments)*, *74*, e50360.

Prajapati, J., Goswami, D., and Rawal, R. M. 2021. Endophytic fungi: a treasure trove of novel anticancer compounds. *Current Research in Pharmacology and Drug Discovery*, *2*, 100050.

Rafat, A., Philip, K., and Muniandy, S. 2012. A novel source of bioactive compounds: Endophytic bacteria isolated from Centella asiatica. *Journal of Pure and Applied Microbiology,* 6(1), 11–20.

Rajendran, L., Ramanathan, A., Durairaj, C. and Samiyappan, R. 2011. Endophytic Bacillus subtilis enriched with chitin offer induced systemic resistance in cotton against aphid infestation. *Archives of Phytopathology and Plant Protection*, *44*(14), 1375–1389.

Ratnaweera, P. B., Williams, D. E., Patrick, B.O., de Silva, E. D., Andersen, R. J. 2015, May. Solanioic acid, an antibacterial degraded steroid produced in culture by the fungus Rhizoctonia solani isolated from tubers of the medicinal plant Cyperus rotundus. *Organic Letters*, *17*(9):207–2077. doi: 10.1021/acs.orglett.5b00596

Semenzato, G., Del Duca, S., Vassallo, A. et al. 2023. Genomic, molecular, and phenotypic characterization of Arthrobacter sp. OVS8, an endophytic bacterium isolated from and contributing to the bioactive compound content of the essential oil of the medicinal plant *Origanum vulgare* L. *International Journal of Molecular Sciences*, *24*(5), 4845.

Shahzad, R. Waqas, M. and Khan, A. L., et al. 2016. Seed-borne endophytic *Bacillus amyloliquefaciens* RWL-1 produces gibberellins and regulates endogenous phytohormones of *Oryza sativa. Plant Physiology and Biochemistry*, *106*, 236–243.

Singh, S., Ghadge, V. A., Kumar, P., et al. 2021. Biodiversity and antimicrobial potential of bacterial endophytes from halophyte *Salicornia brachiata. Antonie Van Leeuwenhoek*, *114*, 591–608.

Singh, M., Naskar, A., Rupashree, A., Rajput, M. and Singh, V. K. 2023. Analysis of endophytic microbes harboring in medicinal plants of Himalayan region with their medicinal properties. *Biocatalysis and Agricultural Biotechnology*, *53*, 102857.

Tidke, S. A., Kumar, R. K., Ramakrishna, D., Kiran, S., Kosturkova, G. and Gokare, R. A. 2017. Current understanding of endophytes: their relevance, importance, and industrial potentials. *IOSR J Biotechnol Biochem*, *3*(3), 43–49.

Vinayarani, G. and Prakash, H. S. 2018. Fungal endophytes of turmeric (*Curcuma longa* L.) and their biocontrol potential against pathogens *Pythium aphanidermatum* and *Rhizoctonia solani. World Journal of Microbiology and Biotechnology*, *34*, 1–17.

Wani, Z. A., Ashraf, N., Mohiuddin, T. and Riyaz-Ul-Hassan, S. 2015. Plant-endophyte symbiosis, an ecological perspective. *Applied Microbiology and Biotechnology*, *99*, 2955–2965.

Wijeratne, E. K., Paranagama, P. A., Marron, M. T., et al. 2008. Sesquiterpene quinones and related metabolites from *Phyllosticta spinarum*, a fungal strain endophytic in *Platycladus orientalis* of the Sonoran Desert. *Journal of Natural Products*, *71*(2), 218–222.

Wulandari, R. S. and Suryantini, R. 2018. Growth of Albizia in vitro: endophytic fungi as plant growth promote of Albizia. *International School Science Research Innovation*, *12*(1), 8.

Xia, Y., Sahib, M. R., Amna, A., Opiyo, S. O., Zhao, Z. and Gao, Y. G. 2019. Culturable endophytic fungal communities associated with plants in organic and conventional farming systems and their effects on plant growth. *Scientific Reports*, *9*(1), 1669.

Zhang, J. Y., Tao, L. Y., Liang, Y. J., et al. 2010. Anthracenedione derivatives as anticancer agents isolated from secondary metabolites of the mangrove endophytic fungi. *Marine Drugs*, *8*(4), 1469–1481.

Zhou, P., Wu, Z., Tan, D., et al. 2017. Atrichodermones A–C, three new secondary metabolites from the solid culture of an endophytic fungal strain, *Trichoderma atroviride*. *Fitoterapia*, *123*, 18–22.

9 Exploring the Potential and Prospects of Fungal Endophytes in the Biosynthesis of High-Value Metabolites

Pragya Tiwari[φ], Mangalam Bajpai[φ], Muhammad Adil, Jyotirmoyee Das and Laurent Dufossé**
[φ] These authors contributed equally.
*Corresponding authors.

9.1 RECENT TRENDS IN DRUG DISCOVERY

Endophytes serve as biological repositories for novel natural compounds, providing fresh opportunities for drug development. Growing in popularity as prime candidates for bioprospecting in the hunt for novel chemical entities are the plant-associated bacteria that inhabit the interior tissues of all species of plants (Staniek et al., 2008; Tiwari et al., 2021b; Tiwari et al,. 2024). Research in endophytic biology was boosted by the discovery of the renowned anticancer medication paclitaxel, which led to potential "drug candidates" which are immune-suppressing, antioxidant and have functions that inhibit neurodegeneration (Staniek et al., 2008; Rashmi & Venkateswara, 2019; Tiwari et al., 2021b). Currently, natural products' influences on clinical applications are important, and the pharmaceutical industry is using high-throughput methods to identify potential lead compounds in plant secondary metabolites (Adil et al., 2023c). The discovery of secondary metabolites with therapeutic value has been made possible by the rich but understudied diversity of endophytic entities and their significant potential to influence the pharmaceutical industry. Bio-prospecting of endophytes has led to the identification of novel, highly valuable metabolites with potential for commercial application (Bose et al., 2022; Tiwari et al., 2023a), such as hinnuliquinone, a potent inhibitor of the HIV-1 protease (Singh et al., 2004), cytonic acids A and B, inhibitors of the human cytomegalovirus protease (Guo et al., 2000), pestacin and isopestacin, antioxidants (Strobel et al., 2002), paclitaxel, antineoplastic agents (Zhang et al., 2009), and lariatins A and B, anti-HIV agents (Matsumoto & Takahashi, 2017).

Endophytes are being investigated more and more as production platforms for bioactive metabolites with a variety of chemical structures because of growing studies on the discovery of natural products from biological species. Research on pharmaceutical metabolites from endophytic sources was revived after taxol was discovered in 1993 (Stierle et al., 1993) from the endophytic fungus *Taxomyces andreanae* and afterward from other endophytic species. Additionally, research has shown that endophytes imitate host metabolites derived from endophytic sources. Furthermore, research has shown that endophytes may create, induce and modify the metabolic chemical entities

 DOI: 10.1201/9781003486275-11

within hosts as well as mimic the host metabolism (Ludwig-Muller, 2015; Venieraki et al., 2017). One of the most promising choices for endophyte-mediated natural product discovery is endophytic fungi. Diverse bioactive metabolites made by endophytic fungi have shown socioeconomic significance (Adil et al., 2023a) and found uses in agriculture (as biofertilizers/biostimulants) (Tiwari et al., 2023a) and the environment (bioremediation) (Tiwari et al., 2023c), as well as biofuels and biocatalysts. As, increasingly, beneficial metabolites with complex environmental ramifications are being discovered, many endophytic fungus strains are being researched (Bajpai et al., 2022), and the relationship between optimal utilization and multifarious applications and constraints is being addressed.

Even though endophytes associated with plants act as "biosynthetic platforms" for pharmaceutically significant secondary metabolites, difficulties with endophyte biology and decreased secondary metabolite production because of repeated subculturing of endophytic strains suggest the need for a more integrated and systematic approach to the exploitation of endophytes in drug discovery and research. To understand how and why endophytes biosynthesize secondary metabolites, one must thoroughly understand plant–endophyte relationships and their dynamics (Meshram & Gupta, 2019). Using genetic engineering, improving growth conditions and media, and co-culturing several endophytic strains is another potential option for strain improvement (Tiwari & Dufossé, 2023b). The "genomic revolution" and recent developments in high-throughput technologies have made significant contributions to the study of natural products and the identification of biosynthetic gene clusters (BGCs) (Zazopoulos et al., 2003; Lautru et al., 2005) (Figure 9.1).

Endophyte genetic engineering is still in its infancy but is frequently seen as a precursor to functional genomics and system biology approaches (Tyo et al., 2007) with the potential for long-term translational success. Endophytes serve as a potential platform for genetic manipulation as they independently biosynthesize secondary metabolites like their host plant (Zazopoulos et al., 2003). Genetic recombination between endophytes and plant hosts has been theorized to contribute to the incorporation of metabolic pathways and genes in endophytes (Tan & Zou, 2000; Tiwari & Bae, 2020).

9.2 ENDOPHYTES: NOVEL RESERVOIRS FOR HIGH-VALUE CHEMICAL ENTITIES

In contemporary medicine, pharmacologically active plants and their bioactive byproducts are employed to treat disease (Pan et al., 2013; Tiwari et al., 2021a). More than 51% of the small-molecule medications that were approved between 1981 and 2014 were made with natural ingredients (Chen et al., 2016). However, the uncontrolled use of natural products has endangered natural resources, particularly plant species, and resulted in a dearth of new bioactive metabolites. Therefore, it is essential to investigate alternate environmentally acceptable methods of producing high-value phytochemicals and plant-associated endophytes, which are turning out to be alluring sources (Berdy, 2005; Adil et al., 2023a). The varied uses of secondary metabolites from endophytes, which include uses in medicinal medications and immune suppressants as well as in agriculture and industry, are largely responsible for their growing significance. Out of 22,500 chemicals generated from microbes, fungi accounted for the bulk (38%) of the bioactive metabolites, including antibiotics (Soliman et al., 2015; Mousa et al., 2016) (Table 9.1).

Bacterial and fungal species that colonize internal plant tissues and finish all or part of their life cycles inside the host plant are known as plant-associated endophytes. Endophytes are ubiquitous and found in almost all vascular plant species. They provide the host plant with several benefits including growth promotion, biotic/abiotic stress tolerance (Rodriguez et al., 2009) and the biosynthesis of bioactive secondary metabolites (Kusari et al., 2012). Protection against diseases (Newman & Cragg, 2020), reduction of abiotic stress in plants (Hamayun et al., 2017) and

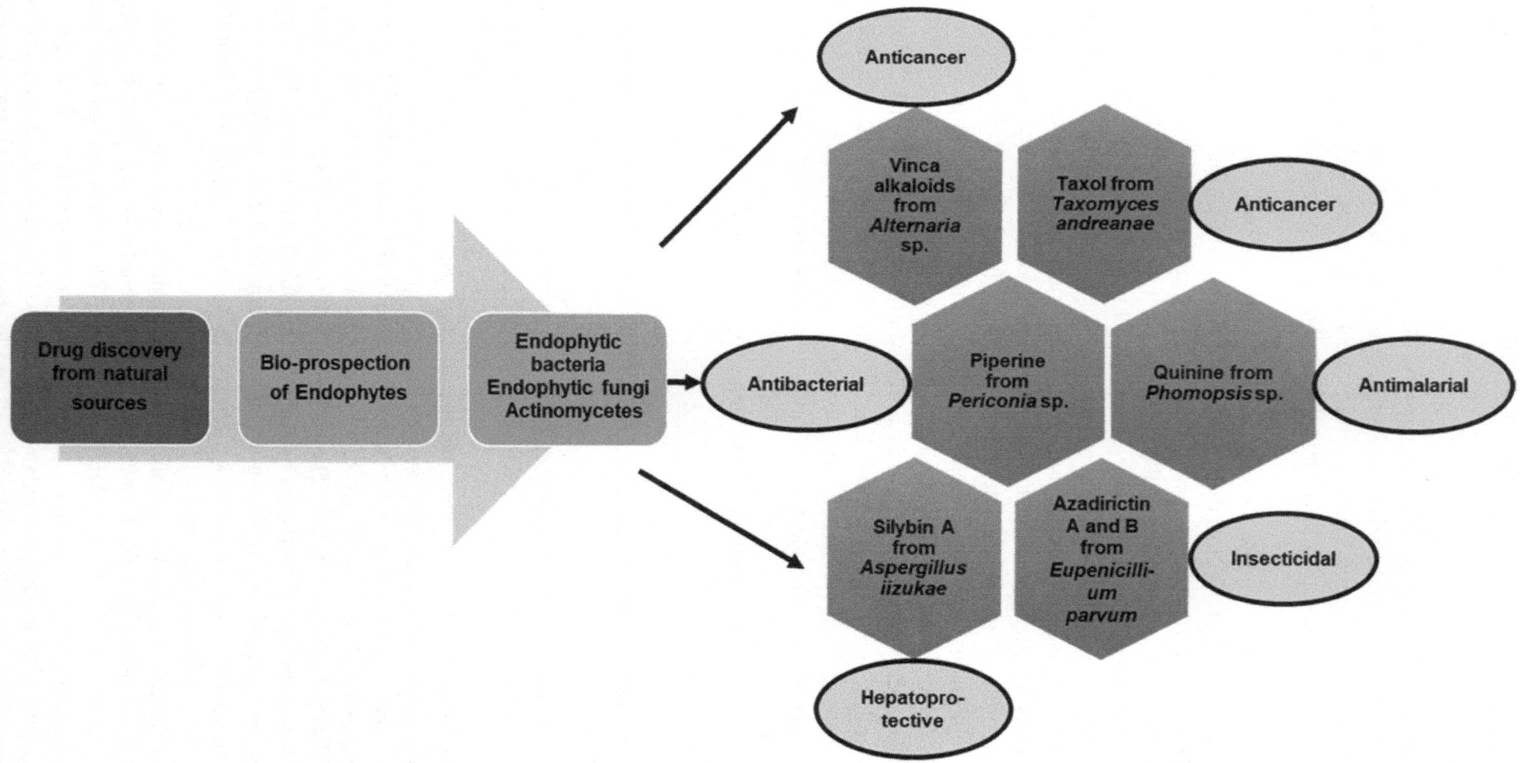

FIGURE 9.1 Bioprospection of endophytes and the identification of novel, highly valuable and commercially relevant metabolites. (Tiwari et al., 2022).

TABLE 9.1
The Multifaceted Applications, Key Examples and Beneficial Outcomes of Fungal Endophytes in the Environment (Tiwari et al., 2022)

Biological application	Fungal endophyte	Plant species	Outcome	References
		Plant Growth Promotion and Agriculture		
Plant growth promotion (PGP)	*Penicillium* sp. *Aspergillus* sp. *Trichoderma gamsii* *Chaetomium globosum* *Fusarium oxysporum* *Ophiosphaerella* sp. *Cochliobolus* sp. *Cladosporium sphaerospermum* *Alternaria* sp., *Fusarium* sp. Fungal endophyte	*Camellia sinensis* *Lens esculenta* *Solanum lycopersicum* *Triticum aestivum* *Glycine max* (L) *S. nigrum* *Hordeum murinum* subsp. *murinum* L.	Mineral-solubilizing function ($Ca_3(PO_4)_2$ and rock phosphate) Solubilization of TCP Siderophore production, HCN, and ammonia production PGP activities Solubilize calcium phosphate Production of phytohormones (gibberellins) IAA production, PGP activities Plant yield increase	(Gupta et al., 2007) (Nath et al., 2012) (Rinu et al., 2014) (Chadha et al., 2015) (Spagnoletti et al., 2017) (Hamayun et al., 2009) (Khan et al., 2015) (Johnson et al., 2014)
Biofertilizers/ biostimulants for crops	*Aspergillus* sp. *Cochliobolus* sp. *Curvularia* sp. *Fusarium equiseti* *A. nidulans*	*S. officinarum* *T. aestivum* *Pisum sativum* *Lycopersicon esculentum*	Phosphorus solubilization Phosphorus solubilization Phosphorus solubilization Plant biomass increase, fruit yield	(Sane & Mehta, 2015) (Spagnoletti et al., 2017) (Sisic et al., 2017) (Xia et al. 2019)
Biocontrol function	*Bacillus subtilis* *Bacillus* sp. *Arthrobacter* sp. *Fusarium verticillioides* *Pseudomonas aeruginosa*	*Oryza sativa* *Piper nigrum* *Zea mays* *L. esculentum*	Biocontrol of *Rhizoctonia solani* Biocontrol of *Phytophthora capsici* Restrict *Ustilago maydes* growth Biocontrol of *Botrytis cinerea*	(Nagendran et al., 2014) (Aravind et al., 2009) (Rodriguez et al., 2012) (Yang et al., 2011)
Biotic/abiotic stress tolerance	*Piriformospora indica* *P. indica* *Proteobacteria* sp. *P. indica* *Pseudomonas* spp.	*Hordeum vulgare* *Brassica rapa* *Raphanus sativus* L. *H. vulgare* *Phragmites australis*	Drought stress tolerance Drought stress tolerance Protection against plant pathogens Biotic/abiotic stress tolerance Protection against plant pathogens	(Ghaffari et al., 2019) (Sun et al., 2010) (Seo et al., 2010) (Johnson, Alex, & Oelmuller, 2014) (White et al., 2017)

(continued)

TABLE 9.1 (Continued)
The Multifaceted Applications, Key Examples and Beneficial Outcomes of Fungal Endophytes in the Environment (Tiwari et al., 2022)

Biological application	Fungal endophyte	Plant species	Outcome	References
Bioremediation	*Mucor* sp. MHR-7 *Aspergillus fumigatus* Fungal endophyte *Neotyphodium coenophialum* *Neotyphodium uncinatum* *Verticillium* sp. *Xylaria* sp. Fungal endophyte *Curvularia* sp. *Neusartorya* sp. *Bjerkandera adusta* *Lasiodiplodia theobromae* *Lindgomycetaceae* *Aspergillus* sp. *Pseudomonas aeruginosa* L10	*Brassica campestris* L. *Lactuca sativa* *Agrostis stolonifera* *Festuca arundinacea* Schreb. *Festuca pratensis* Huds Plants from Ecuadorian Amazon — Mangrove spp. *Sinosenecio oldhamianus* *Boswellia ovalifoliolata* *Aeschynomene fluminensis*	Metal toxicity reduction Chromium detoxification in soil Bioremediation of lead Bioaugmentation, total petroleum hydrocarbons (TPH) and polycyclic aromatic hydrocarbons (PAHs) removal from the soil Degradation of petroleum hydrocarbon Bioremediation of synthetic plastic polymers Heavy metal biosorption Detoxification of triphenylmethane dyes Heavy metal tolerance Heavy metal resistance, bioremediation Degradation of petroleum hydrocarbons	(Zahoor et al., 2017) (Bibi et al., 2018) (Soldi et al., 2020) (Soleimani et al., 2010) (Marin et al., 2018) www.bard.edu/inside/calendar/event/?eid=122431&date=1383843600 (Ling et al., 2016) (Gao et al,. 2020) (Aishwarya et al., 2017) (Pietro-Souza et al., 2020) (Wu et al., 2018)

herbivore deterrents (Panaccione et al., 2014) are a few examples of recent studies into the ecological perspective of secondary metabolite production (Xu et al., 2016). The amazing potential of endophytes in the production of phytochemicals and new platforms for the fabrication of novel antibacterial agents was revealed by a study of the natural product chemistry of endophytes (Tiwari & Bae, 2022).

Endophytes associated with plants have made tremendous progress in the bioprospecting and isolation of bioactive compounds with diverse properties. Reviews of the literature have given important insights into the current and emerging significance of endophytes in drug discovery and research; a few publications worth mentioning are "Discussion on biodiversity of endophytes and its exploitation in drug discovery" (Staniek et al., 2008), "Bioactive metabolites and their role in pharmacological development" (Tiwari et al., 2022) "Newer roles, and actinomycetes endophytes use" (Meshram & Gupta, 2019), "Chemical ecology of endophytes, and its generation of bioactive metabolites" (Tiwari et al., 2022) and "Horizontal gene transfer," as well as its effects on the dissemination of novel features in the synthesis of bioactive metabolites (Deshmukh et al., 2017) providing a thorough understanding of the many metabolites that fungal endophytes produce among other things, and their biological activities. The present review on fungal endophytes provides a detailed insight into the existing and emerging prospects of fungal endophytes as "novel candidates" in drug discovery and research and how the different yield enhancement

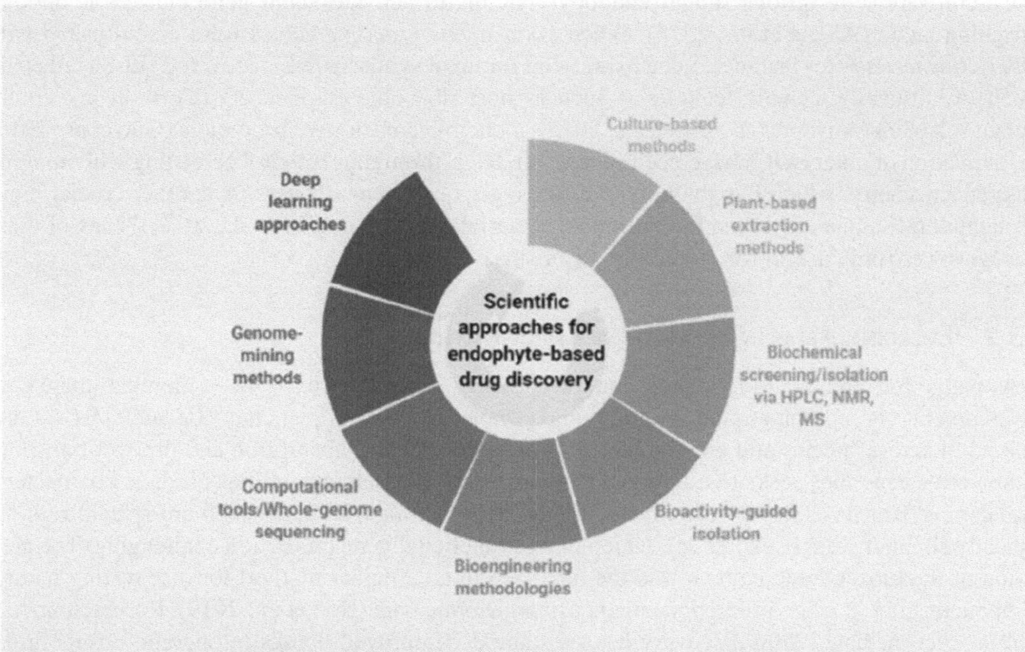

FIGURE 9.2 Conventional and emerging scientific approaches for endophyte-based drug discovery. (Tiwari et al., 2022).

strategies can be adopted to address the associated bottlenecks and enhance bioactive metabolite production (Figure 9.2).

9.3 NATURAL PRODUCT-BASED DRUG DISCOVERY VIA SCIENTIFIC APPROACHES

9.3.1 Traditional Approaches

In the search for valuable metabolites with potential pharmaceutical uses, plant–endophytic interactions have recently been explored more thoroughly. Using deep learning techniques, an intriguing platform has been created for the bioprospecting of regulatory networks, endophytes and the forecasting of novel chemical entities (Aghdam & Brown, 2021). As previously mentioned, plant-associated endophytes have developed a desirable "biosynthetic platform" for the creation of novel bioactive metabolites, although there are several obstacles. Utilizing these biological sources to find new drugs. For instance, most of the knowledge of metabolic networks and secondary metabolite routes is suppressed. These mechanisms must be thoroughly comprehended. Additionally, in silico methods forecast the existence of a large range of metabolites that are silent or inactive in their natural habitats (in plants) (Kusari et al., 2015; Pan et al., 2019). Recent decades have seen reliance on robust, low-throughput methods (Dias et al., 2012) for the discovery of natural products from endophytes, with some existing limitations. These methods essentially included culture-based techniques, plant-based extraction techniques, biochemical screening/isolation using HPLC, NMR and MS, and bioactivity-guided isolation. Sophisticated approaches, including the "one strain, many compounds" (OSMAC) approach, modifying ribosomes, heterologous gene expression and promoter studies, have been extensively investigated to overcome the current issues (Baral et al., 2018).

Co-cultivation of various fungal endophyte strains to activate silent gene clusters is another intriguing tactic (Akone et al., 2018). When *Podocarpus gracilior* leaves were co-cultivated with *Aspergillus terreus*, for instance, gene expression for taxol synthesis was recovered (El-Sayed et al., 2019). Additionally, genetic techniques such as host ribosome engineering (Corre et al., 2008), mutant selection (Anyaogu & Mortensen, 2015), inducible/constitutive promoters (Guo et al., 2015) and activation of quiet BGCs have been successful. High-throughput elicitor screening with imaging mass spectrometry, which has the potential to trigger quiet gene clusters, is another crucial high-throughput technique employed for culturable bacterial endophytes (Xu et al., 2019). Many of these strategies were only useful for culturable organisms.

9.3.2 Emerging Areas in High-Throughput Methods

New paths for endophyte-mediated natural product discovery have been defined thanks to developments in computational biology and whole-genome sequencing. Because BGCs are conserved across species and are connected to regulatory genes, absorption and product transport, comparative genomics seeks to comprehend the diversity of chemical entities produced by bacteria (Aghdam & Brown, 2021). Furthermore, in laboratory settings, BGCs, which are specialized, are typically silent or weakly expressed, which makes functional gene prediction challenging. The activation of the biosynthetic pathways is the next step in an efficient method for discovering natural compounds after in silico predictions made using genomic data (Koo et al., 2019). Research on natural product-mediated drug discovery has substantially improved thanks to ongoing efforts in the creation and upgrade of computer resources. The identification of novel gene clusters has been made easier by databases like ClusterMine 360 (Ichikawa et al., 2013), ClustScan software (Starcevic et al., 2008) and the Database of BIoSynthesis clusters CUrated and InTegrated (DoBISCUIT) (Ichikawa et al., 2013). Additionally, the Minimum Information on Biosynthetic Gene Clusters (MIBiG), a helpful resource, was used to find BGCs using phylogenetic analysis (Epstein et al., 2018). Antibiotics and secondary metabolite analysis shell (antiSMASH) (Wang et al., 2018), Metaflux (Blin et al., 2019) and Reconstruction, Analysis, and Visualization of Metabolic Networks' RAVEN 2.0 software (Latendresse et al., 2012) are additional bioinformatics databases that support the annotation of BGCs. It is also now possible to identify and profile natural product gene clusters across the entire genome. Additionally, network algorithm-based bioinformatics software enhances the predictions of genome mining techniques (Faust & Raes, 2012) and can be linked with metabolic modeling strategies (Stolyar et al., 2007). Through single-cell analysis for endophyte metabolism (Ravikrishnan et al., 2020), bio-kinetic models for the estimation of interspecific interactions among microbes (Mori et al., 2018) and transcriptome approaches for comprehensive understanding (Kuang et al., 2019) the predictive strategies further our understanding of metabolic interactions in microbial cultures.

9.4 HIGH-VALUE METABOLITES FROM FUNGAL ENDOPHYTES

Many studies on the bioprospecting of fungal endophytes capable of generating high-value bioactive compounds such as paclitaxel, vincristine, vinblastine, camptothecin and podophyllotoxin have been reported. However, due to the extensively known genomic instability of endophytes in axenic culture, economic utilization of endophytes for high-value–low-volume plant secondary metabolites remains elusive. While most of the endophyte research focuses on screening endophytes for novel or current high-value biomolecules, very few studies investigate the putative processes of host plant-linked or unique secondary metabolite synthesis in these organisms. The review emphasizes the evidence for and against the presence of host-independent biosynthetic machinery in endophytes by providing an overview of the host–endophyte interaction and its potential influence on the secondary metabolite production capability of endophytes. Endophytes are largely unexplored as valuable sources of medication. They are microorganisms that are usually found in plants. Endophytes are known to create active metabolites with the same function as their hosts, such as taxol, podophyllotoxin, hypericin and azadirachtin, when they are exploited. These metabolites have shown promise in terms of safety and human health problems. We are obliged to make efforts to increase output due to the poor metabolite yield.

9.4.1 Taxol

During the late 1960s, comprehensive screening of plant materials for antineoplastic drugs yielded taxol, a complex polyoxygenated diterpene isolated from the Pacific Yew, *Taxus brevifolia*. Over the last two decades, interest and study on taxol (Snader, 1990; Anon, 1992) have gradually increased to the point that the popular press now appears set to scoop each new advance. What was formerly an esoteric substance of interest only to the most masochistic of synthetic chemists and an equally tiny number of cellular biologists has become one of the few organic molecules identifiable by name to the average citizen, alongside benzene and aspirin. Simultaneously, scientific research on taxol has exploded. Doctors are now investigating its effects on practically every known malignancy. Taxol is being used by biologists to investigate the mechanics of cell function by analyzing the impact of its interactions with the cellular skeletal systems. Synthetic chemists (Bissett & Kaye, 1993) are examining almost every available method for the molecule's production, captivated by its unusual and delicate structure and functioning. Indeed, the demand for taxol has surged so dramatically in the last 5 years that alternatives to *T. brevifolia* extraction (Anjanamurthy & Basavaraju, 1999) are being actively investigated. Because of the fast-developing area of research in the multidimensional study of taxol, those interested in the topic may find it difficult to acquire a reasonable basis of knowledge.

The tale begins in the 1950s, not with *Taxus*, but with the status of cancer therapy research (Wani et al., 1971). The discovery of penicillin and its application in World War II ushered in the "antibiotic era," and many drugs, such as erythromycins and tetracyclines (Miller, 1981; Lenaz & De Furia, 1993), were becoming available in addition to penicillins and sulfa drugs; deaths from infectious diseases of all types, but especially pneumonia, which were a major killer, were on the decline, and this drastically changed mortality patterns, such that heart disease and cancer were now the major killers. There were few big cancer research facilities at the time, and several small groups worked mostly in isolation on various elements of carcinogenesis, etiology and therapy. Goodman et al. on nitrogen mustards in leukemia in 1946 (Chen, 1990), Farber et al. on aminopterin in childhood leukemia in 1948, and Burchenal et al. on 6-mercaptopurine in leukemias in 1953 were the first significant studies on chemotherapeutic drugs used to treat human cancer (Arimondo et al., 2000). These promising discoveries piqued the interest of many chemists, biochemists and biologists in cancer research, but there were insufficient facilities for these researchers to have their compounds examined for anticancer properties. The first source of plant extracts for screening was based on their quick availability as well as the understanding that evaluating plants for anticancer agents was a relatively new topic. As a result, it was decided to begin the collecting effort with native plants from the United States, and an interagency agreement was signed with the United States Department of Agriculture (USDA) (Anon, 1992) to collect and identify plant samples for the screening program. From 1960 through 1981, they worked together. Hartwell also began accumulating literary allusions to plant products as antitumor agents from antiquity to the present, amassing over 3000 citations in the process (Tiwari et al., 2022). These legendary plants were prioritized for collection any time collectors were in locations where they were found, but due to the price and difficulties of establishing expeditions to collect only a few specific plants, these folkloric leads were rarely explored individually.

9.4.2 Podophyllotoxin

Podophyllum lignans (Ayres & Loike, 1990; Bennouna et al., 2008) research is presently focusing on structural optimization to build derivatives with improved pharmacological characteristics and a larger therapeutic scope, as well as the discovery of alternative and renewable podophyllotoxin sources. Podophyllotoxin is a naturally occurring lignan with important antineoplastic and antiviral properties. Podophyllotoxin analogs have advanced to the forefront of several areas of therapeutic and developmental chemotherapy, aided by a detailed understanding of their mechanism of action and facilitated by chemical manipulations that have amplified their bioactivity. Laboratory interactions with other cytotoxic drugs have been used to develop podophyllotoxin-based multidrug regimens (Barret et al., 2002) that are showing significant activity in several malignancies and many of its related analogs will complement conventional pharmaceuticals in disease treatment, prevention and diagnosis, while also adding value to agriculture. Extensive structural modifications of podophyllotoxin were performed to obtain more potent and less toxic antitumor agents, resulting in the widespread clinical introduction of two semisynthetic glycoconjugate analogs of etoposide and teniposide (Gidding et al., 1999), and newer agents with promising preclinical activity are in various stages of clinical evaluation. As our understanding of molecular and biochemical mechanisms of action and resistance grows, additional and improved podophyllotoxin-based techniques for cancer therapy are expected to emerge.

9.4.3 Camptothecin

This chapter discusses recent improvements in the medicinal chemistry of camptothecin, a powerful antitumor antibiotic. Two camptothecin analogs are now authorized for use as anticancer medicines, and numerous others are in clinical development. Camptothecin has a new method of action that involves inhibiting DNA relaxation by DNA topoisomerase I and, more particularly, stabilizing a covalent binary complex produced between topoisomerase I and DNA.

9.4.4 Vinca Alkaloids

Vinca alkaloids are a kind of organic chemical composed of carbon, hydrogen, nitrogen and oxygen that is frequently obtained from plants. Although the term implies alkali, some do not have alkaline qualities. Many toxic alkaloids also have physiological effects that make them useful as medicines (Schutz et al., 2011). The *Madagascar periwinkle* plant yields vinca alkaloids. These are nitrogenous bases that occur naturally or are semisynthetically synthesized from the pink periwinkle plant *Catharanthus roseus* G. Don (Himes, 1991). Vinca alkaloids were discovered for the first time in the 1950s by Canadian scientists Robert Noble and Charles Beer. Medicinal uses of this plant need the examination of these chemicals for hypoglycemic action, which is minor in comparison to their cytotoxic effects (Kumar & Surapaneni, 2001). They have been used to treat diabetes and high blood pressure, as well as acting as disinfectants. Nonetheless, the vinca alkaloids are critical cancer fighters. Vinblastine (VBL), vinorelbine (VRL), vincristine and vindesine (VDS) are the four primary vinca alkaloids in clinical usage, however, only VCR, VBL and VRL are authorized for use in the United States (Himes, 1991). Since 2008, a novel synthetic vinca alkaloid, vinflunine, has been licensed for therapeutic use in Europe (White, 1998; White, 2000). The major mechanisms of vinca alkaloid cytotoxicity include interactions with tubulin and disruption of microtubule function, particularly of microtubules that comprise the mitotic spindle machinery, resulting in metaphase arrest (Baillie et al., 2002). They can, however, perform a variety of additional metabolic actions that may or may not be connected to their effects on microtubules. Many of the effects that do not involve microtubule disruption occur only when cells are treated with clinically insignificant concentrations of vinca alkaloids. Nonetheless, because microtubules are involved in many non-mitotic processes, vinca alkaloids and other antimicrotubule drugs influence both nonmalignant and malignant cells in the nonmitotic cell cycle (Himes, 1991).

9.4.5 Other Metabolites of Pharmacological Importance

Pharmacologically active metabolites can have a major impact on both the therapeutic and harmful effects of medicines. To completely comprehend the mechanism of action of medications, it is necessary to understand the involvement of active metabolites. Active metabolites can potentially be created as standalone medicines. This study explores a range of biotransformation processes that create active metabolites and their structure–activity connections using illustrated examples. Drugs, for example, undergo a variety of biotransformation processes. These responses serve as the body's first line of defense against any invading xenobiotic material, including medications and environmental toxins. The body is equipped with a battery of enzyme systems to catalyze these processes. The biotransformation metabolites are often more water soluble than the parent molecule and hence more easily handled by excretory organs such as the kidney. The transformation of a xenobiotic chemical into numerous products also reduces the total xenobiotic burden. As a result, most medicines' primary clearance mechanism is metabolism (Guengerich, 2001; Ahmad & Beg, 2001; Nassar et al., 2004). Biotransformation processes can be accompanied by a variety of events, such as the synthesis of chemically stable metabolites with no pharmacological or toxicological activity, or the generation of short-lived chemically reactive metabolites with toxicological activity (Aljabali, 2018; Armand et al., 2018). Biotransformation events can also produce chemically stable metabolites with medicinal efficacy (on or off target). Because of the influence of biotransformation events on the overall destiny of pharmaceuticals (e.g., drug clearance and bioavailability), drug discovery and development processes include in vitro (e.g., metabolic stability screening) and in vivo metabolism and pharmacokinetic research (Nassar et al., 2004; Aljabali, 2018; Armand et al., 2018).

9.5 CHALLENGES AND THE WAY FORWARD IN MULTIFACETED APPLICATIONS

Plants have been utilized as medicines since antiquity due to the availability of several phytocompounds that aid in the prevention and treatment of a variety of illnesses and ailments. They are used in a variety of preparations, including herbal tea, extracts, decoctions, infusions, tinctures and powders (Atanasov et al., 2015; David et al., 2015). Initially, people began to eat plants, which were eventually separated as therapeutic plants with specific pharmacological effects (Baker et al., 1995). Historically, the procedures for using a therapeutic herb for a certain condition were mostly dependent on the plant's history, which was recorded in herbals. According to cave art and literature, medicinal plants were widely used as a source of medicinal drugs for the treatment of diseases in ancient healthcare systems such as Ayurveda, traditional Chinese medications and several ancient medications from civilizations such as Mesopotamia, Persia, Sumeria, Egypt and Arabia. More than 700 medicinal plants are recorded as a source of medications in Ayurvedic works of literature, such as the Charaka Samhita and Sushruta Samhita of the first millennium BC (David et al., 2015), and about 7500 plants are believed to be employed in ancient Indian villages as local health traditions.

Plants can produce secondary metabolites because of metabolism, which are isolated as phytochemicals (Henrich & Beutler, 2013). Plants require phytochemicals for protection and maintenance, although they are not considered essential nutrients (Baker et al., 1995; Henrich & Beutler, 2013). These phytochemicals derived from medicinal plants are the sources of many modern natural medicines. Traditional medical systems have many advantages when it comes to curing illnesses; they do, however, use crude phytochemicals that may include harmful compounds (Baker et al.,

1995). Furthermore, advances in the extraction, purification and characterization of phytochemicals have paved the way for the synthesis of several medical medications that outperform older synthetic and conventional pharmaceuticals. Even with current synthetic medications, several drug types from medicinal plants are irreplaceable. Drugs derived from medicinal plants are also bioactive, biocompatible and bioavailable (Adil et al., 2023b; Adil et al., 2023c; Henrich & Beutler, 2013). The most recent in silico methodologies and biological tools aid in maximizing the therapeutic potential of plant extracts, making them an important drug source.

9.6 SUMMARY AND FUTURE PERSPECTIVES

The discovery and commercial production of taxol highlighted the bioprospecting of endophytes in naturally occurring product-mediated drug development and were a turning point in endophyte biology and research. The main pharmacological metabolites, such as CPT, podophyllotoxin, huperzine A and vinca alkaloids (vincristine/vinblastine), constitute a new pharmacological paradigm, to find and use valuable metabolites for therapeutic purposes. The purity of endophyte strains, subculturing issues, low yields and side effects are some of the current and emerging barriers that need to be further addressed for endophytes to be fully utilized as a manufacturing platform for pharmaceutical therapies at an affordable scale. Recent developments in scientific interventions and high-throughput techniques have significantly aided endophyte research, which aims to address these anticipated difficulties (Tiwari et al., 2022). Aiming to improve various endophyte strains and increase the supply of desired metabolites, genetic engineering of endophytes uses gene overexpression and gene constructions, strain mutagenesis and co-cultivation of suitable strains. The bioprospecting of endophytes and their genetic enhancement are a crucial area of research in drug development programs due to the depletion of natural resources and the scarcity of natural products.

ACKNOWLEDGMENTS

The authors acknowledge their respective institutions for research facilities and support.

REFERENCES

Adil, M, P Tiwari, J Chen, and S Kanwal. 2023a. "Major bioactive metabolites and antimicrobial potential of Orchidaceae Fungal endophytes." In: *Advances in Orchid Biology, Biotechnology*, and Omics, Tiwari, P, Chen, J, Eds., Springer Publishers, 2023a. ISBN: 13 978-9819910786.

Adil, M, P Tiwari, J Chen, and S Kanwal. 2023b. "Major bioactive metabolites and antimicrobial potential of Orchidaceae Fungal endophytes." In: *Advances in Orchid Biology, Biotechnology, and Omics*, Tiwari, P., and Chen, J-T, Eds., Springer Publishers, 2023. ISBN: 13 978-9819910786.

Adil, M, P Tiwari, J Chen, and S Kanwal. 2023c. "Plants-derived bioactive compounds as potential ACE-2 inhibitors against SARS-CoV-2 infection." In: *Ethnopharmacology and Drug Discovery for COVID-19: Anti-SARS-CoV-2 Agents from Herbal Medicines and Natural Products*, Chen, J., Ed., CRC Press/Taylor & Francis Group, LLC. ISBN: 9781032347967

Adil, M, P Tiwari, J Chen, and S Kanwal. 2023c. "Therapeutic potential of essential oils against SARS-CoV-2." In: *Ethnopharmacology and Drug Discovery for COVID-19: Anti-SARS-CoV-2 Agents from Herbal Medicines and Natural Products*, Chen, J, Ed., CRC Press/Taylor & Francis Group, LLC. ISBN: 9781032347967

Aghdam, SA, and AMV Brown. 2021. "Deep learning approaches for natural product discovery from plant endophytic microbiomes." *Environ. Microbiome* 16:6.

Ahmad, I, and AZ Beg. 2001. "Antimicrobial, and phytochemical studies on 45 Indian medicinal plants against multi-drug resistant human pathogens." *J. Ethnopharmacol.* 74, no. 2:113–123.

Aishwarya, SANI, N Nagam, T Vijaya, and RV Netala. 2017. Screening and identification of heavy metal-tolerant endophytic fungi Lasiodiplodia theobromae from Boswellia ovalifoliolata an endemic plant of Tirumala hills. *Asian J Pharm. Clin. Res.* 10:488–491.

Akone, SH, C-D Pham, H Chen, ARB Ola, F Ntie-Kang, and Proksch, P. 2018. "Epigenetic modification, co-culture, and genomic methods for natural product discovery." *Phys. Sci. Rev.* 4:1–13.

Aljabali AA. 2018. "Viral nanoparticles: a drug delivery platform." *J. Pharm. Toxicol.* 1, no. 1:1–2.

Anjanamurthy, C, and YB Basavaraju. 1999. "Synthesis of podophyllotoxin and related analogues: part X-tetralone esters as intermediates for the synthesis of analogues of b-apopicropodophyllin." *Ind. J. Chem. B* 38:137–140.

Anon. 1992. "Taxol recommended for second-line therapy in platinum-resistant ovarian cancer." *Cancer Letters* 18, no. 45:1.

Anyaogu, DC, and UH Mortensen. 2015. "Heterologous production of fungal secondary metabolites in Aspergilli." *Front. Microbiol.* no. 6:1–6.

Aravind, R, A Kumar, SJ Eapen, and KV Ramana, 2009. Endophytic bacterial flora in root and stem tissues of black pepper (*Piper nigrum* L.) genotype: isolation, identification, and evaluation against *Phytophthora capsici. Lett. Appl. Microbio.l* 48 (1):58–64.

Arimondo, P, C Boukarim, C Bailly, D Dauzonne, D, and C Monneret. 2000. "Design of two etoposide–amsacrine conjugates: topoisomerase II and tubuline polymerization inhibition and relation to cytotoxicity." *Anticancer Drug Des.* 15:413–421.

Armand, Z, A Abel, RR Hanitra, RD Joseph, and A Mahandrimanana. 2018. "Ethnobotanical, antimicrobial and phytochemical screening of *Euphorbia thymoflia* L. (Kinononono madiniky) to cure the menstrual hemorrhage of a woman Sakalava Bemazava in the Northern Region of Madagascar." *Int J Plant Res.* 8, no. 1:1–7.

Atanasov AG, B Waltenberger, E Pferschy-Wenzig, T Linder, C Wawrosch, P Uhrin, V Temml, L Wang, S Schwaiger, E Heiss, JM Rollinger, D Schuster, JM Breuss, V Bochkov, MD Mihovilovic, B Kopp, R Bauer, VM Dirsch, and H Stuppner. 2015. "Discovery and resupply of pharmacologically active plant-derived natural products: a review." *Biotechnol Adv.* 33:1582–1614.

Ayres, DC, and JD Loike. 1990. *Lignans. Chemical, Biological and Clinical Properties*. Cambridge University Press, Cambridge.

Baillie, TA, MN Cayen, H Fouda, RJ Gerson, JD Green, SJ Grossman, LJ Klunk, B LeBlanc, DG Perkins, LA Shipley 2002. "Drug metabolites in safety testing." *Toxicol. Appl. Pharmacol.* 182, no. 3:188–196.

Bajpai, M, J Das, and P Tiwari. 2022. "Molecular and Biochemical methods for isolation, and characterization of Endophytes." In: *Endophytes: Types, Potential Uses, and Mechanisms of Action*, Tiwari P, Ed., Nova Publishers. ISBN: 979-8-88697-045-6.

Baker, JT, RP Borris, B Carte, GA Cordell, DD Soejarto, GM Cragg, MP Gupta, MM Iwu, DR Madulid, and VE Tyler. 1995. "Natural products drug discovery and development: new perspectives on international collaboration." *J Nat Prod.* 58, no. 9:1325–1357.

Baral, B, A Akhgari, and M Metsä-Ketelä. 2018. "Activation of microbial secondary metabolic pathways: avenues and challenges." *Synth. Syst. Biotechnol.* 3:163–178.

Barret, JM, C Etievant, C Baudouin, K Skov, M Charveron, and BT Hill. 2002. "F 11782, a novel catalytic inhibitor of topoisomerases I and II, induces atypical, yet cytotoxic DNA double-strand breaks in CHO-K1 cells." *Anticancer Res.* 22:187–192.

Bennouna, J, JP Delord, M Campone, and Nguyen L. 2008. "Vinflunine: A new microtubule inhibitor agent." *Clin Cancer Res.* 14:1625–1632.

Berdy, J. 2005. "Bioactive microbial metabolites." *J. Antibiot.* 58:1–26.

Bibi, S, A Hussain, M Hamayun, H Rahman, A Iqbal, M Shah, M Irshad, M Qasim, and B Islam. 2018. Bioremediation of hexavalent chromium by endophytic fungi; safe and improved production of *Lactuca sativa* L. *Chemosphere* 211:653–663.

Bissett, D, and SB Kaye. 1993. "Taxol and Taxotere-current status and future prospects." *Eur. J. Cancer* 29A:1228.

Blin, K, S Shaw, K Steinke, R Villebro, N Ziemert, SY Lee, MH Medema, and T Weber. 2019. "antiSMASH 5.0: Updates to the secondary metabolite genome mining pipeline." *Nucleic Acids Res.* 47:W81–W87.

Bose, Subir Kumar, Mangalam Bajpai, Jyotirmoyee Das, Ashutosh Gautam, and Pragya Tiwari. 2022. "Endophytic actinomycetes: Overview, distribution and multi-faceted applications." In: *Endophytes: Types, Potential Uses, and Mechanisms of Action*, Tiwari, P., Ed., Nova publishers. ISBN: 979-8-88697-045-6.

Chadha, N, R Prasad, and A Varma. 2015. "Plant promoting activities of fungal endophytes associated with tomato roots from central Himalaya, India and their interaction with Piriformospora indica." *Int. J. Pharm. Bio Sci.* 6:333–343.

Chen, L, QY Zhang, M Jia, QL Ming, W Yue, K Rahman, LP, Qin, and T Han. 2016. "Endophytic fungi with antitumor activities: their occurrence and anticancer compounds." *Crit. Rev. Microbiol.* 42:454–473.

Chen, WM. 1990. "Chemical constituents and biological activity of plants of the genus Taxus." *Acta Pharm Sinica* 25:227.

Conway, KR, and CN Boddy. 2012. "ClusterMine360: A database of microbial PKS/NRPS biosynthesis." *Nucleic Acids Res.* 41:402–407.

Corre, C, L Song, S O'Rourke, KF Chater, and GL Challis. 2008. "2-Alkyl-4-hydroxymethylfuran-3-carboxylic acids, antibiotic production inducers discovered by Streptomyces coelicolor genome mining." *Proc. Natl. Acad. Sci. USA* 105:17510–17515.

David, B, JL Wolfender, and DA Dias. 2015. "The pharmaceutical industry and natural products: historical status and new trends." *Phytochem Rev.* 14, no. 2:299–315.

Deshmukh, SK, V Prakash, and N Ranjan. 2017. "Recent advances in the discovery of bioactive metabolites from *Pestalotiopsis*." *Phytochem. Rev.* 16:883–920.

Dias, DA, S Urban, and U Roessner. 2012. "A historical overview of natural products in drug discovery." *Metabolites* 2:303–336.

El-Sayed, ASA, NZ Mohamed, S Safan, MA Yassin, L Shaban, AA, Shindia, GS Ali, and MZ Sitohy. 2019. "Restoring the taxol biosynthetic machinery of *Aspergillus terreus* by *Podocarpus gracilior pilger* microbiome, with retrieving the ribosome biogenesis proteins of WD40 superfamily." *Sci. Rep.* 9:11534.

Epstein, SC, LK Charkoudian, and MH Medema. 2018. "A standardized workflow for submitting data to the minimum information about a biosynthetic gene cluster (MIBiG) repository: Prospects for research-based educational experiences." *Stand. Genomic Sci.* 13:16.

Faust, K, and J Raes. 2012. "Microbial interactions: From networks to models." *Nat. Rev. Microbiol.* 10, no. 8:538–550.

Gao,T, D Qin, S Zuo, Y Peng, J Xu, B Yu, H Song, and J Dong. 2020. Decolorization and detoxification of triphenylmethane dyes by isolated endophytic fungus, Bjerkandera adusta SWUSI4 under non-nutritive conditions. *Bioresour. Bioprocess.* 7:53.

Ghaffari, MR, M Mirzaei, M Ghabooli, B Khatabi, Y Wu, M Zabet-Moghaddam, G Mohammadi-Nejad, PA Haynes, and MR Hajireza. 2019. Root endophytic fungus *Piriformospora indica* improves drought stress adaptation in barley by metabolic and proteomic reprogramming. *Environ. Exp. Bot.* 157, 197–210.

Gidding, CE, SJ Kellie, WA Kamps, and SS de Graaf. 1999. "Vincristine revisited." *Crit Rev Oncol Hematol.* 29:267–287.

Guengerich, FP. 2001. "Common and uncommon cytochrome P450 reactions related to metabolism and chemical toxicity." *Chem. Res. Toxicol.* 14:611–650.

Guo, B, J Dai, S Ng, Y Huang, C Leong, W Ong, and BK Carte. 2000. "Cytonic acids A & B: Novel tridepside inhibitors of hCMV protease from the endophytic fungus Cytonaema species." *J. Nat. Prod.* 63:602–604.

Guo, F, S Xiang, L Li, B Wang, J Rajasärkkä, K Gröndahl-Yli-Hannuksela, G Ai, M Metsä-Ketelä, and K Yang. 2015. "Targeted activation of silent natural product biosynthesis pathways by reporter-guided mutant selection." *Metab. Eng.* 28:134–142.

Gupta, N, J Sabat, R Parida, and D Kerkatta. 2007. "Solubilization of tricalcium phosphate and rock phosphate by microbes isolated from chromite, iron and manganese mines." *Acta Bot. Croat.* 66:197–204.

Hamayun, M, A Hussain, SA Khan, HY Kim, AL Khan, M Waqas, M Irshad, A Iqbal, G Rehman, S, Jan, et al. 2017. "Gibberellins producing endophytic fungus *Porostereum spadiceum* AGH786 rescues growth of salt-affected soybean." *Front. Microbiol.* 8:686.

Hamayun, M, SA Khan, N Ahmad, D-S Tang, S-M Kang, C-I Na, E-Y Sohn, Y-H Hwang, D-H Shin, and B-H Lee. 2009. "Cladosporium sphaerospermum as a new plant growth-promoting endophyte from the roots of Glycine max (L.) Merr." *World J. Microbiol. Biotechnol.* 25, 627–632.

Henrich, CJ, and JA Beutler. 2013. "Matching the power of high throughput screening to the chemical diversity of natural products." *Nat Prod Rep.* 30, no. 10:1284–1298.

Himes, RH. 1991. "Interactions of the *Catharanthus* (Vinca) alkaloids with tubulin and microtubules." *Pharmacol Ther.* 51:257–267.

Johnson, JM, T Alex, and R Oelmuller. 2014. *Piriformospora indica*: the versatile and multifunctional root endophytic fungus for enhanced yield and tolerance to biotic and abiotic stress in crop plants. *J. Trop. Agr.* 52:103–122.

Johnson, LJ, AC de Bonth, LR Briggs, JR Caradus, SC Finch, DJ Fleetwood, LR Fletcher, DE Hume, RD Johnson, AJ Popay, et al. 2013. "The exploitation of Epichloae endophytes for agricultural benefit." *Fungal Diver.* 60, 171–188.

Ichikawa, N, M Sasagawa, M Yamamoto, H Komaki, Y Yoshida, S Yamazaki, and N Fujita. 2013. "DoBISCUIT: A database of secondary metabolite biosynthetic gene clusters." *Nucleic Acids Res.* 41:408–414. doi: 10.1093/nar/gks1177

Khan, AR, I Ullah, M Waqas, R Shahzad, SJ Hong, GS Park, BK Jung, IJ Lee, and JH Shin. 2015. "Plant growth-promoting potential of endophytic fungi isolated from Solanum nigrum leaves." *World J. Microbiol. Biotechnol.* 31, 1461–1466.

Koo, T, J Lee, and S Hwang. 2019. "Development of an interspecies interaction model: An experiment on Clostridium cadaveris and Clostridium sporogenes under anaerobic condition." *J. Environ. Manag.* 237:247–254.

Kuang, X, S Sun, J Wei, Y Li, and C Sun. 2019. "Iso-Seq analysis of the *Taxus cuspidata* transcriptome reveals the complexity of Taxol biosynthesis." *BMC Plant Biol.* 19: 210.

Kumar, GN, and S Surapaneni. 2001. "Role of drug metabolism in drug discovery and development." *Med. Res. Rev.* 21:397–411.

Kusari, P, S Kusari, M Spiteller, and O Kayser. 2015. "Implications of endophyte-plant crosstalk in light of quorum responses for plant biotechnology." *Appl. Microbiol. Biotechnol.* 99:5383–5390.

Kusari, S, C Hertweck, and M Spiteller. 2012. "Chemical ecology of endophytic fungi: Origins of secondary metabolites." *Chem. Biol.* 19:792–798.

Latendresse, M, M Krummenacker, M Trupp, and PD Karp. 2012. "Construction and completion of flux balance models from pathway databases." *Bioinformatics* 28: 388–396.

Lautru, S, RJ Deeth, LM Bailey, and GL Challis. 2005. "Discovery of a new peptide natural product by Streptomyces coelicolor genome mining." *Nat. Chem. Biol.* 1: 265–269.

Lenaz, L, and MD De Furia. 1993. "Taxol: a novel natural product with significant anticancer activity." *Fitoterapia* 64, no. Suppl. 1:27.

Ling, OM, LP Teen, A Mujahid, P Proksch, and M Müller. 2016. Initial screening of Mangrove endophytic fungi for antimicrobial compounds and heavy metal biosorption potential. *Sains Malaysiana.* 45, 1063–1071.

Ludwig-Müller, J. 2015. "Plants and endophytes: Equal partners in secondary metabolite production?" *Biotechnol. Lett.* 37:1325–1334.

Marín, F, H Navarrete, and A Narvaez-Trujillo. 2018. Total petroleum hydrocarbon degradation by endophytic fungi from the Ecuadorian Amazon. *Advan. Microbiol.* 8:1029–1053.

Matsumoto, A, and Y Takahashi. 2017. "Endophytic actinomycetes: Promising source of novel bioactive compounds." *J. Antibiot.* 70:514–519.

Meshram, V, and M Gupta. 2019. "Endophytic fungi: A quintessential source of potential bioactive compounds." In: *Endophytes for a Growing World*, Hodkinson, T.R., Doohan, F.M., Saunders, M.J., Murphy, B.R., Eds., Cambridge University Press, pp. 277–309.

Miller, RW. 1981. "A brief survey of Taxus alkaloids and other taxane derivatives." *J. Nat. Prod.* 43:425.

Mori, T, JKB Cahn, MC Wilson, RA Meoded, V Wiebach, AFC Martinez, EJN Helfrich, A Albersmeier, D Wibberg, S Dätwyler, S, et al. 2018. "Single-bacterial genomics validates rich and varied specialized metabolism of uncultivated *Entotheonella* sponge symbionts." *Proc. Natl. Acad. Sci. USA* 115:1718–1723.

Mousa, WK, C Shearer, V Limay-Rios, CL Ettinger, JA Eisen, and MN Raizada. 2016. "Root hair endophyte stacking in finger millet creates a physicochemical barrier to trap the fungal pathogen *Fusarium graminearum*." *Nat. Microbiol.* 1:16167.

Nagendran, K, G Karthikeyan, PM Faisal, P Kalaiselvi, M Raveendran, K Prabakar, and T Raguchander. 2014. Exploiting endophytic bacteria for the management of sheath blight disease in rice. *Biol. Agric. Hortic.* 30 (1), 8–23.

Nassar, AF, AM Kamel, C Clarimont. 2004. "Improving the decision-making process in the structural modification of drug candidates: enhancing metabolic stability." *Drug Discov. Today* 9, no. 23:1020–1028.

Nath, R, GD Sharma, M Barooah. 2012. "Plant growth promoting endophytic fungi Isolated from tea (*Camellia sinensis*) shrubs of Assam, India." Appl. Ecol. Environ. Res. 13(3): 877–891.

Newman, DJ, and GM Cragg. 2020. "Natural products as sources of new drugs over the nearly four decades from 01/1981 to 09/2019." *J. Nat. Prod.* 83:770–803.

Pan, R, X Bai, J Chen, H Zhang, and H Wang. 2019. "Exploring structural diversity of microbe secondary metabolites using OSMAC strategy: a literature review." *Front. Microbiol.* 10:294.

Pan, SY, SF Zhou, SH Gao, ZL Yu, SF Zhang, MK Tang, JN Sun, DL Ma, YF Han, WF Fong, et al. 2013. "New perspectives on how to discover drugs from Herbal medicines: CAM's outstanding contribution to modern therapeutics." *Evid. Based Complement. Altern. Med.* 2:627375.

Panaccione, DG, WT Beaulieu, and D Cook. 2014. "Bioactive alkaloids in vertically transmitted fungal endophytes." *Funct. Ecol.* 28:299–314.

Pietro-Souza, W, FC Pereira, IS Mello, FFF Stachack, AJ Terezo, CN da Cunha, JF White, H Li, and MA Soares. 2020. Mercury resistance and bioremediation mediated by endophytic fungi. *Chemosphere*. 240, 124874.

Rashmi, M, and SV Venkateswara. 2019. "Secondary metabolite production by Endophytic fungi: The gene clusters, nature, and expression." In: *Endophytes and Secondary Metabolites Reference Series in Phytochemistry*, Jha, S., Ed., pp. 475–490. Springer Nature. ISBN: 978-3-319-90483-2.

Ravikrishnan, A, LM Blank, S Srivastava, and K Raman. 2020. "Investigating metabolic interactions in a microbial co-culture through integrated modeling and experiments." *Comput. Struct. Biotechnol J.* 18:1249–1258.

Rinu, K, P Sati, and A Pandey. 2014. Trichoderma gamsii (NFCCI 2177): A newly isolated endophytic, psychrotolerant, plant growth promoting, and antagonistic fungal strain. *J. Basic Microbiol.* 54:408–417.

Rodriguez Estrada AE, W Jonkers, HC Kistler, and G May. 2012, July. Interactions between Fusarium verticillioides, Ustilago maydis, and Zea mays: an endophyte, a pathogen, and their shared plant host. *Fungal Genet Biol.* 49(7):578–587.

Rodriguez, RJ, JF White, AE Arnold, and RS Redman. 2009. "Fungal endophytes: Diversity and functional roles." *New Phytol.* 182:314–330.

Sane, S. and S Mehta. 2015. Isolation and evaluation of rock phosphate solubilizing fungi as potential biofertilizer. *J Fertil Pestic.* 6:156–160.

Schutz, FA, J Bellmunt, JE Rosenberg, and TK Choueiri. 2011. "Vinflunine: Drug safety evaluation of this novel synthetic vinca alkaloid." *Expert Opin Drug Saf.* 10:645–53.

Seo, WT, WJ Lim, EJ Kim, HD Yun, YH Lee, and KM Cho. 2010. Endophytic bacterial diversity in the young radish and their antimicrobial activity against pathogens. *J. Korean Soc. Appl. Biol. Chem.* 53:493–503.

Singh, SB, JG Ondeyka, N Tsipouras, C Ruby, V Sardana, M Schulman, M Sanchez, F Pelaez, MW Stahlhut, S Munshi, et al. 2004. "Hinnuliquinone, a C2-symmetric dimeric non-peptide fungal metabolite inhibitor of HIV-1 protease." *Biochem. Biophys. Res. Commun.* 324:108–113.

Šiši´c, A, J Ba´canovi´c, and MR Finckh. 2017. Endophytic Fusarium equiseti stimulates plant growth and reduces root rot disease of pea (Pisum sativum L.) caused by Fusarium avenaceum and Peyronellaea pinodella. *Eur. J Plant Pathol.* 148:271–282.

Snader, KM. 1990. "Overview of supply issues, Workshop on Taxol and Taxus: Current and Future Perspectives." NCI, Bethesda, MD, June 26, 1990.

Soldi, E, C Casey, BR Murphy, and TR Hodkinson. 2020. Fungal endophytes for grass-based bioremediation: an endophytic consortium isolated from *Agrostis stolonifera* stimulates the growth of *Festuca arundinacea* in Lead contaminated soil. *J. Fungi* 6 (4):254. https://doi.org/10.3390/jof6040254

Soleimani, M, M Afyuni, MA Hajabbasi, F Nourbakhsh, MR Sabzalian, and JH Christensen. 2010. Phytoremediation of an aged petroleum contaminated soil using endophyte infected and non-infected grasses. *Chemosphere*. 81:1084–1090.

Soliman, SSM, JS Greenwood, A Bombarely, LA Mueller, R Tsao, DD Mosser, and MN Raizada. 2015. "An endophyte constructs fungicide-containing extracellular barriers for its host plant." *Curr. Biol.* 25:2570–2576.

Spagnoletti, F, N Tobar, AF Di Pardo, V Chiocchio,and R Lavado. 2017. Dark septate endophytes present different potential to solubilize calcium, iron and aluminium phosphates. *Appl. Soil Ecol.* 111, 25–32.

Staniek, A, J Woerdenbag, and O Kayser. 2008. "Endophytes exploiting biodiversity for the improvement of natural product-based drug discovery." *J. Plant Interact.* 3:75–98.

Starcevic, A, J Zucko, J Simunkovic, PF Long, J Cullum, and D Hranueli. 2008. "ClustScan: An integrated program package for the semi-automatic annotation of modular biosynthetic gene clusters and in silico prediction of novel chemical structures." *Nucleic Acids Res.* 36:6882–6892.

Stierle, A, G Strobel, and D Stierle. 1993. "Taxol and taxane production by *Taxomyces andreanae*, an endophytic fungus of Pacific yew." *Science* 260:214–216.

Stolyar, S, S Van Dien, KL Hillesland, N Pinel, TJ Lie, JA Leigh, and DA Stahl. 2007. "Metabolic modeling of a mutualistic microbial community." *Mol. Syst. Biol.* 3:92.

Strobel, GA, E Ford, J Worapong, JK Harper, AM Arif, DM Grant, PC Fung, and K Chan. 2002. "Isopestacin, an isobenzofuranone from *Pestalotiopsis microspora*, possessing antifungal and antioxidant activities." *Phytochemistry* 60:179–183.

Sun, C, JM Johnson, D Cai, I Sherameti, R Oelmüller, and B Lou. 2010. *Piriformospora indica* confers drought tolerance in Chinese cabbage leaves by stimulating antioxidant enzymes, the expression of drought-related genes, and the plastid-localized CAS protein. *J. Plant Physiol.* 167:1009–1017.

Tan, RX, and WX Zou. 2000. "Endophytes: a rich source of functional metabolites." *Nat. Prod. Rep.* 18:448–459.

Tiwari, P. 2022. Triterpenoids from *Gymnema sylvestre* R. Br: biological significance in the treatment of Diabetes. Phytochemicals and medicinal plants in food design: strategies and technologies for improved healthcare. In: *Innovations in Agricultural and Biological Engineering*, Goyal, MR, Birwal, P, Mishra, SK, Eds., Apple Academic Press, volume 31, ISBN 9781774639450.

Tiwari, P and H Bae. 2020. "Horizontal gene transfer and endophytes: An implication for the acquisition of novel traits." *Plants* 9:305.

Tiwari, P, and H Bae. 2022. "Endophytic Fungi: key insights, emerging prospects, and challenges in natural product drug discovery." *Microorganisms* 10:360.

Tiwari, P, and L Dufossé. 2023b. "Focus and insights into the synthetic biology-mediated chassis of economically important fungi for the production of high-value metabolites." *Microorganisms* 11:1141.

Tiwari, P, M Adil, and H Bae. 2023. "Endophyte-mediated bioremediation- an efficient biological strategy in ecological subsistence and agriculture." In: *Endophytic and Arbuscular Mycorrhizal Fungi and their Role in Sustainable Agriculture*, Erwin, DJ., Ed., Nova Publishers. ISBN: 979-8-88697-766-0.

Tiwari, P, S Kang, and H Bae. 2023a. "Plant-endophyte associations: Rich yet under-explored sources of novel bioactive molecules and applications." *Microbiological Research* 266:127241.

Tiwari, P, and L Dufossé. 2023b. "Focus and insights into the synthetic biology-mediated chassis of economically important fungi for the production of high-value metabolites." *Microorganisms* 11:1141.

Tiwari, P, Y Srivastava, H Bae. 2021b. "Endophytes: Trend of pharmaceutical design of Endophytes as anti-infective." *Curr Top. Med. Chem.* 21:1572–1586.

Tiwari, P, Y Srivastava, and V Kumar. 2022. "Antimicrobial peptides as effective agents against drug-resistant pathogens." In: *Antimicrobial Resistance*, Kumar, V., Shriram, V., Paul, A., Thakur, M., Eds., Springer, pp. 289–322.

Tiwari, P, SK Bose, K Park, L Dufossé, and M Fouillaud. 2024. "Plant-microbe interactions under the extreme habitats and their potential applications." *Microorganisms* 12:448.

Tiwari, P, A Katyal, MF Khan, GM Ashraf, and K Ahmed. 2019. "Lead optimization resources in drug discovery studies." *Endocr. Metab. Immune Disord.-Drug Targets* 19:754–774.

Tiwari, P, T Khare, V Shriram, H Bae, V Kumar. 2021a. "Exploring synthetic biology strategies for producing potent antimicrobial phytochemicals." *Biotechnol. Adv.* 48:107729.

Tyo, KE, HS Alper, and GN Stephanopoulos. 2007. "Expanding the metabolic engineering toolbox: More options to engineer cells." *Trends Biotechnol.* 25:132–137.

Venieraki, A, M Dimou, and P Katinakis. 2017. "Endophytic fungi residing in medicinal plants have the ability to produce the same or similar pharmacologically active secondary metabolites as their hosts." *Hell. Plant Protect. J.* 10:51–66.

Wang, H, S Marcišauskas, BJ, Sánchez, I Domenzain, D Hermansson, R Agren, J Nielsen, and EJ Kerkhoven. 2018. "RAVEN 2.0: A versatile toolbox for metabolic network reconstruction and a case study on *Streptomyces coelicolor*." *PLoS Comput. Biol.* 14:1–17.

Wani, MC, HL Taylor, ME Wall, P Coggon, and AT McPhail. 1971. "Plant antitumor agents. VI. The isolation and structure of taxol, a novel antileukemic and antitumor agent from *Taxus brevifolia*." *J. Amer. Chem. Soc.* 93:2325.

White, JF, KL Kingsley, KP Kowalski, I Irizarry, A Micci, MA Soares, et al. 2017. Disease protection and allelopathic interactions of seed-transmitted endophytic pseudomonads of invasive seed grass (*Phragmites australis*). *Plant Soil* 422, 195–208.

White, RE. 1998. "Short- and long-term projections about the use of drug metabolism in drug discovery and development." *Drug Metab. Dispos.* 26:1213–1216.

White, RE. 2000. "High-throughput screening in drug metabolism and pharmacokinetic support of drug discovery." *Annu. Rev. Pharmacol. Toxicol.* 40:133–157.

Wu, T, J Xu, W Xie, Z Yao, H Yang, C Sun, and X Li. 2018. *Pseudomonas aeruginosa* L10: A hydrocarbon-degrading, biosurfactant-producing, and plant-growth-promoting endophytic bacterium isolated from a Reed (*Phragmites australis*). *Front. Microbiol.* 9, 1087. https://doi.org/10.3389/fmicb.2018.01087.

Xu, D, BY Zhang, and XL Yang. 2016. "Antifungal monoterpene derivatives from the plant endophytic fungus *Pestalotiopsis foedan*." *Chem. Biodivers.* 13:1422–1425.

Xu, F, Y Wu, C Zhang, KM Davis, K Moon, LB Bushin, and MR Seyedsayamdost. 2019. "A genetics-free method for high-throughput discovery of cryptic microbial metabolites." *Nat. Chem. Biol.* 15:161–168.

Zazopoulos, E, K Huang, A Staffa, W Liu, BO Bachmann, K Nonaka, J Ahlert, JS Thorson, B Shen, and CM Farnet. 2003. "A genomics-guided approach for discovering and expressing cryptic metabolic pathways." *Nat Biotech.* 21:187–190.

Zhang, P, PP Zhou, and LJ Yu. 2009. "An endophytic taxol-producing fungus from Taxus media, *Cladosporium cladosporioides* MD2." *Curr. Microbiol.* 59:227–232.

Zahoor, M. M Irshad, H Rahman, M Qasim, SG Afridi, M Qadir, and A Hussain. 2017. Alleviation of heavy metal toxicity and phytostimulation of *Brassica campestris* L. by endophytic *Mucor* sp. MHR-7. *Ecotoxicol. Environ. Saf.* 142, 139–149.

10 Endophytic Fungi as a Potential Source of Natural Pigments

Madhavi Tiwari and Ashish Saraf*
*Corresponding author.

10.1 INTRODUCTION

Endophytic fungi are a group of microorganisms that live within the tissues of plants without causing apparent harm to their host (Clay, 1988). These fungi establish symbiotic relationships with various plant species and can be found in the roots, stems, leaves and other plant tissues. Unlike pathogenic fungi that cause diseases in plants, endophytic fungi form mutually beneficial associations with their host plants. The relationship between endophytic fungi and plants is generally considered to be symbiotic or mutualistic. In a mutualistic association, both the plant and the fungus benefit (Strobel, 2018, 57). The endophytic fungi receive nutrients and a protected environment inside the plant, while the plants can derive several advantages from their fungal partners. The exact benefits vary depending on the specific fungal species and the host plant, but some common advantages include enhanced growth, increased resistance to environmental stresses and improved tolerance to diseases and pests (Pathak et al., 2022, 2203).

Endophytic fungi play crucial roles in plant ecosystems, forming symbiotic relationships with various plant species (Singh et al., 2021, 197). One primary role of endophytic fungi involves enhancing nutrient uptake efficiency in plants. These fungi can solubilize nutrients, such as phosphorus and nitrogen, making them more accessible to the host plant. This contributes to improved nutrient status and overall growth (Khare et al., 2018, 2732).

Endophytic fungi are gaining recognition as a promising source of natural pigments, offering a diverse array of colors with potential applications in various industries. These fungi, residing within the tissues of plants, have been found to produce pigments such as anthocyanins, carotenoids and melanins, showcasing a wide spectrum of colors (Strobel., 2014). One of the key advantages of exploring endophytic fungi for pigment production is the biological and environmental friendliness associated with natural pigments. Compared to synthetic alternatives, pigments derived from endophytic fungi are considered more environmentally sustainable, with potential benefits (Pimentel et al., 2011, 383–406). Moreover, the stability of natural pigments from endophytic fungi under different conditions, including variations in pH, temperature and light exposure, is noteworthy. This stability is crucial for applications in the food, cosmetics and textiles industries, where consistent color is highly desirable (Sarsaiya et al., 2019, 316–334). Understanding the intricate mechanisms behind pigment production in endophytic fungi provides insights into their ecological roles and the dynamic interactions within plant ecosystems. Additionally, the antioxidant properties of many

DOI: 10.1201/9781003486275-12

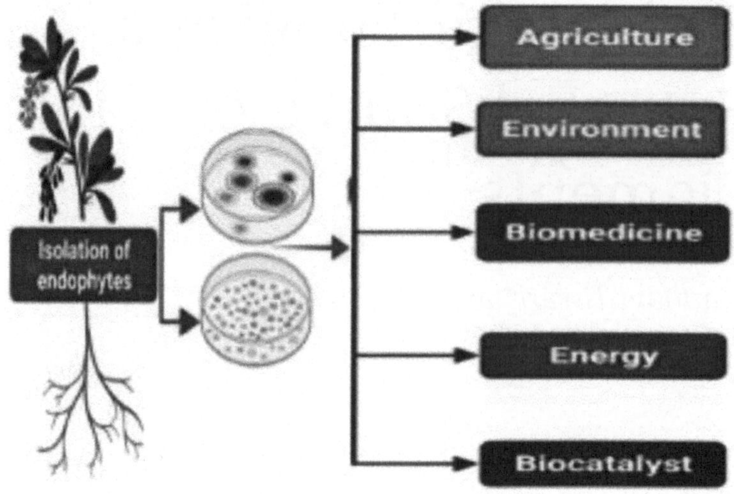

FIGURE 10.1 Applications of endophytes isolated from plants. The diagram shows the process of isolating endophytes and their potential applications in various fields including agriculture, environment, biomedicine, energy and biocatalysis.

pigments produced by endophytic fungi contribute to the potential health benefits of products containing these pigments (Tan & Zou, 2001, 448–459).

The biotechnological potential of endophytic fungi is evident in the cultivation of these fungi in controlled environments, allowing for scalable production of natural pigments. Advances in biotechnological techniques, particularly in fermentation processes, facilitate the efficient extraction and purification of these pigments for commercial use (Schardl et al., 2013, e1003323). The potential applications of natural pigments derived from endophytic fungi span a range of industries, including food, cosmetics, textiles and pharmaceuticals. These pigments can serve as colorants in food and beverages, natural dyes in textiles, and additives in cosmetic products, offering versatile solutions across multiple sectors (Chepkirui et al., 2019). Furthermore, the exploration of endophytic fungi for pigment production contributes to biodiversity studies and the discovery of novel compounds. The unique microbial communities within different plants provide a rich source of untapped bioactive molecules, fostering innovation and supporting the development of environmentally friendly products (Kusari et al., 2012, 107–115).

Fungal pigmentation refers to the ability of fungi to produce pigments that contribute to their coloration. This phenomenon has garnered significant attention due to its relevance in various fields, including medical mycology, agriculture and biotechnology. The escalating concerns surrounding the adverse effects of artificial colorants on consumers and the environment have fueled a growing interest in natural coloring alternatives. This chapter delves into the realm of natural colorants, focusing on the pivotal role of endophytic fungi as key producers. It provides an overview of the major sources of natural colorants, emphasizing the significance of plants and microbes. The exploration of endophytic fungi as primary sources of natural compounds and an overview of pigment production in these fungi underscore their potential as nature's palette for sustainable coloration. The chapter explores the significant role of endophytic fungi in producing a diverse array of pigments with applications ranging from textile dyes to cosmetic formulations and food supplements along with various techniques for extraction and identification and characterization of extracted pigments.

10.2 INCREASING DEMAND FOR NATURAL COLORANTS

Growing consumer awareness and preference for natural and organic products have led to an increased demand for natural colorants (Akhtar et al., 2015, 123–145). Industries such as food, cosmetics and textiles are witnessing a shift toward sustainable and eco-friendly alternatives (Jaiswal & Abu-Ghannam, 2019, 87–100). Synthetic colorants often involve complex chemical processes leading to environmental pollution. Disposal of synthetic colorants contributes to soil and water pollution (MacCan et al., 2007). Some artificial colorants have been associated with adverse health effects, including allergic reactions and hyperactivity in children. Long-term exposure to certain synthetic colorants has raised concerns about potential carcinogenicity (Gao et al., 2020).

Increased awareness among consumers about the environmental impact of synthetic colorants has driven the demand for sustainable alternatives (Harnischfeger & Lohmann, 2018). Consumers are seeking products with minimal ecological footprints. Many companies are adopting sustainability as a core value, leading to the incorporation of natural colorants in their products (Abad-Somovilla & Bermejo-Barrera, 2018). Global initiatives and certifications promoting sustainable practices are gaining traction.

10.3 ENDOPHYTIC FUNGI AND PIGMENT PRODUCTION AS "NATURAL COLORANTS"

Endophytic fungi play a fascinating role in the production of various pigments, contributing to the diversity of colors observed in plants. The production of pigments by endophytic fungi has ecological significance and potential applications in various fields, including agriculture and medicine (Strobel, 2014). Pigments derived from endophytic fungi are often considered environmentally friendly alternatives to synthetic dyes. The biological nature of these pigments aligns with the growing demand for sustainable and eco-friendly products, particularly in industries such as textiles and food (Pimentel et al., 2011, 383–406). The study of pigment production by endophytic fungi enhances our understanding of plant–fungus interactions and has implications for various fields.

This chapter dives into the world of natural colorants, spotlighting plants, microbes and the starring role of endophytic fungi. Key resources like Wrolstad's *Natural Colorants: Chemistry and Biochemistry* and Carle's *Handbook on Natural Pigments in Food and Beverages* lay down the basics of natural colorants (Wrolstad, 2019; Carle & Schweiggert, 2019). Shifting gears to endophytic fungi, those unsung heroes in biodiversity, Strobel and Daisy's look into microbial endophytes and Rodriguez's insights on fungal endophytes show how diverse and important they are in plant ecosystems (Strobel & Daisy, 2003; Rodriguez, 2009). Kusari's exploration of endophytic fungi in *Cannabis sativa* and Pimentel's dive into antimicrobial activity underscore these fungi's potential across various industries (Kusari et al., 2012). Aly's findings on endophytic fungi as sources of antifungal compounds and Strobel's review of the chemistry and biodiversity of fungal endophytes add more weight to their significance (Aly et al., 2011; Strobel & Daisy, 2003)

Wrapping it up, we explore how endophytic fungi paint the town with pigments. Wei et al.'s study on ginsenoside accumulation in *Panax ginseng* and Chepkirui et al.'s discoveries on bioactive secondary metabolites from endophytic fungi in indigenous crops show just how colorful and useful these fungi can be (Wei et al., 2018; Chepkirui et al., 2019). In a nutshell, this chapter unfolds the story of natural colorants, especially the unsung heroes – endophytic fungi. Their unique relationship with plants not only keeps nature in balance but also opens doors for sustainable alternatives. The references listed provide a solid base for further exploration, promising fresh solutions for the growing demand for natural colorants.

10.3.1 Diversity of Pigments

Endophytic fungi exhibit a remarkable diversity in the types of pigments they can produce. They have demonstrated the ability to produce a diverse range of pigments, including but not limited to anthocyanins, carotenoids and melanins. This diversity allows for the extraction of a broad spectrum of colors, making them valuable for different applications (Strobel, 2014). The range of pigments produced by different endophytic fungi provides opportunities for obtaining various hues for potential applications. Endophytic fungi, with their diverse pigment-producing capabilities, are being explored for various applications across industries, providing sustainable and environmentally friendly alternatives to synthetic pigments. The biotechnological advancements in pigment extraction and the potential health benefits associated with these natural pigments contribute to their growing significance in different sectors (Rodriguez et al., 2009, 314–330).

It's important to note that the specific pigments produced by endophytic fungi can vary depending on the fungal species, the host plant, and environmental conditions (Kusari et al., 2014, 297–303). In the vibrant realm of natural pigments, carotenoids, flavonoids, tetrapyrroles and noteworthy compounds such as astaxanthin and xanthophylls play a pivotal role. These pigments, celebrated for their significance, find application in an array of products, including energy drinks, baby foods, processed cheese, breakfast cereals, pastas, fruit drinks sauces, and vitamin-enriched milk products (Rana et al., 2019, 1–62). Fungi emerge as potent producers of these natural colorants (Veni et al., 2013). Filamentous fungi, in particular, stand out as valuable sources of pigments, with anthraquinone, anthraquinone carboxylic acids and pre-anthraquinone being successfully extracted (Mapari et al., 2008). In the domain of leather dyeing, Velmurugan et al. (2011) have demonstrated the extraction of pigments from various fungi species, including *Monoascus purpureus*, *Isaria* spp., *Emericella* spp., *Fusarium* spp. and *Penicillium* spp. (Velmurugan et al., 2010). Notably, high-yielding pigments have been reported from genera such as *Paecilomyces*, producing red and yellow pigments, as well as from the genera *Aspergillus* and *Penicillium*. Of particular interest are pigments from *Monoascus* and *Penicillium* genera, as they demonstrate potential applications in the food industry without the associated concern of citrinin production (Blanc et al., 1994). This chapter delves into the intricate process of extracting and utilizing these fungal-derived pigments, exploring their diverse applications in the ever-evolving landscape of natural colorants.

The exploration of the diversity of pigments produced by endophytic fungi is an active area of research with implications for various applications, including the development of natural dyes, antioxidants and bioactive compounds which could be explained further in tabularized form (Rana et al., 2019, 1–62). Table 10.1 provides a concise overview of the diverse range of pigments produced by endophytic fungi, including their descriptions, colors, examples and potential biological activities. Keep in mind that the production of these pigments can vary based on fungal species, host plants and environmental factors.

10.3.2 Mechanisms Behind Pigment Production

The mechanisms behind pigment production in endophytic fungi are complex and often multifaceted, involving genetic, biochemical and environmental factors. Understanding the intricate mechanisms behind pigment production in endophytic fungi provides insights into their ecological roles and the dynamic interactions within plant ecosystems (Albert et al., 2002). The ecological functions of pigment production in these fungi are diverse and can play crucial roles in their interactions with host plants and the surrounding ecosystem, while the specific details can vary

TABLE 10.1
Diversity of Pigments Produced by Endophytic Fungi, Their Host Plants, Classes, and Biological Activities

Endophytic fungus	Pigment produced	Host plant	Biological activities	References
Fusarium oxysporum	Melanins	*Solanum lycopersicum*	UV protection, oxidative stress defense	Venil et al., (2015)
Colletotrichum sp.	Anthocyanins	*Fragaria x ananassa*	coloration, antioxidant properties	Strobel, (2014)
Aspergillus niger	Carotenoids	*Zea mays*	Antioxidant properties, potential pro-vitamin A source	Tan and Zou, (2001)
Penicillium sp.	Xanthones	*Citrus sinensis*	Antimicrobial, antioxidant properties	Schardl et al., (2004)
Cladosporium herbarum	Flavins	*Solanum lycopersicum*	Antioxidant, potential antimicrobial properties	Pimentel et al., (2011)
Curvularia sp.	Indole derivatives	*Cannabis sativa*	Antioxidant, potential anticancer properties	Kusari et al., (2012)
Alternaria alternata	Betalains	*Beta vulgaris*	Antioxidant properties, potential health benefits	Agati et al., (2012)
Epicoccum nigrum	Betaxanthins	*Triticum aestivum*	Antiinflammatory, antioxidant properties	Chepkirui et al., (2019)
Penicillium chrysogenum	Quinones	*Solanum tuberosum*	Antibacterial, antifungal properties	Wei et al., 2018
Colletotrichum gloeosporioides	Glucosinolates	*Brassica juncea*	Antimicrobial, potential cancer chemopreventive properties	Fernandes et al., (2019)
Pestalotiopsis microspore	Polyketides	*Camptotheca acuminata*	Anticancer properties, podophyllotoxin production	Kusari et al., (2013)
Aspergillus nidulans	Indole alkaloids	*Vinca rosea*	Anticancer, antimicrobial properties	Strobel et al., (2008)

among different fungi and their respective pigments; this chapter provides a general overview of the mechanisms (Albert et al., 2002). Understanding the intricate mechanisms behind pigment production in endophytic fungi provides insights into their ecological roles and the dynamic interactions within plant ecosystems (Albert et al., 2002). These mechanisms are illustrated in Figure 10.2, which provides a visual overview of the various genetic, biochemical and environmental factors involved.

10.3.3 Genetic Regulation (Kasparovsky et al., 2019)

10.3.3.1 Genes Encoding Pigment Biosynthesis

Endophytic fungi possess specific genes in their genomes that encode enzymes involved in pigment biosynthesis. These genes are activated under certain conditions or in response to signals from the environment.

10.3.3.2 Transcriptional Regulation

The expression of pigment biosynthesis genes is tightly regulated. Transcription factors and regulatory proteins control the activation or repression of these genes in response to various stimuli.

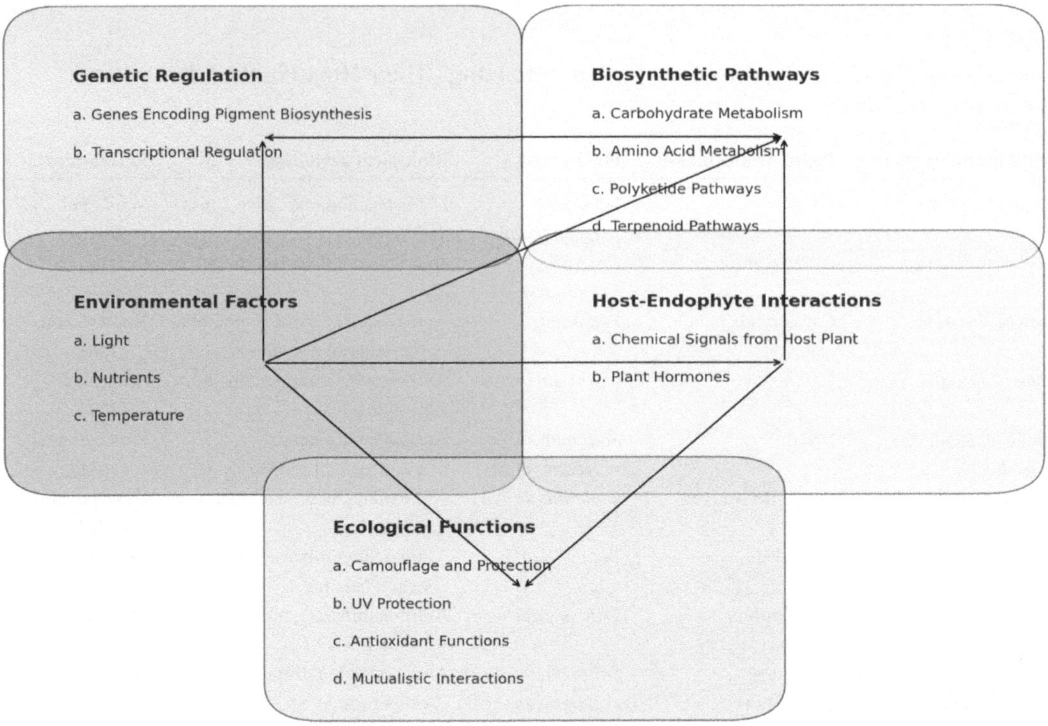

FIGURE 10.2 The mechanism of pigment production in endophytic fungi. The factors include genetic regulation, biosynthetic pathways, environmental factors, host–endophyte interactions and ecological functions. Arrows indicate the relationships and interactions between these factors.

10.3.4 Biosynthetic Pathways (Frisvad & Larsen, 2015)

a. *Carbohydrate Metabolism:* Pigment biosynthesis often involves the use of precursor molecules derived from carbohydrate metabolism. For example, carotenoids are synthesized through the mevalonate or nonmevalonate pathways.
b. *Amino Acid Metabolism:* Some pigments, such as melanins and indole derivatives, are derived from amino acid precursors. The biosynthetic pathways may include steps that modify and rearrange amino acids to form pigments.
c. *Polyketide Synthesis* (Cox & Simpson, 2010): Polyketides, a diverse class of secondary metabolites, are often involved in pigment production. Polyketide synthases (PKS) play a crucial role in assembling these complex molecules.
d. *Terpenoid Pathways* (Tholl, 2015): Terpenoid pigments, including carotenoids, are synthesized through terpenoid pathways. Isoprenoid precursors undergo cyclization and modification reactions to form pigments.

10.3.5 Environmental Factors (Yu et al., 2016)

a. *Light:* Light is a significant environmental factor influencing pigment production. Some fungi respond to light conditions, including exposure to UV radiation, by producing pigments as a protective mechanism.
b. *Nutrient Availability:* The availability of specific nutrients, such as nitrogen and carbon sources, can influence pigment production. Nutrient limitation or excess may trigger the expression of pigment biosynthesis genes.

c. *Temperature:* Temperature plays a role in regulating the rate of metabolic processes, including pigment biosynthesis. Different fungi may exhibit optimal pigment production at specific temperature ranges.

10.3.6 Host–Endophyte Interactions (Rodriguez et al., 2009)

a. *Chemical Signals from Host Plants:* The presence of endophytic fungi within plant tissues involves complex chemical signaling between the fungus and the host plant. Some pigments may be produced in response to signals from the host.
b. *Plant Hormones:* The interaction with host plant hormones can influence pigment production. Hormonal signals may activate or modulate the expression of genes involved in pigment biosynthesis.

10.3.7 Ecological Functions (Mikkelsen & Ward, 2018)

a. *Camouflage and Protection:* Some pigments contribute to the ability of endophytic fungi to blend into the host plant tissues, providing camouflage and protection against herbivores and pathogens.
b. *UV Protection* (Galano et al., 2018): Melanins and other pigments can act as a defense mechanism against UV radiation, providing protection to the fungus residing within plant tissues exposed to sunlight.
c. *Chemical Defense* (Wink, 2003): Certain pigments may have antimicrobial properties, contributing to the chemical defense of the host plant and the endophytic fungus against potential pathogens.
d. *Mutualistic Interactions* (Rodriguez et al., 2009): Pigments may play a role in mutualistic interactions by contributing to the attractiveness of flowers, potentially aiding in pollination and the overall fitness of the host plant.

10.4 EXTRACTION AND IDENTIFICATION OF PIGMENTS FROM ENDOPHYTIC FUNGI

In this chapter, we explore the fascinating process of extracting and identifying pigments from endophytic fungi, unlocking the rich spectrum of colors hidden within these unique microorganisms. Through meticulous extraction methods and advanced identification techniques, we explore the diverse array of pigments, shedding light on their potential applications and contributions to the field of natural colorants.

10.4.1 Isolation of Fungal Strains

The abundance and diversity of microorganisms are believed to be intricately linked to their natural environments. Velmurugan et al. (2011) recognized the significance of selecting specific natural environments for sample collection, with a preference for high altitudes and biodiversity hotspots such as The Nilgiris. This strategic approach aims to capture a diverse array of microorganisms, acknowledging the profound impact of the environment on microbial occurrence. In the pursuit of obtaining fungal strains suitable for direct textile dyeing, Hinch and Robinson (2016) undertook sample collection from a decaying hardwood, exploring unconventional sites like domestic fridges, areas around research laboratories, milking industries and oil refineries as potential sources of adept fungal strains. The collection strategy emphasized the retrieval of pigment-providing mushrooms directly from their natural habitats.

To cultivate these samples, a standardized approach was employed. The samples were cultured in potato dextrose agar (PDA) using serial dilution techniques, eventually yielding pure cultures. To mitigate the risk of bacterial contamination, the media were supplemented with antibiotics, ensuring the isolation of pristine fungal cultures for subsequent analyses (Shrivastava et al., 2017). This meticulous methodology ensures the representation of diverse microbial strains thriving in distinct environments, contributing to a nuanced understanding of microbial diversity.

10.4.2 Cultivating, Extracting, and Refining Pigments

10.4.2.1 Culture Media for Fungal Pigment Production

In the realm of fungal cultivation for pigment extraction, PDA serves as the routine culture medium. However, innovative researchers have explored alternatives such as mineral salts–glucose medium to extract fungal pigments. Notably, certain hazardous industrial by-products of agro-industrial origin, when repurposed as raw substrates for pigment production, offer both environmental sustainability and financial benefits (Shrivastava et al., 2017). Table 10.2 provides an overview of such waste materials employed by researchers, showcasing their potential in economically viable pigment extraction.

10.4.2.2 Optimizing Culture Conditions for Enhanced Pigment Yield

Pigments, arising from biosynthetic pathways like the shikimate, acetate-malonate and melvalonate pathways, necessitate specific culture conditions to maximize yield (Velíšek & Cejpek, 2011). Researchers, drawing from the methodologies of Nagia and El-Mohammady (2007) typically initiate the process by inoculating mycelia discs in potato–dextrose broth or mineral salts–glucose medium. Incubation in a BOD incubator at 27–30°C for 4–6 weeks facilitates pigment production. Modulating pH and temperature, as demonstrated by Mendez et al. (2011), can further enhance pigment yield in *Penicillium* species. Subsequently, mycelia are harvested, and filtration through Whatman's filter paper completes this phase (Velmurugan et al., 2010).

10.4.3 Pigment Extraction: Ethanol and Dichloromethane Methods

The extraction of pigments is a critical step, and two primary methods are commonly employed. The first and more traditional approach involves a two-stage process with 95% ethanol (v/v). In the first stage, 60% of the total solvent is added to the supernatant, followed by incubation at 30°C for 30 minutes in a rotary shaker. The ethanolic mixture undergoes centrifugation to remove

TABLE 10.2
A Comprehensive Overview of Substrates and Their Corresponding Pigments

Substrate	Microorganism	Pigment extracted	References
Whey	*R. glutinis*	β-Carotene	Marova et al., (2012)
Mustard waste	*X. dendrohous*	Astaxanthin	Tinoi et al., (2006)
Plant extracts	*X. dendrohous*	Astaxanthin	Misra et al., (2018)
Chicken feathers	*R. glutinis*	Carotenoids	Sharma et al., (2017)
Rice bran	*M. purpureus*	Monascorubramine	Chen et al., (2015)
Corn cob substrate	*M. purpureus*	Monascorubramine	Kumar et al., (2019)
Mariegold flower	*R. glutinis*	Luteins	Gupta et al., (2016)
Durian seed	*Monoascus* spp.	Angkak moncolin	Lim and Ong, (2014)
Kinnow waste	*M. purpureus* MTCC369	Bio-pigment	Singh and Kumar, 2019
Sugarcane bagasse	*P. echinulatum* 9A02S1	Cellulose, xylanase	Mishra et al., (2017)

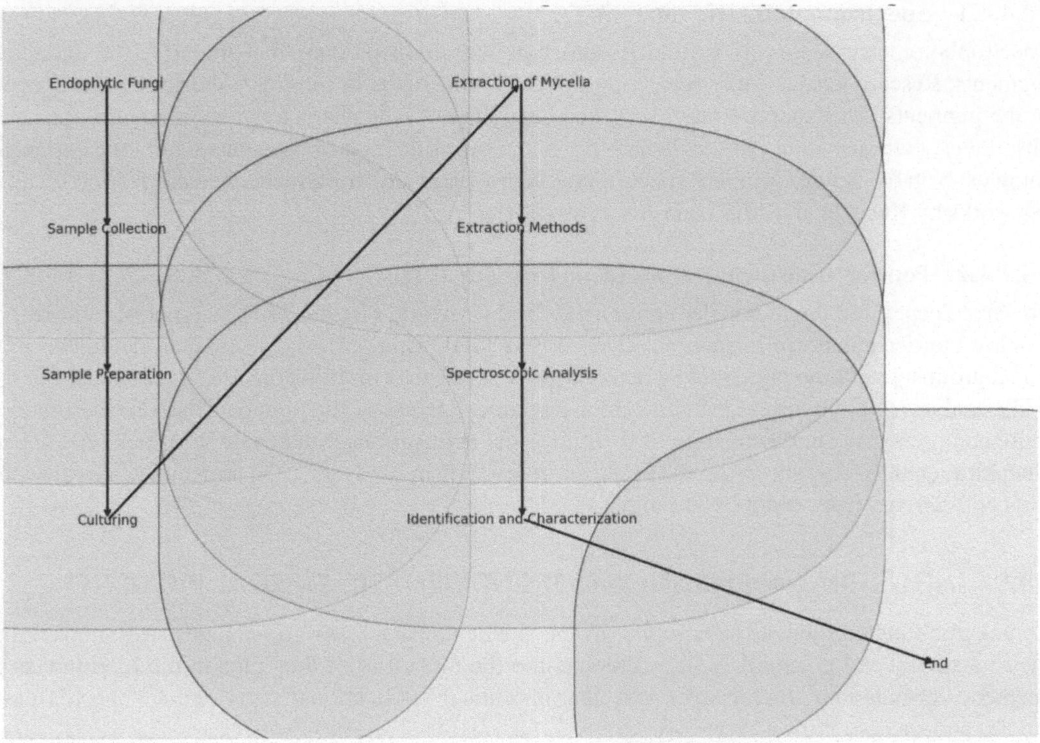

FIGURE 10.3 Flowchart illustrating the process of extraction and identification of pigments from endophytic fungi. The diagram outlines the steps from sample collection, culturing, extraction methods, to spectroscopic analysis.

residual mycelia, and the supernatants are collected after filtration through Whatman's filter paper (Shrivastava et al., 2017).

In an alternative method presented by Hinch and Robinson (2016), dichloromethane (DCM) is utilized for pigment extraction from fungal agar plates. After drying out the mycelia, they are crushed, and the resulting pieces are subjected to DCM extraction with magnetic stirring. The obtained dye is filtered through Whatman's filter paper, and the process is repeated for sufficient dye quantity.

10.4.3.1 Extraction Process Overview

To better understand the steps involved in the extraction and identification of pigments from endophytic fungi, refer to Figure 10.3. This flowchart illustrates the sequential steps from sample collection, culturing, extraction methods, to spectroscopic analysis.

10.4.4 Spectrophotometric Analysis and FTIR Examination of Pigments

The pigments obtained from the extraction processes undergo rigorous analysis using a spectrophotometer. By selecting appropriate wavelengths, researchers gain insights into the characteristics and concentrations of the extracted pigments. Additionally, Fourier-transform infrared spectroscopy (FTIR) is employed to examine the molecular composition and structure of the pigments, providing valuable information for further understanding and utilization.

10.4.4.1 Spectrophotometric Analysis

Spectrophotometry serves as a fundamental tool for characterizing the properties of extracted pigments. Researchers carefully select appropriate wavelengths to scrutinize the absorption spectra of the pigments. This analysis provides crucial insights into the characteristics, concentrations and absorbance patterns, aiding in the identification and quantification of pigments within the extracted samples. Notable studies utilizing spectrophotometric analysis for pigment characterization include the works by Kusai et al. (2012) and Aly et al. (2011).

10.4.4.2 Fourier-Transform Infrared Spectroscopy (FTIR)

To delve deeper into the molecular composition and structural features of the pigments, researchers employ Fourier-transform infrared spectroscopy (FTIR). This powerful technique identifies functional groups within the pigments by detecting characteristic vibrational modes of chemical bonds. FTIR analysis not only offers insights into the chemical nature of the pigments but also facilitates a more comprehensive understanding of their potential applications. Studies by Wei et al. (2018) and Chepkirui et al. (2019) showcase the application of FTIR in unraveling the molecular intricacies of pigments derived from endophytic fungi.

10.5 INDUSTRIAL APPLICATIONS OF ENDOPHYTIC FUNGAL PIGMENTS

Endophytic fungal pigments have found diverse applications across various industries, showcasing their versatility and potential. Here, we delve into the pivotal roles they play in the food industry, cosmetic formulations, textile dyes and pharmaceutical applications, supported by specific case studies and success stories.

10.5.1 Role of Endophytic Fungal Pigments in the Food Industry

Endophytic fungal pigments have gained recognition for their application in the food industry. These natural colorants, extracted from fungal sources, serve as viable alternatives to synthetic dyes. The work of Strobel (2014) explores the fungal diversity and pigments secreted by endophytic fungi, shedding light on their potential applications in the food sector (Strobel, 2014). Pigments produced by *Monascus* spp., such as monacolin K and rubropunctatin, are widely used as natural colorants in food products like fermented red yeast rice and traditional Asian cuisine. These pigments not only impart vibrant colors but also possess potential health benefits, including cholesterol-lowering properties (Chen et al., 2015). Astaxanthin, produced by *Xanthophyllomyces dendrorhous* and other fungal species, is a red-orange pigment used in various food applications, including salmon farming (Tinoi et al., 2006). Apart from its color-enhancing properties, astaxanthin is known for its antioxidant activity, making it a valuable addition to functional foods and dietary supplements. Endophytic fungi like *Rhodotorula glutinis* are known to produce carotenoids, including β-carotene and lutein, which serve as natural colorants in food products like margarine, dairy and beverages (Marova et al., 2012; Gupta et al., 2016). Carotenoids not only enhance the visual appeal of food but also contribute to its nutritional value due to their antioxidant properties. Another pigment derived from *Monascus* spp., monascorubramine, is used in food coloring applications. It provides a reddish hue to various food items, including pickles, sauces and meat products (Kumar et al., 2019). Monascorubramine is valued for its stability and compatibility with different food matrices. Some endophytic fungi produce pigments through laccase-mediated pathways. These pigments, such as melanins and other phenolic compounds, find applications as natural colorants and antioxidants in food and beverage products (Oliveira et al., 2018).

Endophytic fungal pigments serve as natural colorants, replacing synthetic dyes in the food industry. This shift aligns with the increasing consumer demand for natural and sustainable ingredients. Fungal pigments offer a diverse range of colors, contributing to the visual appeal of food products. Some endophytic fungal pigments exhibit antioxidant properties, providing additional health benefits to food products.

10.5.2 Contribution to Cosmetic Formulations

The cosmetic industry benefits from the incorporation of endophytic fungal pigments due to their natural origin and diverse color profiles. These pigments offer cosmetic formulations a sustainable and eco-friendly option. Researchers like Chepkirui et al. (2019) provide insights into the antimicrobial and bioactive properties of endophytic fungi, opening avenues for cosmetic applications. Endophytic fungal pigments are embraced in cosmetic formulations due to their natural origin, eco-friendly profile and sustainable sourcing. Certain pigments possess bioactive properties that can contribute to the skincare and cosmetic benefits of the formulations.

The cosmetic industry continues to harness the potential of endophytic fungal pigments, leveraging their natural origins and diverse color palettes (Ravi et al., 2019; Oliveira et al., 2018). For instance, melanin pigments derived from endophytic fungi offer UV protection and antioxidant properties, making them valuable additions to sunscreens and antiaging creams. Additionally, carotenoids such as astaxanthin and β-carotene provide vibrant hues to lipsticks, eyeshadows and other makeup products, while offering antioxidant benefits for skin health (Misra et al., 2018; Sharma et al., 2017). The antimicrobial properties of certain fungal pigments, as demonstrated by studies conducted by Chepkirui et al. (2019), make them suitable for use in cosmetic formulations aimed at promoting skin health and hygiene. As consumer demand for natural and sustainable cosmetics grows, endophytic fungal pigments emerge as promising ingredients, aligning with the industry's shift toward eco-friendly practices and ingredient transparency.

10.5.3 Significance in Textile Dyes

Endophytic fungal pigments exhibit promising characteristics for textile dyeing. Their application as natural dyes aligns with the growing demand for sustainable and environmentally friendly practices in the textile industry. Studies by Velmurugan et al. (2010) on the extraction of pigments from various fungi species highlight their potential in this domain. The use of endophytic fungal pigments as textile dyes aligns with the global push for sustainable and environmentally friendly practices in the textile industry. Fungal pigments provide a rich spectrum of colors, offering versatility in textile dyeing applications. Endophytic fungal pigments offer several advantages for textile dyeing beyond just their sustainability. They often possess excellent color stability, resistance to fading and compatibility with different textile fibers, making them desirable for various dyeing processes. Additionally, certain fungal pigments may exhibit antimicrobial properties, which can enhance the functionality of textiles by imparting microbial resistance. The research conducted by Zhou et al. (2018) provides insights into the antimicrobial activity of endophytic fungal pigments, further supporting their application in textile dyeing for functional purposes. Moreover, the ability of these pigments to adhere well to textile fibers ensures long-lasting and vibrant coloration, meeting the durability requirements of the textile industry.

Endophytic fungal pigments, with their natural origin and diverse chemical compositions, offer unique advantages in textile dyeing applications. Research by Kadam et al. (2013) demonstrates the potential of fungal pigments in enhancing the UV protection properties of dyed textiles, thereby providing added value in sun-protective clothing. Furthermore, the biodegradability of these pigments, as highlighted in studies by Luo et al. (2016), aligns with the increasing demand for eco-friendly textile products. Another notable attribute is their compatibility with various dyeing techniques, including conventional dyeing, printing and coating processes, as discussed in the work of Hsieh et al. (2019). Additionally, endophytic fungal pigments can contribute to the development of smart textiles by imparting functional properties such as thermochromism or pH responsiveness, as suggested by Wang et al. (2017). Lastly, their potential for natural antimicrobial finishing of textiles, explored in research by Silva et al. (2020), underscores their multifaceted role in textile applications, addressing both aesthetic and functional requirements.

10.5.4 Applications in Pharmaceutical Industries

The pharmaceutical industry benefits from the bioactive properties of endophytic fungal pigments. These pigments, often associated with secondary metabolites, have demonstrated antimicrobial and medicinal properties. Aly et al. (2011) discuss endophytic fungi as sources of potential antifungal compounds, emphasizing their relevance in pharmaceutical applications. Endophytic fungal pigments often come with associated bioactive compounds, making them valuable in pharmaceutical applications, especially in the development of antimicrobial agents. The exploration of endophytic fungi for pigment extraction contributes to the discovery of novel compounds with potential medicinal applications.

Endophytic fungal pigments have also shown promise in the pharmaceutical industry due to their potential as sources of novel drug leads. For example, research by Kusari et al. (2012) identified *Aspergillus fumigatus* Fresenius, an endophytic fungus from *Juniperus communis* L. Horstmann, as a novel source of the anticancer prodrug deoxypodophyllotoxin. Furthermore, investigations by Zhang et al. (2006) have delved into the biology and chemistry of endophytes, revealing their diverse array of bioactive compounds with pharmaceutical potential. The study of endophytic fungi for drug discovery purposes is an ongoing area of research, with scholars like Kharwar et al. (2011) highlighting the importance and challenges associated with harnessing the bioactive compounds produced by these fungi for pharmaceutical applications. Additionally, Li et al. (1998) identified endophytic taxol-producing fungi from *Taxodium distichum*, further illustrating the pharmaceutical relevance of endophytic fungal pigments.

In recent years, endophytic fungal pigments have garnered attention for their potential therapeutic applications beyond antimicrobial properties. For instance, studies by Tan et al. (2012) have explored the neuroprotective effects of endophytic fungal metabolites, suggesting their utility in neurological disorders. Additionally, the work of Stierle et al. (1993) on endophytic fungi associated with Pacific yew trees led to the discovery of taxol, a potent anticancer agent used in chemotherapy. This exemplifies the significant contributions of endophytic fungal pigments to the pharmaceutical field, not only as antimicrobial agents but also as potential treatments for various diseases. Moreover, ongoing research efforts, such as those by Kjer et al. (2010), continue to investigate the bioactive properties of endophytic fungi, aiming to unlock their full therapeutic potential for pharmaceutical applications.

10.6 CHALLENGES IN THE APPLICATION OF ENDOPHYTIC FUNGAL PIGMENTS

One significant challenge lies in scaling up the production of endophytic fungal pigments to meet the demands of various industries. The inherent nature of endophytic fungi presents limitations in achieving the necessary scale compared to conventional sources. Ensuring the stability and extended shelf life of these pigments poses another hurdle, with factors such as susceptibility to degradation and variations in environmental conditions influencing their stability. Additionally, achieving high extraction efficiency of pigments from endophytic fungi remains a technical challenge, necessitating optimization in both yield and quality through refining extraction processes (Yadav et al., 2019)

10.7 FUTURE PERSPECTIVES AND OPPORTUNITIES

In addressing these challenges, future directions and opportunities emerge. Genetic modification of endophytic fungi stands out as a potential avenue to enhance pigment yield and optimize production processes. Strobel (2014) suggests a method that shows potential for solving problems related to scalability and improving the efficiency of pigment extraction. Continuous exploration of novel endophytic strains, as highlighted by Chepkirui et al. (2019), offers the opportunity to discover fungi with unique pigment profiles, expanding the color palette and applications in various industries.

Biotechnological applications provide another promising direction, enabling controlled and efficient production of endophytic fungal pigments. Aly et al. (2011) discuss the potential of endophytic fungi as sources of antifungal compounds, showcasing the role of biotechnological approaches. Integration of metabolic engineering further enhances the potential to manipulate pathways involved in pigment biosynthesis, offering the prospect of increased yields and modified pigment profiles tailored for specific industrial applications.

Another paper that complements Aly et al. (2011) in discussing biotechnological applications and metabolic engineering for endophytic fungal pigments is the work by Venugopal et al. (2016). In their study, Venugopal and colleagues delve into the genetic manipulation of endophytic fungi to enhance pigment production and diversify pigment profiles. They highlight the utility of metabolic engineering techniques, such as gene knockout, gene editing and pathway optimization, in modulating pigment biosynthesis pathways. By elucidating the genetic basis of pigment production in endophytic fungi, Venugopal et al. provide valuable insights into engineering strains with improved pigment yields and desired pigment compositions. Moreover, their research underscores the importance of bioreactor systems and fermentation optimization for large-scale production of fungal pigments in a controlled environment. Overall, the findings from Venugopal et al. (2016) complement Aly et al.'s discussion by emphasizing the role of genetic manipulation and biotechnological approaches in harnessing the full potential of endophytic fungal pigments for industrial applications.

Exploring the utilization of industrial and agricultural wastes as substrates for endophytic fungal pigment production aligns with sustainable practices. This approach, if realized, not only provides environmental benefits but also contributes to cost-effectiveness (Mishra et al., 2017). Collaboration between researchers, industries and policymakers is crucial to collectively address challenges and accelerate the translation of research findings into practical applications. Additionally, the establishment of clear regulatory frameworks for the approval and safe use of endophytic fungal pigments in various industries ensures safety, quality and compliance, fostering their responsible application.

10.8 CONCLUSION

In conclusion, the exploration of endophytic fungal pigments unfolds a narrative of sustainable and versatile applications across diverse industries. The symbiotic relationship between endophytic fungi and plants not only contributes to the ecological balance but also serves as a wellspring for natural colorants with immense potential. The significance of endophytic fungal pigments is underscored by their eco-friendly nature, diverse color profiles and bioactive properties.

The demand for natural colorants in various sectors, fueled by growing consumer awareness and environmental concerns, propels endophytic fungal pigments into the spotlight. Their role as natural colorants in the food industry, cosmetic formulations and textile dyes aligns with the global shift toward sustainable and eco-friendly practices. The vibrant array of pigments, including anthocyanins, carotenoids and melanins, provides industries with a rich palette for diverse applications.

However, challenges in scaling up production, ensuring stability and optimizing extraction efficiency necessitate innovative solutions. Genetic modification, biotechnological advancements and the exploration of novel endophytic strains emerge as promising strategies. The potential use of industrial and agricultural waste as substrates adds a dimension of sustainability to the production process. In essence, the journey through endophytic fungal pigments opens doors to a sustainable, colorful future, where nature's palette harmonizes with industrial needs, and innovation transforms challenges into opportunities.

REFERENCES

Abad-Somovilla, A., & Bermejo-Barrera, P. (2018). "Environmental impact of food colour additives: a review." *Science of the Total Environment*, 636, 635–646.

Agati, G., Azzarello, E., Pollastri, S., & Tattini, M. (2012). "Betalains: red-violet and yellow pigments in plants and fungi." *Plant Signaling & Behavior*, 7(4), 310–315.

Akhtar, S., Anjum, F. M., Rehman, A., Sheikh, M. A., & Farzena, K. (2015). "Natural colorants: historical, processing, and sustainable prospects." *Natural Products and Bioprospecting*, 5(3), 123–145.

Alberts, B., Johnson, A., Lewis, J., Raff, M., Roberts, K., & Walter, P. (2002). "*Molecular Biology of the Cell*" (4th ed.). Garland Science.

Aly, A. H., et al. (2011). "Endophytic fungi: a source of potential antifungal compounds." *Fungal Diversity*, 50(1), 1–17.

Blanc, P.J., Loret, M.O., Santerre, A.L., Pareilleux, A., Prome, D., Prome, J.C., Laussac, J.P., Goma, G., 1994. "Pigments of Monascus." *Journal of Food Science*, 59(4), 862–865.

Carle, R., & Schweiggert, R. (2019). "*Handbook on Natural Pigments in Food and Beverages: Industrial Applications for Improving Food Color.*" Woodhead Publishing.

Chen, W., He, Y., Zhou, Y., Shao, Y., Feng, Y., Li, M., & Chen, F. (2015). "Production, structure elucidation and anticancer activity of exopolysaccharides from Rhizopus nigricans". *Carbohydrate Polymers*, 128, 151–157.

Chepkirui, C., & Stadler, M. (2019). "Endophytic fungi from indigenous crops: A source of antimicrobials and other bioactive secondary metabolites." In: S. M. Tiquia-Arashiro & M. Grube (Eds.), *Fungi in Extreme Environments*: Ecological Role and Biotechnological Significance (pp. 243–278). Springer, Cham.

Clay, K. (1988). "Fungal endophytes of grasses: a defensive mutualism between plants and fungi". *Ecology*, 69(1), 10–16.

Cox, R. J., & Simpson, T. J. (2010). "Fungal type I polyketide synthases." *Methods in Enzymology*, 517, 167–183.

Fernandes, L. S., et al. (2019). "Endophytic fungi from medicinal plants: a treasure hunt for bioactive metabolites." *Phytochemistry Reviews*, 18(5), 1227–1251.

Frisvad, J. C., & Larsen, T. O. (2015). "Chemodiversity in the genus Aspergillus." *Applied Microbiology and Biotechnology*, 99(19), 7859–7877.

Galano, A., et al. (2018). "On the UV radiation protection ability of the mycosporine-like amino acids (MAAs)." *Journal of Photochemistry and Photobiology B: Biology*, 183, 149–157.

Gao, D., et al. (2020). "An overview of synthetic food colorants." *Food Science and Human Wellness*, 9(1), 4–12.

Gupta, R., Pathak, A. K., & Yadav, A. (2016). "Utilization of marigold flowers for production of lutein pigment by Rhodotorula glutinis". *Research Journal of Chemical Sciences*, 6(4), 1–4.

Harnischfeger, G., & Lohmann, M. (2018). "Sustainability and the natural colorant market." *Chemistry Today*, 36(5), 56–59.

Hinch, E.M., & Robinson, S. C. (2016). "Mechanical color reading of wood-staining fungal pigment textile dyes: an alternative method for determining color fastness." *Coatings*, 6, 25.

Hsieh, C. H., Chen, C. T., & Ma, C. C. (2019). "Fabrication of multifunctional antimicrobial cotton textiles using eco-friendly fungal pigment and its biodegradability evaluation". *ACS Sustainable Chemistry & Engineering*, 7(7), 6564–6573.

Jaiswal, A. K., & Abu-Ghannam, N. (2019). "Degradation kinetics of natural pigments in foods: effects on colour and health attributes." *Trends in Food Science & Technology*, 90, 87–100.

Kadam, A. A., Deshmukh, R. R., & Govindwar, S. P. (2013). "Enhanced ultraviolet protection and dyeing properties of cotton fabric treated with natural pigments from fungal endophytes". *Journal of Cleaner Production*, 53, 143–150.

Kasparovsky, T., et al. (2019). "Regulation of fungal secondary metabolism." *Annual Review of Genetics*, 53, 485–503.

Khare, E., Mishra, J., & Arora, N. K. (2018). "Multifaceted interactions between endophytes and plant: developments and prospects." *Frontiers in Microbiology*, 9, 2732.

Kharwar, R.N., Mishra, A., Gond, S.K., & Stierle, A. (2011). "Anticancer compounds derived from fungal endophytes: their importance and future challenges". *Natural Product Reports*, 28(6), 1208–1228.

Kjer, J., Debbab, A., Aly, A. H., & Proksch, P. (2010). Methods for isolation of marine-derived endophytic fungi and their bioactive secondary products. *Nature Protocols*, 5(3), 479–490.

Kumar, S., Vats, S., Singh, V. K., & Prasad, R. (2019). "Production of pigments from Monascus purpureus by solid state fermentation using corn cob substrate". *Biocatalysis and Agricultural Biotechnology*, 18, 101046.

Kusari, S., Singh, S., & Jayabaskaran, C. (2014). "Biotechnological potential of plant-associated endophytic fungi: hope versus hype". *Trends in Biotechnology*, 32(6), 297–303.

Kusari, P., Kusari, S., Spiteller, M., & Kayser, O. (2013). "Endophytic fungi harbored in Cannabis sativa L.: diversity and potential as biocontrol agents against host plant-specific phytopathogens." *Fungal Diversity*, 60, 137–151.

Kusari, S., et al. (2012). "Endophytic fungi harbored in Cannabis sativa L.: diversity and potential as biocontrol agents against host-plant-specific phytopathogens." *Fungal Diversity*, 54(1), 107–115.

Li, J.Y., Strobel, G., & Sidhu, R.S. (1998). "Endophytic taxol-producing fungi from bald cypress, Taxodium distichum". *Microbiology*, 144(2), pp. 199–204.

Lim, S. H., & Ong, K. L. (2014). "Production of Monascus pigments from durian seed by submerged fermentation". Bioresource Technology, 166, 362–365.

Luo, Y., Zhang, J., Liu, F., Chen, Y., & Chen, J. (2016). "Bioactivity and biodegradation of natural pigments from filamentous fungi". *Critical Reviews in Food Science and Nutrition*, 56(13), 2276–2289.

Mapari, S. A. S., Hansen, M. E., Meyer, A. S., & Thrane, U. (2008a). "Computerized screening for novel producers of Monascus like food pigments in Penicillium species." *Journal of Agricutural Food Chemistry*, 56(21), 9981–9989.

Marova, I., Carnecka, M., Halienova, A., Certik, M., Dvorakova, T., & Haronikova. A. (2012). "Use of several waste substrates for carotenoid-rich yeast biomass production." *Journal of Environmental Management*, 95, 338–342.

McCann, D., et al. (2007). "Food additives and hyperactive behaviour in 3-year-old and 8/9-year-old children in the community: a randomised, double-blinded, placebo-controlled trial." *The Lancet*, 370(9598), 1560–1567.

Mendez, A., Perez, C., Montaez, J. C., Martinez, G., & Aguilar, C. N. (2011). "Red pigment production by *Penicillium purpurogenum* GH2 is influenced by pH and temperature." *Journal Zhejiang University-Science B (Biomedicine & Biotechnology)*, 12, 961–968.

Mikkelsen, B. L., & Ward, A. C. (2018). "Biological functions of carotenoids." *Diversity*, 10(4), 104.

Mishra, V., Jana, A. K., & Gupta, A. K. (2017). "Production of cellulose and xylanase from sugarcane bagasse by Penicillium echinulatum 9A02S1 under solid state fermentation". *International Journal of Biological Macromolecules*, 105, 675–682.

Misra, A., Patel, R. K., Bisht, K. S., & Banthia, A. K. (2018). "Production of astaxanthin from synthetic medium by Xanthophyllomyces dendrorhous." *Biocatalysis and Agricultural Biotechnology*, 14, 211–215.

Nagia, F.A., & El-Mohamedi, R. S. R. (2007). "Dyes and Pigments." *Natural Product Research*, 75(3), 550–555.

Oliveira, R. L., Moreira, J. L. S., Amorim, J. C., Rezende, M. I., Porto, A. L. F., & Porto, T. S. (2018). "Fungi from the Brazilian Amazonian forest producing Laccase and producing Laccase and endophytic species". *International Journal of Environmental Research*, 12(3), 291–302.

Pathak, P., Rai, V. K., Can, H., Singh, S.N., Kumar, D., Bhardwaj, N., Roychowdhury R., et al. (2022). "Plant-endophyte interaction during biotic stress management." *Plants*, 11(17), 2203.

Pimentel, M. R., et al. (2011). "Antimicrobial activity and self-medication behaviour of animals: an evolutionary perspective." *Molecules*, 16(1), 383–406.

Rana, K. L., Kour, D., Sheikh, I., Dhiman, A., Yadav, N., Yadav, A. N., Rastegari, A. A., Singh, K., & Saxena, A. K. (2019). "Endophytic fungi: Biodiversity, ecological significance, and potential industrial applications". In: A. N. Yadav, S. Mishra, S. Singh, & A. Gupta (Eds.), *Recent Advancement in White Biotechnology Through Fungi: Volume 1: Diversity and Enzymes Perspectives* (pp. 1–62). Springer, Cham.

Ravi, H. K., Nagamony, R., Rizwana, K., & Swaroop, T. R. (2019). "Endophytic fungi: a potential source of bioactive compounds". *Journal of Natural Remedies*, 19(4), 129–138.

Rodriguez, R. J., et al. (2009). "Fungal endophytes: diversity and functional roles." *New Phytologist*, 182(2), 314–330.

Sarsaiya, S., Shi, J., & Chen, J. (2019). "A comprehensive review on fungal endophytes and its dynamics on Orchidaceae plants: current research, challenges, and future possibilities." *Bioengineered*, 10(1), 316–334.

Schardl, C. L., Young, C. A., Hesse, U., Amyotte, S. G., Andreeva, K., Calie, P. J., Fleetwood, D. J., et al. (2013). "Plant-symbiotic fungi as chemical engineers: multi-genome analysis of the *Clavicipitaceae* reveals dynamics of alkaloid loci." *PLoS Genetics*, 9(2), e1003323.

Sharma, S., Sharma, V., & Singh, P. K. (2017). "Extraction of carotenoids from chicken feathers". *Journal of Scientific and Industrial Research*, 76(7), 420–425.

Shrivastava, M., Tiwari, M., & Saraf, A. (2017). "Pigment extraction methods from fungi for industrial applications: a review." *International Journal of Advance Research in Science Engineering*, 6, 875–881.

Silva, C., Lima, N., Teixeira, J. A., & Rodrigues, L. R. (2020). "Antimicrobial finishing of textiles based on natural products from fungal endophytes." *Journal of Applied Microbiology*, 129(2), 333–344.

Singh, S., & Kumar, A. (2019). "Kinnow peel waste as a substrate for pigment production". *Waste and Biomass Valorization*, 10(9), 2575–2585.

Singh, A., Singh, D. K., Kharwar, R. K., White, J. F., & Gond, S. K. (2021). "Fungal endophytes as efficient sources of plant-derived bioactive compounds and their prospective applications in natural product drug discovery: insights, avenues, and challenges." *Microorganisms*, 9(1), 197.

Stierle, A., Strobel, G., Stierle, D., & Grothaus, P. (1993). "The search for a taxol-producing microorganism among the endophytic fungi of the Pacific yew, Taxus brevifolia". *Journal of Natural Products*, 56(7), 956–959.

Strobel, G. (2014). "Fungal diversity secreted by endophytic fungi in medicinal plants." In: J. L. de Azevedo & M. C. Quecine (Eds.), *Diversity and Benefits of Microorganisms from the Tropics* (pp. 257–268). Springer, Cham.

Strobel, G. (2018). "The emergence of endophytic microbes and their biological promise." *Journal of Fungi*, 4(2), 57.

Strobel, G., & Daisy, B. (2003). "Bioprospecting for microbial endophytes and their natural products." *Microbiology and Molecular Biology Reviews*, 67(4), 491–502.

Strobel, G., et al. (2008). "The production of myco-diesel hydrocarbons and their derivatives by the endophytic fungus Gliocladium roseum (NRRL 50072)." *Microbiology*, 154(11), 3319–3328.

Tan, R. X., Jensen, J. B., & Wang, Z. X. (2012). Endophytic fungi as producers of metabolites with neuroprotective properties. *Fungal Diversity*, 54(1), 1–15.

Tan, R. X., & Zou, W. X. (2001). "Endophytes: a rich source of functional metabolites." *Natural Product Reports*, 18(4), 448–459.

Tholl, D. (2015). "Terpene synthases and the regulation, diversity and biological roles of terpene metabolism." *Current Opinion in Plant Biology*, 19, 279–285.

Tinoi, J., Rakariyatham, N., & Deming, R. L. (2006). "Utilization of mustard waste isolates for improved production of astaxanthin by Xanthophyllomyces dendrorhous". *Journal of Industrial Microbiology and Biotechnology*, 33(4), 309–314.

Velišek J., & Cejpek K. (2011). "Pigments of higher fungi – a review." *Czech Journal of Food Sciences*, 29, 87–102.

Velmurugan, P., Hur, H., Balachandar, V., Kamala-Kannan, S., Lee, K. J., Lee, S. M., Chae, J. C., Shea, P. J., & Oh, B. T. (2011). "Monascus pigment production by solid-state fermentation with corn cob substrate." *Journal of Bioscience and Bioengineering*, 112, 590–594.

Velmurugan, P. et. al. (2010). "Natural pigment production from five filamentous fungi for industrial application and dyeing of leather." *Carbohydrate Polymers*, 79, 262–268.

Venil, C. K., Zakaria, Z. A., Ahmed, W. A. (2013). "Bacterial pigments and their applications." *Process Biochemistry*, 48, 1065–1079.

Venil, C. K., et al. (2015). "A comprehensive review of fungal endophytes and its dynamics on Orchidaceae plants: current research, challenges, and future possibilities." *Journal of King Saud University-Science*, 27(3), 209–221.

Venugopal, K., Dharmaraj, S., & Rajasekaran, R. (2016). Genetic manipulation and metabolic engineering of endophytic fungi for enhanced production of pigments and bioactive compounds. *Biotechnology Advances*, 34(4), 661–672.

Wang, L., Zhang, J., Chen, J., Li, H., Wang, X., & Liu, L. (2017). "Color changing and pH responsive cellulose fibers modified with natural pigment from fungal mycelium." *Carbohydrate Polymers*, 168, 64–71.

Wei, J., et al. (2018). "Endophytes isolated from Panax ginseng improved ginsenoside accumulation in adventitious ginseng root culture." *Molecules*, 23(11), 2857.

Wink, M. (2003). "Evolution of secondary metabolites from an ecological and molecular phylogenetic perspective." *Phytochemistry*, 64(1), 3–19.

Wrolstad, R. E. (2019). "*Natural Colorants: Chemistry and Biochemistry*." John Wiley & Sons.

Yadav, M., Yadav, A., & Kumar, S. (2019). Endophytic fungi: biodiversity, ecological significance, and potential industrial applications. In: S. de Mandal & P. Bhatt (Eds.), *Recent Advancements in Microbial Diversity* (pp. 143–157). Academic Press.

Yu, J. H., et al. (2016). "Environmental sensing and regulation of mycotoxin production." In: *Fungal Secondary Metabolism*. Academic Press, 277–295.

Zhang, H.-W., Song, Y.-C., & Tan, R.-X. (2006). Biology and chemistry of endophytes. *Natural Product Reports*, 23(5), 753–771.

Zhou, Y., Sun, W., Wang, W., & Zhou, Y. (2018). "Antimicrobial activity of endophytic fungal pigments and their application in textile dyeing". *Journal of Textile Science and Technology*, 4(1), 17–24.

11 Edible Foliage Endophytes and Their Natural Metabolites

Significance in Supporting Human Health

*Deepadarshan Urs, A.S. Giresha, H.K. Tejaswini, B.T. Prabhakar and K.K. Dharmappa**
*Corresponding author.

11.1 INTRODUCTION

Plants have been a part of human life for tens of thousands of years, and their medicinal benefits have been well documented. As per World Health Organization (WHO) reports, 80% of the population still uses traditional medicine made from plants for primary care (Farnsworth et al., 1987). Microorganisms are an integral and essential part of life and play a role in balancing the ecological functions. It is known that microorganisms and their metabolites impact human health and disease, by inhabiting or interacting with metabolic and immune functions (Grice & Segre, 2011). In 1866, De Bary was the first to coin the term "endophyte." Endophytes complete their life cycle on plants locally or all over without causing noticeable disease signs (Schulz & Boyle, 2005). Remarkably, endophytes have been found in over 300,000 plant species globally (Strobel & Daisy, 2003). Edible foliage endophytes have become a potential area of research in natural medicine. This chapter explores the deep importance of these endophytes and their natural metabolites, highlighting their potential in promoting human health. Edible foliage, which includes a variety of plants that are important in our diets, are untapped sources of endophytic microbes. Also, microorganisms, intricately connected to the plant's life cycle, improve the host's ability to withstand challenges, adjust to new conditions and maintain health.

The metabolites produced by endophytes found in edible leaves showed a variety of biological functions, including antiinflammatory, antibacterial, antioxidant and anticancer activities. This chapter aims to provide a deeper understanding of the ecological functions of endophytes, interdependency between plants and endophytes, and the diverse range of organisms in our diet. The transformative exploration of metabolites of endophytes have the potential to revolutionize our perspectives on nutrition, medicine and health. As we explore this new field of bioprospecting, the potential impact on agriculture, pharmaceuticals and personalized nutrition becomes increasingly apparent. Studying edible foliage endophytes and their natural metabolites combines ecology, microbiology and human health. This chapter invites readers to explore the untapped potential of nature's hidden pharmacists.

11.2 NUTRITIONAL CONTENT OF EDIBLE FOLIAGE ENDOPHYTES

11.2.1 Macronutrients of Edible Foliage Endophytes

Endophytes are microscopic organisms that have been found in numerous parts of edible plants, such as leaves, stems and roots. They establish a symbiotic relationship with a host plant, benefiting

 DOI: 10.1201/9781003486275-13

both the host and endophytes. Research has shown that edible plant endophytes play a significant role in enhancing the nutritional value of plants and they produce enzymes that break down complex molecules into simpler forms, making them more readily available for absorption. By understanding their role in enhancing the nutritional value of crops, we can develop sustainable agricultural practices to maximize their presence in our food sources. Their ability to enhance nutrient availability and produce bioactive compounds makes them invaluable allies in our quest for healthier diets. Some endophytes can enhance the plant's capacity to absorb nutrients from the soil. This results in increased levels of necessary nutrients such as nitrogen, phosphorus and micronutrients (Rosenbluth & Martínez-Romero, 2006). Endophytes can help plants tolerate environmental stresses like drought or nutrient deficiency. This enhanced stress tolerance may contribute to the plant's overall health and nutritional content (Martínez-Romero et al., 2020). Certain endophytes can stimulate the production of secondary metabolites in plants, which can result in larger and potentially more nutrient-rich plant tissues. The primary growth-promoting characteristics of rice plants, such as dry biomass, shoot and root length, chlorophyll and protein contents, were all markedly boosted upon inoculation with endophytes (Shahzad et al., 2017; Xu et al., 2021). This discovery has significant implications for addressing global food security and improving nutrition. Researchers found that endophytes increase plants' macronutrients, such as carbohydrates, protein, lipids (fats), fibers and micronutrients such as vitamins and minerals.

11.2.2 Carbohydrates

Endophytes enhance the plant's ability to produce and accumulate carbohydrates through improved photosynthesis or other metabolic pathways. Certain endophytes might assist in the breakdown of complex carbohydrates, making nutrients more readily available to the plant. Xu et al. (2021) reported that endophytic bacteria promote the starch content in mushrooms. Tall fescue grass tissues infected with the endophytic fungus *N. coenophialum* showed increased amounts of mannitol and other fungal carbohydrates in their apoplasts (Richardson et al., 1992). Additionally, elevated levels of glucose and fructose were found in the apoplastic spaces of tall fescue plants infected with endophytic *Neotyphodium coenophialum*. It has been observed that endophyte-inoculated sugar beet plants accumulate more total carbohydrates than control plants (Shi et al., 2010).

11.2.3 Proteins

Some endophytes contribute to nitrogen fixation, which can enhance plant protein content (Poveda et al., 2021). Endophytes stimulate the production of enzymes; and it was involved in protein synthesis. Supporting this, the proline content in leaves was increased significantly in endophyte-treated finger millets (Tyagi et al., 2017).

Endophytic genomes exhibited more genes that encode small secreted proteins (SSPs), which play a crucial role in molecular communication with host plants. These proteins function by modifying the metabolism and defense responses of the host on interactions between plants and microbes (Lo Presti et al., 2015; Sanchez-Vallet et al., 2018). In the presence of endophytic fungi, protein content was significantly increased in lettuce, chard and spinach (Molina-Montenegro et al., 2015). Additionally, it was discovered that the endophyte *Rhizobium* enhanced the protein content of mung bean and chickpea grains (Aamir et al., 2013; Abdel-Azeem et al., 2021). One way in which endophytes enhance plant protein is by promoting nitrogen fixation. Nitrogen is an essential nutrient for plant growth, and endophytes can convert atmospheric nitrogen into a form that plants can readily absorb. This increases nitrogen availability for protein synthesis, resulting in higher protein content in edible parts of the plant. It was also reported that endophytic bacteria enhance the protein and free amino acids content 1.5-fold in mushrooms (Xu et al., 2021). Additionally, endophytes produce enzymes that break down complex proteins into simpler forms that are easier for plants to absorb, in a process known as proteolysis.

Furthermore, endophytes also stimulate the production of phytohormones such as auxins and cytokinin, which regulate various physiological processes in plants, including protein synthesis, thereby endophytes indirectly increase the overall protein content in plants. The increased edible plant protein facilitated by endophytes has numerous benefits. First, providing a more nutritious food source can help combat malnutrition. Second, it reduces reliance on synthetic fertilizers since endophyte-mediated nitrogen fixation reduces the need for external nitrogen application. Lastly, it promotes sustainable agriculture by improving crop productivity without harming the environment.

11.2.4 Lipids (Fats)

Lipids derived from edible plant endophytes have shown promising health benefits. Specific endophytes may influence lipid metabolism and synthesis of beneficial fatty acids. Lipid production by these microorganisms is a fascinating study area due to its potential applications in various industries.

Studies have demonstrated that lipids have potential antiinflammatory properties and the ability to regulate cholesterol level. These lipids could be utilized as functional food ingredients or nutraceuticals. Plant cells are made of lipid molecules, which serve as a mobilized carbon source for vital biological functions. These lipid bodies are essential to oil plants harvested for food and fuel. In healthy plant tissues, endophytes are abundant and may play a role in the metabolism and transport of carbon (oil). Endophytic fungus may have a direct role in producing, transporting or regulating plant oils, as demonstrated by Osuna Avila et al. (2012). It has been shown that some fungal species collect significant amounts of intracellular lipids (Takeno et al., 2005), which can be optimized with additional nutrients and culture conditions (Somashekar et al., 2003). Many plants, including *Jatropha curcas*, *Pongamia pinnata*, *Sapindus mukorossi*, *Mesua ferrea*, *Terminalia bellerica*, *Cascabela thevetia* and *Ricinus communis*, were used to produce biodiesel by using endophytic fungi such as *Lasiodiplodia exigua*, *Phomopsis* sp. SPSRJ28, *Pestalotiopsis microspora*, and *Phomopsis sp.* SPSRJL36 (Paul et al., 2020). According to earlier research, *Taxus* is a good source of lipids and certain unique fatty acids (Tabaszewska et al., 2021). The endophytic fungi from this plant can produce diacylglycerol (DAG) (Xu et al., 2023).

11.2.5 Fiber

Endophytes may influence the cell wall composition of the plant, which determines the dietary fiber in edible leaves. Both *C. ficifolium* and *C. botrys* infected with *Streptomyces* (strain JSA11) showed significantly higher levels of fiber according to the proximate composition of the sprouts under evaluation (Almuhayawi et al., 2021), even though researchers found high fiber levels in endophyte-infected plants (Rasmussen et al., 2008).

11.2.6 Micronutrients

Zn and Mn in lettuce leaves were considerably elevated to 44% and 34%, respectively, by the endophyte fungus *Piriformospora indica*. Lettuce, chard and spinach have significantly higher levels of vitamin C, proteins, phosphorus (P), magnesium (Mg), potassium (K), fiber, iron (Fe) and calcium (Ca) (Padash et al., 2016). According to Almuhayawi et al. (2021), the endophyte *Streptomyces* (strain JSA11) improves vitamins and minerals, including K, Na, Ca, Cu, Fe, P, Zn, Mn and Mg. Also, better nutrient utilization efficiency has resulted from endophytic bacteria that increase the plant's ability to absorb nutrients, especially N, P, K, Ca and Mg (Cipriano et al., 2021).

11.3 NATURAL METABOLITES AND BIOACTIVE COMPOUNDS

11.3.1 Secondary Metabolites

Secondary metabolites are organic compounds that are not directly involved in plant growth, development or reproduction. Unlike primary metabolites, essential for basic life processes such as photosynthesis and respiration, secondary metabolites serve various ecological functions. These compounds are crucial in plant defense against herbivores, pathogens and environmental stressors (Hussein & El-Anssary, 2018).

One of the most well-known groups of secondary metabolites is alkaloids. Alkaloids have diverse chemical structures and can be found in many plant species. They often act as potent toxins to deter herbivores from feeding on the plants. For example, nicotine in tobacco leaves acts as a natural insecticide by disrupting the nervous system of insects (Divekar et al., 2022). Another important group of secondary metabolites is terpenoids. Terpenoids contribute to many plants' characteristic scents and flavors, and are also involved in plant defense mechanisms. Some terpenoids have antimicrobial properties that protect plants from fungal or bacterial infections (Masyita et al., 2022). Phenolic compounds are another class of secondary metabolites found in plants. These compounds have antioxidant properties and help protect plants from oxidative stress caused by environmental factors such as UV radiation or pollution (Dai & Mumper, 2010).

11.3.2 Role of Endophytes in the Production of Bioactive Compounds

Endophytes have been found to play a crucial role in producing bioactive compounds in plants. Numerous bioactive metabolites have been reported to be produced by endophytes in a single plant or microbe. These metabolites have been used as a great source of drugs to treat various diseases and have potential uses in the food, cosmetics, medicine and agriculture industries (Godstime et al., 2014; Shukla et al., 2014). Described below are some critical roles of endophytes in the production of bioactive compounds.

11.3.3 Endophytic Fungi Bioactive Compounds

An endophyte that has attracted much attention lately as a potential source of novel bioactive chemicals is endophytic fungus. By establishing colonization sites, battling pathogens, generating antibiotics and inducing resistance mechanisms, endophytic fungi protect plants against illness (Heydari & Pessarakli, 2010). In addition to creating phytohormones that strengthen the host plant's tolerance to a variety of stresses, endophytic fungi can also create pesticides to protect plants from herbivores (Badawy et al., 2021; Abdelaziz et al., 2022).

Bioactive compounds extracted from endophytes often include steroids, xanthones, terpenoids, isocoumarins, phenols, tetralones, benzopyranones and enniatines (Manganyi & Ateba, 2020). Furthermore, it has been reported that specific endophytic fungi generate 12 metabolites similar to their host plants (Singh et al., 2021). Three endophytic bacteria, *Diaporthe phaseolorum*, *Aspergillus fumigatus* and *Aspergillus versicolor* were recently isolated from tomato by Rashid (2021). Additionally, they extracted 42 bioactive compounds from *A. versicolor*, 36 bioactive compounds from *D. phaseolorum* and 38 bioactive compounds from *A. fumigatus*. Table 11.1 shows some significant bioactive compounds from endophytic fungi and their biological activity.

11.4 HUMAN HEALTH IMPLICATIONS

11.4.1 Immune System Support

The impact of edible leaf endophytes on the immune system is an intriguing field investigating the complex connection between plants, their accompanying microbes and human well-being. There has

TABLE 11.1
Endophytic Fungi with Their Bioactive Compounds and Biological Activities

Endophyte	Bioactive compound	Bioactivities	References
Alternaria sp.	Berberine	Antibiotic, antidiabetic, antihypertensive, antiproliferative hepatoprotective, hypolipidemic, vasodilator	(Duan et al., 2009)
Apiospora montagnei	Epiepoxydon	Anticancer	(Klemke et al., 2004)
Aspergillus niger	Rubrofusarin B	Anticancer	(Song et al., 2004)
Aspergillus parasiticus	Sequoiatones A	Anticancer	(Stierle et al., 1999)
Aspergillus sp.	6-Methyl-1,2,3-trihydroxy-7, 8-cyclohepta-9,12-diene-11-one-5,6,7,8-tetralene-7-acetamide	Anticancer	(Budhiraja et al., 2012)
Berkleasmium sp.	Diepoxin, Palmarumycin C11–C12, Cladospirone B, Palmarumycin C6.1,4,7β-trihydroxy-8-(spirodioxy-10,80-naphthyl)-7, 8-dihydronaphthalene, palmarumycin	Anticancer	(Shan et al., 2014)
Cephalotheca faveolata	Radicicol	Anticancer	(Turbyville et al., 2006)
Chaetomium sp., *Chaetomium chiversii*, *Chaetomium cupreum*, *Chaetomium globosum*	Radicicol, Cochliodinol, Mollicellin O, Mollicellin H, Mollicellin I, Altenuene, 4 -epialtenuene, aureonitolic acid, cochliodinol, isocochliodinol, indole-3-carboxylic acid, cyclo(alanyltryptophane), orsellinic acid, chaetoglobosin, ergosta-5,7,22-trien-3beta-ol, chaetopyranin	Anticancer	(Turbyville et al., 2006), (Debbab et al., 2007) (Zhao et al., 2016) (Wang et al., 2006)
Cladosporium cladosporioides	Aconitine	Anticancer, antiinflammatory, antineuralgic, cardiotoxic	(Yang et al., 2012)
Cytospora sp.	Cytoskyrin A, B	Anticancer	(Brady et al., 2001)
Emericella nidulans	Emindole DA	Anticancer	(Kralj et al., 2006)
Eutypella sp.	Eutypellin A	Anticancer	(Isaka et al., 2009)
Fusarium chlamydosporium	Fusarithioamide A	Anti-microbial	(Ibrahim et al., 2016)
Fusarium fujikuroi	5-Hydroxy 2(3H)-benzofuranone, Harpagoside	Anti-fungal	(Mousa et al., 2016)
Fusarium oxysporum	Beauvericin, bikaverin, cyclosporine, vincristine	Anticancer	(Zhan et al., 2007) (Jia et al., 2016)
Guignardia sp., *Halorosellinia* sp.	Anthracenedione	Anticancer	(Zhang et al., 2010)
Hyalodendriella sp.	Hyalodendriol CL	larvicidal	(Mao et al., 2017)
Hypoxylon truncatum	Daldinone	Anticancer	(Gu et al., 2006)

TABLE 11.1 (Continued)
Endophytic Fungi with Their Bioactive Compounds and Biological Activities

Endophyte	Bioactive compound	Bioactivities	References
Lachnum palmae	Isocoumarin derivatives	Cytotoxic	(Zhou et al., 2017)
Lopherdermium nitens	Sesquiterpenoids, pyrenophorin	Antifungal	(McMullin et al., 2015)
Mycelia sterilia	Spiromamakone A	Anticancer	(van der Sar et al., 2006)
Nodulisporium sp.	Sporothriolide	Antiphytopathogenic	(Cao et al., 2020)
P. citrinum	15-Dimethyl-2-epi-fumiquinazoline A, deoxytryptoquialanone, Citrinadin A, Chrysogenamide A	Antioxidant	(Huang et al., 2007)
Penicillium oxalicum	Gymnemagenin	Antidiabetic	(Parthasarathy & Sathiyabama, 2014)
Periconia atropurpurea	Periconicin A–B	Anticancer	(Teles et al., 2006)
Pestalotiopsis photiniae	Photinides A–F	Anticancer	(Ding et al., 2006)
Pestalotiopsis theae	Pestalotheol A, B, C and D	Neuroprotective	(Lin et al., 2008)
Phialophor amustea	Phialomustin C and D	Antimicrobial and cytotoxic	(Nalli et al., 2015)
Phomopsis sp.	Phomoxanthone A, Phomoxanthone B	Anticancer	(Isaka et al., 2001)
Phyllosticta spinarum	Tauranin, (+)-(5 S,10 S)-40–hydroxy methylcyclozonarone, 3-ketotauranin, 3alpha-hydroxytauranin, 12-hydroxytauranin, phyllospinarone	Anticancer	(Wijeratne et al., 2008)
Talaromyces purpureogenus	Talaromyolide K	Antiviral	(Cao et al., 2020)
Taxomyces andreanae	Taxol	Anticancer	(Stierle et al., 1993)
Thielavia subthermophila	Emodin	Anticancer	(Kusari et al., 2009)
Trichoderma brevicompactum	Trichodermin	Antifungal	(Shentu et al., 2014)
Trichoderma koningiopsis	Koningiopisin C		(Shi et al., 2017)
Xylaria sp.	Eremophilanolides, griseofulvin, cytochalasin	Anticancer	(Isaka et al., 2010) (Oliveira et al., 2011)

been an increasing interest in studying the potential of endophytes, namely those present in edible leaves, to improve the human immune system (Lu et al., 2021). The function of endophytes in these plants extends beyond fundamental nourishment. One study indicated that specific endophytes contain bioactive chemicals that can have immunomodulatory effects when ingested (Khare et al., 2018).

Edible foliage endophytes can impact the immune system by fostering a harmonious gut microbiota (Martínez-Romero et al., 2020). The gut microbiota has a vital function in controlling immunological responses, and changes have been associated with different immune-related disorders. When consumed, edible foliage endophytes can interact with the gut microbiota, stimulating the proliferation of advantageous bacteria and influencing the overall microbial community (Xu et al.,

2023). Consequently, this can benefit the immune system by improving its ability to respond to harmful microorganisms and maintaining a balanced immunological state. In addition, several edible plant's internal microorganisms produce chemical compounds that can modify the immune system (Flandroy et al., 2018). Bioactive substances possess properties that can reduce inflammation (Giresha et al., 2022), combat oxidative stress and inhibit the growth of microorganisms. These effects collectively enhance the immune system's ability to protect it against pathogens. Some endophytic fungi have been discovered to synthesize chemicals that possess antiinflammatory properties, which could potentially decrease the likelihood of developing chronic inflammatory disorders (Chen et al., 2020).

Studies indicate that specific endophytes have the potential to enhance the function of immune cells, including macrophages and lymphocytes, by increasing their capacity to identify and eradicate harmful microorganisms (Mengistu, 2020; Provenza & Villalba, 2010). The direct contact between endophytes and immune cells emphasizes the complex relationship between plant-associated microbes and the human immune system (Mishra et al., 2021).

The study of edible foliage endophytes and their impact on the immune system is still in its nascent phase, and further investigation is required to understand the precise mechanisms and potential health advantages comprehensively. Furthermore, plant genetics, ambient conditions and the individual's overall diet can all influence the diversity and functionality of endophytes. The impact of edible leaf endophytes on the immune system holds great potential for further investigation in human health. The relationship between plants and our immune system presents new opportunities to utilize the advantages of these microorganisms to enhance a solid and harmonious immunological response, leading to improved health.

11.4.2 Potential for Disease Prevention and Immune Response Enhancement

There is increasing interest in investigating the potential of edible leaf endophytes for disease prevention and improving the human immune response. Numerous edible plants contain endophytes that synthesize antimicrobial chemicals. These chemicals can inhibit the proliferation and dissemination of detrimental diseases, such as bacteria and fungi (Zhong et al., 2017). By consuming these plants, humans can derive advantages from the antimicrobial characteristics of the endophytes, hence diminishing the likelihood of contracting infectious diseases (Haque et al., 2015). Edible foliage endophytes often release bioactive compounds with antioxidant and antiinflammatory activities (Singh et al., 2021). These substances can benefit overall health by reducing oxidative stress and inflammation, both associated with various disorders, including chronic diseases (Zengin et al., 2019). Edible plants are susceptible to infection by foodborne pathogens. Endophytes can help the plant's defensive mechanisms, which makes it more resistant to infections. Humans may obtain indirect protection against foodborne infections if they consume such plants (Rizzo et al., 2021).

11.4.3 Antiinflammatory Effects

Role of natural metabolites in reducing inflammation: The interplay of edible leaves and endophytes has attracted interest for its possible medicinal implications, particularly in inflammation reduction. Inflammation is an intricate physiological reaction the body triggers in response to detrimental stimuli, such as pathogens or tissue damage (Dharmappa, 2022; Giresha et al., 2022). Although acute inflammation is a normal and defensive mechanism, chronic inflammation can result in a range of health problems, such as autoimmune disorders, cardiovascular illnesses and specific forms of cancer (Narayanappa et al., 2022; Sophiya et al., 2022). Hence, discovering natural and sustainable methods to regulate inflammation holds significant significance.

Research has been conducted on the health advantages of edible foliage, which refers to various plant leaves suitable for human consumption. Many edible plants include endophytes that produce bioactive substances possessing antiinflammatory functions (Abdel-Azeem et al., 2021). These bioactive chemicals can modify the immune response and assist in regulating inflammatory pathways within the human body. Many studies have found endophytes in edible leaves that demonstrate antiinflammatory properties (Weber et al., 2004; Maharjan et al., 2020; Chen et al., 2021). They synthesize secondary metabolites, including alkaloids, phenols and flavonoids, demonstrating anti-inflammatory and antioxidant characteristics. These chemicals can interact with cellular signaling pathways and prevent the generation of proinflammatory molecules, decreasing the overall inflammatory response.

An instance of an ingestible plant containing antiinflammatory endophytes is basil (*Ocimum basilicum*). Basil is a widely favored herb utilized in diverse global culinary traditions, and its leaves have endophytes that synthesize chemicals renowned for their antiinflammatory properties (Kamelnia et al., 2023). These substances may function by inhibiting the action of inflammatory enzymes and cytokines and reducing inflammation. It is crucial to acknowledge that although the potential of edible foliage endophytes in lowering inflammation is encouraging, additional research is necessary to comprehend the exact mechanisms involved and to create optimal dietary recommendations.

To summarize, edible foliage endophytes are a natural and sustainable reservoir of antiinflammatory substances. Incorporating various consumable plants into one's diet could enhance general well-being and potentially mitigate the likelihood of chronic inflammatory ailments. Continuing study in this area is expected to reveal fresh knowledge about the complex connection between edible plants and their endophytes, leading to novel nutritional approaches that can enhance human well-being.

11.5 FUTURE DIRECTIONS FOR RESEARCH

Despite positive results, various challenges limit the advancement of research in this field. Inadequate understanding of the diversity of edible foliage endophytes, the complicated interplay between endophytes and their host plants, and the variety of metabolite production pose significant obstacles. There need to be standardized methodologies for isolating, characterizing and evaluating compounds' bioactivity. By using advanced metagenomic techniques to comprehensively characterize the diversity of edible foliage endophytes and their genetic potential for metabolite production, as well as investigating the intricate relationships between edible plants and their endophytes to understand better how these interactions affect metabolite production and bioactivity. Functional genomics tools can be utilized to decipher the genetic basis of metabolite synthesis, leading to the targeted manipulation of endophytes for enhanced metabolite production and developing bioinformatics tools for efficient data analysis and integration of multi-omics datasets to uncover novel bioactive compounds and their potential health benefits. The study of edible foliage endophytes and their natural metabolites shows excellent potential in promoting human health. By overcoming current problems and using new methods, scientists may be able to fully exploit these microorganisms, which would lead to significant breakthroughs and improvements in health and nutrition.

11.6 CONCLUSION

Studying the edible leaf endophytes and their natural metabolites is a potential area of research that can help us better comprehend the complex connection between plants and human health. This chapter explored endophytes' essential functions in improving edible plants' nutritional content and bioactivity. This could lead to new ways of enhancing human well-being. This chapter highlights the importance of edible foliage endophytes and their metabolites as beneficial resources for promoting human health. To fully understand the potential benefits these mutually beneficial interactions offer society, it is crucial to be involved in interdisciplinary collaboration and ongoing research around the globe.

REFERENCES

Aamir, M., Aslam, A., Khan, M. Y., Jamshaid, M. U., Ahmad, M., Asghar, H. N., & Zahir, Z. A. (2013). Co-inoculation with Rhizobium and plant growth promoting rhizobacteria (PGPR) for inducing salinity tolerance in mung bean under field condition of semi-arid climate. *Asian Journal of Agriculture Biology, 1*(1), 7–12.

Abdel-Azeem, A. M., Yadav, A. N., Yadav, N., & Sharma, M. (2021, December 2). *Industrially Important Fungi for Sustainable Development*. Springer Nature.

Abdelaziz, A. M., El-Wakil, D. A., Attia, M. S., Ali, O. M., Abdelgawad, H., & Hashem, A. H. (2022, May 6). Inhibition of Aspergillus flavus growth and Aflatoxin production in Zea mays L. using endophytic Aspergillus fumigatus. *Journal of Fungi, 8*(5), 482. https://doi.org/10.3390/jof8050482

Almuhayawi, M. S., Abdel-Mawgoud, M., Al Jaouni, S. K., Almuhayawi, S. M., Alruhaili, M. H., Selim, S., & AbdElgawad, H. (2021, December 13). Bacterial endophytes as a promising approach to enhance the growth and accumulation of bioactive metabolites of three species of Chenopodium Sprouts. *Plants, 10*(12), 2745. https://doi.org/10.3390/plants10122745

Badawy, A. A., Alotaibi, M. O., Abdelaziz, A. M., Osman, M. S., Khalil, A. M. A., Saleh, A. M., Mohammed, A. E., & Hashem, A. H. (2021, June 29). Enhancement of seawater stress tolerance in Barley by the endophytic fungus Aspergillus ochraceus. *Metabolites, 11*(7), 428. https://doi.org/10.3390/metabo11070428

Brady, S. F., Singh, M. P., Janso, J. E., & Clardy, J. (2001, February 27). ChemInform abstract: Cytoskyrins A and B, new BIA active Bisanthraquinones Isolated from an Endophytic Fungus. *ChemInform, 32*(9). https://doi.org/10.1002/chin.200109224

Budhiraja, A., Nepali, K., Sapra, S., Gupta, S., Kumar, S., & Dhar, K. L. (2012, March 30). Bioactive metabolites from an endophytic fungus of Aspergillus species isolated from seeds of Gloriosa superba Linn. *Medicinal Chemistry Research, 22*(1), 323–329. https://doi.org/10.1007/s00044-012-0032-z

Cao, X., Shi, Y., Wu, S., Wu, X., Wang, K., Sun, H., He, S., Dickschat, J. S., & Wu, B. (2020, July). Polycyclic meroterpenoids, talaromyolides E – K for antiviral activity against pseudorabies virus from the endophytic fungus Talaromycespurpureogenus. *Tetrahedron, 76*(30), 131349. https://doi.org/10.1016/j.tet.2020.131349

Chen, Y., Zhang, L., Zou, G., Li, C., Yang, W., Liu, H., & She, Z. (2020, April). Anti-inflammatory activities of alkaloids from the mangrove endophytic fungus Phomopsis sp. SYSUQYP-23. *Bioorganic Chemistry, 97*, 103712. https://doi.org/10.1016/j.bioorg.2020.103712

Chen, Y., Zou, G., Yang, W., Zhao, Y., Tan, Q., Chen, L., Wang, J., Ma, C., Kang, W., & She, Z. (2021, January 24). Metabolites with anti-Inflammatory activity from the mangrove endophytic fungus Diaporthe sp. QYM12. *Marine Drugs, 19*(2), 56. https://doi.org/10.3390/md19020056

Cipriano, M. A. P., Freitas-Iório, R. D. P., Dimitrov, M. R., de Andrade, S. A. L., Kuramae, E. E., & Silveira, A. P. D. D. (2021, February 25). Plant-growth endophytic bacteria improve nutrient use efficiency and modulate Foliar N-Metabolites in sugarcane seedling. *Microorganisms, 9*(3), 479. https://doi.org/10.3390/microorganisms9030479

Dai, J., & Mumper, R. J. (2010, October 21). Plant Phenolics: Extraction, analysis and their antioxidant and anticancer properties. *Molecules, 15*(10), 7313–7352. https://doi.org/10.3390/molecules15107313

Debbab, A., Edrada, R., Wray, V., Müller, W., Hakiki, A., Mosaddak, M., Proksch, P., & Ebel, R. (2007). New bioactive metabolites isolated from the endophytic fungus Stemphylium globuliferum isolated from Mentha pulegium growing in Morocco. *Planta Medica, 73*(09), SL_016. https://doi.org/10.1055/s-2007-986734

Dharmappa, K. K. (2022). Medicinal and nutritional aspects of wild edible fruits from Western Ghat of India. *International Journal of Green Pharmacy (IJGP), 16*(4). http://dx.doi.org/10.22377/ijgp.v16i4.3345

Ding, G., Song, Y. C., Chen, J. R., Xu, C., Ge, H. M., Wang, X. T., & Tan, R. X. (2006, February 1). Chaetoglobosin U, a Cytochalasan Alkaloid from Endophytic Chaetomium globosum IFB-E019. *Journal of Natural Products, 69*(2), 302–304. http://dx.doi.org/10.1021/np050515+

Divekar, P. A., Narayana, S., Divekar, B. A., Kumar, R., Gadratagi, B. G., Ray, A., Singh, A. K., Rani, V., Singh, V., Singh, A. K., Kumar, A., Singh, R. P., Meena, R. S., & Behera, T. K. (2022, February 28). Plant secondary metabolites as defense tools against herbivores for sustainable crop protection. *International Journal of Molecular Sciences, 23*(5), 2690. https://doi.org/10.3390/ijms23052690

Duan L.I., Liwei G., & Hong Y. (2009). Isolation and identification of producing endophytic fungi of berberine from the plant *Phellodendronamurense*. *Journal of Anhui Agriculture Science*; 22:7.

Farnsworth, N., Akerele, O., & Bingel, A. (1987, May). Medicinal plants in therapy. *Journal of Ethnopharmacology*, *19*(3), 336. https://doi.org/10.1016/0378-8741(87)90016-x

Flandroy, L., Poutahidis, T., Berg, G., Clarke, G., Dao, M. C., Decaestecker, E., Furman, E., Haahtela, T., Massart, S., Plovier, H., Sanz, Y., & Rook, G. (2018, June). The impact of human activities and lifestyles on the interlinked microbiota and health of humans and of ecosystems. *Science of the Total Environment*, *627*, 1018–1038. https://doi.org/10.1016/j.scitotenv.2018.01.288

Giresha, A. S., Urs, D., Manjunatha, J. G., Sophiya, P., Supreetha, B. H., Jayarama, S., & Dharmappa, K. K. (2022, May 10). Group IIA secreted phospholipase A2 inhibition by elemolic acid as a function of anti-inflammatory activity. *Scientific Reports*, *12*(1). https://doi.org/10.1038/s41598-022-10950-1

Godstime, O. C., Enwa, F. O., Augustina, J. O., & Christopher, E. O. (2014). Mechanisms of anti-microbial actions of phytochemicals against enteric pathogens – a review. *Journal of Pharmaceutical, Chemical and Biological* Science, *2*, 77–85.

Grice, E. A., & Segre, J. A. (2011, March 16). The skin microbiome. *Nature Reviews Microbiology*, *9*(4), 244–253. https://doi.org/10.1038/nrmicro2537

Gu, W., Ge, H. M., Song, Y. C., Ding, H., Zhu, H. L., Zhao, X. A., & Tan, R. X. (2006, December 24). Cytotoxic Benzo[j]fluoranthene metabolites from Hypoxylontruncatum IFB-18, an endophyte of Artemisia annua. *Journal of Natural Products*, *70*(1), 114–117. https://doi.org/10.1021/np0604127

Haque, M. A., Lee, J. H., & Cho, K. M. (2015, October). Endophytic bacterial diversity in Korean kimchi made of Chinese cabbage leaves and their anti-microbial activity against pathogens. *Food Control*, *56*, 24–33. https://doi.org/10.1016/j.foodcont.2015.03.006

Heydari, A., & Pessarakli, M. (2010, May 1). A review on biological control of fungal plant pathogens using microbial antagonists. *Journal of Biological Sciences*, *10*(4), 273–290. https://doi.org/10.3923/jbs.2010.273.290

Huang, W. Y., Cai, Y. Z., Hyde, K. D., Corke, H., & Sun, M. (2007, February 21). Endophytic fungi from Nerium oleander L (Apocynaceae): Main constituents and antioxidant activity. *World Journal of Microbiology and Biotechnology*, *23*(9), 1253–1263. https://doi.org/10.1007/s11274-007-9357-z

Hussein, R., & El-Anssary, A. (2018, January 1). *Plants Secondary Metabolites: The Key Drivers of the Pharmacological Actions of Medicinal Plants*. IntechOpen. doi: 10.5772/intechopen.76139.

Ibrahim, S. R. M., Elkhayat, E. S., Mohamed, G. A. A., Fat'hi, S. M., & Ross, S. A. (2016, October). Fusarithioamide A, a new anti-microbial and cytotoxic benzamide derivative from the endophytic fungus Fusarium chlamydosporium. *Biochemical and Biophysical Research Communications*, *479*(2), 211–216. https://doi.org/10.1016/j.bbrc.2016.09.041

Isaka, M., Chinthanom, P., Boonruangprapa, T., Rungjindamai, N., & Pinruan, U. (2010, February 15). Eremophilane-Type Sesquiterpenes from the Fungus Xylaria sp. BCC 21097. *Journal of Natural Products*, *73*(4), 683–687. https://doi.org/10.1021/np100030x

Isaka, M., Jaturapat, A., Rukseree, K., Danwisetkanjana, K., Tanticharoen, M., & Thebtaranonth, Y. (2001, July 13). Phomoxanthones A and B, novel Xanthone Dimers from the endophytic fungus Phomopsis species. *Journal of Natural Products*, *64*(8), 1015–1018. https://doi.org/10.1021/np010006h

Isaka, M., Palasarn, S., Lapanun, S., Chanthaket, R., Boonyuen, N., & Lumyong, S. (2009, September 9). γ-Lactones and ent-Eudesmane Sesquiterpenes from the endophytic fungus Eutypella sp. BCC 13199. *Journal of Natural Products*, *72*(9), 1720–1722. https://doi.org/10.1021/np900316x

Jia, M., Chen, L., Xin, H. L., Zheng, C. J., Rahman, K., Han, T., & Qin, L. P. (2016, June 9). A friendly relationship between endophytic fungi and medicinal plants: A systematic review. *Frontiers in Microbiology*, *7*, 906. https://doi.org/10.3389/fmicb.2016.00906

Kamelnia, E., Mohebbati, R., Kamelnia, R., El-Seedi, H. R., & Boskabady, M. H. (2023). Anti-inflammatory, immunomodulatory and antioxidant effects of *Ocimum basilicum* L. and its main constituents: A review. *Iranian Journal of Basic Medical Science*, *26*(6), 617–627. https://doi.org/10.22038%2FIJBMS.2023.67466.14783

Khare, E., Mishra, J., & Arora, N. K. (2018, November 15). Multifaceted interactions between endophytes and plant: Developments and prospects. *Frontiers in Microbiology*, *9*. https://doi.org/10.3389/fmicb.2018.02732

Klemke, C., Kehraus, S., Wright, A. D., & König, G. M. (2004, May 27). New secondary metabolites from the marine endophytic fungus Apiosporamontagnei. *Journal of Natural Products*, *67*(6), 1058–1063. https://doi.org/10.1021/np034061x

Kralj, A., Kehraus, S., Krick, A., Eguereva, E., Kelter, G., Maurer, M., Wortmann, A., Fiebig, H. H., & König, G. M. (2006, June 23). Arugosins G and H: Prenylated Polyketides from the marine-derived fungus Emericellanidulans var. acristata. *Journal of Natural Products*, *69*(7), 995–1000. https://doi.org/10.1021/np050454f

Kusari, S., Zühlke, S., Košuth, J., Čellárová, E., & Spiteller, M. (2009, September 11). Light-independent metabolomics of endophytic Thielaviasubthermophila provides insight into microbial Hypericin Biosynthesis. *Journal of Natural Products*, *72*(10), 1825–1835. https://doi.org/10.1021/np9002977

Lin, Z., Wen, J., Zhu, T., Fang, Y., Gu, Q., & Zhu, W. (2008, February). Chrysogenamide A from an endophytic fungus associated with Cistanche deserticola and its neuroprotective effect on SH-SY5Y cells. *The Journal of Antibiotics*, *61*(2), 81–85. https://doi.org/10.1038/ja.2008.114

Lo Presti, L., Lanver, D., Schweizer, G., Tanaka, S., Liang, L., Tollot, M., Zuccaro, A., Reissmann, S., & Kahmann, R. (2015, April 29). Fungal effectors and plant susceptibility. *Annual Review of Plant Biology*, *66*(1), 513–545. https://doi.org/10.1146/annurev-arplant-043014-114623

Lu, H., Wei, T., Lou, H., Shu, X., & Chen, Q. (2021, September 1). A critical review on communication mechanism within plant-endophytic fungi interactions to cope with biotic and Abiotic stresses. *Journal of Fungi*, *7*(9), 719. https://doi.org/10.3390/jof7090719

Maharjan, S., Lee, S. B., Kim, G. J., Cho, S. J., Nam, J. W., Chin, J., & Choi, H. (2020, February 19). Isolation of unstable isomers of Lucilactaene and evaluation of anti-inflammatory activity of secondary metabolites produced by the endophytic fungus Fusarium sp. QF001 from the Roots of Scutellariabaicalensis. *Molecules*, *25*(4), 923. https://doi.org/10.3390/molecules25040923

Manganyi, M. C., & Ateba, C. N. (2020, December 6). Untapped potentials of endophytic fungi: A review of novel bioactive compounds with biological applications. *Microorganisms*, *8*(12), 1934. https://doi.org/10.3390/microorganisms8121934

Mao, Z., Lai, D., Liu, X., Fu, X., Meng, J., Wang, A., Wang, X., Sun, W., Liu, Z. L., Zhou, L., & Liu, Y. (2017, January 5). Dibenzo-α-pyrones: A new class of larvicidal metabolites against *Aedesaegypt*i from the endophytic fungus Hyalodendriellasp. Ponipodef12. *Pest Management Science*, *73*(7), 1478–1485. https://doi.org/10.1002/ps.4481

Martinez-Romero, E., Aguirre-Noyola, J. L., Bustamante-Brito, R., González-Román, P., Hernández-Oaxaca, D., Higareda-Alvear, V., Montes-Carreto, L. M., Martínez-Romero, J. C., Rosenblueth, M., & Servín-Garcidueñas, L. E. (2020, December 15). We and herbivores eat endophytes. *Microbial Biotechnology*, *14*(4), 1282–1299. https://doi.org/10.1111/1751-7915.13688

Masyita, A., Mustika Sari, R., Dwi Astuti, A., Yasir, B., Rahma Rumata, N., Emran, T. B., Nainu, F., & Simal-Gandara, J. (2022, March). Terpenes and terpenoids as main bioactive compounds of essential oils, their roles in human health and potential application as natural food preservatives. *Food Chemistry: X*, *13*, 100217. https://doi.org/10.1016/j.fochx.2022.100217

McMullin, D. R., Green, B. D., & Miller, J. D. (2015, December). Antifungal sesquiterpenoids and macrolides from an endophytic Lophodermium species of Pinus strobus. *Phytochemistry Letters*, *14*, 148–152. https://doi.org/10.1016/j.phytol.2015.10.006

Mengistu, A. A. (2020, July 30). Endophytes: Colonization, behaviour, and their role in defense mechanism. *International Journal of Microbiology*, *2020*, 1–8. https://doi.org/10.1155/2020/6927219

Mishra, S., Bhattacharjee, A., & Sharma, S. (2021, March 4). An ecological insight into the multifaceted world of plant-endophyte association. *Critical Reviews in Plant Sciences*, *40*(2), 127–146. https://doi.org/10.1080/07352689.2021.1901044

Molina-Montenegro, M. A., Oses, R., Torres-Díaz, C., Atala, C., Zurita-Silva, A., & Ruiz-Lara, S. (2015, August 18). Root-endophytes improve the ecophysiological performance and production of an agricultural species under drought condition. *AoB PLANTS*, *8*. https://doi.org/10.1093/aobpla/plw062

Mousa, W., Schwan, A., & Raizada, M. (2016, September 3). Characterization of antifungal natural products isolated from endophytic fungi of Finger Millet (Eleusine coracana). *Molecules*, *21*(9), 1171. https://doi.org/10.3390/molecules21091171

Nalli, Y., Mirza, D. N., Wani, Z. A., Wadhwa, B., Mallik, F. A., Raina, C., chaubey, A., Riyaz-Ul-Hassan, S., & Ali, A. (2015). Phialomustin A–D, new anti-microbial and cytotoxic metabolites from an endophytic fungus, Phialophoramustea. *RSC Advances*, *5*(115), 95307–95312. https://doi.org/10.1039/c5ra18121f

Narayanappa, N., Deepadarashan, D., Sathisha, A. D., Krishna, R. H., Rajkumar, S. M., & Dharamppa, K. K. (2022, November 14). Evaluation of anti-inflammatory activity of Garcinia indica, a wild edible fruit by inhibiting secretory phospholipase A2 group IIA enzyme from human pleural fluid. *Biomedicine*, *42*(5), 1058–1064. https://doi.org/10.51248/.v42i5.2173

Oliveira, C. M., Regasini, L. O., Silva, G. H., Pfenning, L. H., Young, M. C., Berlinck, R. G., Bolzani, V. S., & Araujo, A. R. (2011, June). Dihydroisocoumarins produced by Xylaria sp. and Penicillium sp., endophytic fungi associated with Piper aduncum and Alibertia macrophylla. *Phytochemistry Letters*, *4*(2), 93–96. https://doi.org/10.1016/j.phytol.2010.11.003

Osuna Avila, P., Barrow, J. R., Lucero, M. E., & Aaltonen, R. E. (2012, March). Relationship between plant lipid bodies and fungal endophytes. *Terra Latinoamericana*, *30*(1), 39–45.

Padash, A., Shahabivand, S., Behtash, F., & Aghaee, A. (2016, December). A practicable method for zinc enrichment in lettuce leaves by the endophyte fungus Piriformospora indica under increasing zinc supply. *Scientia Horticulturae*, *213*, 367–372. https://doi.org/10.1016/j.scienta.2016.10.040

Parthasarathy, R., & Sathiyabama, M. (2014, February 5). Gymnemagenin-producing endophytic fungus isolated from a medicinal plant Gymnemasylvestre R.Br. *Applied Biochemistry and Biotechnology*, *172*(6), 3141–3152. https://doi.org/10.1007/s12010-014-0754-0

Paul, S., Bhagobaty, R., Nihalani, M., & Joshi, S. (2020, November). Characterization of oleaginous endophytic fungi of biodiesel plants as potential biofuel minifactories. *Biomass and Bioenergy*, *142*, 105750. https://doi.org/10.1016/j.biombioe.2020.105750

Poveda, J., Eugui, D., Abril-Urías, P., & Velasco, P. (2021, August 4). Endophytic fungi as direct plant growth promoters for sustainable agricultural production. *Symbiosis*, *85*(1), 1–19. https://doi.org/10.1007/s13199-021-00789-x

Provenza, F., & Villalba, J. (2010, April). The role of natural plant products in modulating the immune system: An adaptable approach for combating disease in grazing animals. *Small Ruminant Research*, *89*(2–3), 131–139. https://doi.org/10.1016/j.smallrumres.2009.12.035

Rashid, T. S. (2021, March). Bioactive metabolites from tomato endophytic fungi with antibacterial activity against tomato bacterial spot disease. *Rhizosphere*, *17*, 100292. https://doi.org/10.1016/j.rhisph.2020.100292

Rasmussen, S., Parsons, A. J., Fraser, K., Xue, H., & Newman, J. A. (2008, January 24). Metabolic profiles of Loliumperenne are differentially affected by nitrogen supply, carbohydrate content, and fungal endophyte infection. *Plant Physiology*, *146*(3), 1440–1453. https://doi.org/10.1104/pp.107.111898

Richardson, M. D., Chapman, G. W., Hoveland, C. S., & Bacon, C. W. (1992, July). Sugar alcohols in endophyte-infected tall fescue under drought. *Crop Science*, *32*(4), 1060–1061. https://doi.org/10.2135/cropsci1992.0011183x003200040045x

Rizzo, D. M., Lichtveld, M., Mazet, J. A. K., Togami, E., & Miller, S. A. (2021, March 31). Plant health and its effects on food safety and security in a one health framework: Four case studies. *One Health Outlook*, *3*(1). https://doi.org/10.1186/s42522-021-00038-7

Rosenblueth, M., & Martínez-Romero, E. (2006, August). Bacterial endophytes and their interactions with hosts. *Molecular Plant-Microbe Interactions®*, *19*(8), 827–837. https://doi.org/10.1094/mpmi-19-0827

Sanchez-Vallet, A., Fouché, S., Fudal, I., Hartmann, F. E., Soyer, J. L., Tellier, A., & Croll, D. (2018, August 25). The genome biology of effector gene evolution in Filamentous plant pathogens. *Annual Review of Phytopathology*, *56*(1), 21–40. https://doi.org/10.1146/annurev-phyto-080516-035303

Schulz, B., & Boyle, C. (2005, June). The endophytic continuum. *Mycological Research*, *109*(6), 661–686. https://doi.org/10.1017/s095375620500273x

Shahzad, R., Waqas, M., Khan, A. L., Al-Hosni, K., Kang, S. M., Seo, C. W., & Lee, I. J. (2017, June). Indoleacetic acid production and plant growth promoting potential of bacterial endophytes isolated from rice (Oryza sativa L.) seeds. *Acta BiologicaHungarica*, *68*(2), 175–186. https://doi.org/10.1556/018.68.2017.2.5

Shan, T., Tian, J., Wang, X., Mou, Y., Mao, Z., Lai, D., Dai, J., Peng, Y., Zhou, L., & Wang, M. (2014, September 19). Bioactive Spirobisnaphthalenes from the endophytic fungus Berkleasmium sp. *Journal of Natural Products*, *77*(10), 2151–2160. https://doi.org/10.1021/np400988a

Shentu, X., Zhan, X., Ma, Z., Yu, X., & Zhang, C. (2014). Antifungal activity of metabolites of the endophytic fungus Trichoderma brevicompactum from garlic. *Brazilian Journal of Microbiology*, *45*(1), 248–254. https://doi.org/10.1590/s1517-83822014005000036

Shi, Y., Lou, K., & Li, C. (2010, April 20). Growth and photosynthetic efficiency promotion of sugar beet (Beta vulgaris L.) by endophytic bacteria. *Photosynthesis Research*, *105*(1), 5–13. https://doi.org/10.1007/s11120-010-9547-7

Shi, X. S., Wang, D. J., Li, X. M., Li, H. L., Meng, L. H., Li, X., Pi, Y., Zhou, X. W., & Wang, B. G. (2017). Anti-microbial polyketides from Trichoderma koningiopsis QA-3, an endophytic fungus obtained from the medicinal plant Artemisia argyi. *RSC Advances*, *7*(81), 51335–51342. https://doi.org/10.1039/c7ra11122c

Shukla, S. T., Habbu, P. V., Kulkarni, V. H., Jagadish, K. S., Pandey, A. R., & Sutariya, V. N. (2014). Endophytic microbes: A novel source for biologically/pharmacologically active secondary metabolites. *Asian Journal of Pharmacology Toxicology, 2*, 1–16.

Singh, A., Singh, D. K., Kharwar, R. N., White, J. F., & Gond, S. K. (2021, January 19). Fungal endophytes as efficient sources of plant-derived bioactive compounds and their prospective applications in natural product drug discovery: Insights, avenues, and challenges. *Microorganisms*, *9*(1), 197. https://doi.org/10.3390/microorganisms9010197

Somashekar, D., Venkateshwaran, G., Sambaiah, K., & Lokesh, B. (2003, July). Effect of culture conditions on lipid and gamma-linolenic acid production by mucoraceous fungi. *Process Biochemistry*, *38*(12), 1719–1724. https://doi.org/10.1016/s0032-9592(02)00258-3

Song, Y., Li, H., Ye, Y., Shan, C., Yang, Y., & Tan, R. (2004, December). Endophytic naphthopyrone metabolites are co-inhibitors of xanthine oxidase, SW1116 cell and some microbial growths. *FEMS Microbiology Letters*, *241*(1), 67–72. https://doi.org/10.1016/j.femsle.2004.10.005

Sophiya, P., Urs, D., K. Lone, J., Giresha, A. S., Krishna Ram, H., Manjunatha, J. G., El-Serehy, H. A., Narayanappa, M., Shankar, J., Bhardwaj, R., Ahmad Guru, S., & Dharmappa, K. K. (2022, October 17). Quercitrin neutralizes sPLA2IIa activity, reduces the inflammatory IL-6 level in PC3 cell lines, and exhibits anti-tumor activity in the EAC-bearing mice model. *Frontiers in Pharmacology*, *13*, 996285. https://doi.org/10.3389/fphar.2022.996285

Stierle, A. A., Stierle, D. B., & Bugni, T. (1999, July 1). Sequoiatones A and B: Novel antitumor metabolites isolated from a Redwood endophyte. *The Journal of Organic Chemistry*, *64*(15), 5479–5484. https://doi.org/10.1021/jo990277l

Stierle, A., Strobel, G., & Stierle, D. (1993, April 9). Taxol and Taxane production by Taxomycesandreanae, an endophytic fungus of Pacific Yew. *Science*, *260*(5105), 214–216. https://doi.org/10.1126/science.8097061

Strobel, G., & Daisy, B. (2003, December). Bioprospecting for microbial endophytes and their natural products. *Microbiology and Molecular Biology Reviews*, *67*(4), 491–502. https://doi.org/10.1128/mmbr.67.4.491-502.2003

Tabaszewska, M., Rutkowska, J., Skoczylas, U., Słupski, J., Antoniewska, A., Smoleń, S., Łukasiewicz, M., Baranowski, D., Duda, I., & Pietsch, J. (2021, January 30). Red Arils of Taxus baccata L. A new source of valuable fatty acids and nutrients. *Molecules*, *26*(3), 723. https://doi.org/10.3390/molecules26030723

Takeno, S., Sakuradani, E., Murata, S., Inohara-Ochiai, M., Kawashima, H., Ashikari, T., & Shimizu, S. (2005, January). Molecular evidence that the rate-limiting step for the biosynthesis of arachidonic acid in Mortierellaalpina is at the level of an elongase. *Lipids*, *40*(1), 25–30. https://doi.org/10.1007/s11745-005-1356-6

Teles, H. L., Sordi, R., Silva, G. H., Castro-Gamboa, I., Bolzani, V. D. S., P fenning, L. H., Magalhães de Abreu, L., Costa-Neto, C. M., Young, M. C. M., & Araújo, N. R. (2006, December). Aromatic compounds produced by Periconiaatropurpurea, an endophytic fungus associated with Xylopiaaromatica. *Phytochemistry*, *67*(24), 2686–2690. https://doi.org/10.1016/j.phytochem.2006.09.005

Turbyville, T. J., Wijeratne, E. M. K., Liu, M. X., Burns, A. M., Seliga, C. J., Luevano, L. A., David, C. L., Faeth, S. H., Whitesell, L., & Gunatilaka, A. A. L. (2006, February 1). Search for Hsp90 inhibitors with potential anticancer activity: Isolation and SAR studies of radicicol and Monocillin I from two plant-associated fungi of the Sonoran desert. *Journal of Natural Products*, *69*(2), 178–184. https://doi.org/10.1021/np058095b

Tyagi, J., Varma, A., & Pudake, R. N. (2017, July). Evaluation of comparative effects of arbuscular mycorrhiza (Rhizophagusintraradices) and endophyte (Piriformospora indica) association with finger millet (Eleusine coracana) under drought stress. *European Journal of Soil Biology*, *81*, 1–10. https://doi.org/10.1016/j.ejsobi.2017.05.007

van der Sar, S. A., Blunt, J. W., & Munro, M. H. G. (2006, April 14). Spiro-Mamakone A: A unique relative of the Spirobisnaphthalene class of compounds. *Organic Letters*, *8*(10), 2059–2061. https://doi.org/10.1021/ol060434k

Wang, S., Li, X. M., Teuscher, F., Li, Diesel, A., Ebel, R., Proksch, P., & Wang, B. G. (2006, October 18). Chaetopyranin, a Benzaldehyde derivative, and other related metabolites from Chaetomium globosum, an endophytic fungus derived from the Marine Red Alga Polysiphonia urceolata. *Journal of Natural Products*, *69*(11), 1622–1625. https://doi.org/10.1021/np060248n

Weber, D., Sterner, O., Anke, T., Gorzalczancy, S., Martino, V., & Acevedo, C. (2004). Phomol, a new anti-inflammatory metabolite from an endophyte of the medicinal plant Erythrina crista-galli. *The Journal of Antibiotics*, *57*(9), 559–563. https://doi.org/10.7164/antibiotics.57.559

Wijeratne, E. M. K., Paranagama, P. A., Marron, M. T., Gunatilaka, M. K., Arnold, A. E., &Gunatilaka, A. A. L. (2008, February 1). Sesquiterpene Quinones and related metabolites from Phyllostictaspinarum, a Fungal strain endophytic in Platycladusorientalis of the Sonoran desert. *Journal of Natural Products*, *71*(2), 218–222. https://doi.org/10.1021/np070600c

Xu, M., Bai, H. Y., Fu, W. Q., Sun, K., Wang, H. W., Xu, D. L., Dai, C. C., & Jia, Y. (2021, January). Endophytic bacteria promote the quality of Lyophyllumdecastes by improving non-volatile taste components of mycelia. *Food Chemistry*, *336*, 127672. https://doi.org/10.1016/j.foodchem.2020.127672

Xu, W., Bi, H., Peng, H., Yang, L., He, H., Fu, G., Liu, Y., & Wan, Y. (2023, January 14). Fermentative production of Diacylglycerol by endophytic fungi screened from Taxus chinensis var. mairei. *Foods*, *12*(2), 399. https://doi.org/10.3390/foods12020399

Xu, Y., Zhu, M., Feng, Y., & Xu, H. (2023, October). Panax notoginseng-microbiota interactions: From plant cultivation to medicinal application. *Phytomedicine*, *119*, 154978. https://doi.org/10.1016/j.phymed.2023.154978

Yang, K., Liang, J., Li, Q., Kong, X., Chen, R., & Jin, Y. (2012, December 27). Cladosporium cladosporioides XJ-AC03, an aconitine-producing endophytic fungus isolated from Aconitum leucostomum. *World Journal of Microbiology and Biotechnology*, *29*(5), 933–938. https://doi.org/10.1007/s11274-012-1246-4

Zengin, G., Stefanucci, A., Rodrigues, M. J., Mollica, A., Custodio, L., Aumeeruddy, M. Z., & Mahomoodally, M. F. (2019, January). Scrophularia lucida L. as a valuable source of bioactive compounds for pharmaceutical applications: In vitro antioxidant, anti-inflammatory, enzyme inhibitory properties, in silico studies, and HPLC profiles. *Journal of Pharmaceutical and Biomedical Analysis*, *162*, 225–233. https://doi.org/10.1016/j.jpba.2018.09.035

Zhan, J., Burns, A. M., Liu, M. X., Faeth, S. H., & Gunatilaka, A. A. L. (2007, February 1). Search for Cell Motility and Angiogenesis Inhibitors with Potential Anticancer Activity: Beauvericin and Other Constituents of Two Endophytic Strains of Fusarium oxysporum. *Journal of Natural Products*, *70*(2), 227–232. https://doi.org/10.1021/np060394t

Zhang, J. Y., Tao, L. Y., Liang, Y. J., Chen, L. M., Mi, Y. J., Zheng, L. S., Wang, F., She, Z. G., Lin, Y. C., To, K. K. W., & Fu, L. W. (2010, April 23). Anthracenedione Derivatives as Anticancer Agents Isolated from Secondary Metabolites of the Mangrove Endophytic Fungi. *Marine Drugs*, *8*(4), 1469–1481. https://doi.org/10.3390/md8041469

Zhao, S. S., Zhang, Y. Y., Yan, W., Cao, L. L., Xiao, Y., & Ye, Y. H. (2016, December 22). Chaetomium globosum CDW7, a potential biological control strain and its antifungal metabolites. *FEMS Microbiology Letters*, *364*, fnw287. https://doi.org/10.1093/femsle/fnw287

Zhong, L. Y., Zou, L., Tang, X. H., Li, W. F., Li, X., Zhao, G., & Zhao, J. L. (2017, March 7). Community of endophytic fungi from the medicinal and edible plant Fagopyrum tataricum and their anti-microbial activity. *Tropical Journal of Pharmaceutical Research*, *16*(2), 387. https://doi.org/10.4314/tjpr.v16i2.18

Zhou, J., Li, G., Deng, Q., Zheng, D., Yang, X., & Xu, J. (2017, October 31). Cytotoxic constituents from the mangrove endophytic Pestalotiopsis sp. induce G0/G1 cell cycle arrest and apoptosis in human cancer cells. *Natural Product Research*, *32*(24), 2968–2972. https://doi.org/10.1080/14786419.2017.1395431

12 Biological Activity of Endophytic Community Residing in *Aegle marmelos*

*Anagha Narale and Rachana Pachori**
*Corresponding author.

12.1 INTRODUCTION

Bioactive compounds are secondary metabolites or phytochemicals synthesized by metabolic processes that can be used in the treatment of different ailments. Plants are considered as a chief source for the isolation of biologically active compounds. However, recent studies have revealed that microbial populations residing in the internal tissues of plants can produce these bioactive compounds and also provide beneficial effects to the host plant (Sorrenti et al., 2023). Microorganisms that inhabit internal plant tissues forming a mutualistic association with the host are known as endophytes. These are endosymbionts, which may be bacteria, fungi or actinomycetes, that can be isolated from any part of the plant (Gouda et al., 2016). Previous studies have pointed out the presence of more than one endophyte in a single host plant (Strobel & Daisy, 2003). According to many ancient literatures of India, diverse plant species possess multiple medicinal properties, of which *Aegle marmelos* was extensively used. This plant has been studied by many researchers mostly for the therapeutic properties of extracts isolated from its fruits and leaves. However, very few studies have been reported to date focusing on the diversity of endophytes present in this plant as well as the bioactive compounds synthesized by them. Hence, this chapter concentrates on endophytes and bioactive compounds produced from them with special reference to *A. marmelos*.

12.2 *AEGLE MARMELOS*

Aegle marmelos, known as Bael in India, is considered one of the most sacred trees that is used in the rituals for Lord Shiva. The tree itself is also worshiped due to some regional beliefs. It has great importance in the diets of populations residing in the northern parts of India. *Aegle marmelos* is mostly found in the Indian subcontinent, i.e. India, Nepal, Sri Lanka, Myanmar and Thailand. It is also found in other regions of world and is recognized by different names. For example, it is called Bael in India. Bilva, Shivaphala in Sanskrit, Bel in Assamese and Marathi, Vilvamarum in Tamil and Belo in Oriya, Bela in Spanish, Be Li in Sinhalese and Matoom in Thai (Sarkar et al., 2020). Bael fruit, wood apple, stone apple, wood stone, bitter apple, elephant apple and Bengal quince are some of the common names used for this fruit. *Aegle marmelos* belongs to the Rutacea family and is a deciduous plant that has leaves in sets of threes, twos or singles (Orwa et al., 2009). It has a short trunk with thick, soft, flaky bark, and spines over spreading branches. The plant can grow in tropical regions with minimum availability of water. *Aegle marmelos* can grow in temperatures between –6°C and 45°C and can bear fruits in regions with a long dry season (Pushpakumara, 2007). The fruit of *A. marmelos* is mostly included in regular diets in the form of jams, pickles, salads, etc. (Figure 12.1).

 DOI: 10.1201/9781003486275-14

FIGURE 12.1 *Aegle marmelos* (Bael) tree with light green-colored ball-shaped fruit in the inset.

Aegle marmelos is a plant with high medicinal importance. The use of the plant in therapeutic procedures as well as medicinal doses is significantly found in the ancient medicinal literature of India (Dwivedi & Dwivedi, 2007). It is used to treat different disorders and diseases such as diarrhea, neurological disorders, inflammation, vomiting, maladsorption, heart diseases, dyspepsia, edema, swollen joints, to control asthma, and to heal wounds, high blood pressure, typhoid, etc. (Chanda et al., 2008; Parichha, 2004). *Aegle marmelos* possesses different therapeutic properties such as antimicrobial, antioxidant, anticancer, antidiabetic, antiinflammatory, immunomodulatory, chemo-protective, antigenotoxic, etc. (Rahman & Parvin, 2014). Different bioactive compounds have been found to be present in *Aegle marmelos* including alkaloids, coumarins, tannins, carotenoids, terpenoids, steroids, saponins, polysaccharides, phenols, etc. (Dhankhar et al., 2011; Mujeeb et al., 2014).

12.3 WHAT ARE ENDOPHYTES?

In 1886, Anton de Barry, a German botanist and the father of plant pathology, first used the term "endophytes" (Muthu et al., 2022). Petrini (1986) used the definition of Anton de Barry to define endophytes as "the organism residing in the plant tissue" (Petrini, 1991). Earlier, the term endophyte was concerned with the fungal species inhabiting plant tissues. However, discoveries and studies have stated that along with fungi, bacteria can also reside in plant tissues with a mutualistic association (Hardoim et al., 2015). Hallmann et al. (1997) defined endophytes as bacteria residing in plants without visibly harming the host and which can be isolated from surface disinfected parts of the plants. Many studies have revealed that endophytic microorganisms showed some similarities with the rhizospheric community of microorganisms that are present near the roots of the plants. However, the mutualistic relationship between the host and the endophytes is beneficial for both the microorganism and the plant for defense against plant pathogens. Along with this mutualistic association, endophytes exhibit plant growth-promoting activities as well (Wippel, 2023). They also exhibited stimulation of the host immune system and survival during abiotic stress conditions (Teixeira et al., 2019). Although millions of endophytes exist very few are being studied. This requires exploring diverse endophytes for their potential characteristics (Joseph & Priya, 2011). Recently, more research has been focused on the exploration of secondary metabolites produced by endophytes which may be beneficial for therapeutic purposes (Xia et al., 2022).

Fungal endophytes are majorly classified into two classes: (1) clavicipitaceous and (2) non-clavicipitaceous, of which non-clavicipitaceous has diverse forms of endophytic fungal communities. Non-clavicipitaceous endophytes can be isolated from all terrestrial land plants. They can be distinguished into three major classes based on the transmission route, colonizing part, etc. (Rodriguez et al., 2009). Actinomycetes also inhabit plants, internal tissues to reside as endophytes. They also form a symbiotic association with the host plant and produce biologically active compounds (Saini et al., 2022). Bacteria can also colonize all parts of plants, forming mutualistic associations with them. These are classified as bacterial endophytes (Kandel et al., 2017).

There are over 300,000 million plant species, and all of them are hosts to more than one endophyte (Strobel & Daisy, 2003). Endophytes have been studied for their characteristic features, revealing that they synthesize various bioactive secondary metabolites that are used by the host plants for different activities such as plant growth, mitigation of biotic and abiotic stress and defense mechanisms, etc. (Singh et al., 2017). Endophytes may have unique biological systems or genetic makeup, which can be the source of these bioactive compounds (Strobel, 2003). However, these metabolites are also found to be beneficial in treating various therapeutic diseases. These secondary metabolites are then considered as bioactive compounds. Drug resistance by pathogens is a major emerging problem faced by humans, which requires the discovery of therapeutic compounds, which can be accomplished by endophytic compounds. Biologically active compounds produced by endophytes have promising health-healing potentials and are safe for use by humans (Sorrenti et al., 2023; Ancheeva et al., 2020). Endophytic fungi are recognized as the major source of biologically active compounds, having therapeutic properties such as antibiotics, antioxidants, anticancer compounds, antimicrobial volatiles, etc. (Singh & Kumar, 2023; Kaul et al., 2012; Huang et al., 2007). Endophytes produce diverse forms of compounds with distinctive and unique structures, such as alkaloids, saponins, tannins, taxols, flavonoids, terpenoids, steroids, phenolic acids, etc. Isopestacin isolated from endophytic fungi shows efficient antifungal and antioxidant activities (Gouda et al., 2016; Sorrenti et al., 2023). Hence, these bioactive compounds are in demand by pharmaceutical companies.

12.4 ENDOPHYTES OF *AEGLE MARMELOS*

The endophytic community present in plants remains underexplored for its therapeutic content. Different strains of endophytes have been recognized by studies conducted on *A. marmelos*. Gond et al. (2007) stated that the inner bark of *A. marmelos* harbors a greater population of endophytic fungi as compared to the other parts of the plant. They also reported the presence of diversified fungal species of endophytes, most of which belong to the deuteromycotina genera. One of the isolated endophytes of *Fusarium* sp. has been found to colonize the plant dominantly as compared to other species like *Aspergillus* sp., *Drechslera* sp., *Alternaria* sp., *Rhizoctonia* sp., *Nigrospora* sp. and *Stenella* sp. Along with these endophytes, they also found two other members of endophytes belonging to the Ascomycetous genera: *Chaetomium globosum* and *Emericella* spp. A similar study was then conducted by Mani et al. (2015). In this study, they found that endophytic fungi like *Alternaria alternata*, *Alternaria citrimacularis* and *Curvularia australiensis* showed significant antimicrobial activity. Patil et al. (2015) isolated *Aspergillus flavus* from *Aegle marmelos* and reported it as a potential source of pharmaceutically important bioactive compounds. *Bartalinia robillardoides* is a strain of endophytic fungus isolated from *Aegle marmelos*, as reported by Gangadevi and Muthumary (2008). Gupta et al. (2014) studied the lipase activity of *Aegle marmelos* exhibited by the endophytic fungi. In their study they found that prominent lipase activity was shown by *Fusarium incarntum*, an endophyte isolated and identified from *Aegle marmelos*. In most of the medicinal plants. Arora et al. (2016) isolated *Xylaria psidii*, an endophyte from *A. marmelos* (L.) Corr. Actinomycetes isolated from *Aegle marmelos* were also reported to produce a significant amount of bioactive compounds. Saini et al. (2022) studied the activity of *Microbiospora* sp. to test its enzyme inhibitory activity. Bacterial endophytes isolated from *A. marmelos* were also seen to possess various potential bioactive properties (Londhe & Berde, 2015). Myint isolated 11 strains of bacterial endophytes from the leaves of *A. marmelos*. They were identified as belonging to the genera *Streptococcus*, *Corynebacterium*, *Pseudomonas*, *Aeromonas*, *Bacillus* and *Azomonas*.

12.5 ENDOPHYTIC BIOACTIVE COMPOUNDS REPORTED IN *AEGLE MARMELOS*

Aegeline, marmeline, coumarin, imperatorin and xanthotoxol are the bioactive compounds found in *A. marmelos* majorly contributing to the antidiabetic, anticancerous, antifertility, antimicrobial and immunogenic activities (Pathirana et al., 2020). Similarly, Mani et al. (2018) reported *Curvularia australiensis* and *Alternaria citrimacularis* as the fungal endophytes isolated from *A. marmelos*, which are capable of synthesizing various bioactive compounds. They predominantly produce phenols and flavonoids that show prominent antioxidant properties. *Curvularia australiensis* endophytic fungi mostly found in the leaves of *Aegle marmelos* produce flavonoids (a dimer of epicatechin) that can be used to treat cervical cancers and inflammation, and also can be used for pain relief and antipyretic activities (Mani et al., 2021; Monika et al., 2023). *Aspergillus flavus* isolated from *A. marmelos* was reported to produce bioflavonoid rutin, which showed extensive antioxidant activities (Patil et al., 2015). *Fusarium incarntum* isolated from *Aegle marmelos* was studied for pancreatic lipase inhibitory activity for antiobesity compounds. However, the exact compound exhibiting pancreatic lipase inhibitory activity by *Fusarium incarntum* remains to be identified (Gupta et al., 2014). *Bartalinia robillardoides* isolated from *A. marmelos* was reported to produce taxols, a compound that can be used in cancer treatments (Gangadevi & Muthumary, 2008). Arora et al. (2016) isolated two potential bioactive compounds from *Xylaria psidii* that showed cytotoxic activity against pancreatic cancerous cells. *Alternaria alternata* species of endophytic fungi isolated from *A. marmelos* produce different bioactive compounds such as steroids, tannins and glycosides that exhibit antimicrobial activities (Eram et al., 2018) (Table 12.1).

TABLE 12.1
Bioactive Compounds Isolated from Endophytic Microorganisms Present in *Aegle marmelos* and Their Therapeutic Properties

S. no.	Bioactive compounds	Endophytes	Bioactivity	References
1.	Aegeline	ND	Antioxidant activity, antidiabetic activity	Pathirana et al., 2020
2.	Marmeline	ND	Anticancer, antibacterial, antiinflammatory activities	Pathirana et al., 2020; Shinde et al., 2014

(continued)

TABLE 12.1 (Continued)
Bioactive Compounds Isolated from Endophytic Microorganisms Present in *Aegle marmelos* and Their Therapeutic Properties

S. no.	Bioactive compounds	Endophytes	Bioactivity	References
3.	Coumarin	*Microbiospora* sp.	Antiinflammatory, antimicrobial, antioxidant activities	Pathirana et al., 2020; Srinivasa et al., 2022; Saini et al., 2022
4.	Xanthotoxol	ND	Anticancer, antiinflammatory, antioxidant activities	Pathirana et al., 2020
5.	Bioflavonoid Rutin	*Aspergillus flavus*	Antioxidant activities	Patil et al., 2015
6.	Taxols	*Bartalinia robillardoides*, *Xylaria psidii*	Anticancer activities	Gangadevi & Muthumary, 2008; Arora et al., 2016
7.	Flavonoids (a dimer of epicatechin)	*Alternaria citrimacularis*, *Aspergillus flavus*, *Curvularia australiensis*, *Microbiospora* sp.	Anticancer, antiinflammatory, pain relief, antipyretic, antioxidant activities	Patil et al., 2015; Mani et al., 2018; Mani et al., 2021; Monika et al., 2023; Saini and Gangwar, 2022
8.	Steroids	*Alternaria alternata*	Antimicrobial activities	Eram et al., 2018
9.	Phenols	*Curvularia australiensis*, *Alternaria citrimacularis*, *Microbiospora* sp.	Anticancer, antiinflammatory, pain relief, antipyretic, antioxidant, antimutagenic activities	Mani et al., 2021; Monika et al., 2023; Saini and Gangwar, 2022
10.	Glycosides	*Alternaria alternata*	Antimicrobial activities	Eram et al., 2018
11.	Imperatorin	ND	Antioxidant, antidiabetic activities.	Pathirana et al., 2020
12.	Tannins	*Alternaria alternate*	Antimicrobial activities	Eram et al., 2018
13.	Betalains	*Microbiospora* spp.	Antioxidant, antidiabetic activities	Saini and Gangwar, 2022
14.	Vitamins	*Microbiospora* spp.	Antioxidant activities	Saini and Gangwar, 2022
15.	Cellulolytic enzymes	*Aspergillus oryzae*, *Trichoderma harzianum*	Breakdown of cellulose, fermentation	Badiyaa et al., 2018
16.	Gallic acid	ND	Antioxidant, antidiabetic activities	Ahmad et al., 2021

* ND: not determined.

12.6 CONCLUSION

Aegle marmelos has huge applications in therapeutic areas. Most of the studies on *Aegle marmelos* have been conducted over aqueous or solvent extracts of leaves and fruits. However, recently studies were also conducted on the endophytic community present in *Aegle marmelos*. These studies revealed that there is a diverse population of endophytic communities residing in *A. marmelos* involving bacteria, actinomycetes and fungi. Prominently, fungal endophytes were found as compared to bacterial and actinomycetal endophytes. Furthermore, reports on endophytic fungi for the production of secondary metabolites that have different biological activities like antimicrobial, anticancer, antioxidant and antiinflammatory activities pave the way for therapeutic applications. However, there is a need to explore the endophytic population present in *Aegle marmelos* from various geographical locations, since there are many endophytes that remain to be studied for their potential to produce bioactive compounds with possible therapeutic applications.

REFERENCES

Ahmad, Wasim, Mohd Amir, Adil Ahmad, Abuzer Ali, Amena Ali, Shadma Wahab, Harshita Abul Barkat et al. "*Aegle marmelos* leaf extract phytochemical analysis, cytotoxicity, in vitro antioxidant and antidiabetic activities." *Plants* 10, no. 12 (2021): 2573.

Ancheeva, Elena, Georgios Daletos, and Peter Proksch. "Bioactive secondary metabolites from endophytic fungi." *Current Medicinal Chemistry* 27, no. 11 (2020): 1836–1854.

Arora, Divya, Nisha Sharma, Venugopal Singamaneni, Vishal Sharma, Manoj Kushwaha, Vidushi Abrol, Santosh Guru et al. "Isolation and characterization of bioactive metabolites from Xylaria psidii, an endophytic fungus of the medicinal plant *Aegle marmelos* and their role in mitochondrial dependent apoptosis against pancreatic cancer cells." *Phytomedicine* 23, no. 12 (2016): 1312–1320.

Badiyaa, Pradeep Kumar, Sai Praneeth Thotaa, Sandeep Yerrama, Praveen V. Vadlanib, Pallavi Vedantamd, Sai Sathish Ramamurthya, Nageswara Rao Golakotia, Robin Sharmae, and BS Vijaya Kumare. "Endophytic Fungi from *Aegle marmelos* Plant: a Potent and Innovative Platform for Enhanced Cellulolytic Enzyme Production." *Journal of Technology Innovations in Renewable Energy* 7 (2018): 7–18.

Chanda, R., A. Ghosh, T. Mitra, J. P. Mohanty, N. Bhuyan, and G. Pawankar. "Phytochemical and pharmacological activity of *Aegle marmelos* as a potential medicinal plant: an overview." *The Internet Journal of Pharmacology* 6, no. 1 (2008): 3.

Dwivedi, Girish, and Shridhar Dwivedi. "Sushruta-the clinician-teacher par excellence." *Indian Journal of Chest Diseases and Allied Sciences* 49, no. 4 (2007): 243.

Eram, Daniya, Manoj-Kumar Arthikala, Govindappa Melappa, and Gustavo Santoyo. "Alternaria species: endophytic fungi as alternative sources of bioactive compounds." *Italian Journal of Mycology* 47 (2018): 40–54.

Gangadevi, V., and J. Muthumary. "Taxol, an anticancer drug produced by an endophytic fungus Bartalinia robillardoides Tassi, isolated from a medicinal plant, *Aegle marmelos* Correa ex Roxb." *World Journal of Microbiology and Biotechnology* 24 (2008): 717–724.

Gond, S. K., V. C. Verma, A. Kumar, V. Kumar, and R. N. Kharwar. "Study of endophytic fungal community from different parts of *Aegle marmelos* Correae (Rutaceae) from Varanasi (India)." *World Journal of Microbiology and Biotechnology* 23 (2007): 1371–1375.

Gouda, Sushanto, Gitishree Das, Sandeep K. Sen, Han-Seung Shin, and Jayanta Kumar Patra. "Endophytes: a treasure house of bioactive compounds of medicinal importance." *Frontiers in Microbiology* 7 (2016): 1538.

Gupta, Mahiti, Sanjai Saxena, and Dinesh Goyal. "Lipase inhibitory activity of endophytic fungal species of *Aegle marmelos*: a bioresource for potential pancreatic lipase inhibitors." *Symbiosis* 64 (2014): 149–157.

Hallmann, J., A. Quadt-Hallmann, W. F. Mahaffee, and J. W. Kloepper. "Bacterial endophytes in agricultural crops." *Canadian Journal of Microbiology* 43, no. 10 (1997): 895–914.

Hardoim, Pablo R., Leonard S. Van Overbeek, Gabriele Berg, Anna Maria Pirttilä, Stéphane Compant, Andrea Campisano, Matthias Döring, and Angela Sessitsch. "The hidden world within plants: ecological and evolutionary considerations for defining functioning of microbial endophytes." *Microbiology and Molecular Biology Reviews* 79, no. 3 (2015): 293–320.

Huang, Wu-Yang, Yi-Zhong Cai, Jie Xing, Harold Corke, and Mei Sun. "A potential antioxidant resource: endophytic fungi from medicinal plants." *Economic Botany* 61, no. 1 (2007): 14–30.

Joseph, Baby, and R. Mini Priya. "Bioactive compounds from endophytes and their potential in." *American Journal of Biochemistry and Molecular Biology* 1, no. 3 (2011): 291–309.

Kandel, Shyam L., Pierre M. Joubert, and Sharon L. Doty. "Bacterial endophyte colonization and distribution within plants." *Microorganisms* 5, no. 4 (2017): 77.

Kaul, Sanjana, Suruchi Gupta, Maroof Ahmed, and Manoj K. Dhar. "Endophytic fungi from medicinal plants: a treasure hunt for bioactive metabolites." *Phytochemistry Reviews* 11 (2012): 487–505.

Londhe, Aditee A., and Chanda Parulekar Berde. "Study of endophytic bacteria from *Aegle marmelos*." *World Journal of Pharmaceutical Research* 4, no. 12 (2015): 1013–1021.

Mani, Vellingiri Manon, A. J. P. G. Soundari, and Kathirvel Preethi. "Enzymatic and phytochemical analysis of endophytic fungi on *Aegle marmelos* from Western Ghats of Tamil Nadu, India." *International Journal of Life Science Pharma Research* 8, no. 1 (2018): L1–L8.

Mani, Vellingiri Manon, Arockiam Jeyasundar Parimala Gnana Soundari, Balamuralikrishnan Balasubramanian, Sungkwon Park, Utthapon Issara, Kathirvel Preethi, and Wen-Chao Liu. "Evaluation of dimer of epicatechin from an endophytic fungus Curvularia australiensis FC2AP on acute toxicity levels, anti-inflammatory and anti-cervical cancer activity in animal models." *Molecules* 26, no. 3 (2021): 654.

Mani, Vellingiri Manon, Arockiamjeyasundar Parimala Gnana Soundari, Damodharan Karthiyaini, and Kathirvel Preethi. "Bioprospecting endophytic fungi and their metabolites from medicinal tree *Aegle marmelos* in Western Ghats, India." *Mycobiology* 43, no. 3 (2015): 303–310.

Monika, S., M. Thirumal, and P. R. Kumar. "Phytochemical and biological review of *Aegle marmelos* Linn." *Future Science OA* 9, no. 3 (2023): FSO849.

Mujeeb, Farina, Preeti Bajpai, and Neelam Pathak. "Phytochemical evaluation, antimicrobial activity, and determination of bioactive components from leaves of *Aegle marmelos*." *BioMed Research International* 2014 (2014): 497606.

Muthu Narayanan, Manjula, Norhayati Ahmad, Pooja Shivanand, and Faizah Metali. "The role of endophytes in combating fungal-and bacterial-induced stress in plants." *Molecules* 27, no. 19 (2022): 6549.

Orwa, C., A. Mutua, R. Kindt, R. Jamnadass, and A. Simons. "Agroforestree Database: a tree reference and selection guide. Version 4." *Version 4.* (2009).

Parichha, Saswati. "Bael (*Aegle marmelos*): nature's most natural medicinal fruit." *Orissa Review* 9 (2004): 16–17.

Pathirana, Chamila Kumari, Terrence Madhujith, and Janakie Eeswara. "Bael (*Aegle marmelos* L. Corrêa), a medicinal tree with immense economic potentials." *Advances in Agriculture* 2020 (2020): 1–13.

Patil, M. P., R. H. Patil, and V. L. Maheshwari. "Biological activities and identification of bioactive metabolite from endophytic Aspergillus flavus L7 isolated from *Aegle marmelos*." *Current Microbiology* 71 (2015): 39–48.

Petrini, Orlando. "Fungal endophytes of tree leaves." In Andrews, J.H., Hirano, S.S. (eds) *Microbial ecology of leaves*, pp. 179–197. New York, NY: Springer New York, 1991. https://doi.org/10.1007/978-1-4612-3168-4_9

Pushpakumara, D. K. N. G. "Beli: *Aegle marmelos* L. Correa: In: Pushpakumara DKNG, Gunasena HPM, Sing VP, editors. Underutilized fruit trees in Sri Lanka: Voulme 1: World Agroforestry Centre." South Asia Office, New Delhi, India, (2007): 249–276.

Rahman, Shahedur, and Rashida Parvin. "Therapeutic potential of *Aegle marmelos* (L.)-An overview." *Asian Pacific Journal of Tropical Disease* 4, no. 1 (2014): 71–77.

Rodriguez, R. J., J. F. White Jr, A. E. Arnold, and A. R. Redman. "Fungal endophytes: diversity and functional roles." *New phytologist* 182, no. 2 (2009): 314–330.

Saini, Preeti, and Madhurama Gangwar. "In vitro enzyme inhibitory and free radical scavenging potentials of an *Aegle marmelos* endophytic actinomycete extract." *Research Journal of Recent Sciences, ISSN* 2277 (2022): 2502.

Sandeep Dhankhar, S. Ruhil, M. Balhara, Seema Dhankhar, and A. K. Chhillar. "*Aegle marmelos* (Linn.) Correa: a potential source of Phytomedicine." *Journal of Medicinal Plant Research* 5, no. 9 (2011): 1497–1507.

Sarkar, Tanmay, Molla Salauddin, and Runu Chakraborty. "In-depth pharmacological and nutritional properties of bael (*Aegle marmelos*): a critical review." *Journal of Agriculture and Food Research* 2 (2020): 100081.

Shinde, P. B., S. D. Katekhaye, M. B. Mulik, and K. S. Laddha. "Rapid simultaneous determination of marmelosin, umbelliferone and scopoletin from *Aegle marmelos* fruit by RP-HPLC." *Journal of Food Science and Technology* 51 (2014): 2251–2255.

Singh, Vivek Kumar, and Awanish Kumar. "Secondary metabolites from endophytic fungi: production, methods of analysis, and diverse pharmaceutical potential." *Symbiosis* 90 (2023): 111–125.

Singh, Monika, Ajay Kumar, Ritu Singh, and Kapil Deo Pandey. "Endophytic bacteria: a new source of bioactive compounds." *3 Biotech* 7 (2017): 1–14.

Sorrenti, Valeria, Ilaria Burò, Valeria Consoli, and Luca Vanella. "Recent advances in health benefits of bioactive compounds from food wastes and by-products: biochemical aspects." *International Journal of Molecular Sciences* 24, no. 3 (2023): 2019.

Srinivasa, Chandrashekar, Govindappa Mellappa, Shashank M. Patil, Ramith Ramu, Bhargav Shreevatsa, Chandan Dharmashekar, Shiva Prasad Kollur, Asad Syed, and Chandan Shivamallu. "Plants and endophytes–a partnership for the coumarin production through the microbial systems." *Mycology* 13, no. 4 (2022): 243–256.

Strobel, Gary A. "Endophytes as sources of bioactive products." *Microbes and Infection* 5, no. 6 (2003): 535–544.

Strobel, Gary, and Bryn Daisy. "Bioprospecting for microbial endophytes and their natural products." *Microbiology and Molecular Biology Reviews* 67, no. 4 (2003): 491–502.

Teixeira, Paulo José PL, Nicholas R. Colaianni, Connor R. Fitzpatrick, and Jeffery L. Dangl. "Beyond pathogens: microbiota interactions with the plant immune system." *Current Opinion in Microbiology* 49 (2019): 7–17.

Wippel, Kathrin. "Plant and microbial features governing an endophytic lifestyle." *Current Opinion in Plant Biology* 76 (2023): 102483.

Xia, Yandong, Junang Liu, Cang Chen, Xiuli Mo, Qian Tan, Yuan He, Zhikai Wang, Jia Yin, and Guoying Zhou. "The multifunctions and future prospects of endophytes and their metabolites in plant disease management." *Microorganisms* 10, no. 5 (2022): 1072.

13 Mushrooms as a Source of Better Food and Health in Northern Pakistan

Shakil Ahmad Zeb, Shujaul Mulk Khan, Abdullah Abdullah, Zeesahn Ahmad, Muhammad Shakeel Khan, Tehsin Ullah and Husain Badshah*
*Corresponding author.

13.1 INTRODUCTION: HUMAN FOOD SYSTEMS

The production, distribution and motivations behind our food choices need to be discussed to understand the importance of the food system. It connects both the food economy view point and the food chain's linear paradigm. The supply of food for human consumption is guaranteed by a complex web of interconnected processes, activities and organizations that make up the human food system. Every aspect of food is covered by this complex system, which includes production, processing, distribution, marketing, consumption and waste management. A complex web of interdependent players and components, such as farmers, processors, distributors, merchants, consumers and waste management systems compose the network (Tansey & Worsley, 2014). The food chain is connected to three distinct facets (1) biological (the ecological sustainability of the living systems that provide food); (2) economic and political (the authority and dominance that various factions possess over the various components of the system); and (3) social and cultural aspects (including interpersonal relationships, communal values and food-related cultural traditions).

The nature and direction of changes in the food system are important because they are dynamic and significantly affect the food industry. The food industry is a part of society and a larger economy; hence, the trends, values and actions in the industry are reflected in its products. On the other hand, determining how to improve the efficiency of food systems and address problems related to these systems may provide insight for the rest of the economy and society.

13.2 MUSHROOMS

The term mushroom comes from the French Mousseron (Mucerom), which is derived from mousse, moss (rapid-growing) (Ramsbottom, 1954). Even though the caps of some mushrooms are only a few millimeters long, Rinaldi and Tyndalo (1972) defined them as the fruiting bodies of organisms known as higher fungi or macromycetes (Macromycetes = large fungi). Mushrooms are gill-bearing fleshy agarics that are either edible or lethal. A fleshy structure with a cap and gills for producing spores on the underside is what is commonly used to identify mushrooms, which belong to the *Agaricus* or other families of fungi. Specifically, the term "mushroom" refers to the "agaric" mushroom that is grown for commercial purposes (Dickinson & Lucas, 1979; Philips, 1981). The agaric fungus's fruiting body develops an umbrella-like structure with a central stipe supporting a pileus or cap. On its lower surface, the cap has multiple gills or lamellae arranged radially (Webster, 1970). A mushroom is a large fungus that can grow above or below the ground and has an apparent fruiting

DOI: 10.1201/9781003486275-15

body. It is large enough to be seen on its own and can be manually harvested (Chang & Miles, 1989). Miller (1984) described mushrooms as gilled mushrooms that are both edible and lethal. Mushrooms were classified as edible, nonedible, toxic or nonpoisonous by Purukayastha and Chandra (1985). Agarics and other fleshy species of fungi from different families are particularly included in this category. According to the *Dictionary of Fungi*, a mushroom is any fungus with edible properties that resembles agarics or is closely related to the Agaricaceae family. Based on historical evidence, it appears that humans consumed wild mushrooms, probably as a means of survival. But because of their acknowledged nutritional advantages, mushrooms have become extremely popular in modern diets (Bano, 1976). Pacioni's 1985 taxonomy distinguishes between "toadstools," which are toxic or inedible species, and "mushrooms," which are only edible species. There are many different types of habitats where mushrooms can be found, including forests and desert ecosystems. Fungi are heterotrophic creatures that range from unicellular to multicellular (Sanders, 2001; Stajich et al., 2009; Allen et al., 2003).

13.3 DIVERSITY OF MUSHROOMS

The western Himalayan broadleaf forests have the greatest macrofungal diversity (36%), followed by the northwestern thorn scrub forests (25%). Furthermore, Himalayan subtropical pine woods exhibit abundant varieties of macrofungi with 17% taxonomic representation, while a 13% distribution shown by the western Himalayan subalpine conifer forests. The Baluchistan xeric woods and Karakoram West Tibetan Plateau alpine steppe have a 5% or lower distribution. On the other hand, Thar Desert, Registan North Pakistan sandy desert, South Iran semi-desert, Nubo-Sindian desert, Arabian Sea mangroves, Indus River Delta, Sulaiman range alpine meadows, East Afghan montane conifer forests, Rann of Kutch seasonal marsh and Himalayan northwestern alpine scrub and meadows have 0–2% macromycete records (Aman et al., 2022).

13.4 MUSHROOMS AND HUMAN NUTRITION

The relationship between mushrooms and humans is captivating because mushrooms have been increasingly used as food and medicine over the last few years (Rathore et al., 2017). Studies on the cultural revolution show that mushrooms have been an essential part of the human food system since we were hunters and gatherers (Wani et al., 2010). Edible mushrooms are a source of numerous nutraceutical compounds, including dietary fiber, peptides, terpenes, mineral elements, alcohols, glycoproteins, unsaturated fatty acids and antioxidants, such as ascorbic acid, tocopherols and phenolic compounds. This has been confirmed by numerous researchers (Pardeshi & Pardeshi, 2009). These mushrooms are therapeutically valuable, because they contain specific bioactive compounds, which can be used for immune system strengthening for the treatment and prevention of life-threatening conditions such as cancer, heart disease, hypertension and cerebral stroke. Other known benefits of mushrooms include hepatoprotective, antidiabetic, antiinflammatory, antitumor, antiviral, antibacterial, hypolipidemic, antithrombotic and hypotensive effects (Wasser & Weis, 1999). Mushrooms are a valuable food for the coming generations, providing not only high-quality protein but also therapeutic benefits. They are known to help in curing various diseases, including tumors, cancers, and nervous disorders. Fully utilizing this amazing gift from nature and advancing it from the farm to the fork is a formidable task.

13.4.1 Morels

Morel mushrooms are edible mushrooms valued for their meaty texture, subtle flavor and rich, characteristic aroma in many different cuisines (especially *Morchella* spp., Pezizales, Ascomycota).

Gastronomists highly value these mushrooms for their desirable quality of taste (Tietel & Masaphy, 2018). All over the world, morels are eaten as food. In India and Tibet, these plants are cooked with vegetables and are regarded as not suitable with fish or meat (Ajmal et al., 2015). The distinctive conic head of morels, which resembles a honeycomb due to a network of ridges and pits forming its cap, is what makes them distinctive. They are mostly found in areas that have been devastated by fire. The various habitats of these plants include roads and road cuts, swampy ground and excavations, and lightly burned grassy areas (Huffman, 2008; Negi, 2006). Fruiting bodies of the genus *Morchella* (morels) are remarkably morphological in shape, mature and immature in color, taste and edibleness (Masaphy, 2010), and the varieties of species found in the *Morchella* genus have different colors, chemical compositions and levels of bioactivity. Collectors of morels differentiate between yellow morels, such as *M. esculenta*, *M. deliciosa* and *M. crassipes*, and black morels, such as *Morchella angusticeps*, *M. conica*, *M. vulgaris* and *M. elata* (Kuo, 2005). Because of their high consumption level, the nutritional and phytochemical compositions of the yellow and black morels *Magna esculenta* and *Magna conica* have been the subject of most related research (Richard et al., 2015).

13.4.2 Himalaya

The name Himalaya is derived from Sanskrit and refers to "abodes of snow." The Himalayas are an Asian mountain range that acts as a natural barrier between the plains of the Indian subcontinent and the Tibetan Plateau. The Himalayas feature more than 100 peaks that are more than 7,200 meters (23,600 feet) above sea level. Mount Everest is the highest peak in the range. The Himalayas cross five countries, i.e. India, Nepal, Bhutan, Pakistan and China. The Tibetan Plateau to the north, the Indo-Gangetic Plain to the south, and the Hindu Kush and Karakoram ranges to the northwest are all within the Himalayan range. The combined drainage basin of the Tsangpo-Brahmaputra, Indus and Ganges rivers, three of the world's major rivers, is home to approximately 600 million people; 53 million people live in the Himalayas. Some parts of the Himalayas, including the Galyat tract, Margalla Hills and Kaghan Valley, stretch to the Pakistani provinces of Punjab and Khyber Pakhtunkhwa. Murree is a mountain resort city that is situated in the Galyat region of the PirPanjal range, District Punjab, Pakistan. Its altitude ranges between 520 m–2,380 m. It is a famous hill station located northeast of Islamabad at a distance of 30 km. Due to their beautiful scenery, these hill stations attract visitors, both foreign and domestic, especially during the summer months (Figure 13.1).

FIGURE 13.1 Beautiful views of the study area of Murree Forest Division.

13.4.3 Monsoon

According to the Köppen climate classification, Murree has a subtropical highland climate (Cwb) that is influenced by monsoons. It has chilly, snowy winters and comparatively cool summers with heavy precipitation, particularly in July and August. There are two peaks in the yearly average precipitation of 1,573 mm (61.9 inches), one in the winter and one in the summer. In summer, cool light rain is common and frequent in the study area, while in winter, it is covered by a thick blanket of snow (Ali et al., 2020).

13.5 ECOLOGICAL ZONES

The study area is located in two ecological zones: (1) the Chir zone (subtropical forest) and (2) the blue pine or quail zone (moist temperate coniferous forest). The study area consisted of mainly *Pinus wallichiana* (blue pine), *Pinus roxburghii* (pinechir), *Quercus incana* W. Bartram, Indian olive (*Olea europaea* subsp. *cuspidata*) and *Aesculus indica* (chestnut) (Table 13.1).

TABLE 13.1
Different Morels showing Ethno-Mycological Activity

S. No.	Botanical name	Ethno-mycology
1	*Agaricus bisporus*	Increases the enzymatic activities of serum, biochemical levels, antioxidant and lipid contents can protect the liver and kidneys while also having hepatoprotective, immunostimulatory and antitumor effects as well as antiaging properties
2	*Agaricus sylvaticus*	Regulates diabetes, strengthens the liver, acting as an antidote and enhancing pancreatic functions
3	*Coprinellus disseminates*	Flu, fever, cough, wound healing, skin tonic, and diuretic
4	*Morchella costata* Pers	Reduces oxidation stress and inhibits growth of bacteria
5	*Morchella conica* Fries	Body tonic, antitumor, relieves muscle soreness, eye irritation, and boosts immunity
6	*Morchella crassipes* (Vent.) Pers.	Stomach and intestinal tonic, liver tonic, aids in breathing, and acts as a diuretic
7	*Morchella deliciosa* Fr.	Antiinflammatory, blood purifiers, liver tonic, antitumor, and a treatment for diabetes, colds, digestive problems and reproductive issues
8	*Morchella elata*	In the human body, cofactors for enzymes involved in cellular substrate metabolism include antioxidants, vital minerals and vitamins needed for neurotransmission and the hematopoiesis process, which produces blood cells
9	*Morchella esculenta* Pers	As a vegetable, laxative and body tonic, it has potential benefits for muscle pain, fatigue and even anticancer properties
10	*Morchella galilaea*	Shows anticancer activity in animals
11	*Morchellarotunda* (Fr.) Boud.	Reduces oxidation stress and inhibits growth of bacteria. Used for its antimicrobial and antioxidant activities,
12	*Morchella smithiana*	Used as food and medicine
13	*Lactarius piperatus*	Reduce diabetic, cirrhosis and cancer symptoms
14	*Morchella tridentina* Bres.	Utilized in medications to treat heart conditions, cancer, edema and breast cancer (petunidin, veprisinium and visnagin)
15	*Morchella vulgaris*	Antibacterial, antioxidant. Nephroprotective against drug-induced kidney damage

(*continued*)

TABLE 13.1 (Continued)
Bioactive Compounds Isolated from Endophytic Microorganisms Present in *Aegle marmelos* and Their Therapeutic Properties

S. No.	Botanical name	Ethno-mycology
16	*Pleurotus ostreatus* (Jacq.)	Protective against cardiovascular disease, hypertension, elevated cholesterol, with antimicrobial, antioxidant and antidiabetic properties
17	*Volvariella volvacea*	Prevents anemia and strengthens the bones
18	*Pleurotus cornucopiae*	Used for its antitumor and antiviral activities
19	*Coprinus comatus*	Used for its antitumor and antiviral activities
20	*Boletus edulis*	Antioxidant, antimicrobial, antiviral and antiinflammatory
21	*Tricholomaa cerbum*	Reduces the risk of coronary heart diseases, also used as an antioxidant
22	*Pleurotus eryngii*	Enhances the level of high-density lipoprotein, glycogen and cholesterol
23	*Cantharellus cibarius*	Good source of polysaccharides like chitin and chitosan, lowers the risk of developing certain cancers and also reduces inflammation
24	*Agaricus campestris*	Enhances the secretion of insulin and used in the treatment of ulcers
25	*Calocybe indica*	Used for its antiaging, antioxidant and neuroprotective activities

13.6 THREATS TO MUSHROOM BIODIVERSITY

The biodiversity of mushrooms is threatened by a number of factors, including habitat loss, pollution, climate change and overharvesting. Nearly 50% of the species evaluated in a study assessing the threat status of fungi are in danger, mostly as a result of habitat loss. The lack of appropriate organic material for the growth of mushrooms is a result of various activities, such as deforestation, land degradation and the conversion of forest areas into agricultural land. For mushrooms, pollution, which includes the use of fertilizers containing fungicides, is a major risk factor. By changing the humidity and temperature, climate change has an additional effect on mushrooms. Additionally, overharvesting caused by the growing demand for wild edible mushrooms has resulted in a decrease in the quantity and variety of mushroom fruiting bodies in traditional consumption hubs. To preserve the biodiversity of mushrooms, these difficulties highlight the urgent need for conservation efforts (Koune, 2001; Mortimer et al., 2012).

13.7 MUSHROOMS AND HUMAN NUTRITION

Mushrooms of Pakistan are highly priced delicious foods which have been used by a large number of people for more than 2000 years (Figure 13.2). They are abundant in nutrients and have therapeutic value, and help to create a healthy diet because they contain a high concentration of proteins, mineral and vitamins (Shah et al., 2004). Mushrooms are non-timber forest products that are important for their medicinal and nutritional properties. They supply a variety of biologically active substances that can support immune system vigor and provide protection from carcinogens (Ramesh & Pattar, 2010).

These plants have more than 100 medicinal properties. Mushrooms are primarily used medicinally for their antiinflammatory, antitumor, anticancer, antiallergic, antidiabetic, immunomodulatory, cardiovascular-protective, anticholesterolemic, antifungal, antibacterial, antiviral, antiparasitic and detoxification properties (Finimundy et al., 2013; Chang & Wasser, 2012; Zhang et al., 2011; Yu et al., 2009). Apart from a few primary metabolites like oxalic acid, peptides and proteins, mushrooms may contain other antimicrobial compounds. These compounds primarily include secondary metabolites

FIGURE 13.2 Morale and mushroom species (A and B) collected from Murree Forest Division.

such as terpenes, steroids, anthraquinones, derivatives of benzoic acid and quinolones. Mushrooms have been utilized by various civilizations for centuries to support the immune system and as a means of preventing and treating illnesses, thanks to their antineoplastic and antiinflammatory properties (Alves et al., 2012). Badshah et al. (2015) reported that another edible and antibacterial mushroom species that grows wild in southern Waziristan is *Astraeus hygrometricus* (Diplocystaceae). Due to their desirable qualities, such as their ease of culture and variety of useful activities, humans have been cultivating and eating mushrooms as a type of macrofungi for generations (Bhambri et al., 2022).

13.8 MUSHROOM PRODUCTS

In general, canning, drying and certain technologies for value addition are used to store mushrooms for a long time. Drying appears to be the most successful method for extending shelf life and ensuring dispersion of the many methods used to preserve mushrooms. Different food items are dried using different drying procedures. Each method has benefits and drawbacks (Arumuganathan et al., 2010). Studies contrasting conventional sun drying with alternative drying methods reveal that using a solar dryer resulted in a significant decrease in the drying time and a noticeable boost in the product's texture and color quality and flavor. The most common method for processing mushrooms is canning. In recent years, mushroom preservation by canning has become much more sophisticated (Wakchaure et al., 2010).

An essential component of many culinary preparations, such as quick soups, snack spices, pasta, casseroles, meat and rice meals, is derived from dehydrated mushrooms (Tuley, 1996; Gothandapani et al., 1997). Mehta et al. (2011) created a number of *Agaricus bisporus* products, including ready-to-eat mushroom curries, chips, candy mushrooms, ketchup mushrooms, jam and soup-powder mushrooms. Sharma et al. (1991) also efficaciously prepared biscuits from mushrooms (Figure 13.3). Ulziijargal et al. (2013) studied the quality of bread enriched with mushroom mycelia and discovered that replacing 5% of wheat flour with mycelium powder had no negative effect on the textural profile of bread. Noodles fortified with mushrooms were created by Desayi (2012), who discovered that among the various treatments, the 10% mushroom powder and 0.2% vanilla flavor had the highest ratings for organoleptic parameters such as color and appearance, crispness, flavor, taste and overall acceptability, even after 30 days of storage. These findings suggest that medicinal mushrooms have great potential to treat various health problems. Fresh or dried mushrooms can be used to create value-added goods (Figure 13.4).

FIGURE 13.3 Different mushroom products: (A) mushroom biscuits; (B) mushroom nuggets; (C) mushroom pancakes.

FIGURE 13.4 (A) Jam produced from mushroom; (B) pickle produced from mushroom.

13.9 MUSHROOM MARKETING

A significant component of non-timber forest products (NTFPs) is wild edible mushrooms (WEMs). More than 80 nations across the world are engaged in the commercial harvesting of WEMs and therapeutic mushrooms (Pieroni et al., 2005; Hussain & Sher 2022). Approximately two billion USD of mushrooms were traded internationally in 2003 (Yun & Hall, 2004). *Morchella* spp. are highly sought among WEMs; as a result, they are widely collected and traded in foreign marketplaces. Researchers have collected numerous *Morchella* species from different regions of Pakistan (Ajmal et al., 2015; Badshah et al., 2015; Hamayun et al., 2006; Sher et al., 2015; Sultana & Qureshi, 2007). However, commonly collected species include *M. conica*, *M. esculenta* and *M. aungusticepes* (Hamayun et al., 2006). Khyber Pakhtunkhwa's overall annual productivity, which comprises more than 60% of the country's total production, was 500 tons (fresh morels converted into 55–65 tons dry mass) (Hamayun et al., 2006). The Swat district of Khyber Pakhtunkhwa, Pakistan, produced 50 million PKRs (Pakistani Rupees) in total earnings from the morel business (Sher, 2013).

13.10 TRADE AND LOGISTICAL NETWORK

A high level of production has led to the establishment of several local businesses in metropolitan marketplaces, while many more are operating in neighboring forest towns and villages. The data revealed that market value continued to be quite erratic, reliant on demand and highly variable. The resource supply chain is shown as follows: resource collectors; local stores; traders' representatives; stockers; major trade centers in major marketplaces in Dir (Dir Upper & Dir Lower) Chitral, Swat, Palas valley, Peshawar, Lahore and Karachi; and consumer nations. At various points throughout the supply chain, value-added operations, including cleaning/scrubbing, drying, cutting/grading and packing of the harvested mushrooms, were completed.

13.11 POLICIES AND MANAGEMENT FOR CONSERVATION

The present study's recommendations for designing policies relating to the sustainable collection and use of WEMs will be very helpful. The principles delineated here are equally relevant to a range of stakeholders, including forest owners, users, resource traders and local forestry/management organizations, with regard to the conservation of morel resources, including species diversity as well as genetic diversity.

1. You can export, buy, sell and gather useful mushrooms with government authorization.
2. Conducting training seminars for stakeholders, particularly collectors, may be organized to familiarize them with current science-based management and conservation techniques.
3. To avoid soil compaction, raking and trampling difficulties, regulations on access and harvester limit incursion to morel yielding stands should be rigorously implemented.
4. To protect the microclimate from mushroom contamination, accompanying vegetation and canopy cover are needed.
5. Young-stage draining is prohibited (for nonsporulating sporocarps), although picking up specimens up to a certain size may be permitted.
6. To maintain resource availability, harvesting should be performed in different mushroom-producing areas.
7. Before any management actions, such as fire setting, trash buildup and forest thinning, can be taken, prior consent must be acquired.
8. Never remove a mushroom from its natural habitat; instead, cut the fruiting bodies off above the soil line and leave the base portion alone.

REFERENCES

Ajmal, M., Akram, A., Ara, A., Akhund, S., & Nayyar, B. G. (2015). *Morchella esculenta*: An edible and health beneficial mushroom. *Pakistan Journal of Food Sciences, 25*(2), 71–78.

Ali, S., Khalid, B., Kiani, R. S., Babar, R., Nasir, S., Rehman, N., ... & Goheer, M. A. (2020). Spatio- temporal variability of summer monsoon onset over Pakistan. *Asia-Pacific Journal of Atmospheric Sciences, 56*, 147–172.

Allen, T. R., Millar, T., Berch, S. M., &Berbee, M. L. (2003). Culturing and direct DNA extraction find different fungi from the same ericoid mycorrhizal roots. *New Phytologist, 160*(1), 255–272.

Alves, M. J., Ferreira, I. C., Dias, J. (2013). A review on antifungal activity of mushroom (basidiomycetes) extracts and isolated compounds. *Current Topics Medicinal Chemisrty, 13*, 2648–2659

Aman, N., Khalid, A. N., & Moncalvo, J. M. (2022). A compendium of macrofungi of Pakistan by ecoregions. *MycoKeys, 89*, 171–233.

Arumuganathan, T., Amnikantan, M. R., Indurani, C., Rai, R. D., & Kamal, S. (2010). Texture and quality parameters of oyster mushroom as influenced by drying methods. *International Agrophysics, 24*, 339–342.

Badshah, H., Ullah, F., Khan, M. U., Mumtaz, A. S., & Malik, R. N. (2015). Pharmacological activities of selected wild mushrooms in South Waziristan (FATA), Pakistan. *South African Journal of Botany, 97*, 107–110.

Bano, Z. (1976). The nutritive value of mushrooms. In First Symposium on survey and cultivation of edible mushrooms of India, Regional Research Laboratory, Jammu (pp. 148–169).

Bhambri, A., Srivastava, M., Mahale, V. G., Mahale, S., & Karn, S. K. (2022). Mushrooms as potential sources of active metabolites and medicines. *Frontiers in Microbiology*, *13*, 837266–837266.

Chang, S. T., & Miles, G. P. (1989). Recent trends in world production of cultivation edible mushroom. *Mushroom Journal*, *504*, 15–17.

Chang, S. T., Wasser, S. P. (2012). The role of culinary-medicinal mushrooms on human welfare with a pyramid model for human health. *International Journal Medicinal Mushrooms*, *14*(2), 95–134.

Desayi, D. (2012). Development and sensory evaluation of mushroom fortified noodles. *International Journal of Food, Agriculture and Veterinary Sciences*, *2*(2), 187–189.

Dickinson, C., & Lucas, J. (1979). *Encyclopedia of Mushrooms*. Orchid Publication, London.

Finimundy, T. C., Gambato, G., Fontana, R., et al. (2013). Aqueous extracts of Lentinula edodes and Pleurotussajor-caju exhibit high antioxidant activity and promising in vitro antitumour activity. *Nutrition Research*, *33*(1), 76–84.

Gothandapani, L., Parvathi, K., & Kennady, Z. J. (1997). Evaluation of different methods of drying on the quality of oyster mushroom (Pleurotus sp.). *Drying Technology*, *15*(6–8), 1995–2004.

Hamayun, M., Khan, S. A., Ahmad, H., Shin, D. H., & Lee, I. J. (2006). Morel collection and marketing: A case study from the Hindu-Kush Mountain region of Swat, Pakistan. *Lyonia, 11*, 7–13.

Huffman, D. M. (2008). *Mushrooms and Other Fungi of the Midcontinen-tal United States*. University of Iowa Press.

Hussain, S., & Sher, H. (2022). Indigenous ecological knowledge and wild harvesting of morel mushrooms: The resource productivity and marketing in Swat, Pakistan. *Environment, Development and Sustainability*, 25(2), 1773–1791.

Koune, J. P. (2001). *Threatened Mushrooms in Europe* (Vol. 122). Council of Europe: Strasbourg, France.

Kuo, M. (2005). *Morels*. Ann Arbor, MI: The University of Michigan Press.

Masaphy, S. (2010). Biotechnology of morel mushrooms: Successful fruiting body formation and development in a soilless system. *Biotechnology Letters, 32*, 1523–1527.

Mehta, B. K., Jain, S. K., Sharma, G. P., Doshi, A., & Jain, H. K. (2011). Cultivation of button mushroom and its processing: An Techno –Economic feasibility. *International Journal of Advanced Biotechnology and Research*, 2(1), 201–207.

Miller, O. K. (1984). *Mushrooms of North America*. E.P. Duttom Publishing Company Inc., NewYork.

Mortimer, P. E., Karunarathna, S. C., Li, Q., Gui, H., Yang, X., Yang, X., ... & Hyde, K. D. (2012). Prized edible Asian mushrooms: Ecology, conservation and sustainability. *Fungal Diversity*, *56*, 31–47.

Negi, C. S. (2006). Morels (Morchella spp.) in Kumaun Himalaya. *Natural Product Radiology*, *5*, 306–310.

Ramesh, C. H., & Pattar, M. G. (2010). Antimicrobial properties, antioxidant activity and bioactive compounds from six wild edible mushrooms of western ghats of Karnataka, India. *Pharmacognosy Research*, *2*(2), 107.

Pardeshi, B. M., & Pardeshi, P. M. (2009). The edible medicinal mushrooms as supportive natural nutrients: Study of nonvolatile mineral contents of some edible medicinal mushrooms from India; eastern remedies for modern western maladies. In *Proceedings of the 5th international medicinal mushroom conference* (pp. 514–518). Mycological Society of China, Nantong, China.

Philips, R. (1981). *Mushrooms and other Fungi of Great Britain and Europe*. Pan Book Ltd., London.

Pieroni, A., Nebel, S., Santoro, R. F., & Heinrich, M. (2005). Food for two seasons: Culinary uses of noncultivated local vegetables and mushrooms in a south Italian village. *International Journal of Food Sciences and Nutrition, 56*(4), 245–272.

Purkayastha, R. P., & Chandra, A. (1985). *Manual of Indian Edible Mushrooms*. Jagendra Book Agency, New Delhi, India.

Ramsbottom, J. (1954). *Mushrooms & Toadstools*. Collins 14 ST James's Place London.

Rathore, H., Prasad, S., & Sharma, S. (2017). Mushroom nutraceuticals for improved nutrition and better human health: A review. *PharmaNutrition*, *5*(2), 35–46.

Richard, F., Bellanger, J.-M., Clowez, P., Hansen, K., O'Donnell, K., Urban,A., Sauve, M., Courtecuisse, R., & Moreau, P.-A. (2015). True morels (Morchella, Pezizales) of Europe and North America: Evolutionary relationships inferred from multilocus data and a unified taxonomy. *Mycologia, 107*, 359–382.

Rinaldi, A., & Tyndalo, V. (1972). *The Complete Book of Mushrooms*. Crescent Book, New York.

Sanders, W. B. (2001). Lichens: The interface between mycology and plant morphology: Whereas most other fungi live as an absorptive mycelium inside their food substrate, the lichen fungi construct a plant-like body within which photosynthetic algal symbionts are cultivated. *AIBS Bulletin, 51*(12), 1025–1035.

Shah, Z. A., Ashraf, M., Ishtiq, C. (2004). Comparative study on cultivation and yield performance of oyster mushroom (Pleurotusostreatus) on different substrates (Wheat straw, Leaves, saw dust). *Pakistan Journal of Nutrition, 3*, 158–160.

Sharma, R. C., Jandaik, C. L., & Bhardwaj, S. R. (1991). Enrichment of cookies/biscuits with mushroom prepared from unmarketable mushroom portions. *Advances in Mushroom Sciences*, 102.

Sher, H. (2013). *Collection and Marketing of High Value Medicinal and Aromatic Plants from District Swat, Pakistan* (Vol. 11). International Food Policy Research Institute (IFPRI), Washington, D.C.

Sher, H., Aldosari, A., & Bussmann, R. W. (2015). Morels of Palas valley, Pakistan: A potential source for generating income and improving livelihoods of mountain communities. *Economic Botany, 69*(4), 345–359.

Stajich, J. E., Berbee, L. M., Blackwell, M., Hibbett, S. D., James, Y. T., Spatafora, W. J., & Taylor. (2009). The fungi. *Current Biology, 19*(18), R840–R845.

Sultana, K., & Qureshi, R. A. (2007). Distribution of medicinally important mushrooms of mountainous/Northern areas of Pakistan. *Plant Pathology Journal, 6*(2), 183–186.

Tansey, G., & Worsley, A. (2014). *The Food System*. Routledge, London.

Tietel, Z., & Masaphy, S. (2018). True morels (Morchella)—nutritional and phytochemical composition, health benefits and flavour: A review. *Critical Reviews in Food Science and Nutrition, 58*(11), 1888–1901.

Tuley, L. (1996). Swell time for dehydrated vegetables. *International Food Ingredients, 4*(1), 23–27.

Ulziijargal, E., Yang, J. H., Lin, L. Y., Chen, C. P., & Mau, J. L. (2013). Quality of bread supplemented with mushroom mycelia. *Food Chemistry, 138*(1), 70–76.

Wakchaure, G. C., Shirur, M., Manikandan, K., & Rana, L. (2010). Development and evaluation of oyster mushroom value added products. *Mushroom Research, 19*(1), 40–44.

Wani, B. A., Bodha, R. H., & Wani, A. H. (2010). Nutritional and medicinal importance of mushrooms. *Journal of Medicinal Plants Research, 4*(24), 2598–2604.

Wasser, S. P., & Weis, A. L. (1999). Medicinal properties of substances occurring in higher basidiomycetes mushrooms: Current perspectives. *International Journal Of Medicinal Mushrooms, 1*(1), 31–62.

Webster, J. (1970). *Introduction to Fungi*. Cambridge University Press, Bently House, London.

Yu, S., Weaver, V., Martin, K., & Cantorna, M. T. (2009). The effects of whole mushrooms during inflammation. *BMC Immunology, 10*, Article 12.

Yun, W., & Hall, I. R. (2004). Edible ectomycorrhizal mushrooms: Challenges and achievements. *Canadian Journal of Botany, 82*(8), 1063–1073.

Zhang, L., Fan, C., Liu, S., Zang, Z., & Jiao, L. (2011). Chemical composition and antitumour activity of polysaccharide from Inonotusobliquus. *Journal of Medicinal Plants Research, 5*(7), 1251–1260.

14 Exploring the Green Gold Rush

Commercializing Medicinal Plant Endophyte-Derived Products

*Shalini Pandey, Richa Shahwal, Arpita Srivastava, Soniya Goyal, Mahiti Gupta and Arunima Sur**
*Corresponding author.

14.1 INTRODUCTION

The realm of medicinal plant endophytes stands as an emerging frontier in scientific inquiry and industrial innovation, offering a treasure trove of opportunities for diverse applications. The symbiotic relationship between plants and the microorganisms residing within their tissues has sparked profound interest and intensive exploration in recent years (Lu et al., 2021). This exploration isn't merely confined to the scientific community but extends fervently into industrial sectors seeking novel, sustainable and commercially viable products (Diab et al., 2021). Within the intricate tapestry of nature, these often-overlooked microorganisms, dwelling silently within plants, weave an elaborate array of bioactive compounds (Dwibedi et al., 2022). It is within this biological symbiosis that an abundance of bioactive molecules, exhibiting remarkable therapeutic properties, awaits discovery. The quest for these compounds has sparked a surge of interest, not only for their potential in pharmaceuticals but also in various other industrial sectors (Lu et al., 2021). The global landscape for products derived from medicinal plant endophytes has witnessed an upsurge in demand, buoyed by an increasingly discerning consumer base seeking sustainable and natural alternatives. This burgeoning market has not only piqued the interest of industries but has also prompted significant research endeavors aimed at unraveling the hidden mechanisms and untapped potential residing within these microscopic inhabitants of plants (Aye et al., 2019; Zaferanloo, 2014).

As cultivation and fermentation techniques for these endophytes continue to be optimized, ensuring a consistent and scalable supply of valuable bioactive compounds, the stage is set for an era of expansion and innovation within this domain. Concurrently, as our understanding of these endophytes deepens, so does our responsibility to navigate the regulatory landscape and implement stringent quality control measures to ensure the safety and efficacy of derived products (Talukdar et al., 2021). Moreover, the utilization of medicinal plant endophytes aligns harmoniously with sustainable practices, promoting biodiversity conservation and reducing reliance on synthetic chemicals (Chem et al., 2017). This symbiotic relationship between nature's microcosm and industrial applications not only holds promise for innovation but also for fostering environmentally friendly practices (Bu et al., 2023).

In recent years, the study of endophytes microorganisms that reside within the tissues of plants – has emerged as a focal point of scientific curiosity and industrial significance. This burgeoning field has garnered substantial attention due to the inherent potential of these microscopic inhabitants to yield a plethora of bioactive compounds with diverse applications across various sectors (Pal

 DOI: 10.1201/9781003486275-16

et al., 2019). The importance and relevance of studying endophytes in contemporary research have been underscored by several pivotal factors, each contributing to the burgeoning interest and exploration within this domain. Endophytes have unveiled a hidden realm within the botanical world, presenting an extensive reservoir of bioactive compounds. These compounds, ranging from alkaloids and flavonoids to terpenoids and beyond, possess remarkable therapeutic properties (Adinortey & N'guessan, 2022). Their discovery presents a novel avenue for drug development, offering potential solutions to combat diverse ailments that afflict human health, including infectious diseases, cancer, inflammation and more. The exploration of these microbial symbionts has unveiled nature's hidden pharmacy, stimulating a quest for innovative pharmaceuticals and biotechnological breakthroughs. In an era marked by a resounding call for sustainability and natural alternatives, endophytes have emerged as allies in the pursuit of eco-friendly solutions. Their ability to produce bioactive compounds within the natural confines of plant tissues aligns harmoniously with the growing consumer demand for ethically sourced, environmentally sustainable products. This quest extends beyond pharmaceuticals, permeating industries such as agriculture, cosmetics and beyond, fostering a paradigm shift toward nature-inspired, sustainable practices. Beyond their implications for human health and industrial innovation, endophytes have implications for agriculture (Alvin & Alfonsus, 2016). These microorganisms influence plant health, enhancing nutrient uptake, bolstering resilience against diseases and mitigating environmental stresses. Understanding and harnessing the potential of endophytes offer prospects for revolutionizing agricultural practices, promising increased crop yields, reduced dependence on synthetic chemicals and ultimately contributing to global food security. The recent surge in studying endophytes is underpinned by technological advancements. Sophisticated tools in genomics, metabolomics and bioinformatics have revolutionized our ability to decipher the complexities of endophytic communities (Mili et al., 2021a). This has led to profound insights into microbial interactions within plants, shedding light on intricate ecological relationships and prompting novel avenues for scientific exploration.

In light of these pivotal factors, the recent emphasis on studying endophytes represents not only a scientific pursuit but also a gateway to transformative applications across multiple domains. This introduction sets the stage for delving deeper into the multifaceted realms of endophytic research, exploring their implications for human health, industry, agriculture and ecological understanding in the sections that follow (Manam, n.d.). This chapter aims to delve into the multifaceted aspects of the burgeoning market for commercial products derived from medicinal plant endophytes. It will illuminate their significance across diverse industrial sectors, explore regulatory considerations and underscore the pivotal role they play in promoting sustainable practices. By examining these elements, we aim to offer insights into both the potential and challenges that lie ahead in this burgeoning domain (Rochín-Hernández et al., 2022).

14.2 UNDERSTANDING MEDICINAL PLANT ENDOPHYTES

Medicinal plants are known to host a wide variety of endophytic microorganisms, which play a crucial role in the production of secondary metabolites within the plant. Secondary metabolites refer to organic compounds synthesized by plants that serve purposes other than primary metabolic functions like growth or reproduction. Instead, these compounds often possess biological activities that can be beneficial for humans or other organisms (Mili et al., 2021b). Endophytic microorganisms residing within medicinal plants contribute significantly to the biosynthesis of these secondary metabolites. Through intricate biochemical processes, these microorganisms collaborate with the plant to produce a diverse array of bioactive compounds. This symbiotic relationship between medicinal plants and their endophytes underscores the importance of microorganisms in enhancing the medicinal properties of plants and highlights their potential for the discovery of new therapeutic agents. Instead, these bacteria contribute positively to the growth of host plants and bolster their resilience against various pathogens and environmental stresses. Additionally, they have the capacity to modulate the synthesis of secondary metabolites within plants, which often possess significant medicinal properties and exert various biological effects. The applications of endophytic bacteria span across diverse sectors including agriculture, medicine and other industries. In agriculture, they are utilized for enhancing plant growth, fortifying resistance against both biotic and abiotic stresses and facilitating the production of metabolites with medicinal potential (Pirttilä AM, 2008). The distribution and population structure of endophytic bacteria are influenced by factors such as the genetic characteristics and health status of their host plants, as well as the ecological conditions of the surrounding environment. A comprehensive understanding of bacterial endophytes is essential for maximizing their efficacy and leveraging their benefits effectively. By unraveling the intricacies of their interactions with host plants, we can harness their potential more efficiently. Moreover, applying this knowledge to medicinal plants holds promise for improving both yield and quality. Therefore, continued research into bacterial endophytes is crucial for unlocking their full potential and advancing sustainable agricultural practices and medicinal plant production (Russo ML, 2016). The quality and yield of medicinal plants are profoundly influenced by various environmental factors, including temperature, illumination, moisture, soil conditions and the presence of soil fauna (Namdar et al., 2019). However, there is a growing recognition that the relationships between medicinal plants and specific bacterial endophytes also play a significant role in shaping plant health and productivity (Ek-Ramos et al., 2019). These symbiotic relationships, which often develop over the long term, can enhance plant growth and prove particularly beneficial in agricultural settings (Compant et al., 2010). Bacterial endophytes can be found in various parts of plants, including flowers, leaves, roots, seeds and stems (Qin et al., 2009;Elmagzob et al., 2019). Colonization by these bacteria occurs at different stages of plant development, with some establishing stable communities even at the seed stage (Mocali et al., 2003). Endophytes utilize the plant's internal environment, known as the endosphere, as a specialized niche, providing them protection from external environmental fluctuations (Senthilkumar et al., 2011). The evolution of these symbiotic relationships necessitates close coordination of physiology, structure and life cycles between the host plant and the endophytes, resulting in mutual benefits for both parties (Saikkonen et al., 2004; Reinhold-Hurek & Hurek, 2011). These mutualistic interactions serve crucial roles in terrestrial ecosystems: host plants offer shelter and support to endophytes, which reciprocate by promoting plant growth through mechanisms such as nitrogen fixation, phosphorus enrichment and phytohormone synthesis (Borah et al., 2019). Additionally, since pathogenic microorganisms and endophytic bacteria occupy the same niche within plants, inoculation with endophytic bacteria can serve as an effective method for biological control of pathogens (Senthilkumar et al., 2011). Bacterial endophytes isolated from medicinal plants have the capability to produce bioactive metabolites and can significantly influence the production of secondary metabolites by their host plants. The host–endophyte relationship is dynamic and flexible, with endophytic bacteria adjusting their gene expression or producing different metabolites in response to subtle changes in host plant growth, and vice versa (Ek-Ramos et al., 2019). Despite their critical role in the plant micro-ecosystem, the relationship between bacterial endophytes and their host plants remains incompletely understood, highlighting the need for further research in this area (Tiwari et al., 2013). The community of bacterial endophytes is intricately influenced by both biotic and abiotic factors, which collectively shape their species composition, community structure, diversity and functions (Walitang et al., 2018). Environmental factors play a pivotal role not only in determining the distribution of medicinal plants but also in dictating the species of bacterial endophytes capable of colonizing the host plant throughout its life cycle. Additionally, endophyte diversity is significantly impacted by various host plant characteristics, including genotype (Walitang et al., 2018), tissue type (Dai et al., 2014), growth stage (Vendan et al., 2010) and health status.

14.2.1 Factors that Affect Endophytic Bacteria

14.2.1.1 Environment and Host Plant Environment Effects on Community Structure of Endophytic Bacteria

The community structure of endophytic bacteria is intricately influenced by various external factors, encompassing seasonality, altitude, latitude, longitude and soil conditions (Chiellini et al., 2014). For instance, in mulberry (*Morus* L.), Actinobacteria and Proteobacteria were the dominant endophytic bacteria during spring, whereas only Proteobacteria were prevalent during fall. Furthermore, spring endophytes exhibited greater diversity and species richness compared to their autumn counterparts. In another study involving *Pyrus ussuriensi*, the community structure of endophytic bacteria exhibited substantial correlations with soil carbon, nitrogen, pH and temperature. Notably, an increase in nitrogen content was associated with enhanced richness and diversity of root endophytic bacteria (Yarte et al., 2020). Moreover, research by Xu et al. (2014) demonstrated strong correlations between soil water content, annual precipitation and endophytic bacterial RFLP (restriction fragment length polymorphism) type variation in *Caragana jubata* and *Oxytropis ochrocephala*. Additionally, latitude, longitude, soil nitrogen content and soil potassium content also influenced the variation in endophytic bacterial communities. These findings underscore the intricate interplay between environmental parameters and the structure of endophytic bacterial communities in diverse plant species. The dynamic changes in bacterial endophyte community structure in response to shifting environmental conditions suggest that the maintenance of a diverse endophytic community may serve as a physiological strategy for plants to adapt to their surroundings. Understanding the intricate relationships between plants and their endophytic bacteria under varying environmental contexts holds significance for elucidating plant adaptation mechanisms and optimizing agricultural practices.

14.2.1.2 Host Plant Effects on Endophytic Bacteria

The relationship between host plants and specific endophytic bacterial isolates can exhibit characteristics of either parasitism or mutualism, contingent upon factors such as host genotype, tissue type and health status (Cope-Selby et al., 2017). Numerous studies have demonstrated that endophytic bacterial communities are significantly influenced by their host plants. For instance, in the Patagonian ecosystem, bacterial abundance and diversity were found to be higher on *Hieracium pilosella* compared to *Gaultheria mucronata* (Zhang et al., 2019). This observation is consistent with the notion that the diverse physiological structures, metabolites and growth habits of different plants can influence their capacity to recruit various endophytic bacteria (Kawaguchi & Minamisawa, 2010; Campisano et al., 2014). Furthermore, the health status of the host plant plays a pivotal role in endophyte colonization. For example, the bacterial endophytic community of *Paullinia cupana* with asymptomatic anthracnose was predominantly composed of Firmicutes, whereas plants with symptomatic anthracnose exhibited a prevalence of Acidobacteria. These findings suggest that pathogen-mediated surface damage to the host plant may facilitate the establishment of certain endophytes while disrupting the original stable microbial ecological environment (Hallmann et al., 1998). In addition, molecular techniques have been employed to investigate cultivable bacteria isolated from different compartments of medicinal plants such as *Echinacea angustifolia* and *Echinacea purpurea*. (Ahmad et al., 2021; Bramhchari et al., 2019). The results revealed distinct bacterial communities between the two plant species and among different plant compartments, indicating strong selective pressures and low degrees of strain sharing within plant tissues (Chiellini et al., 2014). These differences may arise from variations in nutritional and environmental conditions or from phytochemical and anatomical features that establish specific ecological niches for endophytes. Overall, endophytic bacteria demonstrate the ability to adjust their structure and diversity in response to various factors, including plant genotype, organ type, health status and growth stage, in order to secure a consistent supply of nutrients. Moreover, studies in olive systems have shown that

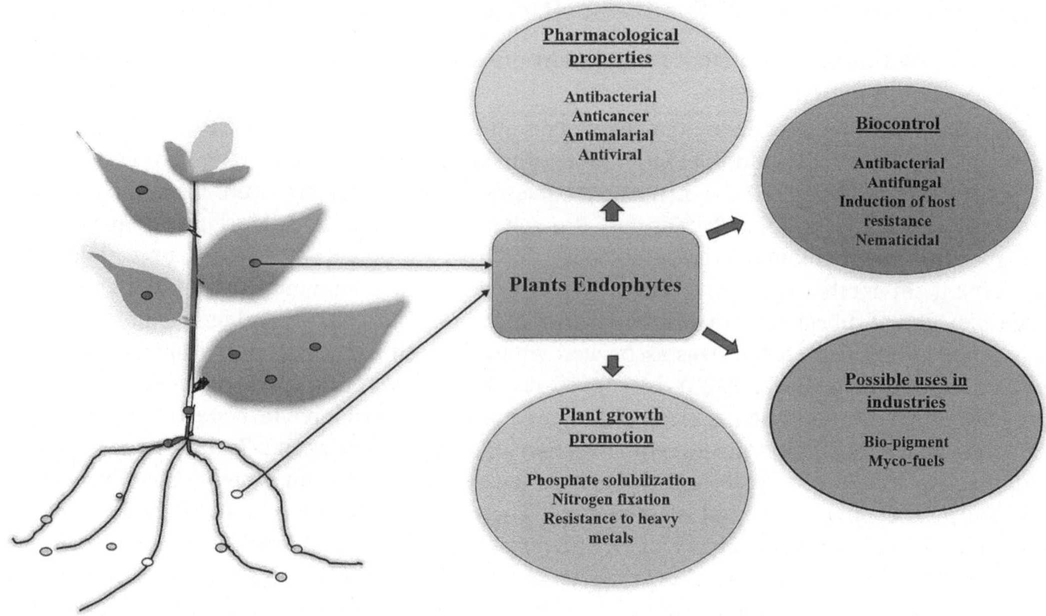

FIGURE 14.1 Medicinal plant endophytes and their applications in various aspects.

the belowground communities of endophytic bacteria are predominantly influenced by the cultivar genotype, highlighting the significant role of plant genotype in shaping belowground bacterial communities (Fernandez-Gonzalez et al., 2019). These findings underscore the complex and dynamic nature of the interactions between plants and their endophytic bacterial communities, which ultimately influence plant health and productivity. Researchers are actively delving into the intricate interactions between medicinal plants and endophytes to unlock the therapeutic potential harbored by these microorganisms. Through the isolation and study of endophytes sourced from medicinal plants, scientists endeavor to unearth novel compounds possessing medicinal properties, elucidate their mechanisms of action, and potentially develop new drugs or herbal remedies. To achieve these aims, a myriad of techniques is employed, including culture-based methods, molecular biology, metagenomics and metabolomics, allowing for a comprehensive exploration of endophyte diversity within medicinal plants and their potential contributions to drug discovery and plant health (Figure 14.1) (Johnston-Monje & Raizada, 2011).

14.3 MARKET TRENDS AND GROWTH FACTORS

Recent research has illuminated significant market trends and growth factors in the dynamic field of endophytes. One conspicuous trend driving this domain is the escalating demand for sustainable agricultural practices, which has precipitated a notable shift toward bio-based solutions. In response to this demand, endophytes have emerged as promising alternatives to conventional synthetic fertilizers and pesticides, offering a spectrum of benefits including the promotion of plant growth, augmentation of nutrient uptake and fortification against pathogens and environmental stressors. Additionally, the burgeoning biopharmaceutical industry has kindled keen interest in endophyte-derived bioactive compounds, encompassing antimicrobial agents and immunomodulators, thereby catalyzing intensive research endeavors and commercialization efforts in this realm (Suryanarayanan TS, 2005).

Moreover, the advent of technological advancements in microbial genomics and metagenomics has ushered in a new era of exploration into the diversity and functional capacities of endophytes

TABLE 14.1
Bioactive Natural Compounds from Endophytic Microorganisms

Biological activity	Bioactive compound	Endophytic microorganisms	Host plants
Antibacterial	Pestacin	*Penicillium microspore*	*Ephedra fasciculata*
Antibacterial	Brefeldin A	*Aspergillus clavatus*	*Taxus mairei*
Antifungal	Clavatol	*Aspergillus clavatonanicus*	*Taxus mairei*
Antiviral	Mullein	*Penicillium janczewskii*	*Prumnopitys andina*
Anticancer	Nigerasperone	*Aspergillus niger*	*Colpomenia sinuosa*
Antifungal	Phenolics, flavonoids.	*A. alternata*	*Azadirachta indica, A. Juss*
Antibacterial	Outovirin C	*P. raciborskii*	*Rhododendron tomentosum Harmaja*
Antibacterial	Trichodermin	*Trichoderma brevicompactum*	*Allium sativum*
Antiviral	Griseofulvin	*Xylaria* sp.	*Alibertia macrophylla*
Antibacterial	Tauranin	*Phyllosticta spinarum*	*Platycladus orientalis*
Antifungal	Herbarin	*Syncephalastrum* sp.	*Acacia nilotica*

(Hussain H, 2015; Kusari et al. 2011). These advancements have substantially enhanced our ability to decipher novel strains and unearth bioactive compounds, thereby broadening the horizon of possibilities in endophyte research. Regulatory support for biopesticides and biostimulants, coupled with collaborative initiatives and partnerships spanning academia, research institutions and industry stakeholders, has further propelled innovation and spurred market expansion within the endophyte sector (Rukachaisirikul V, 2013). Furthermore, the advent of emerging markets in developing countries has presented lucrative opportunities for endophyte-based solutions to address pressing challenges in agriculture, health care and beyond. These nascent markets offer fertile ground for the adoption and implementation of endophyte-derived products, thereby contributing to the alleviation of agricultural constraints and healthcare disparities. These discernible trends collectively underscore the growing recognition of endophytes' potential benefits and their increasingly indispensable role in fostering sustainability across diverse sectors including agriculture, biopharmaceuticals and beyond (Madasi et al., 2021). The trajectory of market trends and growth factors in the endophyte domain is underpinned by a confluence of factors including heightened awareness of their merits, continual technological innovations, supportive regulatory frameworks and collaborative synergies within the scientific and industrial communities (Supratman U, 2021). Some of the important bioactive compounds derived from endophytes are listed in Table 14.1.

14.4 CULTIVATION AND FERMENTATION TECHNIQUES

Various strategies are employed to augment the biosynthesis of secondary metabolites in endophytic fungi. These methods encompass optimization techniques, co-cultivation approaches, epigenetic modifications and molecular methodologies. Each of these strategies offers distinct avenues for enhancing the production of bioactive compounds within endophytic fungi, thereby enriching their potential for medicinal and industrial applications.

14.4.1 Optimization Method

The one strain, many compounds (OSMAC) approach stands as a significant method for amplifying secondary metabolites derived from fungal endophytes or generating diverse compounds through modifications in cultivation conditions. These alterations encompass adjustments in parameters like temperature, light exposure, pH levels, agitation and the composition of the growth medium, ultimately influencing the production of secondary metabolites (Hewage RT, 2014). The OSMAC approach involves modifying cultivation conditions, including altering media composition (such as carbon and nitrogen sources, metal ions), physical parameters (temperature, pH, aeration) or introducing specific molecules (such as enzyme activators/inhibitors, $MgSO_4$, NaCl), to induce the production of novel secondary metabolites or enhance the yield of existing ones (Pan R, Bai X, 2019). Different secondary metabolites (including four chaetomugilins, two malformins and seven chaetoglobosins) were extracted from a fungal endophyte (*Chaetomium* sp.) obtained from *Astragalus membranaceus* using the OSMAC strategy. This strategy involved altering the growth media during fungal cultivation to induce the production of these compounds (Zhang H, 2021).

14.4.2 Co-Culture Method

The co-culture method involves combining two or more microorganisms within the same growth medium to increase the production of secondary metabolites or trigger the expression of cryptic genes (Wakefeld J, 2017). This strategy, termed co-culture, entails the construction of an artificial microbial community aimed at activating biosynthetic gene clusters, ultimately leading to the synthesis of novel compounds. Co-culture techniques can be categorized into interactions between fungus and bacterium, fungus and fungus, or bacterium and bacterium, all of which contribute to the enhancement or generation of new compounds and the activation of silent gene clusters within microorganisms (Peng X-Y, 2021).

14.4.3 Molecular-Based Approaches

This method is employed to activate dormant biosynthetic gene clusters (BGCs) and improve the production of secondary metabolites through various approaches, including genetic engineering and targeted exploration of the strain's genome for specific gene clusters using bioinformatics tools. Pathway-specific regulatory genes, whether situated within or outside specific BGCs, play a crucial role (Wiemann P, Keller, 2013). These genes can either activate or suppress the biosynthesis of secondary metabolites, with transcription factors regulating the expression of genes involved in secondary metabolite production (Macheleidt J, 2016). Different strategies are utilized to awaken dormant BGCs and enhance secondary metabolite production in fungi, such as amplifying the expression of transcription factors, eliminating transcriptional repressors and substituting native promoters with either inducible or constitutive promoters (Figure 14.2) (Skellam E, 2019).

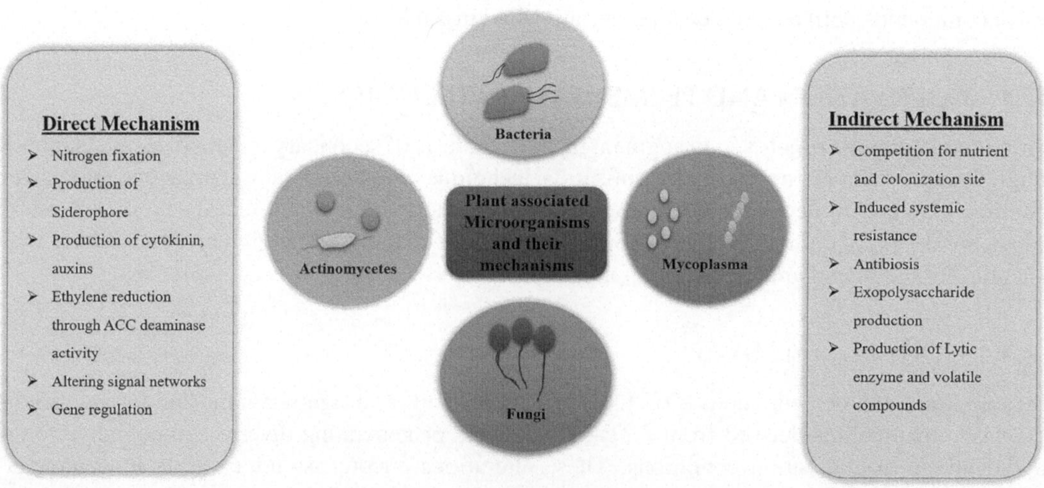

FIGURE 14.2 Various types of plant-associated microorganisms and their different mechanisms.

14.5 INDUSTRIAL APPLICATIONS AND SIGNIFICANCE OF ENDOPHYTE-DERIVED PRODUCTS

In recent years there has been a great surge in exploring endophytes for pharmaceutical, agricultural and environmental remediation purposes. Some of the important studies are listed in Table 14.2.

14.5.1 Pharmaceutical Industry

14.5.1.1 Antibiotics

Compounds derived from endophytes possessing antimicrobial properties present promising opportunities for the development of novel antibiotic medications. The wide range of bioactive compounds synthesized by endophytes serves as a reservoir of innovative molecules that could potentially combat drug-resistant pathogens. Several active substances with antimicrobial effects have been isolated from various sources. These include linoleic acid, R-glycerol monolinoleate, bisdethio-(bis-methyl-thio)-gliotoxin, fumiquinazoline-F, fumiquinazoline-D, deoxy-thymidine, cerebroside A and (Z, Z).)-N, N0-[1-[(4-hydroxy-phenyl)-methylene]-2-[(4-methoxy-phenyl)-methylene]-1,2-ethanediyl]-bisformamide among others, extracted from endophytic fungi *A. fumigatus*. *L. cubeba* harbors a diverse fungal endophyte population with antimicrobial properties. Notably, SF22 (*C. globosum*) and SF14 (*P. minioluteum*) exhibit antipathogenic fungal properties,

TABLE 14.2
Patented Natural Products Produced from Plant Endophytes

Patent No.	Endophyte	Host	Patent particulars	References
CN1421522A	*Alternaria* sp.	*Erigeron* sp.	Synthesized cerrenin D, utilized in the production of anticancer medications	Chen et al., 2003
US6911338B2	*Muscodor* sp.	*Cinnamomum zeylanicum*, *Grevillea pteridifolia*	Volatile organic antibiotics synthesized for targeting pathogens for combating infections	Strobel & Ezra, 2005
US20090142816A1	*Gliocadium* sp.	*Eucryphia cordifolia*	The synthesis of volatile compounds for the generation of biofuels	Strobel, 2009
CN101701230A	*Fusarium proliferatum*	Mangrove	Anthraquinone compound, known for its anticancer properties, through the implementation of various vaccination techniques	Zhou et al., 2010
CN102190614A	*Aspergillus oryzae*	Red algae *Heterosiphonia* sp.	Employing diterpenoid alkaloid as an effective agent for insecticide purposes	Ji & Qiao, 2011
CA2766412A1	Endophyte from *Pinus strobus*	*Pinus strobus*	Antifungal metabolites	Miller et al., 2012
US20130302480A1	*Muscodor crispans*	*Ananas ananassoides*	Manufacturing versatile compounds with applications spanning agriculture, industry, construction, pharmaceuticals or personal care products	Gandhi et al., 2013
CN104059044A	*Trichoderma* sp./ *Penicillium* sp.	Mangrove	Synthesizing a xanthone derivative for its utilization as a microbial pesticide and fungicide	Li & Ding, 2014
CN109456191A	*Cerrena* sp.	*Pogostemon cablin*	The utilization of cerrenin D in drug formulation with antitumor properties	Liu et al., 2019

potentially leading to the discovery of novel natural antimicrobial compounds in endophytes (Wu et al., 2019). Anofnic acid, isolated from endophytic *A. tubenginsis* by Elkhouly et al. (2021), demonstrates promising antimicrobial activity against Gram-positive and Gram-negative bacteria as well as unicellular fungi, along with the ability to inhibit biofilm formation. Additionally, Mohamed et al. (2020) identified aspergillethers A and B as new diaryl ether derivatives from *A. versicolor*, which display potent antimicrobial activity against various microorganisms such as *S. aureus*, *B. cereus*, *E. coli*, *C. albicans* and *Geotrichium candidum* (Falade & Ekundayo, 2021). Trichocadinins B–G and new cadinane-type sesquiterpene derivatives isolated from *T. virens*, sourced from *Artemisia argyi*, exhibit antibacterial and antifungal properties. Furthermore, Zhao et al. (2019) discovered three new derivatives of α-pyrone from *P. ochrochloron*, isolated from *Taxus* media, demonstrating antimicrobial activity against several pathogenic bacterial and fungal strains, with MIC values ranging from 12.5 to 100 μg/mL. Additionally, patents have been filed for innovations in algal control, such as the utilization of diterpene alkaloid-like compounds (CN102190699A). These patents also encompass developments in antibacterial, antibiotic, antimicrobial and bacteriostatic applications.

14.5.1.2 Anticancer Agents

Certain endophyte-derived products exhibit cytotoxic properties, suggesting their potential as candidates for developing anticancer drugs. These compounds merit investigation for their ability to inhibit tumor cell growth and induce apoptosis. Particularly noteworthy are *Aspergillus* endophytes found in plants of the genera *Taxus* and *Torreya*, which are known for their applications in obtaining high yields of paclitaxel or its precursors, such as baccatin III and cephalomannine, due to their anticancer activity (Blunt et al., 2010; Latz et al., 2018). Li et al. (2020) isolated eight new compounds from the co-culturing of *Armillaria* sp. and *Epicoccum* sp., with only one compound demonstrating anticancer activity. *Eurotium* sp., an endophytic fungus isolated from *C. longa* rhizomes, is involved in producing the asparaginase anticancer enzyme, which serves as a therapeutic agent in leukemia treatment (Ferdous et al., 2019; Li et al., 2020). The N97 endophytic fungal extract was found to inhibit the proliferation of HeLa cells and induce apoptosis in this investigation (Majoumouo et al., 2020). However, patents have been developed over time to include methods and procedures for producing and recovering these compounds of interest. The primary applications of these patents include providing metabolites that serve as precursors of bioactive molecules (such as baccatin III and cephalomannine in CN103194502A) and those that can be utilized as anticancer and antitumor agents.

14.5.2 Agricultural Sector

14.5.2.1 Plant Growth-Promoting Agents

Certain endophytes have been identified for their ability to enhance nutrient absorption, stimulate root growth and improve stress resistance in plants (Patil et al., 2016). This trait makes them invaluable for developing plant growth-promoting agents, thereby enhancing crop yield and resilience. These agents, produced by plant endophytes, play a critical role in augmenting plant growth, nutrient uptake and stress tolerance. Endophytic fungi (EF) have been extensively utilized as biotic elicitors in plant cells, tissues and organ cultures to stimulate growth and metabolite production. For instance, strains such as DO14 (*Pestalotiopsis* sp.), DO18 (*Talaromyces* sp.), DO19 (*Xylariaceae* sp.) and DO120 (*Hypoxylon* sp.) have been utilized to obtain various extracts from four endophytic fungi, which when applied individually to their host *Dendrobium catenatum* resulted in significant increases in biomass yields and active contents after co-culturing. These findings suggest that fungal elicitor PPFs (protein-polysaccharide fractions) can be harnessed to produce high-quality *D. catenatum* seedlings in industrial settings and serve as a valuable microbial fertilizer resource for other plant developments (Zhu et al., 2018). Li et al. (2012) found that the crude oligosaccharides obtained from partially acid hydrolyzed cell wall fragments of the isolated EF (Dzf17) demonstrated a significant ability to boost diosgenin synthesis in *D. zingiberensis* seedling cultures when employed as elicitors. Additionally, certain endophytic bacteria, particularly those belonging to the genera *Rhizobium* and *Bradyrhizobium*, exhibit nitrogen-fixing capabilities. These bacteria establish symbiotic associations with leguminous plants like soybeans and peas, converting atmospheric nitrogen into ammonia, which serves as a nitrogen source for the host plant. By supplementing plants with nitrogen, these endophytes promote robust growth and higher yields, especially in nitrogen-deficient soils.

14.5.2.2 Biopesticides

Endophytes play a crucial role in enhancing plant resilience not only against pathogens but also against pests. Plants emit volatile organic compounds (VOCs) with various pesticidal properties, including toxicity, repellence or attraction to pests like caterpillars and spiders. Endophytes contribute to VOC production in plants, often by activating the jasmonic pathway (JA). By stimulating this pathway and producing key compounds, endophytes boost plants' resistance to pests by promoting the synthesis of antiherbivore VOCs (Eid et al., 2019). For instance, commercially available ryegrass strains such as AR1, AR37, NEA2 and NEA4 are engineered with endophyte-associated traits (EATs), resulting in improved yield, increased lateral root growth by up to 92%, and a notable rise in the number of tillers, thus aiding in plant persistence (Brosset & Blande 2022). Moreover, these strains exhibit heightened resistance to pests, showing significant reductions in damage caused by weevils (by approximately 75%) and black beetles (by around 60%) (Card et al., 2016; Datta et al., 2022; Shrivastava et al., 2010).

14.5.3 Bioremediation

Certain endophytes possess the ability to degrade pollutants and contaminants in both soil and water, offering promising potential for bioremediation applications in industrial settings, thus advancing sustainable and eco-friendly technologies (Cain et al., 2020). *Cyperus laxus* has been effectively employed for phytoremediation of hydrocarbon-contaminated soil through the inoculation with endophytic fungi (EF), as demonstrated by Escalante-Espinosa et al. (2005). Notably, among the targeted contaminants was phenanthrene, a recognized environmental carcinogen, which showed enhanced degradation in plants treated with EF, suggesting a viable approach for remediation efforts. Active groups in the fungal cell wall, such as carboxyl, hydroxyl and hydrosulfide, play a pivotal role in interacting with heavy metal ions, leading to their precipitation on the surface. Fungi employ various metal tolerance mechanisms, including adsorption, immobilization, complexing and quanta-valence shifting, as elucidated by Collin et al. (2003) and Gadd (2007). Furthermore, Sun et al. (2020) demonstrated the efficacy of Pb-resistant endophytic fungi, particularly *Trichoderma asperellum* (SD-5), in enhancing the growth and biomass of ryegrass (*Lolium perenne*) under Pb stress conditions. This enhancement extended to improving photosynthesis, as evidenced by increased leaf chlorophyll and total protein content.

Additionally, the expression of key genes involved in antioxidant defense mechanisms, such as SOD, POD, GPX and APX, was influenced. The presence of Pb-resistant organisms facilitated increased Pb extraction from the soil, thereby contributing to improved remediation outcomes. Endophytic fungi residing within plant roots can play a role in rhizoremediation, a process where contaminants in the rhizosphere are degraded or immobilized. For instance, endophytic fungi like *Trichoderma* spp. have been demonstrated to enhance the degradation of hydrocarbons and pesticides in the rhizosphere by producing enzymes capable of breaking down these pollutants. Additionally, these fungi can promote plant growth and stimulate root exudation, facilitating the degradation of contaminants in the soil (Li et al., 2012).

14.6 CONCLUSION

The exploration of medicinal plant endophytes presents a promising avenue for commercial opportunities in various industries. Our investigation has revealed a wealth of potential applications, ranging from pharmaceuticals to agriculture and beyond. Through harnessing the symbiotic relationship between plants and their endophytes, we can unlock a treasure trove of bioactive compounds with therapeutic potential. Furthermore, the scalability and sustainability of endophyte-derived products offer a distinct advantage in today's environmentally conscious market. By reducing reliance on traditional extraction methods and minimizing ecological impact, these products align with the growing demand for eco-friendly solutions.

However, challenges such as regulatory hurdles, standardization of production processes and ensuring fair benefit-sharing with indigenous communities must be addressed to fully realize the commercial potential of endophyte-derived products. Collaboration between researchers, industry stakeholders and policymakers will be essential in navigating these complexities and fostering a thriving marketplace for green gold. In essence, the future of medicinal plant endophytes is bright, with ample opportunities for innovation and economic growth. By leveraging nature's pharmacy, we can not only develop novel therapeutics but also contribute to sustainable development and biodiversity conservation on a global scale. The journey ahead may be challenging, but the rewards are boundless for those willing to embark on this green revolution.

REFERENCES

Adinortey, M. B., & N'guessan, B. B. (2022). H+/K+-ATPase Inhibitors from Plants: A Potential Source for Drug Discovery. In Vijay Kumar, S., Pradeep K. Agarwal, & Subhadeep Chattopadhyay (Eds.), *Natural Products as Enzyme Inhibitors* (pp. 125–136). Springer Nature Singapore. https://doi.org/10.1007/978-981-19-0932-0_5

Ahmad, N., Bhat, M. Y., Wani, A. H., & Peer, L. A. (2021). Rhizosphere mycobiome diversity of medicinal plants: A review. *Journal of Plant Science Research*, 37(1), 175–187. https://doi.org/10.32381/JPSR.2021.37.01.21

Alvin, A. (2016). Bioprospecting for Anti-Tuberculosis Endophytic Natural Products from Indonesian Traditional Medicinal Plants. *UNSW Sydney*, [Doctoral dissertation]. https://doi.org/10.26190/unsworks/18860.

Aye, M. M., Aung, H. T., Sein, M. M., & Armijos, C. (2019). A review on the phytochemistry, medicinal properties and pharmacological activities of 15 selected Myanmar medicinal plants. *Molecules*, 24, 293.

Blunt, J., Cole, T., Munro, M., Sun, L., Weber, J.-F.R., Ramasamy, K., Abu Bakar, H., & Abdul Majeed, A. B. B. (2010, 3 June). Bioactive Compounds Derived from Endophytic Aspergillus Fungus Strain Isolated from Garcinia Scortechinii. WO Patent 2010062159 A1.

Borah, P., Saikia, K., Hrishikesh Hazarkia, H., & Ragesh, G. (2019). Exploring potential of bacterial endophytes in disease management of horticultural crops. *Current Horticulture*, 7(2), 3237.

Bramhachari, P. V., Anju, S., Sheela, G. M., Komaraiah, T. R., Venkataiah, P., & Prathyusha, A. M. V. N. (2019). Secondary Metabolites from Marine Endophytic Fungi: Emphasis on Recent Advances in Natural Product Research. In *S. R. S. M. Rao & S. B. Yadav* (Eds.), *Advances in Endophytic Fungal Research: Present Status and Future Challenges* (pp. 339–350). Springer Nature. https://doi.org/10.1007/978-3-030-03589-1_15.

Brosset, A., & Blande, J. D. (2022). Volatile-mediated plant–plant interactions: Volatile organic compounds as modulators of receiver plant defence, growth, and reproduction. *Journal Experimental Botany*, 73, 511–528.

Bu, Z., Li, W., Liu, X., Liu, Y., Gao, Y., Pei, G., Zhuo, R., Cui, K., Qin, Z., Zheng, H., Wu, J., Yang, Y., Su, P., Cao, M., Xiong, X., Liu, X., & Zhu, Y. (2023). The rice endophyte-derived α-Mannosidase ShAM1 degrades host cell walls to activate DAMP-triggered immunity against disease. *Microbiology Spectrum*, 11(3), e04824–22.

Cain, J. W., Miller, K. I., Kalaitzis, J. A., Chau, R., & Neilan, B. A. (2020). Genome mining of a fungal endophyte of Taxus yunnanensis (Chinese yew) leads to the discovery of a novel azaphilone polyketide, lijiquinone. *Microbial Biotechnology*, 13(5), 1415–1427. https://doi.org/10.1111/1751-7915.13568.

Card, S., Johnson, L., Teasdale, S., & Caradus, J. (2016). Deciphering endophyte behaviour: The link between endophyte biology and efficacious biological control agents. *FEMS Microbiology Ecology*, 92(8), fiw114.

Chemat, F., Rombaut, N., Meullemiestre, A., Turk, M., Perino, S., Tixier, A. S., & Vian, M. A. (2017). Review of Green Food Processing techniques. Preservation, transformation, and extraction. *Innovative Food Science and Emerging Technologies*, 41, 357–377.

Chen, Y., Yang, L., Li, S., Li, Z., Zhou, B., & Zhang, Q. (2003). Erigeron Breviscapus-Endogenic Fungus. CN Patent 1421522 A.

Chiellini, C., Iannelli, R., Lena, R., Gullo, M., & Petroni, G. (2014). Bacterial community characterization in Paper Mill White Water. *BioResources*, 9(2), 25412559.

Collin, H. C., Andersen, R. A., & Steinnes, E. (2003). Isolation and N-terminal sequencing of a novel cadmium-binding protein from Boletus edulis. *Journal of Physics IV France*, 107(1):311–314. https://doi.org/10.1051/jp4:20030304.

Compant, S., Clement, C., & Sessitsch, A. (2010). Plant growth-promoting bacteria in the rhizoand endosphere of plants: their role, colonization, mechanisms involved and prospects for utilization. *Soil Biology and Biochemistry*, 42(5), 669678. https://doi.org/10.1016/j.soilbio.2009.11.024.

Campisano, A., Antonielli, L., Pancher, M., Yousaf, S., Pindo, M., & Pertot, I. (2014). Bacterial endophytic communities in the grapevine depend on pest management. *PLoS ONE*, 9(11), e112763.

Cope-Selby, N., Cookson, A., Squance, M., Donnison, I., & Flavell, R. (2017). Endophytes as potential biocontrol agents: Mechanisms, applications, and their limitations. *Biocontrol Science and Technology*, 27(8), 967982. https://doi.org/10.1080/09583157.2017.1356544.

Dai, W., Schmid, M. F., King, J. A., & Chiu, W. (2014). Identifying the assembly pathway of cyanophage inside the marine bacterium using electron cryo-tomography. *Microbial Cell,* 1(1). doi:10.15698/mic2014.01.125.

Datta, D., Behera, L., Chaudhary, V. T., Kumar, S., & Bisen, K. (2022). Endophytes: Rendering Systemic Resistance to Plants. In *Rhizosphere Microbes: Biotic Stress Management* (pp. 175–195), Singh, U.B., Sahu, P.K., Singh, H.V., Sharma, P.K., Sharma, S.K., Eds., Springer Nature Singapore, 2022.

Diab, M. K., Mead, H. M., Khedr, M. A., Nae, M. S., Abu-Elsaoud, A. M., & Ahmed El-Shatoury, S. (2021). Artemisia-derived Actinobacteria Producing Inhibitors of Spodoptera Littoralis via Molecular Modeling Studies. *Research Square*. https://doi.org/10.21203/rs.3.rs-878185/v1

Dwibedi, V., Rath, S. K., Joshi, M., Kaur, R., Kaur, G., Singh, D., Kaur, G., & Kaur, S. J. (2022). Microbial Endophytes: Application Towards Sustainable Agriculture and Food Security. In *Applied Microbiology and Biotechnology* (Vol. 106, Issue 17, pp. 5359–5384). Springer Science and Business Media Deutschland GmbH. https://doi.org/10.1007/s00253-022-12078-8.

Eid, A. M., Salim, S. S., Hassan, S. E. -D., Ismail, M. A., & Fouda, A. Role of Endophytes in Plant Health and Abiotic Stress Management. In *Microbiome in Plant Health and Disease* (pp. 119–114). Springer, 2019.

Ek Ramos, M. J., Gomez-Flores, M. J., Alonso, R., Orozco-Flores, A., Rodriguez-Padilla, C., Gonzalez-Ochoa, G., & Tamez-Guerra, P. (2019). Bioactive products From plant-endophytic gram-positive bacteria. *Frontiers in Microbiology,* 10. doi:10.3389/fmicb.2019.00463.

Elkhouly, H. I., Hamed, A. A., El Hosainy, A. M., Ghareeb, M. A., & Sidkey, N. M. (2021). Bioactive secondary metabolite from endophytic Aspergillus tubenginses ASH4 isolated from Hyoscyamus muticus: Antimicrobial, antibioflm, antioxidant and anticancer activity. *Pharmacognosy Journal,* 13(2), 434.

Elmagzob, A. A. H., Ibrahim, M. M., & Zhang, G. F. (2019). Seasonal diversity of endophytic bacteria associated with Cinnamomum camphora (L.) Presl. *Diversity,* 11(7), 112. doi:10.3390/d11070112.

Falade, A. O., Adewole, K. E., & Ekundayo, T. C. (2021). Therapeutic potentials of endophytes for healthcare sustainability. *Egyptian Journal of Basic and Applied Sciences*, 8(1), 117–135.

Ferdous, K. J., Afroz, F., Islam, M. R., et al. (2019). Isolated endophytic fungi from the plant Curcuma longa and their potential bioactivity—a review. *Pharmacology Pharmacy*, 10, 244–270. https://doi.org/10.4236/pp.2019.105021.

Fernandez-Gonzalez, A. J., Martínez-Hidalgo, P., Cobo-Diaz, J. F., Villadas, P. J., Martinez-Molina, E., Toro, N., & Velazquez, E. (2019). The rhizosphere microbiome of burned holm-oak: Potential role of bacterial endophytes in the recovery of burned forest soils. *Frontiers in Microbiology*, 10, 760. https://doi.org/10.3389/fmicb.2019.00760.

Gadd, G. M. (2007). Geomycology: Biogeochemical transformations of rocks, minerals, metals and radionuclides by fungi, bioweathering and bioremediation. *Mycological Research*, 111, 3–49. https://doi.org/10.1016/j.mycres.2006.12.001.

Gandhi, N. R., Palmer Skebba, V., & Strobel, G.A. (2013) Antimicrobial Compositions and Related Methods of Use. U.S. Patent 20130302480 A1.

Hallmann, J., Quadt-Hallmann, A., Mahaffee, W. F., & Kloepper, J. W. (1998). Bacterial endophytes in agricultural crops: Potential role in disease suppression. *Plant and Soil*, 207(1), 119131. https://doi.org/10.1023/A:1026208325972.

Hewage, R. T., & Perera, H. (2014). Endophytic fungi and their applications in agriculture. *Journal of Applied Microbiology*, 117(1), 2132. https://doi.org/10.1111/jam.12515.

Hussain, H., Wang, S., Zuo, Z., & Zhang, H. (2015). Endophytes: A treasure for plant growth promotion and biocontrol. *Microbiological Research*, 178, 1932. https://doi.org/10.1016/j.micres.2015.04.001.

Ji, N., & Qiao, M. (2011) Agricultural Application of Diterpenoid Alkaloid of Endophytic Fungi of Algae. CN Patent 102190614 A.

Johnston-Monje, D., & Raizada, M. N. (2011). Plant and Endophyte Relationships: Nutrient Management. In *M. Moo-Young* (Ed.), *Comprehensive Biotechnology, Second Edition* (Vol. 4, pp. 713–727). Elsevier Inc.

Kawaguchi, M., & Minamisawa, K. (2010). Plant-microbe communications for symbiosis. *Plant and Cell Physiology*, 51(9), 13771380. https://doi.org/10.1093/pcp/pcq120.

Kusari, S., Singh, S., & Jayabaskaran, C. (2014). Biotechnological potential of plant-associated endophytic fungi: Hope versus Hype. *Trends in Biotechnology*, 32, 297–303.

Latz, M. A. C., Jensen, B., Collinge, D. B., & Jørgensen, H. J. L. (2018). Endophytic Fungi as Biocontrol Agents: Elucidating Mechanisms in Disease Suppression. In *Plant Ecology and Diversity* (Vol. 11, Issues 5–6, pp. 555–567). Taylor and Francis Ltd. https://doi.org/10.1080/17550874.2018.1534146.

Li, C., Ding, W. X. (2014). Derivative, Its Preparation Method by Fermentation, Extraction and Recrystallization, and Application Thereof in Preparing Microbicide for Controlling Phytopathogenic Fungi. CN Patent 104059044 A.

Li, P., Lou, J., Mou, Y., et al. (2012). Efects of oligosaccharide elicitors from endophytic Fusarium oxysporum Dzf17 on diosgenin accumulation in Dioscorea zingiberensis seedling cultures. *Journal Medicinal Plants Research*, 6, 5128–5134. https://doi.org/10.5897/JMPR12.120.

Li, H. Y., Wei, D., Shen, M., & Zhou, Z. P. (2012). Endophytes and their role in phytoremediation. *Fungal Diversity*, 54:11–18.

Li, H-T., Tang, L-H., Liu, T., Yang, R-N., Yang, Y-B., Zhou, H., & Ding, Z-T. (2020). Protoilludane-type sesquiterpenoids from Armillaria sp. by co-culture with the endophytic fungus Epicoccum sp. associated with Gastrodia elata. *Bioorganic Chemistry,* 95, 103503. https://doi.org/10.1016/j.bioorg.2019. 103503.

Li, Y., Zhang, X., & Liu, Q. (2020). Plant endophytes and their role in sustainable agriculture: A review. *Frontiers in Plant Science*, 11, 603680.

Lu, H., Wei, T., Lou, H., Shu, X., & Chen, Q. (2021). A Critical Review on Communication Mechanism Within Plant-Endophytic Fungi Interactions to Cope with Biotic and Abiotic Stresses. In *Journal of Fungi,* (Vol. 7, Issue 9). MDPI. https://doi.org/10.3390/jof7090719.

Liu, H., Zhang, W., Tan, H., Chen, Y., Li, S., & Li, H. (2019) Compound Cerrenin D, Preparing Method Thereof, and Application Thereof in Preparing Antitumor Drug. CN Patent 109456191 A.

Macheleidt, J., Müller, M., & Schuster, A. (2016). Biosynthesis of secondary metabolites in fungal endophytes: Implications for plant-microbe interactions. *Fungal Biology Reviews*, 30(1), 5058. https://doi.org/10.1016/j.fbr.2016.03.002.

Madasi, A., Ajmeera, A., Renuka, G., Vemireddy, B., & Vanteru, K. R. (2021). Production of extracellular enzymes, antimicrobial and other agriculturally important metabolites by fungal endophytes of Litsea glutinosa (Lour.) C.B. Rob. a Medicinal Plant. *Journal of Pure and Applied Microbiology*, 15(4), 2317–2328. https://doi.org/10.22207/JPAM.15.4.56

Majoumouo, M. S., Tincho, M. B., Kouipou Toghueo, R. M., et al. (2020). Cytotoxicity potential of endophytic fungi extracts from Terminalia catappa against human cervical cancer cells. *Journal of Toxicology,* 1, 8871152. https://doi.org/10.1155/2020/8871152.

Mili, C., Roy, S., & Tayung, K. (2021a). Endophytic Fungi of Wild and Domesticated Crop Plants and Their Prospect for Applications in Sustainable Agriculture. In *Endophytes* (pp. 21–35). Springer Singapore. https://doi.org/10.1007/978-981-15-9371-0_2.

Mocali, S., Bertelli, E., Di Cello, F., et al. (2003). Fluctuation of bacteria isolated from elm tissues during different seasons and from different plant organs. *Research in Microbiology*, 154(2), 10514. doi:10.1016/S0923-2508(03)00031-7.

Miller, J. D., Adams, G. W., & Sumarah, M. (2012). Antifungal Metabolites from Fungal Endophytes of Pinus Strobus. CA Patent 2766412 A1.

Mohamed, G. A., Ibrahim, S. R. M., & Asfour, H. Z. (2020). Antimicrobial metabolites from the endophytic fungus *Aspergillus versicolor*. *Phytochemistry Letters*, 35, 152. https://doi.org/10.1016/j.phytol.2019.12.003.

Namdar, D., Klaghofer, N. S., Petrini, L., Rivera, T., Koltai, H., & Lewinsohn, E. (2019). Endophytic colonization of cannabis (Cannabis sativa L.) roots by Piriformospora indica, a mutualistic root-colonizing fungus. *Symbiosis,* 78(2), 145–153. https://doi.org/10.1007/s13199-019-00603-2.

Nath, A., Raghunatha, P., & Joshi, S. R. (2012). Diversity and biological activities of endophytic fungi of Emblica officinalis, an ethnomedicinal plant of India. *Mycobiology*, *40*(1), 8–13. https://doi.org/10.5941/MYCO.2012.40.1.008.

Pal, D., Nayak, A. K., & Saha, S. (2019). Cellulose-Based Hydrogels: Present and Future. In *Natural Bio-active Compounds: Volume 1: Production and Applications* (pp. 285–332). Springer Singapore. https://doi.org/10.1007/978-981-13-7154-7_10.

Pan, R., & Bai, X. (2019). Endophytic fungi in plants and their biotechnological applications: A review. *Applied Microbiology and Biotechnology*, 103(3), 11151133. https://doi.org/10.1007/s00253-019-09871-5.

Patil, R. H., Patil, M. P., & Maheshwari, V. L. (2016). Bioactive Secondary Metabolites From Endophytic Fungi: A Review of Biotechnological Production and Their Potential Applications. In *Studies in Natural Products Chemistry* (Vol. 49, pp. 189–205). Elsevier B.V. https://doi.org/10.1016/B978-0-444-63601-0.00005-3.

Peng, X.-Y., Wang, C., & Li, X. (2021). Endophytes from medicinal plants and their biotechnological applications in health and agriculture. *Frontiers in Plant Science*, 12, 655366. https://doi.org/10.3389/fpls.2021.655366.

Qin, S., Chen, H. H., Klenk, H. P., Zhao, G. Z., Li, J., Xu, L. H., Li, W. J. (2009). Title: Glycomyces scopariae sp. nov. and Glycomyces mayteni sp. nov., isolated from medicinal plants in China. *International Journal of Systemic and Evolutionary Microbiolgy*, 59, 10231027.

Rai, N., Kumari Keshri, P., Verma, A., Kamble, S. C., Mishra, P., Barik, S., Kumar Singh, S., & Gautam, V. (2021). Plant Associated Fungal Endophytes as a Source of Natural Bioactive Compounds. In *Mycology* (Vol. 12, Issue 3, pp. 139–159). Taylor and Francis Ltd. https://doi.org/10.1080/21501203.2020.1870579.

Reinhold-Hurek, B., & Hurek, T. (2011). Living inside plants: Bacterial endophytes. *Current Opinion in Plant Biology,* 14(4), 435–443. https://doi.org/10.1016/j.pbi.2011.04.004.

Rochín-Hernández, L. S., Rochín-Hernández, L. J., & Flores-Cotera, L. B. (2022). Endophytes, a Potential Source of Bioactive Compounds to Curtail the Formation–Accumulation of Advanced Glycation End Products: A Review. In *Molecules* (Vol. 27, Issue 14). MDPI. https://doi.org/10.3390/molecules27144469.

Rukachaisirikul, V., Phongpaichit, S., & Teerawutgulrag, A. (2013). Endophytic fungi from medicinal plants: A review of their bioactive metabolites and biotechnological applications. *Journal of Applied Microbiology*, 115(6), 15521571. https://doi.org/10.1111/jam.12324.

Russo, M. L., Pelizza, S. A., Cabello, M. N., & Scorsetti, A. C. (2016). Endophytic fungi from *Solanum* species and their role as biocontrol agents against *Helicoverpa gelotopoeon* (Lepidoptera: Noctuidae). *Biological Control,* 97, 34–43. https://doi.org/10.1016/j.biocontrol.2016.03.004

Saikkonen, K., Wäli, P. R., Helander, M., & Faeth, S. H. (2004). Evolution of endophyte-plant symbioses. *Trends in Plant Science*, 9(6), 27580. doi:10.1016/j.tplants.2004.04.005.

Senthilkumar, M., Anandham, R., Madhaiyan, M., Venkateswaran, V., & Sa, T. (2011). Endophytic bacteria: Perspectives and applications in agricultural crop production. In *Bacteria in Agrobiology: Crop Ecosystems* (pp. 6196). doi:10.1007/978-3-642-18357-7_3.

Shrivastava, G., Rogers, M., Wszelaki, A., Panthee, D. R., & Chen, F. (2010). Plant volatiles-based insect pest management in organic farming. *Critical Reviews in Plant Sciences*, *29*(2), 123–133. https://doi.org/10.1080/07352681003617483.

Skellam, E., Lefebvre, C., & Vannette, R. L. (2019). Plant endophytes: Biodiversity and ecological roles in sustainable agriculture. *Frontiers in Microbiology*, 10, 1550. https://doi.org/10.3389/fmicb.2019.01550.

Strobel, G. (2009). Gliocladium Isolate C-13 and Methods of Its Use for Producing Volatile Compounds and Hydrocarbons. U.S. Patent 20090142816 A1.

Strobel, G. A., & Ezra, D. (2005). Application of Volatile Antibiotics and Non-Volatile Inhibitors from Muscodor Spp. to Control Harmful Microbes in Human and Animal Wastes. U.S. Patent 6911338 B2.

Sun, X., Sun, M., Chao, Y., et al. (2020). Alleviation of lead toxicity and phytostimulation in perennial ryegrass by the Pb-resistant fungus Trichoderma asperellum SD-5. *Functional Plant Biology*, 48, 333–341. https://doi.org/10.1071/FP20237.

Supratman, U., Yuliana, M., & Santoso, E. (2021). Plant endophytes and their potential for the production of bioactive compounds: A review. *Journal of Fungi*, 7(10), 788. https://doi.org/10.3390/jof7100788.

Suryanarayanan, T. S., Venkateswarlu, B., & Kaveriappa, K. M. (2005). Endophytic fungi of plants: A review. *Fungal Diversity*, 16, 1–16. https://doi.org/10.1007/s13225-005-0033-4.

Talukdar, R., Wary, S., Hajowary, R., Sarma, A., & Tayung, K. (2021). Antimicrobial Activity of Endophytic Fungi Isolated from Some Selected Ethnomedicinal Plants of Assam, India. In *Endophytes: Potential Source of Compounds of Commercial and Therapeutic Applications* (pp. 91–102). Springer Singapore. https://doi.org/10.1007/978-981-15-9371-0_6.

Tiwari, R., Awasthi, A., Mall, M., Shukla, A. K., Srinivas, K. V. N. S., Kv, S., & Kalra, A. (2013). Bacterial endophyte-mediated enhancement of in planta content of key terpenoid indole alkaloids and growth parameters of Catharanthus roseus. *Industrial Crops and Products,* 43, 306310. doi:10.1016/j.indcrop.2012.07.045.

Vendan, R. T., Yu, Y., Lee, S., & Rhee, Y. (2010). Diversity of endophytic bacteria in ginseng and their potential for plant growth promotion. *Journal of Microbiology*, 48, 559565. doi: 10.1007/s12275-010-0082-1.

Wakefield, J. J., & McAllister, T. A. (2017). Endophytic bacteria: Potential roles in promoting plant growth and enhancing plant stress tolerance. *Applied Soil Ecology*, 116, 110. https://doi.org/10.1016/j.apsoil.2017.03.015.

Walitang, D., Kim, C-Gi., Kim, K., Kang, Y., Kim, Y. K., & Sa, T. (2018). The Influence of host genotype and salt stress on the seed endophytic community of salt-sensitive and salt-tolerant rice cultivars. *BMC Plant Biology*, 18(1). doi:10.1186/s12870-018-1261-1.

Wiemann, P., & Keller, N. P. (2013). Inhibition of fungal secondary metabolism and its potential for agricultural applications. *Applied Microbiology and Biotechnology*, 97(6), 14451453. https://doi.org/10.1007/s00253-013-4854-9.

Wu, F., Yang, D., Zhang, L., et al. (2019). Diversity estimation and antimicrobial activity of culturable endophytic fungi from Litsea cubeba (Lour.) Pers. in China. *Forests*, 10, 33. https://doi.org/10.3390/f10010033.

Xu, J., Fan, Q., & Yang, L. (2014). Endophytic fungi from Rhizomes of Paris polyphylla and their antibacterial activity. *Evidence-Based Complementary and Alternative Medicine*, 2014, 1–8. doi:10.1155/2014/276701.

Yarte, T., Li, M., & Nair, M. G. (2020). Endophytes from traditional medicinal plants: A review on their secondary metabolites, bioactivities, and potential applications. *Journal of Fungi*, 6(4), 255. https://doi.org/10.3390/jof6040255.

Zaferanloo, B. (2014). Endophytes from Australian native plants as novel sources of bioactive compounds for industrial, environmental and medicinal applications. PhD thesis, RMIT University, Melbourne, Australia.

Zhang, H., Zhang, Y., & Zhou, Y. (2021). Endophytic fungi from medicinal plants: A rich source of bioactive compounds and their potential applications. *Frontiers in Microbiology*, 12, 726412. https://doi.org/10.3389/fmicb.2021.726412.

Zhang, J., Liu, Y., Zhang, N., Hu, B., Jin, Z., & Zhou, X. (2019). Plant endophytes: A sustainable source of antifungal agents. *Frontiers in Microbiology*, 10, 1068. https://doi.org/10.3389/fmicb.2019.01068.

Zhao, S., Chen, S., & Li, X. (2019). Endophytic fungi from medicinal plants and their bioactive metabolites. *Frontiers in Microbiology*, 10, 1653. https://doi.org/10.3389/fmicb.2019.01653.

Zhou, X., Zhang, Y., Yu, C., Kang, L., Sun, X., & Zou, W. (2010). Method for Enhancing Yield of Antitumor Anthraquinone-like Compounds from Fusarium Proliferatum by Adaption of Various Inoculation Methods. CN Patent 101701230 A.

Zhu, B., Wu, L., Wan, H., et al. (2018), Fungal elicitors stimulate biomass and active ingredients accumulation in Dendrobium catenatum plantlets. *Biologia*, 73, 917–926. https://doi.org/10.2478/s11756-018-0091-9.

Unit 3

Endophytic Fungi for Sustainable Agricultural Practices

15 Plant-Associated Fungi
Functional Role in Plant Development and Mitigation of Biotic and Abiotic Stresses

Prasann Kumar and Joginder Singh*
*Corresponding author.

15.1 INTRODUCTION

Since fungi are everywhere, they have complex relationships with plants that affect their growth, well-being and environmental resilience. Plant–fungi interactions range from beneficial and symbiotic to disease causing. These interactions have shown their importance in shaping plants' ecological and physiological dynamics and their effects on agriculture and the environment. This chapter covers the complex world of plant fungi, including their roles in plant growth and development and their significant contributions to biotic and abiotic stress reduction (Kumar et al., 2011a; Sinam et al., 2023; Kumar & Choudhury, 2024a). We study plants' complex relationships with fungi to understand their mechanisms and effects on agriculture, ecosystem dynamics and plant physiology. The history and significance of plant-associated fungi are of great academic interest. Plant-associated fungi interact in many ways. Symbiotic mycorrhizal fungi are an example of mutually beneficial interactions. Plant–fungi interactions have fascinated researchers for decades. The connections under consideration are diverse and functionally versatile. They affect plant health and efficiency in natural and agricultural ecosystems. Mutualistic symbioses, especially mycorrhizal ones, are prominent in plant–fungi interactions. Most plant roots form symbiotic relationships with mycorrhizal fungi, which benefits both parties. Symbiotic relationships between fungi and plants increase nutrient absorption, especially phosphorus. The plant feeds the fungus photosynthetic products. This symbiotic relationship increases plant root system size, helping them thrive in nutrient-poor environments. Thus, this adaptation is crucial to plant survival and success in diverse ecological settings. Endophytic fungi live in plant tissues and are commensal without harming them. These fungi significantly affect plant resilience to drought and salinity. They also synthesize bioactive compounds to boost the plant's defenses against herbivores and pathogens. Endophytes can help farmers manage crops sustainably. Note that not all plant–fungi interactions are mutualistic. Many fungal species cause plant diseases, resulting in significant financial losses in the agricultural sector. Understanding the mechanisms behind these pathogenic associations is crucial to developing effective disease management strategies. This chapter examines plant–fungi interactions in detail. It examines mycorrhizal, endophytic and pathogenic relationships, focusing on their functions. This study examines how mycorrhizal fungi help plants absorb nutrients, focusing on phosphorus and other essential elements. Fungi play a crucial role in nutrient cycling, which improves ecosystem health and fecundity. The chapter also examines how fungi boost plant pathogen resistance, reducing disease incidence. This study also investigates mycoparasites, which parasitize other fungi, as a biocontrol method in agriculture. We also examine how fungal symbiosis can improve plant resilience to drought, salinity and extreme temperatures (Kumar et al., 2017, 2019; Das et al., 2022; Sharma & Kumar, 2023; Kumar & Choudhury, 2024c; Srivastav et al., 2024). The chapter also explains these stress-relieving mechanisms. The practical

DOI: 10.1201/9781003486275-18

implications of plant–fungi interactions in agriculture dominate this chapter. This study describes the use of mycorrhizal fungi and endophytes for sustainable crop production to increase resilience to environmental stressors, and also discusses the drawbacks of using plant-associated fungi in agriculture and the environment. Additionally, we examine this field's prospects. Emerging trends and new methods for optimizing these interactions are also described. The chapter covers the complex world of plant-associated fungi and their effects on plant development, stress relief and agricultural sustainability. This knowledge will affect how we use these interactions to improve ecosystems, agriculture and global food security (Kumar & Mistri, 2020; Kumar et al., 2021b, 2022c).

15.2 MYCORRHIZAL ASSOCIATIONS: THE SYMBIOTIC RELATIONSHIPS BELOW THE SURFACE

Mycorrhizal associations demonstrate the remarkable collaborative dynamics of nature, modeling plant–fungal symbiosis. Plants and fungi share resources and help in this complex symbiotic relationship. Fungi help plants obtain phosphorus, while plants provide photosynthetic products. Different mycorrhizal associations have evolved to fit specific ecological niches and plant–fungi interactions. Arbuscular and ectomycorrhizae are the main types of mycorrhizae. Plant roots and fungi form arbuscular mycorrhizae underground. This network helps plants and soil absorb and exchange nutrients. Glomeromycota arbuscular mycorrhizal fungi are widespread and essential to nutrient acquisition. They penetrate the cellular structure of most terrestrial plant roots and form complex networks that increase root space in the soil (Kumar et al., 2018c; Kumar & Dwivedi, 2020; Kumar et al., 2021b; Singh et al., 2024). The arbuscular mycorrhiza (AM) association involves developing extensively branched arbuscules in plant root cells to transfer nutrients efficiently. This chapter examines AM symbiosis mechanisms and benefits to both organisms, focusing on phosphorus absorption. Ectomycorrhizae are essential to forest ecosystems. Ectomycorrhizal associations interact closely with tree root tips in boreal and temperate forests. Ectomycorrhizal fungi cover root tips with a mantle and extend hyphae into the soil. This setup maximizes nitrogen and phosphorus absorption. Endophytic fungi, which live in plant tissues without harming them, are fascinating. Fungi found in nearly all plants examined so far are diverse. Endophytic microorganisms contribute to plant health by increasing tolerance to nonliving environmental factors and protecting against pests and diseases. Endophytic fungi may help plants resist drought, salinity and heavy metal pollution. Endophytic interactions in plant defense mechanisms are intriguing. Specific endophytes produce herbivore- and pathogen-repellent bioactive compounds. Thus, these endophytes defend their plant hosts innately. Not all plant–fungi interactions are harmonious. Pathogenic fungi cause plant diseases, which cost agriculture money and disrupt ecosystems. Pathogenic fungi invade and harm plant hosts by releasing harmful substances and forming niche infection structures. Various defense mechanisms have evolved to protect plants from fungal pathogens. Physical barriers, chemical reactions and systemic resistance are among these defenses. Understanding these mechanisms is crucial to developing effective disease management strategies. Two organisms work together to obtain nutrients in mutualistic symbiosis. Mutualistic symbiosis involves resource exchange that benefits both plants and fungi (Kumar et al., 2017; Dey et al., 2023b; Kumar & Kumar, 2023; Sharma et al., 2023b). The symbiotic relationship between mycorrhizal fungi and plants improves phosphorus acquisition. Mutualistic fungi boost the plant's pathogen defenses. This study examines how these fungi induce plant defense responses to reduce disease. Mycorrhizal associations and other mutualistic interactions can help plants tolerate drought, salinity and temperature extremes. Plant–fungi interactions demonstrate ecological relationships' complexity. Mycorrhizal associations, endophytic fungi, pathogenic interactions and mutualistic symbioses help plants thrive and adapt to different environmental systems. By examining these interactions, we can better understand the natural world's synergistic capabilities and how to use these interdependencies to improve agricultural sustainability, ecosystem vitality and global food security (Table 15.1).

TABLE 15.1
Symbiotic Partnerships Below Ground: The Mycorrhizal Connections

Plant Species	Fungal Partner	Mycorrhizal Type	Description
Pine (*Pinus* spp.)	*Amanita muscaria*	Ectomycorrhizae	*Amanita* helps improve nutrient uptake in pine trees
Legumes (Fabaceae)	*Rhizophagus irregularis*	Arbuscular mycorrhizae	Enhances nitrogen fixation in legume plants
Orchids (Orchidaceae)	*Tulasnella* sp.	Orchid mycorrhizae	*Tulasnella* assists in nutrient absorption for orchids
Beech (*Fagus* spp.)	*Lactarius deliciosus*	Ectomycorrhizae	*Lactarius* aids in nutrient exchange for beech trees
Grasses (Poaceae)	*Glomus* spp.	Arbuscular mycorrhizae	*Glomus* enhances nutrient uptake in grasses
Oak (*Quercus* spp.)	*Suillus* spp.	Ectomycorrhizae	*Suillus* improves nutrient absorption for oak trees
Heath (Ericaceae)	*Rhizoscyphus ericae*	Ericoid mycorrhizae	*Rhizoscyphus* aids in nutrient acquisition for heath
Willows (*Salix* spp.)	*Paxillus involutus*	Ectomycorrhizae	*Paxillus* helps willows with nutrient absorption
Alder (*Alnus* spp.)	*Frankia* spp.	Actinorhizal	*Frankia fixes* nitrogen for alder trees
Sugar maple (*Acer saccharum*)	*Laccaria bicolor*	Ectomycorrhizae	*Laccaria* aids in nutrient transfer for sugar maple
Eucalyptus (*Eucalyptus* spp.)	*Pisolithus tinctorius*	Ectomycorrhizae	*Pisolithus* facilitates nutrient uptake in eucalyptus
Alder buckthorn (*Rhamnus frangula*)	*Diversispora epigaea*	Arbuscular mycorrhizae	*Diversispora* aids in nutrient exchange for alder buckthorn
Douglas fir (*Pseudotsuga menziesii*)	*Rhizopogon* spp.	Ectomycorrhizae	*Rhizopogon* improves nutrient absorption for Douglas fir
Blueberry (*Vaccinium* spp.)	*Hymenoscyphus* spp.	Ericoid mycorrhizae	*Hymenoscyphus* enhances nutrient acquisition for blueberries
Aspen (*Populus* spp.)	*Laccaria laccata*	Ectomycorrhizae	*Laccaria* aids in nutrient transfer for aspen trees
Sweet fern (*Comptonia peregrina*)	*Frankia* spp.	Actinorhizal	*Frankia* fixes nitrogen for sweet fern
Sagebrush (*Artemisia* spp.)	*Glomus aggregatum*	Arbuscular mycorrhizae	*Glomus* improves nutrient uptake in sagebrush
Pine barrens sandwort (*Moehringia* spp.)	*Cenococcum geophilum*	Ectomycorrhizae	*Cenococcum* enhances nutrient absorption for sandworts
Banana (*Musa* spp.)	*Gigaspora* spp.	Arbuscular mycorrhizae	*Gigaspora* aids in nutrient exchange for banana plants
Pinyon pine (*Pinus edulis*)	*Scleroderma* spp.	Ectomycorrhizae	*Scleroderma* assists in nutrient uptake for pinyon pine
Truffles (*Tuber* spp.)	*Terfezia* spp.	Ectomycorrhizae	*Terfezia* forms symbiotic relationships with truffles
Common myrtle (*Myrtus communis*)	*Glomus mosseae*	Arbuscular mycorrhizae	*Glomus mosseae* enhances nutrient uptake in common myrtle
Hazel (*Corylus* spp.)	*Inocybe* spp.	Ectomycorrhizae	*Inocybe* assists in nutrient absorption for hazel trees
American beech (*Fagus grandifolia*)	*Amanita fulva*	Ectomycorrhizae	*Amanita fulva* aids in nutrient transfer for American beech

(continued)

TABLE 15.1 (Continued)
Symbiotic Partnerships Below Ground: The Mycorrhizal Connections

Plant Species	Fungal Partner	Mycorrhizal Type	Description
Kiwi fruit (*Actinidia* spp.)	*Glomus intraradices*	Arbuscular mycorrhizae	*Glomus intraradices* improves nutrient exchange in kiwi fruit
American sycamore (*Platanus occidentalis*)	*Pisolithus tinctorius*	Ectomycorrhizae	*Pisolithus* enhances nutrient uptake for sycamore trees
White pine (*Pinus strobus*)	*Laccaria laccata*	Ectomycorrhizae	*Laccaria* assists in nutrient absorption for white pine
Liverwort (*Marchantia* spp.)	*Glomus spores*	Arbuscular mycorrhizae	*Glomus* spores facilitate nutrient exchange in liverworts
Ponderosa pine (*Pinus ponderosa*)	*Rhizopogon* spp.	Ectomycorrhizae	*Rhizopogon* improves nutrient transfer for ponderosa pine
Cotton (*Gossypium* spp.)	*Glomus* spp.	Arbuscular mycorrhizae	*Glomus* enhances nutrient uptake in cotton plants
White clover (*Trifolium repens*)	*Glomus etunicatum*	Arbuscular mycorrhizae	*Glomus etunicatum* aids in nutrient uptake for white clover
Black truffle (*Tuber melanosporum*)	*Rhizopogon* spp.	Ectomycorrhizae	*Rhizopogon* forms symbiotic relationships with black truffles
Red pine (*Pinus resinosa*)	*Suillus* spp.	Ectomycorrhizae	*Suillus* improves nutrient exchange for red pine
Lupine (*Lupinus* spp.)	*Rhizophagus* spp.	Arbuscular mycorrhizae	*Rhizophagus* enhances nitrogen fixation in lupine plants
Red alder (*Alnus rubra*)	*Elaphomyces* spp.	Ectomycorrhizae	*Elaphomyces* assists in nutrient transfer for red alder
Wild blue phlox (*Phlox divaricata*)	*Tomentella* spp.	Ectomycorrhizae	*Tomentella* aids in nutrient absorption for wild blue phlox
Horse chestnut (*Aesculus hippocastanum*)	*Amanita* spp.	Ectomycorrhizae	*Amanita* forms symbiotic relationships with horse chestnut
Milkweed (*Asclepias* spp.)	*Gigaspora margarita*	Arbuscular mycorrhizae	*Gigaspora margarita* enhances nutrient exchange in milkweed
Norway spruce (*Picea abies*)	*Lactarius quietus*	Ectomycorrhizae	*Lactarius quietus* aids in nutrient uptake for Norway spruce
40. Stonecrop (*Sedum* spp.)	*Glomus* spp.	Arbuscular mycorrhizae	*Glomus* improves nutrient absorption in stonecrop plants

15.3 NUTRIENT UPTAKE AND NUTRIENT CYCLING: THE VITAL ROLES OF FUNGI IN ECOSYSTEM NUTRIENT DYNAMICS

In the complex network of ecological interactions, fungi play vital roles in nutrient cycling and plant absorption. This chapter discusses arbuscular mycorrhizal fungi (AMF), fungal endophytes and fungi-mediated decomposition in nutrient acquisition and cycling. Glomeromycota arbuscular mycorrhizal fungi help plants acquire nutrients and are widespread. The literature strongly suggests that mycorrhizal fungi help plants absorb nutrients, particularly phosphorus. The nutrient uptake mechanism AMF form symbiotic relationships with plant roots. This includes the formation of highly branched arbuscules in plant root cells and mycorrhizal networks that expand root systems. AMF improves plant phosphorus acquisition, which is a notable contribution. Endophytic fungi help host plants absorb nutrients, especially nitrogen. Endophytic fungi that produce bioactive compounds protect host plants from pathogens. Fungi help ecosystems cycle nutrients by decomposing organic matter (Dwivedi & Kumar, 2011; Kumar et al., 2018e; Kumar & Dwivedi, 2018; Kumar & Naik,

2020; Kotia et al., 2021). The above process corrupts complex organic compounds into simpler ones, reducing nutrient absorption and recycling. The primary decomposers in ecosystems are fungi and bacteria. These organisms produce enzymes that decompose complex organic compounds like cellulose and lignin, releasing nutrients. Fungi decompose organic matter, releasing carbon, nitrogen and phosphorus into the environment. Fungi help plants and soil microbes exchange nutrients through nutrient cycling. This section discusses how fungi actively participate in nutrient cycling in terrestrial ecosystems, emphasizing their importance to ecosystem health. Arbuscular mycorrhizal fungi and fungal endophytes decomposition processes are crucial to nutrient uptake and cycling in ecological systems, and play a key role in agriculture. Microorganisms improve nutrient uptake, plant health and nutrient cycling, sustaining ecosystems and agricultural systems. By understanding these mechanisms and their ecological importance, we can better understand fungi's complex nutrient dynamics in nature (Table 15.2).

TABLE 15.2
The Crucial Contribution of Fungi in Ecosystem Nutrient Dynamics: Nutrient Uptake and Cycling

Fungal Species	Ecosystem Role	Nutrient Dynamics	Description
Mycorrhizal fungi	Nutrient uptake	Phosphorus	Mycorrhizal fungi form symbiotic relationships with plants, aiding in phosphorus uptake
Saprophytic fungi	Decomposition	Carbon	Saprophytic fungi break down organic matter, releasing carbon back into the ecosystem
Lichen	Nitrogen fixation	Nitrogen	Lichens contain nitrogen-fixing cyanobacteria, contributing to nitrogen cycling
Endophytic fungi	Plant protection	Phosphorus	Endophytic fungi can enhance phosphorus availability for host plants
Truffle fungi	Symbiotic mycorrhiza	Macronutrients	Truffle fungi have mycorrhizal associations and influence nutrient dynamics in forests
Yeast	Fermentation	Carbon	Yeasts are involved in carbon cycling through fermentation processes
White rot fungi	Lignin decomposition	Carbon, phosphorus	White rot fungi break down lignin, releasing carbon and phosphorus
Brown rot fungi	Cellulose decomposition	Carbon	Brown rot fungi specialize in breaking down cellulose, contributing to carbon cycling
Chytrid fungi	Aquatic decomposers	Nitrogen	Chytrid fungi decompose organic matter in aquatic ecosystems, affecting nitrogen levels
Arbuscular mycorrhizae	Nutrient exchange	Nitrogen, phosphorus	Arbuscular mycorrhizal fungi facilitate nutrient exchange between plants and the soil
Glomeromycota	Arbuscular mycorrhiza	Phosphorus	Glomeromycota fungi form arbuscular mycorrhizal associations, enhancing phosphorus uptake by plants
Mycoparasitic fungi	Predatory fungi	Nitrogen	Mycoparasites attack other fungi, releasing nitrogen in the process
Ascomycota	Decomposers	Carbon, nitrogen	Ascomycota fungi decompose organic matter, releasing carbon and nitrogen
Ectomycorrhizal fungi	Nutrient transfer	Macronutrients	Ectomycorrhizal fungi transfer nutrients between trees in forest ecosystems

(*continued*)

TABLE 15.2 (Continued)
The Crucial Contribution of Fungi in Ecosystem Nutrient Dynamics: Nutrient Uptake and Cycling

Fungal Species	Ecosystem Role	Nutrient Dynamics	Description
Zygomycota	Decomposers	Carbon, nitrogen	Zygomycota fungi break down organic matter, cycling carbon and nitrogen
Wood decay fungi	Lignin decomposition	Carbon, phosphorus	Wood decay fungi break down lignin, releasing carbon and phosphorus in forests
Orchid mycorrhizae	Nutrient acquisition	Macronutrients	Orchid mycorrhizae assist orchids in acquiring nutrients from the soil
Marine fungi	Decomposers	Marine nutrients	Marine fungi break down organic matter in ocean ecosystems, influencing nutrient dynamics
Rhizoctonia sp.	Pathogenic fungi	Nutrient release	*Rhizoctonia* fungi cause diseases in plants, releasing nutrients upon host decomposition
Endomycorrhizal fungi	Nutrient exchange	Macronutrients	Endomycorrhizal fungi facilitate nutrient exchange between plants and the soil in diverse ecosystems
Ectendomycorrhizal fungi	Nutrient exchange	Macronutrients	Ectendomycorrhizal fungi combine features of ectomycorrhizal and arbuscular mycorrhizal associations to facilitate nutrient exchange
Geosiphon pyriforme	Nitrogen fixation	Nitrogen	*Geosiphon* is a nitrogen-fixing fungus involved in nutrient cycling in select ecosystems
Rust fungi	Pathogenic fungi	Nutrient release	Rust fungi are plant pathogens that release nutrients when they infect and break down host tissues
Decomposer mushrooms	Litter decomposers	Carbon, nitrogen	Decomposer mushrooms break down organic matter in forests, cycling carbon and nitrogen
Ecteomycorrhizal fungi	Macronutrient transfer	Macronutrients	These fungi transfer essential macronutrients between plants and the soil
Chytridiomycota	Amphibian pathogens	Nutrient cycling	Chytrid fungi affect amphibians, influencing nutrient dynamics in aquatic ecosystems
Mycoparasitic *Trichoderma* sp.	Predatory fungi	Nutrient release	*Trichoderma* mycoparasites attack other fungi, releasing nutrients as they consume their hosts
Rusts and smuts	Plant pathogens	Nutrient release	Rusts and smuts are plant pathogens that release nutrients when infecting and breaking down host tissues
Saccharomyces cerevisiae	Fermentation	Carbon	*Saccharomyces cerevisiae* yeast species are involved in carbon cycling through fermentation
Glomus intraradices	Arbuscular mycorrhiza	Macronutrients	*Glomus intraradices* enhances the uptake of essential macronutrients in plants
Mycoparasitic *Hypomycetes*	Predatory fungi	Nutrient release	*Hypomycetes* species parasitize other fungi, releasing nutrients as they consume their hosts
Glomus clarum	Arbuscular mycorrhiza	Macronutrients	*Glomus clarum* enhances the uptake of essential macronutrients in plants
Ectomycorrhizal *Amanita*	Nutrient transfer	Macronutrients	Ectomycorrhizal *Amanita* fungi transfer macronutrients between trees in forests
Arthrobotrys spp.	Carnivorous fungi	Nitrogen	*Arthrobotrys* species capture and consume nematode prey, releasing nitrogen in the process

TABLE 15.2 (Continued)
The Crucial Contribution of Fungi in Ecosystem Nutrient Dynamics: Nutrient Uptake and Cycling

Fungal Species	Ecosystem Role	Nutrient Dynamics	Description
Penicillium spp.	Decomposers	Carbon, nitrogen	*Penicillium* fungi decompose organic matter, contributing to carbon and nitrogen cycling
Colletotrichum spp.	Plant pathogens	Nutrient release	*Colletotrichum* species cause plant diseases, releasing nutrients during host decomposition
Glomerellales fungi	Endophytes	Macronutrient exchange	*Glomerellales* fungi form endophytic associations, aiding in nutrient exchange between plants
Beauveria spp.	Entomopathogenic fungi	Nutrient cycling	*Beauveria* fungi infect and consume insects, contributing to nutrient dynamics
Cryptococcus spp.	Decomposers	Carbon, nitrogen	*Cryptococcus* fungi decompose organic matter, influencing carbon and nitrogen cycling
Rust and smut fungi	Plant pathogens	Nutrient release	Rust and smut fungi are plant pathogens that release nutrients when they infect and break down host tissues

15.4 PLANT DEFENSE MECHANISMS: SAFEGUARDING ECOLOGICAL INTEGRITY

Plants have developed many defenses against fungal pathogens due to their stationary lifestyle. This chapter covers plant defense mechanisms, including fungal-induced resistance, mycoparasite biocontrol and plant defense responses. Fungal-induced pathogen resistance is a type of plant defense which is fascinating because of the induced resistance (Kumar et al., 2020, 2022a; Devi & Kumar, 2023, 2023b; Devi et al., 2023c; Sharma et al., 2023a; Kumar & Choudhury, 2024b; Kumar & Singh, 2024). After exposure to specific microorganisms, plants become more pathogen-resistant. Induced systemic resistance (ISR) occurs when nonpathogenic organisms, such as bacteria or fungi, stimulate the plant's immune system, causing a defensive response. This section examines ISR's mechanisms and how it helps plants resist fungal pathogens. After a pathogenic infection, plants develop systemic acquired resistance (SAR). This chapter examines SAR's complex signaling pathways and molecular mechanisms, including salicylic acid's role. Cross-protection occurs when plants infected with a less aggressive pathogen develop resistance to a more aggressive strain. We describe cross-protection's disease management potential. Mycoparasites can parasitize pathogenic fungi despite their adverse effects on plants. Mycoparasites include many pathogenic fungi-targeting fungal species. This section discusses *Trichoderma* species and *Clonostachys* mycoparasites' fungal pathogen-fighting strategies (Kumar et al., 2018a, 2018b; Goud et al., 2022; Dey et al., 2023c; Saini et al., 2023). This chapter examines how mycoparasites parasitize pathogenic fungi. This includes lytic enzyme synthesis and host cell apoptosis. Biocontrol strategies for plant diseases using mycoparasites have great potential. This discussion explores mycoparasites' potential use in agriculture for sustainable disease management. To defend against fungal pathogens, plants launch a complex defense system. Understanding these defense mechanisms helps plants resist fungal diseases. Plants have complex fungal pathogen recognition mechanisms. This chapter discuss pattern recognition receptors, effector-triggered immunity and phytohormones in signaling pathways. Plants activate immune responses when they detect pathogens, triggering defensive mechanisms. These processes include developing antimicrobial compounds, strengthening cellular walls and initiating a hypersensitive reaction to contain the infection. Plant tissues can respond systemically to localized pathogen attacks. This chapter discusses SAR ecology and agriculture. Complex defense mechanisms have evolved from plants' dynamic interactions with fungal pathogens. This dynamic interaction relies on fungal-induced resistance, mycoparasite biocontrol and plant defense responses. Understanding and using these mechanisms allows the development of new methods to improve plant health, reduce fungal diseases and advance sustainable agriculture.

15.5 ABIOTIC STRESS TOLERANCE: FUNGI AS ALLIES IN THE BATTLE AGAINST ENVIRONMENTAL CHALLENGES

Abiotic stresses like drought, heavy metal contamination, salinity and extreme temperatures can severely impact plant growth and survival in the ever-changing world of plants. This chapter discusses how fungi help plants tolerate drought, heavy metal stress and salinity and temperature extremes. Abiotic stresses like drought threaten agriculture and ecosystems. Mycorrhizal symbiosis has become a powerful ally for plants in their fight against water scarcity. For drought resilience and mycorrhizal associations, AMF mycorrhizal fungi are known to improve drought tolerance. This section discusses drought resilience mechanisms such as improved water uptake, plant turgor and stress response adaptation. AMF creates extensive hyphal networks that extend into the soil, acting as roots. Plants can absorb more water from a larger soil volume with these networks. The chapter discusses how mycorrhizal symbiosis upregulates aquaporins and modulates stress-responsive genes to improve drought tolerance. Environmental issues like heavy metal contamination affect plant health and ecosystem stability worldwide. Fungi have shown remarkable metal stress relief in plants. Hyperaccumulating fungi can learn to store heavy metals in their cells (Pankaj et al., 2012a; Kumar et al., 2021c, 2022b, 2022a; Devi & Kumar, 2023; Hidangmayum et al., 2023; Kumari et al., 2023; Kumar et al., 2024; Kumar & Dey, 2024). This section discusses hyperaccumulators' ecological importance and phytoremediation potential. Fungi can sequester or chelate heavy metals in their mycelium. This prevents toxic metals from entering plant tissues, protecting plant health. The chapter examines how mycorrhizal associations and other fungal interactions can improve heavy metal stress tolerance. Some mechanisms reduce metal uptake and metal-induced oxidative stress. Salinity stress and extreme cold and heat can harm plants. Some fungi can reduce the effects of abiotic stresses. Certain fungi improve ion homeostasis and minimize salt accumulation in plant tissues, increasing salinity tolerance. We study these effects and mechanisms on agricultural applications. Grass fungal endophytes can withstand extreme temperatures. This section discusses how fungal endophytes help plants survive cold and heat stress through biochemical and physiological changes. The chapter discusses the implications of fungal-mediated tolerance to abiotic stresses and how fungi maintain ecosystem resilience in changing environmental conditions. Abiotic stress tolerance is crucial to plant survival and productivity. Fungi strengthen plants against drought, heavy metal contamination, salinity and temperature extremes through symbiotic and endophytic relationships. In a world facing environmental changes, we can develop innovative, sustainable agriculture and ecosystem management strategies by understanding and harnessing fungal-assisted stress tolerance mechanisms (Kumar et al., 2018d; Naik & Kumar, 2020; Kumar et al., 2021a; Reddy et al., 2022; Kumar et al., 2023b; Raj et al., 2024).

15.6 AGRICULTURAL APPLICATIONS: HARNESSING THE FUNGAL ALLIES FOR SUSTAINABLE CROP PRODUCTION

Agriculture, the backbone of the global food system, faces many challenges, from feeding a growing population to achieving sustainability. This project has relied on fungi to improve crop productivity, environmental sustainability and agricultural ecosystem adaptability. This chapter discusses farm fungi, focusing on mycorrhizal, endophytic and bioinoculant uses. Sustainable agriculture research has concentrated on arbuscular AMF mycorrhizal fungi. The main reason is their ability to improve crop productivity, nutrient absorption and health. Agricultural soil mycorrhizal associations affect plant nutrient uptake, especially phosphorus. This section discusses how mycorrhizal fungi increase crop root spacing, improving nutrient acquisition from a larger soil volume and crop yields. Mycorrhizal fungi improve plant nutrient utilization, reducing the need for chemical fertilizers. Nutrient runoff reduction and pollution reduction promote sustainable agriculture. Mycorrhizal associations can help crops resist drought and pests. Fungal endophytes are essential to sustainable agriculture because they inhabit plant tissues without harming them. These substances boost plant growth, stress tolerance and herbivore and pathogen defense. Endophytic fungi improve nutrient uptake, phytohormone production and abiotic stress mitigation. This section examines how endophytes improve crop health and productivity (Kumar et al., 2019; Aley et al., 2022; Das et al., 2022; Devi et al., 2023a; Kumar, 2023; Jagadesh et al., 2024). Fungal endophytes increase plant resilience to drought, salinity and extreme temperatures. This study examines stress mitigation mechanisms and agricultural applications. Pest and pathogen biocontrol uses endophytic fungi that synthesize bioactive compounds to deter herbivores and pathogens. Using mycorrhizal fungi, endophytes, and other beneficial fungi can improve crop resilience and reduce the environmental impact of agriculture. Fungal bioinoculants help crops resist drought, salinity and extreme temperatures. This section examines bioinoculants' practical applications in sustainable crop management. Fungal bioinoculants reduce agricultural chemical inputs like fertilizers and pesticides. Nutrient uptake, use efficiency and pest and pathogen resistance can achieve this. Fungal bioinoculants can help farmers adopt sustainable farming practices. Mycorrhizal fungi can fix nitrogen in the soil, increasing nitrogen content and reducing the need for nitrogen fertilizers. Fungi help sustain farming, boost crop resilience and reduce environmental impacts. Mycorrhizal fungi, fungal endophytes and bioinoculants can help us create a more efficient, productive and sustainable farming future. These symbiotic fungi can improve farming practices, food sustainability and preserve the Earth's finite resources.

15.7 CHALLENGES AND FUTURE PROSPECTS: NAVIGATING THE PATH OF PLANT-ASSOCIATED FUNGI IN AGRICULTURE

Plant-associated fungi's role in sustainable and resilient farming has grown in light of global agricultural and environmental issues. However, using these extraordinary fungi is difficult. This chapter discusses the challenges of harvesting plant-associated fungi and the potential for research and applications in this rapidly evolving field. To maximize their potential, plant-associated fungi must overcome their drawbacks. Many plant–fungus interactions require high specificity and host compatibility. The current situation limits the crops that can benefit from specific fungal associations, emphasizing the need to understand host specificity better. Commercial production and use of fungal bioinoculants are challenging to scale, and producing high-quality inoculum on a large scale remains difficult. Fungi can create unexpected ecological effects in local ecosystems. Promoting responsible and sustainable agriculture requires understanding the environmental impacts of fungal applications. Fungi face pests, pathogens and biotic and abiotic stresses. These factors affect bioinoculation and biocontrol efficacy. Regulatory approval and regulation of fungal bioinoculants can be complicated and time-consuming. Sustainable practices require regulatory streamlining to accelerate adoption. Emerging trends and innovative approaches dominate the current discussion. Innovations and trends in plant-associated fungi research can expand the field and overcome challenges, such as scientific research using molecular and omics methods. Genomic, transcriptomic and proteomic tools have improved the molecular understanding of plant–fungus interactions. These tools help find and grow superior fungal strains. Synthetic biology can improve fungi's bioinoculation and biocontrol properties, making them more useful. Genetic modification or selective breeding can create fungi with enhanced stress tolerance or nutrient acquisition. Precision agriculture uses sensor-based monitoring and bioinoculant application to optimize fungal interventions in agricultural fields. These methods allow farmers to farm cheaply and sustainably. Microbiome research examines the plant microbiome's fungal community to find complex interconnections and synergistic relationships. These complex interactions may improve sustainable agriculture. Current biological control research focuses on fungal biocontrol agents. Researchers are exploring new ways to enhance these agents, including using fungal nanoparticles or formulations. Future research and applications of plant-associated fungi in agriculture are promising. Fungal bioinoculants, mycorrhizal associations and endophytic fungi can revolutionize agriculture by promoting sustainability and environmental responsibility. Fungi's role in crop resilience to climate change-induced pressures like drought and extreme temperatures offers a chance to develop climate-resilient crops. Fungi provide ecosystem services beyond agriculture, such as soil health, carbon sequestration and water purification. As the world faces environmental issues, these services will become more critical. Fungi bioremediation of polluted soils and water bodies reduces heavy metal and organic pollution (Kumar et al., 2013; Kumari et al., 2022; Kumar et al., 2023a; Sharma et al., 2023d; Kumar & Choudhury, 2024c; Srivastav et al., 2024). Mycology can benefit from interdisciplinary collaboration with agronomists, geneticists and ecologists. This collaborative approach could boost innovation and progress in this field. A multidisciplinary approach will help understand plant–fungus interactions. Utilizing plant-associated fungi is difficult, but new patterns, advancements and opportunities allow this to take place more easily. Plant-associated fungi in agriculture can help solve global food security, sustainability and environmental issues. Accepting and using these opportunities while overcoming constraints will unlock these extraordinary fungal companions' full potential in agriculture and environmental preservation.

15.8 SIGNIFICANT INSIGHTS INTO THE FUNCTIONAL ROLES OF FUNGI ASSOCIATED WITH PLANTS

The complex symbiotic relationship between plants and fungi in agriculture and environmental management reveals functional roles that can promote sustainable plant growth and reduce stress. In conclusion, this study illuminates the significance of these interactions, their agricultural applications and their effects on plant development and stress reduction. Plants and fungi form fascinating mutualistic, endophytic and pathogenic relationships. Mycorrhizal, endophytic and fungal biocontrol agents are essential for resilient and sustainable agriculture. The reciprocal transfer of nutrients, stress relief and pathogen protection give plants the tools to thrive in changing ecosystems. The topic is agricultural innovation and its effects on crop yields and farming sustainability. Plant-associated fungi in agriculture are not just a fad but a transformative step toward sustainable and resilient food production. Mycorrhizal associations improve nutrient absorption, reduce fertilizer use and protect crops from drought and pests. Fungal endophytes boost plant growth, stress tolerance and herbivore defense. Fungal bioinoculants reduce the ecological footprint and ensure agricultural sustainability by replacing chemical inputs with environmentally friendly alternatives. Plant–fungi interactions occur within ecological frameworks where microorganisms, soil quality and ecosystem services converge. Understanding complex molecular, genetic and community connections opens the door to environmentally friendly agriculture and management (Kumar et al., 2017; Karam et al., 2023; Kumar & Kumar, 2023; Sharma & Kumar, 2023; Upadhyay et al., 2023; Kumar & Choudhury, 2024d). Plant microbiomes reveal interactions between diverse microorganisms, suggesting new ways to improve plant health and ecosystem sustainability. Plant-associated fungi enhance abiotic stress tolerance, making them crucial in climate change. Fungi may help crops resist drought, extreme temperatures and salinity. These organisms' ability to increase nutrient acquisition and water efficiency makes them vital partners in climate-resilient agriculture, ensuring food security in an uncertain future. Fungi associated with plants have many uses beyond agriculture, including environmental conservation. Fungal bioinoculants reduce soil pollution by addressing heavy metal and organic pollutants.

Bioremediation can address ecological contamination, supporting ecosystem restoration, conservation of natural resources, soil health and a more sustainable planet. Interdisciplinary synergy involves combining academic disciplines to solve a problem or research area. Studying plant-associated fungi requires collaboration and multidisciplinarity. Mycologists, agronomists, geneticists and ecologists must work together to maximize these symbiotic fungal organisms' potential. Integrating multiple disciplines improves understanding of plant–fungi interactions, generates new ideas and promotes sustainable agriculture and environmental stewardship. Plant development, sustainability and stress mitigation have become increasingly important in recent years. This chapter investigates future sustainable plant development and stress mitigation. These issues are urgent due to global population growth. The functional roles of plant-associated fungi open up a promising future. Sustainable plant development and stress mitigation are now achievable goals. Symbiotic fungi improve food security, reduce environmental impacts and promote climate-resilient agriculture. This exploration has significant implications for sustainable plant development and stress mitigation (Kumar et al., 2011b; Chakraborty et al., 2021; Devi et al., 2023b; Dey et al., 2023b; Sharma et al., 2023b; Devi et al., 2024). The potential is vast, opportunities are plentiful and the challenges are manageable. Agriculture is about integrating productivity and sustainability. Using plant-associated fungi boosts crop productivity and reduces the agricultural environmental impact. By improving nutrient uptake, stress tolerance and pest resistance, farming methods can meet the needs of a growing global population without the environmental impacts of conventional farming. Fungi have effects beyond agriculture. Ecological conservation, bioremediation and ecosystem restoration depend on fungi. These services help create a more sustainable environment and improve resource use. Interdisciplinary collaboration is crucial to this venture. Integrating knowledge from different disciplines helps understand complex plant–fungi interactions and develop new sustainable agriculture and environmental stewardship methods. Fungi's roles in plants show resilience, adaptability and a mutually beneficial symbiotic relationship. The authors envision a society where sustainable plant development and stress mitigation are practical and applicable. In this global context, plants and fungi work together to solve our most pressing problems and build a sustainable, resilient future (Pankaj et al., 2012b; Pathak et al., 2017; Kumar & Pathak, 2019; Yomso et al., 2021; Dey et al., 2023a; Sharma et al., 2023b, 2023c).

15.9 CONCLUSION

The complex plant–fungi network is essential in agriculture and environmental management. This study on fungal allies' roles in plant development and stress reduction concludes with several key findings demonstrating their vital role in creating a more sustainable and resilient global environment. A complex interaction network of mutualistic, endophytic and pathogenic relationships between plants and fungi underpins terrestrial ecosystems. Mycorrhizal associations improve nutrient uptake. However, endophytic fungi increase stress tolerance and herbivore resistance. These partnerships demonstrate the power of natural collaboration. Fungi on plants transform agriculture. These organisms can increase nutrient absorption, reduce chemical use and help crops resist drought and pests, reshaping sustainable agricultural practices. These fungi offer a path to progress in a world with competing food security and environmental concerns. Fungi perform bioremediation, improving soil health and ecosystem restoration outside of agriculture. Fungal bioinoculants in bioremediation may help reduce heavy metal and organic pollutant contamination and restore human-damaged ecosystems. Climate change and abiotic stresses have made climate-resilient agriculture crucial. Thus, fungi's contributions are more important. The solutions aim to make crops more resilient to water shortages, high temperatures and salinity stress. These fungal allies offer hope for climate-resilient agriculture in an uncertain future. The complex and diverse study of plant-associated fungi requires an interdisciplinary approach involving mycologists, agronomists, geneticists and ecologists. Collaboration across disciplines helps these fungal allies realize their full potential. This improves plant–fungi understanding, advancing sustainable agriculture and environmental stewardship. This study shows that plant-associated fungi promote sustainable plant growth and reduce stress. They model cooperation and harmony that other natural systems could follow. Plant-associated fungi offer hope and a pragmatic path forward in a world with an expanding population, a changing climate and a deteriorating environment. These organisms go beyond biological curiosity to help us create a sustainable, resilient and harmonious global environment. Their presence is essential for plant growth and provides resilience against biotic and abiotic stresses. Metamorphosis, adjustment and mutually beneficial symbiosis are the story of plant-associated fungi. This story encourages people to accept fungal allies and recognize their contributions to a sustainable and resilient future. It emphasizes the interconnectedness of humans and the natural world and the potential for collective efforts to overcome obstacles and create a harmonious coexistence where plants, fungi and all organisms thrive.

REFERENCES

Aley P, Singh J, Kumar P. (2022). Chapter 23 – Adapting the Changing Environment: Microbial way of life. In: Kumar A, Singh J, Ferreira LFRBT-MUCC (eds). Microbiome Under Changing Climate Implications and Solutions. Woodhead Publishing, pp 507–525.

Chakraborty S, Kumar P, Sanyal R, et al. (2021). Unravelling the regulatory role of miRNAs in secondary metabolite production in medicinal crops. *Plant Gene* 27:100303. https://doi.org/https://doi.org/10.1016/j.plgene.2021.100303

Das T, Saha SC, Sunita K, et al. (2022). Promising botanical-derived monoamine oxidase (MAO) inhibitors: Pharmacological aspects and structure-activity studies. *South African J Bot* 146:127–145. https://doi.org/https://doi.org/10.1016/j.sajb.2021.09.019

Devi P, Kumar P. (2023a). Role of Oxidative Enzymes in the Alleviation of Oxidative Stress in Plant System. In: Arti Gupta and Pankaj Kumar (eds.), *Microbial Oxidative Enzymes: Biotechnological Applications*. De Gruyter, Department of Agronomy, School of Agriculture, Lovely Professional University, Phagwara, Punjab, 144411, India, pp 109–128.

Devi P, Kumar P. (2023b). Effect of Microbial Oxidative Enzymes on Soil Health. In: Arti Gupta and Pankaj Kumar (eds.), *Microbial Oxidative Enzymes: Biotechnological Applications*. De Gruyter, Department of Agronomy, School of Agriculture, Lovely Professional University, Phagwara, Punjab, 144411, India, pp 37–61.

Devi P, Dey SR, Kumar P. (2023a). Integration of Water Resources Management in Rural Areas. In: Sughosh Madhav, Arun Lal Srivastav, Sylvester Chibueze Izah, Eric van Hullebusch (eds.), *Water Resources Management for Rural Development: Challenges and Mitigation*. Elsevier, Department of Agronomy, School of Agriculture, Lovely Professional University, Punjab, Phagwara, India, pp 139–152.

Devi P, Dey SR, Saini L, et al. (2023b). Toward Sustainable Agriculture: Strategies Involving Phytoprotectants Against Reactive Oxygen Species. In: Faizan, M., Hayat, S., Ahmed, S.M. (eds.), *Reactive Oxygen Species: Prospects in Plant Metabolism*. Springer Nature, Department of Agronomy, School of Agriculture, Lovely Professional University, Phagwara, India, pp 229–247.

Devi P, Dogra P, Kumar P, Singh J. (2024). Global Status of Biomass Energy Programmes-Challenges and Roadmap. In: Maulin P Shah, Pardeep Kaur (eds.), *Biomass Energy for Sustainable Development*. CRC Press, Lovely Professional University, Phagwara, India, pp 394–414.

Devi P, Kumar P, Sharma K, Singh J. (2023c). Chapter 10 – Is there a Relationship between Myconanotechnology and Sustainable Development? In: Kuddus M, Ahmad IZ, Hussain CMBT-M and A of N in B (eds) *Progress in Biochemistry and Biotechnology*. Academic Press, pp 175–207.

Dey SR, Devi P, Kumar P. (2023a). Organic Contaminants in Aquatic Environments: Sources and Impact Assessment. In: Arun Lal Srivastav, Lata Rani, Jyotsna Kaushal, Tien Duc Pham (eds.), *Metal Organic Frameworks for Wastewater Contaminant Removal*. Wiley, Department of Agronomy, School of Agriculture, Lovely Professional University, Punjab, 144411, India, pp 299–317.

Dey SR, Kumar P, Singh J. (2023b). Growth of mustard under open field conditions with salicylic acid and antimicrobial Thiourea. *Biopestic Int* 19:183–192. https://doi.org/10.59467/BI.2023.19.183

Dey SR, Sharma M, Kumar P. (2023c). Effects and Responses of Chromium on Plants. In: Kumar, N., Walther, C., Gupta, D.K. (eds.), *Environmental Science and Engineering*. Springer Science and Business Media Deutschland GmbH, Department of Agronomy, School of Agriculture, Lovely Professional University, Punjab, Phagwara, 144411, India, pp 385–427.

Dwivedi P, Kumar P, (2011). Anti-diabetic medicinal plants and their conservation: Waging green war on diabetes. *Med Plants* 3:181–189. https://doi.org/10.5958/j.0975-4261.3.3.031

Goud EL, Singh J, Kumar P. (2022). Chapter 19 – Climate Change and their Impact on Global Food Production. In: Kumar A, Singh J, Ferreira LFRBT-MUCC (eds.), Microbiome Under Changing Climate Implications and Solutions. Woodhead Publishing, pp 415–436.

Hidangmayum A, Dwivedi P, Kumar P, Upadhyay SK. (2023). Seed Priming and Foliar application of Chitosan Ameliorate drought stress responses in Mungbean genotypes through modulation of Morpho-physiological attributes and increased antioxidative defense mechanism. *J Plant Growth Regul* 42:6137–6154. https://doi.org/10.1007/s00344-022-10792-1

Jagadesh M, Dash M, Kumari A, et al. (2024). Revealing the hidden world of soil microbes: Metagenomic insights into plant, bacteria, and fungi interactions for sustainable agriculture and ecosystem restoration. *Microbiol Res* 285. https://doi.org/10.1016/j.micres.2024.127764

Karam N, Kumar P, Anand YR, Choudhury D. (2023). Management of rice blast-A review of the strategies available. *Biopestic Int* 19:107–116. https://doi.org/10.59467/BI.2023.19.107

Kotia A, Rutu P, Singh V, et al. (2021). Rheological Analysis of Rice Husk-starch Suspended in Water for Sustainable Agriculture Application. In: *2nd International Conference on Functional Materials, Manufacturing and Performances, ICFMMP 2021*. Elsevier Ltd, School of Mechanical Engineering, Lovely Professional Unievrsity, Punjab, Phagwara, 144401, India, pp 1962–1966.

Kumar D, Rameshwar SD, Kumar P. (2019). Effect of intergated application of inorganic and organic fertilizers on the roots of chickpea. *Plant Arch* 19:857–860.

Kumar P. (2023). Degradation of Xenobiotics by bacteria and fungi: An overview. *Biopestic Int* 19:117–132. https://doi.org/10.59467/BI.2023.19.117

Kumar P, Choudhury D. (2024a). Hydroponic Removal of Organic Contaminants from Water. In: Kumar, N. (ed.), *Springer Water*. Springer Nature, Department of Agronomy, School of Agriculture, Lovely Professional University, Punjab, Phagwara, 144411, India, pp 143–164.

Kumar P, Choudhury D. (2024b). Bioremediation of Wastewater Using Hydroponics. In: Kumar, N. (ed.), *Springer Water*. Springer Nature, Department of Agronomy, School of Agriculture, Lovely Professional University, Punjab, Phagwara, 144411, India, pp 27–64.

Kumar P, Choudhury D. (2024c). Microbial Native Soil Bacteria Against Cadmium Toxicity. In: Jha, A.K., Kumar, N. (eds.), *Springer Water*. Springer Nature, Department of Agronomy, School of Agriculture, Lovely Professional University, Punjab, Phagwara, 144411, India, pp 193–226.

Kumar P, Choudhury D. (2024d). Toxicity of Rhizospheric Cadmium Contaminated Soil and Its Phytoremediation. In: *Springer Water*. Springer Nature, Department of Agronomy, School of Agriculture, Lovely Professional University, Punjab, Phagwara, 144411, India, pp 227–264.

Kumar P, Devi P, Dey SR. (2021a). Chapter 6 – Fungal Volatile Compounds: A Source of Novel in Plant Protection Agents. In: Kumar A, Singh J, Samuel JBT-V and M of M (eds). Academic Press, pp 83–104.

Kumar P, Dey SR. (2024). Hydroponics Phytoremediation: An Overview. In: Kumar, N. (ed.), *Springer Water*. Springer Nature, Department of Agronomy, School of Agriculture, Lovely Professional University, Punjab, Phagwara, 144411, India, pp 361–396

Kumar P, Dey SR, Choudhury D. (2024). Effectiveness of Cadmium on Biochemical Shift of Pea Plant Treated with Mycorrhiza and Putrescine. *Nat Environ Pollut Technol* 23:275–285. https://doi.org/10.46488/NEPT.2024.v23i01.023

Kumar P, Dwivedi P. (2018). Putrescine and glomus mycorrhiza moderate cadmium actuated stress reactions in zea mays l. By means of extraordinary reference to sugar and protein. *Vegetos* 31:74–77. https://doi.org/10.5958/2229-4473.2018.00076.9

Kumar P, Dwivedi P. (2020). Lignin estimation in sorghum leaves grown under hazardous waste site. *Plant Arch* 20:2558–2561.

Kumar P, Goud EL, Devi P, et al. (2022a). Heavy Metals: Transport in Plants and Their Physiological and Toxicological Effects. In: Kundan Kumar, Sudhakar Srivastava (ed.), *Plant Metal and Metalloid Transporters*. Springer Nature, Department of Agronomy, School of Agriculture, Lovely Professional University, Punjab, Jalandhar, India, pp 23–54.

Kumar P, Goud EL, Devi P, Koul B. (2022b). Metal Pollutants in the Environment. In: Maulin Shah (ed.), *Environmental Microbiology: Emerging Technologies*. De Gruyter, Department of Agronomy, School of Agriculture, Lovely Professional University, Jalandhar, Punjab, 144411, India, pp 291–323.

Kumar P, Harshavardhan M, Kumar PS, et al. (2018a). Effect on chlorophyll a/b ratio in cadmium contaminated maize leaves treated with putrescine and mycorrhiza. *Ann Biol* 34:281–283.

Kumar P, Koul B, Sharma M. (2022c). Phytoremediation of Heavy Metals. In: Jitendra Kumar, Shweta Gaur, Prabhat Kumar Srivastava, Rohit Kumar Mishra, Sheo Mohan Prasad, Devendra Kumar Chauhan (eds.), *Heavy Metals in Plants: Physiological to Molecular Approach*. CRC Press, Department of Agronomy, School of Agriculture, Lovely Professional University, Punjab, Jalandhar, 144411, India, pp 369–388.

Kumar P, Kumar S, Harshavardhan M, et al. (2018b). Evaluation of plant height and leaf length of sorghum grown under different sources of nutrition. *Ann Biol* 34:284–286.

Kumar P, Kumar T, Singh S, et al. (2020). Potassium: A key modulator for cell homeostasis. *J Biotechnol* 324:198–210.

Kumar P, Latha DH, Singh J. (2023a). Rhizoremediation of Organic Emerging Soil Contaminants: Green Technology. In: Pradeep Verma, Maulin P. Shah (eds.), *Clean Technologies Toward a Sustainable Future: Physicochemical, Biochemical and Biotechnological Approaches*. IWA Publishing, Department of Agronomy, School of Agriculture, Lovely Professional University, Phagwara, Punjab, 144411, India, pp 211–229.

Kumar P, Mandal B, Dwivedi P. (2011a). Heavy metal scavenging capacity of Mentha spicata and Allium cepa. *Med Plants* 3:315–318. https://doi.org/10.5958/j.0975-4261.3.4.053

Kumar P, Mandal B, Dwivedi P. (2013). Combating heavy metal toxicity from hazardous waste sites by harnessing scavenging activity of some vegetable plants. *Vegetos* 26:416–425. https://doi.org/10.5958/j.2229-4473.26.2.106

Kumar P, Mandal B, Dwivedi P. (2011b). Heavy metals scavenging of soils and sludges by ornamental plants. *J Appl Hortic* 13:144–146.

Kumar P, Mistri TK. (2020). Transcription factors in SOX family: Potent regulators for cancer initiation and development in the human body. *Semin Cancer Biol* 67:105–113. https://doi.org/https://doi.org/10.1016/j.semcancer.2019.06.016

Kumar P, Naik M. (2020). Biotic symbiosis and plant growth regulators as a strategy against cadmium and lead stress in chickpea. *Plant Arch* 20:2495–2500.

Kumar P, Pathak S. (2019). Responsiveness index of sorghum (Sorghum bicolor (L.) Moench) grown under cadmium contaminated soil treated with putrescine and mycorrhiza. *Bangladesh J Bot* 48:139–143.

Kumar P, Pathak S, Amarnath KS, et al. (2018c). Effect of growth regulator on morpho-physiological attributes of chilli: A case study. *Plant Arch* 18:1771–1776.

Kumar P, Pathak S, Kumar M, Dwivedi P. (2018d). Role of secondary metabolites for the mitigation of cadmium toxicity in sorghum grown under mycorrhizal inoculated hazardous waste site. *Biotechnol Approaches Med Aromat Plants Conserv Genet Improv Util* 199–212.

Kumar P, Saini L, Sharma M. (2023b). Assessment of the Phytochemical Constituents and Metabolites in the Medicinal Plants and Herbal Medicine Used in the Treatment and Management of Respiratory Diseases. In: Izah, S.C., Ogwu, M.C., Akram, M. (eds.), *Reference Series in Phytochemistry*. Springer Science and Business Media B.V., Department of Agronomy, School of Agronomy, Lovely Professional University, Phagwara, India.

Kumar P, Sharma K, Saini L, Dey SR. (2021b). Chapter 8 – Role and Behavior of Microbial Volatile Organic Compounds in Mitigating Stress. In: Kumar A, Singh J, Samuel JBT-V and M of M (eds.), Volatiles and Metabolites of Microbes. Academic Press, pp 143–161.

Kumar P, Singh BN, Dwivedi P. (2017). Plant Growth Regulators, Plant Adaptability and Plant Productivity: A Review on Abscisic Acid (ABA) Signaling in Plants Under Emerging Environmental Stresses. In: Divya Pandey, Abhijit Sarkar (eds.), *Sustaining Future Food Security in Changing Environments*. Nova Science Publishers, Inc., Department of Plant Physiology, Institute of Agricultural Sciences, Banaras Hindu University, Varanasi, India, pp 81–97.

Kumar P, Singh J. (2024). Hydroponic: An Eco-friendly Future. In: Kumar, N (ed.), *Springer Water*. Springer Nature, Department of Agronomy, School of Agriculture, Lovely Professional University, Punjab, Phagwara, 144411, India, pp 223–267.

Kumar P, Yumnam J, Kumar PS, et al. (2018e). Cadmium induced changes in germination of maize seed treated with mycorrhiza. *Ann Agri Bio Res* 23:169–170.

Kumar V, Dwivedi P, Kumar P, et al. (2021c). Mitigation of heat stress responses in crops using nitrate primed seeds. *South African J Bot* 140:25–36. https://doi.org/https://doi.org/10.1016/j.sajb.2021.03.024

Kumar VA, Kumar P. (2023). Revolutionizing agriculture with satellite technology for farmers: A review. *Biopestic Int* 19:97–106. https://doi.org/10.59467/BI.2023.19.97

Kumari P, Kumar P, Singh J. (2023). Toxic Contamination of Aquatic Habitat and its Remediation. In: Pravat Kumar Shit, Dilip Kumar Datta, Biswajit Bera, Aznarul Islam, Partha Pratim Adhikary (eds.), *Spatial Modeling of Environmental Pollution and Ecological Risk*. Elsevier, Department of Agronomy, School of Agriculture, Lovely Professional University, Punjab, Phagwara, India, pp 209–220.

Kumari P, Singh J, Kumar P. (2022). Chapter 21 – Impact of Bioenergy for the Diminution of an Ascending Global Variability and Change in the Climate. In: Kumar A, Singh J, Ferreira LFRBT-MUCC (eds.), Microbiome Under Changing Climate Implications and Solutions. Woodhead Publishing, pp 469–487.

Naik M, Kumar P. (2020). Role of growth regulators and microbes for metal detoxification in plants and soil. *Plant Arch* 20:2820–2824.

Pankaj PK, Kumar P, Nigam RC, Mishra PK. (2012a). Monitoring and surveillance of synthetic pyrethroids and organophosphate in different brands of soft drinks. *J Chem Pharm Res* 4:3939–3943.

Pankaj PK, Kumar P, Nigam RC, Mishra PK. (2012b). To studies organochlorine insecticides in different brands of cold drinks. *J Chem Pharm Res* 4:3934–3938.

Pathak S, Kumar P, Mishra PK, Kumar M. (2017). Mycorrhiza assisted approach for bioremediation with special reference to biosorption. *Pollut Res* 36:329–332.

Raj A, Goutam E, Kumar P, Singh J. (2024). Polyphenol Oxidase: An Enzyme of Plants. In: Pradeep Verma, Komal Agrawal, Maulin P. Shah (eds.), *Polyphenol Oxidases: Function, Wastewater Remediation, and Biosensors*. De Gruyter, Department of Agronomy, School of Agriculture, Lovely Professional University, Phagwara, Punjab, 144411, India, pp 105–128.

Reddy SS, Kumar P, Dwivedi P. (2022). Heavy Metal Transporters, Phytoremediation Potential, and Biofortification. In: Kumar, K., Srivastava, S. (eds.), *Plant Metal and Metalloid Transporters*. Springer, pp 387–405.

Saini L, Sharma K, Kumar P. (2023). Water Efficiency in Agriculture as a Vital Approach Toward Water Management in Rural Areas. In: *Water Resources Management for Rural Development: Challenges and Mitigation*. Elsevier, Department of Agronomy, School of Agriculture, Lovely Professional University, Punjab, Phagwara, India, pp 127–138.

Sharma K, Kumar P. (2023). Microbe-mediated impact on cell membrane stability and leaf pigmentation in pearl millet under txic mercury stress in soil. *Biopestic Int* 19:217–228. https://doi.org/10.59467/BI.2023.19.217

Sharma K, Devi P, Dey SR, Kumar P. (2023a). Mercury Phytovolatilization: An Overview of the Mechanism and Mitigation. In: *Role of Green Chemistry in Ecosystem Restoration to Achieve Environmental Sustainability*. Elsevier, Department of Agronomy, School of Agriculture, Lovely Professional University, Punjab, Phagwara, India, pp 325–331.

Sharma K, Devi P, Kumar P. (2023b). Metal-Organic Frameworks for Wastewater Contaminants Removal. In: *Metal Organic Frameworks for Wastewater Contaminant Removal*. Wiley, Department of Agronomy, School of Agriculture, Lovely Professional University, Punjab, Phagwara, 144411, India, pp 95–117.

Sharma K, Devi P, Kumar P, et al. (2023c). Hazardous Phytotoxic Nature of Reactive Oxygen Species in Agriculture. In: Faizan, M., Hayat, S., Ahmed, S.M. (eds.), *Reactive Oxygen Species: Prospects in Plant Metabolism*. Springer Nature, Department of Agronomy, School of Agriculture, Lovely Professional University, Phagwara, India, pp 135–146.

Sharma M, Dey SR, Kumar P. (2023d). Antioxidant Defence: A Key Mechanism of Chromium Tolerance. In: Kumar, N., Walther, C., Gupta, D.K. (eds.), *Environmental Science and Engineering*. Springer Science and Business Media Deutschland GmbH, Department of Agronomy, School of Agriculture, Lovely Professional University, Punjab, Phagwara, 144411, India, pp 91–116.

Sinam V, Kumar P, Singh J. (2023). Bioremediation and Ecorestoration Strategies of Aquatic Environment. In: Pravat Kumar Shit, Dilip Kumar Datta, Biswajit Bera, Aznarul Islam, Partha Pratim Adhikary (eds.), *Spatial Modeling of Environmental Pollution and Ecological Risk*. Elsevier, Department of Agronomy, School of Agriculture, Lovely Professional University, Punjab, Phagwara, India, pp 483–499.

Singh J, Kumar P, Siddique A. (2024). Biochar-based organic amendments on soil health, nutrient status and quality of potato (Solanum tuberosum). *Plant Sci Today* 11:102–108. https://doi.org/10.14719/pst.2706

Srivastav AL, Patel N, Rani L, et al. (2024). Sustainable options for fertilizer management in agriculture to prevent water contamination: a review. *Environ Dev Sustain* 26:8303–8327. https://doi.org/10.1007/s10668-023-03117-z

Upadhyay SK, Devi P, Kumar V, et al. (2023). Efficient removal of total arsenic (As3+/5+) from contaminated water by novel strategies mediated iron and plant extract activated waste flowers of marigold. *Chemosphere* 313:137551. https://doi.org/10.1016/j.chemosphere.2022.137551

Yomso J, Kumar P, Siddique A. (2021). Effect of cadmium induced stress on morphological growth and yield attributes of rice (Oryza sativa l.). *Res Crop* 22:246–250. https://doi.org/10.31830/2348-7542.2021.064

16 Endophytic Fungi
Plants' Hidden Armor Against Abiotic Stress

*Gurleen Kaur Sodhi and Sanjai Saxena**
*Corresponding author.

16.1 INTRODUCTION

The impact of climate change has become evident, and this trend is expected to continue in the future. It is speculated that regardless of the mitigation efforts, the historical emissions will likely increase the planet's temperature beyond the 2°C mark, viz., considered dangerous. Moreover, considering the changing weather patterns, global temperatures are expected to increase by 4°C by the year 2100 (Thuiller, 2007; New et al., 2011; Raftery et al., 2017; Lyon et al., 2022). Since agriculture and climate change are inextricably linked, these abrupt changes in climatic conditions endanger global food security (Malhi et al., 2021; Shahzad et al., 2021). A report by the Food and Agriculture Organization (FAO) states that at the ongoing greenhouse emissions rate, a decline in the production of major crops such as wheat (5–50%) and rice (20–30%) is foreseen by the year 2100 (Arora, 2019; Raza et al., 2021). Therefore, the current trends may result in an unprecedented increase in crop losses, thereby driving up food costs and making it more challenging to meet the global food demands of the growing population.

The immobile nature of plants makes them vulnerable to various climate conditions. Extreme temperature conditions coupled with drought, salinity and heavy metal contamination are some of the primary abiotic stress conditions endured by plants (Imran et al., 2021). Heavy metals' contamination of arable land is a global concern as it reduces the plant's nutrition uptake ability, affecting photosynthetic processes and eventually causing retardation (Li et al., 2019a; Khan et al., 2021). Likewise, extreme temperature conditions also affect the plant processes. An increase in temperature beyond a critical threshold for extended periods hinders the countless biochemical reactions inside plants, whereas low temperatures or cold stress cause ice crystal formation (Bita & Gerats, 2013; Ritonga & Chen, 2020). Such conditions lead to the accumulation of reactive oxygen species (ROS) in plants, which cause oxidative stress, a phenomenon persistently observed in other abiotic stress conditions (Kapoor et al., 2019). The drought conditions caused by extreme temperatures and other climatic conditions curtail plants' water uptake and essential nutrients. In contrast, salinity stress disrupts ionic homeostasis, reducing plant productivity (Ma et al., 2020).

However, recent advances in our understanding of the multiprocess nature involved in the responses of plants to abiotic stress conditions provide an edge in enhancing crop productivity via different approaches. One such approach is utilizing the symbiotic association of endophytic fungi and plants (Kaddes et al., 2019; Baron & Rigobelo, 2022). The term "endophytic fungi" describes the ubiquitous mycological diversity in healthy plant tissues without causing any harmful effects. Their intricate association with plants makes it difficult to accurately determine their origin (Faeth, 2002; Stone et al., 2004). For improved understanding, endogenous and exogenous hypotheses about the genesis of endophytic fungi were proposed by Caiyi et al. (2004) and Li and Hu (2005),

DOI: 10.1201/9781003486275-19

respectively. As per the endogenous hypothesis, the endophytes arise from the chloroplast and mitochondria, whereas the exogenous hypothesis contends that endophytes enter plants from the outside. Furthermore, the transmission of endophytic fungi can either occur horizontally through asexual conidia or sexual spores, or vertically through penetration of the host embryo by fungal hyphae.

The relationship between endophytic fungi and the host plant is particularly close-knit, where the hyphae are affixed to the plant cells but do not overburden them. Although little is understood about the endophyte–plant association, the relationship is predominantly mutualistic (Saikkonen et al., 2004). Furthermore, the universal phenomenon of horizontal gene transfer enables them to mimic secondary bioactive compounds produced by plants (Tiwari & Bae, 2020). The endophytic fungi have been reported to improve the overall plant health, especially in stress conditions (Dubey et al., 2020; Poveda et al., 2021; Sodhi & Saxena, 2023a). For instance, the synthesis of phytohormones of all five classes is a significant attribute exhibited by the endophytic fungi. These signaling molecules exhibit a wide array of functions ranging from root formation seed and pollen tube development to aiding the plant to adapt to stress conditions (Checker et al., 2018; Yan et al., 2019; Rhaman et al., 2020). Besides phytohormone production, the endophytic fungi from various genera, such as *Penicillium* and *Aspergillus*, have been reported for mineral solubilization such as phosphate, potassium and calcium (Nath et al., 2012; Adhikari & Pandey, 2019). By solubilizing these minerals, the endophytic fungi make them readily available for plants, where they help stimulate the growth of roots and regulate stomatal opening (Lay-Pruitt et al., 2022). Additionally, other plant growth-promoting attributes such as siderophore, ammonia and extracellular lytic enzyme production have also been reported from endophytic fungi (Bilal et al., 2018; Ripa et al., 2019; Chand et al., 2020; Khan et al., 2022; Sodhi & Saxena, 2023b).

Owing to these critical characteristics, this chapter discusses the recent advances in ameliorating abiotic stress conditions using endophytic fungi in different plants. The growth-promoting effects under four major abiotic stress conditions, namely, heavy metals, extreme temperatures, drought and salinity, are discussed in greater detail.

16.2 MECHANISMS OF PLANT DEFENSE AND ENDOPHYTIC FUNGI AGAINST ABIOTIC STRESS

Abiotic stress can be defined as the detrimental effect seen as a result of inanimate factors on living beings. As discussed earlier, osmotic stress is a secondary stress produced due to abiotic stress conditions (Nadarajah, 2020; Zhang et al., 2022). Moreover, one of the first signs of a stressed plant is the generation of ROS. The ROS act as crucial signaling molecules under normal conditions; however, the stress conditions disrupt the ROS generation balance inside the plants, thereby leading to excessive accumulation (Hasanuzzaman et al., 2020; Sachdev et al., 2021). In order to thrive under stress conditions, plants change their physiological, biochemical and molecular mechanisms. However, associated with beneficial endophytic fungi, a plant's capability to tolerate stress is enhanced because of these microorganisms' plant growth promotion potential (Figure 16.1). For instance, the antioxidant potential of endophytic fungi allows the plant to mitigate the stress caused by ROS effectively. The antioxidant potential also reduces the peroxidation of lipids in plants (Ismail et al., 2020; Hamayun et al., 2021a; Sodhi & Saxena, 2023c). Abiotic stress conditions activate functional and regulatory genes in plants, where the former encodes for osmoprotectants heat shock proteins, and the latter encodes various protein kinases and transcription factors. As a result, plant cells start imbibing Ca^{2+} as a secondary messenger from intracellular and extracellular sources to direct the stress within the nucleus, which then leads to an alteration in gene expression (Nguyen et al., 2018; Li, 2019c; Saijo & Loo, 2020). Changes are also observed at the phenotypic level; for instance, under salinity stress conditions, the roots absorb salts but inhibit their translocation to other parts of the plant.

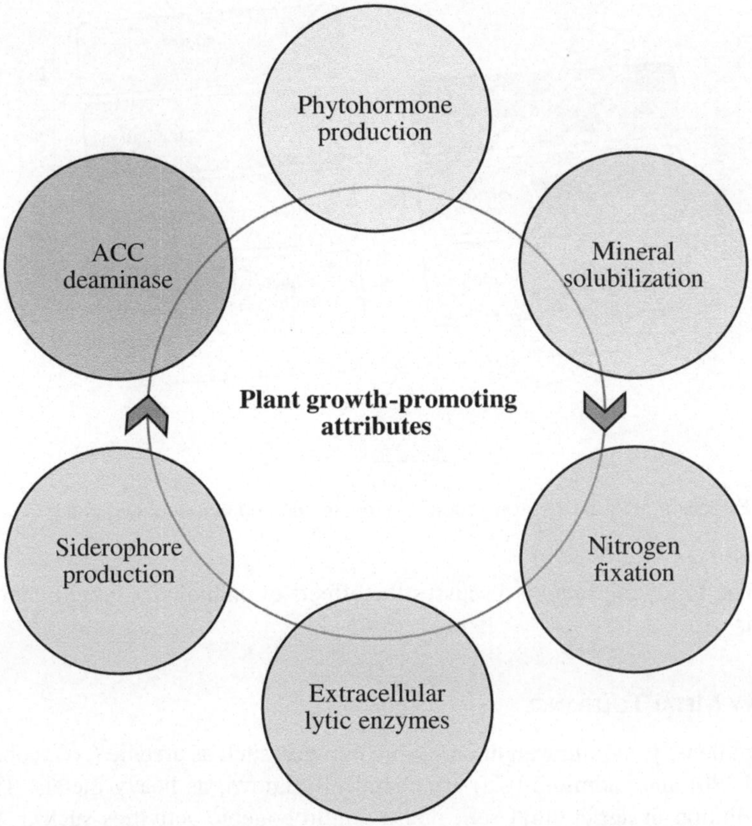

FIGURE 16.1 Plant growth-promoting attributes of endophytic fungi.

On the other hand, endophytic fungi can solubilize minerals such as zinc and phosphate, improving the plant's ability to withstand stress (Adhikari & Pandey, 2019; Chand et al., 2020). These microorganisms can also hold on to the heavy metals, which reduces the negative impact of stress on the host plant (Ripa et al., 2019; Hussain et al., 2022; Shukla et al., 2022). Likewise, in plants at cellular levels, the absorbed salt ions are translocated to the vacuoles, which causes osmotic disbalance and eventually, the cells dehydrate (Parihar et al., 2015; Isayenkov & Maathuis, 2019). In order to maintain the osmotic conditions and keep the metabolic pathways running, the plants then start absorbing proline and glycine betaine. Studies show the capability of endophytic fungi to enhance the accumulation of osmolytes in plants, which helps achieve homeostasis. This also improves the water uptake of the plants, which is beneficial, especially under drought stress (Azizi et al., 2021; He et al., 2022). In addition, the endophytic fungi are excellent producers of phytohormones, which regulate plant growth in unfavorable environments. Furthermore, endophytic fungi have also been observed producing enzymes such as 1-aminocyclopropane-1-carboxylate (ACC) deaminase. Rather than being converted into ethylene, the ACC produced during stress conditions is converted into α-ketobutyrate and ammonia by the action of this enzyme. The inhibition of ethylene production protects plants from detrimental damage (Rauf et al., 2021; Wang et al., 2022; Sodhi & Saxena, 2023d). Endophytic fungi also impart stress tolerance by synthesizing antistress compounds and regulating the host plant's system (Schulz et al., 1999; Schulz et al., 2002). Consequently, endophyte-associated plants can adapt better to abiotic stress conditions

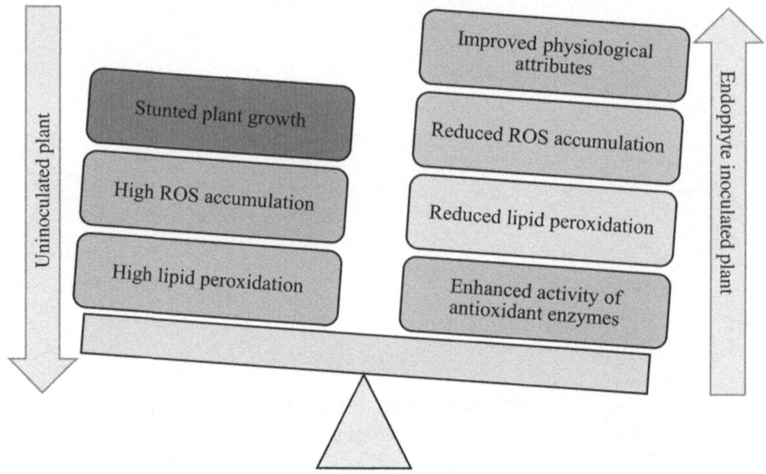

FIGURE 16.2 Effects of abiotic stress on endophyte inoculated and uninoculated plants.

(Figure 16.2). The following section discusses the effects of endophytic fungi under major abiotic stress conditions.

16.2.1 Heavy Metal Tolerance

Metals with high density, atomic weight or atomic number, such as arsenic (Ar), copper (Cu), mercury (Hg), lead (Pb) and cadmium (Cd) are chemically known as heavy metals. The increase in heavy metal pollution in agricultural soil, due to anthropogenic activities such as metal mining, smelting, leaching and utilization of pesticides and fertilizers, has become a severe threat to the environment and food security (Mishra et al., 2019; Hembrom et al., 2020). The cosmopolitan nature of microorganisms, such as endophytic fungi, allows them to inhabit and adapt to extreme environmental conditions (Bang et al., 2018; Khalid et al., 2021). Due to this intrinsic property and the symbiotic association with plants, endophytic fungi can help enumerate heavy metals. For instance, Kushwaha et al. (2022) evaluated the plant growth-promoting (PGP) attributes of *Serendipita* under Ar stress. Ar is generally toxic to plants and inhibits root proliferation. The endophyte exhibited growth under 1 mM Ar concentration and could solubilize phosphate under 5 mM Ar concentration. *S. indica* accumulated 37.12 and 5.67 times higher phosphate levels under 2.5- and 5-mM Ar concentrations, suggesting that the Ar stress enhanced phosphate accumulation. Further elevated expression of ACPase enzyme and exopolyphosphatase (PPX) enzyme under Ar stress was seen. Similarly, elevated levels of the Si *PT* gene were also observed. This study highlights the neutralized effects of arsenic stress and PGP attributes of *S. indica*.

In a similar study, the endophytic fungi *Serendipita indica* was tested for its Cd phytoremediation potential on two flowering plants, *Alyssum montanum* L. and *Helianthus annuus* L. Cd contamination in soil is attributed to mining, smelting practices and extensive utilization of fertilizers. (Mahar et al., 2020). The surface sterilized seeds were exposed to four different Cd concentrations for 28 days. The inoculated seedlings of both plants witnessed an increase in plant length, root and shoot dry matter in *H. annuus*. Further, it increased chlorophyll content, *Fv*/*F*m ratio and electron transport rate (ETR) in *H. annuus*. Meanwhile, *Fv*/*F*m and ETR indicate the maximum quantum efficiency and electron transport rate of the photosystem II (Rastogi et al., 2021).

An increase in the bioaccumulation factor of roots by 16.7–35.6% in *A. montanum* and 20.6–45.9% in *H. annuus* was observed, whereas the bioaccumulation factor of shoots decreased by

19.9–31.1% in *A. montanum* and 37–48.7% in *H. annuus*. Another critical parameter, translocation factor (TF) decreased in *A. montanum* and *H. annuus* in inoculated plants. TF indicates the plant's ability to translocate heavy metals for phytoextraction, and the results demonstrate the enhanced ability in inoculated plants. The results observed in this study indicated that the endophyte may be appropriate for immobilization of Cd in the roots of plants (Shahabivand et al., 2020).

Similarly, El-Shafey et al. (2021) evaluated the effect of three endophytic fungi, namely, *Talaromyces versatilis*, *Emericella nidulans* and *Aspergillus niger* isolated from *Pelargonium graveolens*, under Cd stress. Enhanced leaf biomass of geranium plant on inoculation of *T. versatilis* and *A. niger* was seen. Likewise, *E. nidulans* and *A. niger* inoculation significantly alleviated Cd accumulation in the leaves, whereas *A. niger* also increased the mineral content of K (potassium), Ca (calcium), Mg (magnesium), P (phosphorus), Na (sodium), Fe (iron), Cu (copper) and Zn (zinc) (Fageria, 2001).

As mentioned in the earlier sections, the production of ROS is considered the first sign of stress conditions in the plants. A decrease in hydrogen peroxide (H_2O_2) level under Cd stress by ~ 2.2 and 45% on inoculation of *T. versatilis* and *E. nidulans* was noted. Further, lipid peroxidation decreased on inoculation of *T. versatilis*, *E. nidulans* and *A. niger*. Lipid peroxidation is the deleterious process of oxidative degeneration of lipids accessed by malonaldehyde (MDA), a frequently used biomarker of lipid injury in plants. On evaluation, it was recorded that *T. versatilis*, *E. nidulans* and *A. niger* inoculation enhanced the activity of superoxide dismutase (SOD), catalase (CAT), peroxidase (POD), ascorbate peroxidase (APX), monodehydroascorbate reductase (MDHAR), dehydroascorbate reductase (DHAR) and glutathione peroxidase (GPX). To understand the detoxification mechanism of endophytes, phytochelatins (PCs), metallothioneins (MTs) and glutathione-S-transferase (GST) were studied. Under Cd-stress conditions, the highest activity was observed in *A. niger*, indicating the efficient detoxification of heavy metals (Rizwan et al., 2019).

Tabande et al. (2022) evaluated the efficiency of *S. indica* on *Medicago sativa*, to alleviate ZnO-NPs stress. The inoculated plants observed enhanced root and shoot biomass and chlorophyll and carotenoid contents. Further, the inoculated plants exhibited a reduction in the accumulation of ZnO-NPs by ~ 27–29% in the shoots, along with enhanced accumulation by ~ 23–70% in the roots. The TF also decreased, indicating reduced translocation of ZnO-NPs from roots to shoots and leaves. The endophyte-inoculated plant also exhibited a significant decrease in MDA content and H_2O_2 content. This decrease in oxidative damage was coupled with increased SOD, CAT and POD activity, highlighting the synergistic impact of endophyte inoculation under stress conditions.

Another heavy metal, copper (Cu), is known for inhibiting plants' vital process of photosynthesis. Ragasa et al. (2021) isolated endophytic fungi from *Solanum nigrum* L. The isolates were subjected to Cu tolerance under different concentrations. The tolerant isolate IB-SN1W, identified as *Fusarium oxysporum*, was further analyzed using metatranscriptomics to identify the potential genes involved in Cu tolerance. As a result, 23% of the contigs specific in Cu-treated *F. oxysporum* were expressed, and genes involved in Cu import polyamine transport were upregulated. Enhanced oxidoreductase activities enabled the endophyte to mitigate Cu stress. The expression of a gene *FOXG_02993*, analogous to yeast activation protein 1 (YAP1), was observed in Cu-treated *F. oxysporum*. YAP is a Basic Leucine Zipper Domain (bZIP) protein subfamily. The YAP-1 is specifically activated under oxidative stress conditions, which could be one of the reasons for the Cu tolerance attribute of *F. oxysporum*.

Pb has become one of the top soil contaminants due to rapid industrialization and advanced agricultural practices. Endophytic fungi were isolated from *Chenopodium album* L. The isolates were subjected to Pb stress ranging between 300–1700 mg/L, and tolerant isolate #15, identified as *Trametes hirsute*, was selected for further studies. On inoculation the isolate enhanced the shoot FW, shoot dry weight (DW), root FW and root DW. This study observed an increase in Pb concentration in the shoots and roots (Malik et al., 2020).

Because of global climatic conditions, more than one type of heavy metal in the soil is dominant. However, the stress tolerance virtue of endophytic fungi extends to more than one heavy metal; Wang et al. (2021) evaluated the effect of *Epichloë* sp. on *Festuca sinensis*, an herbaceous perineal grass native to China, under both Cd and Zn stress conditions. On inoculation, the plant height remained constant under Cd stress, whereas an increase of ~ 28% under Zn stress was seen. Similarly, the tiller number increased under Cd and Zn stress. Further, the TF value for both Cd and Zn stress remained high. Additionally, no increase in phytohormone content under Cd stress was seen except in the case of abscisic acid (ABA) content.

In another study, Yu et al. (2021) evaluated *Exophiala pisciphila*, a dark septate endophyte (DSE) isolated from *Arundinella bengalensis*. The endophyte was pretreated with melatonin, a naturally occurring enzyme that scavenges free radicals. The endophyte was subjected to Cd, Zn, Pb and Cu stress for two days, and the oxygen free radical (OFR) concentration decreased by 33.31, 27.64 and 60.91%, respectively. However, no significant difference under Cu stress was observed. On transcriptomics analysis of the endophyte, three melatonin biosynthetic proteins, EpTDC1, EpSNAT1 and EpASMT1, were identified and overexpressed in *Arabidopsis thaliana*. Overall, the results warrant the role of melatonin in other DSEs for phytoremediation.

El-Shahir et al. (2021) evaluated the phytoremediation potential of *Talaromyces pinophilus* on *Triticum aestivum*. The endophyte *T. pinophilus* was isolated from *Rosmarinus officinalis*. It was further observed that the relative water content (RWC) of the shoot, RWC of the root, shoot length and root length also increased. Similarly, the phytohormone-producing endophytic fungi *Paecilomyces formosus* LHL10 and *Penicillium funiculosum* LHL06 were co-cultured to evaluate their phytoremediation potential under multi-metal stress involving 20 mg/kg each of Al (aluminum), Cr (chromium), Ni, Cu, Cd and Pb. Both these isolates exhibited GA production, which was further enhanced by co-culturing. Similarly, co-inoculation also increased IAA production. Furthermore, a significant reduction in multi-metal toxicity on co-inoculation was observed. Co-inoculation enhanced the antioxidant enzyme activity of SOD, CAT and POD. Moreover, a reduction in lipid peroxidation by 61.34 was seen. The amino acid content of serine, glutamic acid, glycine, methionine, lysine, arginine and proline also increased (Bilal et al., 2021).

In another study, Hassan et al. (2022) evaluated the phytoremediation potential of *Alocasia calidora*. A set of 13 endophytic fungi were used to prepare a consortium for inoculation in host plants. In inoculated plants, the root length, shoot length, root FW, shoot FW, root DW and shoot DW increased. Further, the metal accumulation statistics of the plant revealed that the biological absorption coefficient (BAC) for all heavy metals was between 0.5–1, whereas the bioconcentration factor (BCF) was <1. The consortia application also enhanced other biochemical parameters of the plants, such as the chlorophyll content antioxidant enzyme content of POD, CAT and APX. The findings highlight the potential of fungal consortia for enhancing the growth and tolerance of the plant under multi-metal stressed conditions. Various other endophytic fungi and their phytoremediation potentials are shown in Table 16.1.

16.2.2 Extreme Temperature Tolerance

Extreme temperature conditions seen as a consequence of ongoing climate change are another threat to sustainable agriculture. Changes in plant processes, such as photosynthesis, restriction of critical enzymes and impeding nutrition uptake, are some obstructions plants face when the temperatures rise or fall beyond the optimum conditions (Hatfield & Prueger, 2015). Endophytic fungi have shown potential to aid plants in easing these conditions. For instance, Ali et al. (2018) isolated endophytic fungi CpE, identified as *Thermomyces lanuginosus*, from *Gymnanthemum amygdalinum*. The isolate was tested on *Cucumis sativus* under 40°C heat stress. The inoculated plants observed a slow reduction in the photosynthetic processes by maintaining the photosystem II (PSII) apparatus. PSII,

TABLE 16.1
Effect of Endophytic Fungi on Different Plants Under Heavy Metal Stress

Sr. no.	Endophyte	Plant	Response	Stress	References
1.	*Aspergillus welwitschiae*	*Glycine max* L. (soybean)	Enhanced the production of indole acetic acid (54.83 μg/mL as compared to control 15.56 μg/mL), solubilized inorganic phosphate, and production of stress-related metabolites	Cr-VI and As-V in the form of $K_2Cr_2O_7$ and Na_3AsO_4 up to 1200 μg/mL	Hussain et al., 2022
2.	*Epichloë bromicola*	*Hordeum brevisubulatum* (wild barley)	Reduced Cd content to protect the physiological functions of plants	50–100 μM Cd	Zhai et al., 2022
3.	*Epichloë* sp.	*Lolium perenne* L. (perennial ryegrass)	Increased accumulation of heavy metals in aerial parts of the host plants	20 mg/kg Cd^{2+}, 700 mg/kg Pb^{2+} and Cu^{2+}	Żurek et al., 2022
4.	*Serendipita indica*	*Solanum lycopersicum* (tomato)	Increased the arsenic bioaccumulation factor by 1.48 times in root and reduces the arsenic translocation factor by 2.96 times from root to shoot and 13.6 times from root to fruit compared to noninoculated plants	Arsenic stress	Shukla et al., 2022
5.	*Piriformospora indica*	*Artemisia annua* L. (sweet wormwood)	Increased chlorophyll content and water potential, antioxidant enzyme activities. Restricted Cd translocation to aerial parts	40–120 mg/kg Cd	Khalid et al., 2022
6.	*Mucor circinelloides*, *Curvularia lunata* and *Clonostachys rosea*	*Glycine max* L. (soybean)	Enhanced shoot biomass, reduced Cd concentration in roots and shoots. Higher bioaccumulation, immobilization potential and reduced translocation	0.3–5 mg/kg Cd	Wang et al., 2022
7.	*Aspergillus flavus*	*Solanum lycopersicum* (tomato)	Remediated Cd and Cr toxicity by reducing their translocation from the roots to the upper part of the plant	100–500 μg/mL Cd and Cr	Aziz et al., 2021b
8.	*Microdochium bolleyi*	*Hordeum vulgare* L. (barley)	Enhanced shoot length, root length, dry root biomass, dry shoot biomass, chlorophyll content	10–60 mg/kg Cd	Shadmani et al., 2021a

(continued)

TABLE 16.1 (Continued)
Effect of Endophytic Fungi on Different Plants Under Heavy Metal Stress

Sr. no.	Endophyte	Plant	Response	Stress	References
9.	*Piriformospora indica*	*Nicotiana tabacum* L. (tobacco)	Increased Cd accumulation in roots and decreased Cd accumulation in leaves, upregulation of antioxidant enzymes	0–2.5 mg/L Cd	Su et al., 2021a
10.	*Beauveria bassiana*, *Rhodotorula mucilaginosa* and *Metschnikowia pulcherrima*	*Glycine max* (soybean) and *Hordeum vulgare* (barley)	The endophytes were able to remove Cd up to 59% with bioaccumulation of 2.3–11.9 mg/g	100 µg/mL Cd	Ignatova et al., 2021
11.	*Alternaria alternata*, *Microdochium bolleyi*, *Bipolaris zeicola*, *Alternaria* sp., *Pleosporales* sp., *Fusarium redolens*, *Fusarium tricinctum*, *Fusarium monliforme*, *Clonostachys rosea* and *Epicoccum nigrum*	*Hordeum vulgare* L. (barley)	Exhibited tolerance to Cd stress, improved chlorophyll content, root weight, plant length, fresh weight, and dry weight of plants	0.90 mg/kg Cd	Shadmani et al., 2021b
12.	*Aspergillus niger*	*Solanum lycopersicum* (tomato)	Reduced translocation to aerial parts of the plant, enhanced IAA, proline, flavonoids, phenols, CAT and ascorbic acid oxidase, sugar and protein content. Induction of stress response genes *SlGSH1* and *SlPCS1*	100–500 µg/g Cd and Cr	Aziz et al., 2021a
13.	*Falciphora oryzae*	*Oryza sativa* (rice)	Reduced Cd translocation by sequestering it in plant vacuoles. Identification of a *SNARE* Syntaxin 1 gene involved in vacuole enlargement	25–75 µM Cd	Su et al., 2021b
14.	*Aspergillus niger* and *Penicillium chrysosporium*	*Vicia faba* L. (broad bean)	Inhibited the transport of metal and decreased electrolyte leakage and lipid peroxidation	100 mg/L Cd and 200 mg/L Pb	El-Mahdy et al., 2021
15.	*Acrocalymma vagum* and *Scytalidium lignicola*	*Medicago sativa* (lucerne) and *Ammopiptanthus mongolicus* (Mongolian redfin)	Enhanced antioxidant enzymes, biomass, nitrogen and organic carbon content	10–50 mg/L Cd	Hou et al., 2020
16.	*Exophiala pisciphila*	*Zea mays* (maize)	Increased chlorophyll contents in biomass, and length, surface area, volume, branches, tips number, and mass density of maize roots	50 mg/kg Cd	He et al., 2020

located in the thylakoid membrane, plays a crucial role in photosynthesis, and the disorganization of the reaction centers during stress conditions gravely affects the plants.

Likewise, the net photosynthetic rate (Pn) also exhibited a slow reduction, implicating lower inhibition of Rubisco activase and higher assimilation of carbon dioxide (CO_2). An increase in saponin, flavonoid, soluble and insoluble carbohydrate and soluble protein content on inoculation of *T. lanuginosus* under stress conditions was also observed. Similarly, the antioxidant enzyme activities of SOD, CAT and POD increased. However, no significant increase in phenyl ammonia lyase (PAL) activity was observed. The enhanced antioxidant enzyme activity helps the plants by increasing the membrane stability during oxidative damage. Additionally, reduced stomatal opening and rate of transpiration in inoculated plants indicate the ability of plants to maintain the water content and avoid a hydraulic failure. Therefore, enhancement in the morphological features in inoculated plants, such as the root length, leaf FW/DW ratio and leaf water content, increased by 34.6, 62.46 and 62.58%, respectively. This morphological adaptation seen on inoculation of the endophyte is an essential trait in desert ecosystems.

Plants growing in extreme environments such as hot deserts are expected to harbor an affluent diversity of endophytic fungi. Sangamesh et al. (2018) isolated endophytic fungi from 15 naturally occurring plant species of the Thar desert. Of the total isolates, 315 were obtained from stems, 114 from leaves, 71 from roots and seven from thorns. Nine isolates exhibited tolerance to high temperatures of 45°C. Out of the nine isolates, LAS-6, identified as *Chaetomium* sp., was selected for further evaluation as inoculum on IR-64 variety of rice. IR-64 is a salt-sensitive rice variety; on inoculation with LAS-6, enhanced root and shoot growth was observed under heat-stress conditions. Furthermore, the survival rate of inoculated seedlings increased by ~ 17.6 and 19.2% under 40 and 45°C. Similarly, isolates ACJ-2, ACJ-5, SAP-3, SAP-6 and LAS-4, identified as *Aspergillus* sp., were used as an inoculum in IR-64 under PEG-8000 induced drought stress conditions. Polyethylene glycol (PEG) induces artificial drought conditions as it reduces water absorption and causes plant desiccation. As a result, enhanced root length in inoculated seedlings was observed compared to uninoculated seedlings. However, no significant difference in shoot growth was seen. Overall, the results depicted the colonization of endophytic fungi in rice seedlings followed by an enumeration of heat and drought stress conditions.

In a similar study, endophytic fungi were isolated from *Cullen plicata*. The isolate identified as *Thermomyces lanuginosus* was used as an inoculum on the host plant under heat stress conditions (45°C). The Pn increased under heat stress in the inoculated plants, whereas a decline was observed in the uninoculated plants. At high temperatures, effective carbon metabolism is affected in the plants, because of which a decrease in the Pn is observed. However, endophyte inoculation exhibits significant improvement in Pn under heat-stress conditions. Furthermore, the stomatal conductance and transpiration rate increased in uninoculated plants under stress conditions. High temperatures directly affect Calvin cycle enzymes, increasing the transpiration rate. Increases in physical parameters such as root length, shoot length, root FW, root DW, shoot FW, shoot DW, leaf area, leaf number and fruit number were seen. Similarly, the antioxidant enzyme activities of PAL, CAT and POD enhanced by ~ 40, 128.57 and 127.27%, respectively. Since antioxidants are highly effective in mitigating oxidative damage, enhanced activity on endophyte inoculation further increases the chances of plant survival (Ali et al., 2019).

In another study, Ismail et al. (2020) isolated endophytic fungi from *Sonchus asper*. The isolate SonchL-7, identified as *Aspergillus niger*, was preliminarily tested on rice seedlings, where an increase in chlorophyll content, shoot length, root length, shoot FW, root FW, shoot DW and root DW by ~ 4.9, 11.7, 21.9, 11.9, 61.8, 30.5 and 15.7% was observed. Further, testing of the isolate on sunflower and soybean seedlings under 45°C heat stress conditions was carried out. In inoculated sunflower seedlings and soybean seedlings under stress conditions, the chlorophyll content, shoot length, root length, shoot FW, root FW, shoot DW and root DW were also enhanced. The enhanced chlorophyll content could be attributed to the endophyte's ability to uptake magnesium (Mg) from the surroundings. Besides the enhanced growth features, *A. niger* inoculation reduced H_2O_2 generation and MDA content under stress conditions was also recorded. The high MDA content further destabilizes the cell membrane, disturbing the state of cell electrolytes and eventually causing cell death.

On analysis of the antioxidant enzyme concentration under stress conditions, it was observed that *A. niger* enhanced the activity of CAT, POD, ascorbic acid oxidase (AAO), SOD and glutathione reductase (GR). On the contrary, the abscisic acid (ABA) content was reduced. Furthermore, the spectrophotometric analysis revealed an increase in total lipid, protein and soluble sugar content. The enhanced nutritional values indicate the potential of *A. niger* as a biofertilizer to enumerate heat stress. In a similar study, endophytic fungi were isolated from *Verbena officinalis* (Hamayun et al., 2021b). The isolate VerR-20, identified as *Gliocladium cibotii*, was tested on rice seedlings for preliminary analysis. The isolate was further tested on soybean and sunflower seedlings under 40°C stress conditions. The results indicated the protective role of *G. cibotii* under heat stress conditions.

Besides heat stress, low-temperature/cold stress conditions also bring about detrimental plant changes. They cause a decrease in enzyme activity, disrupt nutrient intake and, in severe cases, cause plant death. In their study, Li et al. (2021) evaluated the effect of the endophytic fungi *Piriformospora indica* on *Musa acuminata*, under cold stress (4°C) conditions. Inoculated plants exhibited less severe symptoms with reduced electrolyte leakage (EL). EL is a ubiquitous phenomenon triggered during stress conditions and used as a measure of stress conditions in plants. A rapid decline in Fv/Fm value was observed in the uninoculated plants, whereas no significant difference in Fv/Fm values of inoculated plants was seen. Similarly, the efficient quantum yield (Y(II)), ETR and the photochemical and nonphotochemical quenching coefficient (qP and qN) of inoculated plants were higher.

The antioxidant enzyme activities of SOD, CAT and GR increased. Furthermore, the MDA and H_2O_2 contents decreased. The transcriptomics gene expression revealed upregulation of relative expression of various genes studied. Additionally, the inducer of CBF expression 1 (*ICE1*) was downregulated. *ICE1* co-activates a signaling cascade that results in delayed flowering. The results indicate enhanced tolerance to low-temperature conditions on inoculation of endophytic fungi.

For a better understanding of the ability of endophytic fungi to confer tolerance, Jiang et al. (2021) evaluated the expression profile of *Piriformospora indica* inoculated and uninoculated *Arabidopsis thaliana*. The differential gene expression (DEG) identified 634 genes, of which 193 were upregulated, and 441 were downregulated in inoculated plants under low temperature (4°C) stress. The plants recognize freezing stress in the plasma membrane and activate ABA-responsive genes. Several ABA-responsive DEGs, including *AT1G67080*, were identified. DEGs for other hormones, such as *AT5G56300* for gibberellic acid methyltransferase 2, *AT3G25290* for auxin-responsive protein, *AT2G34600* for jasmonic acid (JA) and *AT4G39780* for ethylene functions were also identified. Low-temperature conditions usually cause a temperature shock, increasing cytosolic calcium levels. In this context, DEGs *AT1G73805* for calmodulin-binding protein, *AT3G59530* for calcium-dependent phosphotriesterase, *AT3G22910* for calcium-dependent cation/H^+ exchanger 19 and *AT1G62810* and *AT3G06130* for potential calcium transporter gene were identified. Furthermore, to mitigate ROS production, the production of antioxidants is vital, and DEGs *AT4G25100* for SOD, *AT1G20620* for CAT and *AT1G62180* for phosphoadenosine phosphosulfate reductase were identified. Overall augmentation of the metabolic processes and protein phosphorylation was observed in inoculated plants, suggesting the ability of endophytic fungi to target new genes involved in low-temperature tolerance. Various endophytic fungi exhibiting tolerance to extreme temperature conditions are shown in Table 16.2.

TABLE 16.2
Effect of Endophytic Fungi on Different Plants Under Extreme Temperature Stress

Sr. no.	Endophyte	Plant	Response	Stress	References
1.	*Piriformospora indica*	*Vitis vinifera* (grapevine)	Enhanced nutrient accumulation, increase in osmoregulants, antioxidant enzymes activity and plant growth regulators ABA and polyamines	4°C low temperature stress	Karimi et al., 2022
2.	*Serendipita indica*	*Musa acuminata* (banana)	Inoculated plants under stress conditions expressed 278 differentially abundant miRNAs including 93 novel miRNAs, significantly associated with nutrient sources uptake, secondary metabolism, growth regulators, auxin-responsive factor, signal transduction pathways	45°C high temperature stress	Bodjrenou et al., 2022
3.	*Stemphylium solani*	*Helianthus annuus* L. (sunflower) and *Glycine max* L. (soybean)	Enhanced chlorophyll contents, total biomass and plant height, CAT, SOD, POD, GR, AAO. Inhibition of ROS, lipid peroxidation proline and ABA	40°C high temperature stress	Ismail et al., 2022
4.	*Penicillium rubens* and *P. bialowienzense*	*Vaccinium corymbosum* (highbush blueberry)	Enhanced gene expression of the Late Embryogenesis Abundant protein (*LEA1*), Fv/Fm, fruit diameter and fruit fresh weight and low oxidative stress	2°C low temperature stress	Acuña-Rodríguez et al., 2022
5.	*Penicillium glabrum*	*Helianthus annuus* L. (sunflower) and *Glycine max* L. (soybean)	Enhanced total biomass, plant height and chlorophyll contents, CAT, POD, AAO, GR, SOD. Significantly curtailed ROS and lipids peroxidation, ABA and proline concentrations	40°C high temperature stress	Hamayun et al., 2021a
6.	*Epichloë* sp.	*Festuca sinensis* (tall fescue)	Enhanced root metabolic activity	–7 to 14°C low temperature stress irrigation with Ca $(NO_3)_2$	Zhou et al., 2021
7.	*Paecilomyces formosus* and *Penicillium funiculosum*	*Glycine max* L. (soybean)	Increase in photosynthetic activity, glutathione, catalase and SOD. Reduced lipid peroxidation, metal accumulation and translocation by downregulating heavy metal ATPase gene (*GmHMA13*, *GmHMA14* and *GmHMA18*) expression	1 mM of each Ni, Cd, and Al, 10% PEG-8000 at 45°C	Bilal et al., 2020

(continued)

TABLE 16.2 (Continued)
Effect of Endophytic Fungi on Different Plants Under Extreme Temperature Stress

Sr. no.	Endophyte	Plant	Response	Stress	References
8.	*Aspergillus niger*	*Helianthus annuus* L. (sunflower) and *Glycine max* L. (soybean)	Reduced ABA content, enhanced ascorbic acid oxidase (AAO), CAT, proline, phenolics, flavonoids, sugars, proteins, lipids, chlorophyll content, shoot and root lengths, fresh and dry biomass	40°C high temperature stress	Ismail et al., 2020a
9.	*Aspergillus violaceofuscus*	*Helianthus annuus* L. (sunflower) and *Glycine max* L. (soybean)	Reduced levels of ROS species, ABA, proline and an overall improved nutritional value	40°C high temperature stress	Ismail et al., 2020b
10.	*Piriformospora indica*	*Arabidopsis thaliana* (thale cress)	Reduced H_2O_2 and MDA levels. Enhanced brassinolide (BR), ABA levels, and the transcript levels of freezing-stress related genes (*CBFs*, *CORs*, *BZR1*, *SAG1* and *PYL6*)	−6°C low temperature stress	Jiang et al., 2020
11.	*Aspergillus flavus*	*Helianthus annuus* L. (sunflower) and *Glycine max* L. (soybean)	Reduced ABA level, proline, and enhanced phenols, flavonoids, CAT and AAO	40°C high temperature stress	Hamayun et al., 2019
12.	*Aspergillus japonicus*	*Helianthus annuus* L. (sunflower) and *Glycine max* L. (soybean)	Enhanced biomass, CAT, AAO, phenolic, flavonoids, soluble sugars, proteins and lipids	40°C high temperature stress	Hamayun et al., 2018

16.2.3 Drought Tolerance

The event of prolonged water shortages is known as drought, and depending upon the underlying causes, it can sometimes last from a few days to years. Plants are gravely affected by drought conditions since they need water to intake nutrients from the soil and transport them throughout the plant (Byers et al., 2020; Del, 2021). A noteworthy improvement in plants inoculated with endophytic fungi under drought-stress conditions has been seen. For instance, Llorens et al. (2019) isolated endophytic fungi from *Aegilops sharonensis*. The isolates #13237 and #14005, identified as *Acremonium sclerotigenum* and *Sarocladium implicatum*, respectively, were used as an inoculum on *Triticum aestivum*, under water-devoid conditions. As a rich source of fiber, wheat is one of the most widely consumed staple crops. *A. sclerotigenum* and *S. implicatum* inoculation enhanced the plant height, plant FW, leaf width, leaf area, leaf DW and root.

Stunted plant height, and a decrease in leaf size and leaf number are some effects seen in plants under drought conditions. The *A. sclerotigenum* and *S. implicatum* inoculated plants also increased RWC and membrane stability index. The results indicate better water retention in inoculated plants and less damage to the plasma membrane of the cells. Under stress conditions the plants exhibited a decrease in MDA, proline, ABA, JA, caffeic and ferulic content. As discussed earlier, MDA is a naturally occurring lipid peroxidation product. Plants having lower levels of MDA content are more tolerant to drought conditions. Furthermore, reduced accumulation of proline, an osmolyte, and ABA, JA and phytohormones indicate the effective adaptive response of inoculated plants to stress conditions.

In another study, He et al. (2019) evaluated three DSEs' effects on *Glycyrrhiza uralensis* growth. Of the three isolates, inoculation of *Acrocalymma vagum* increased shoot biomass, root biomass, total biomass and root-to-shoot ratio. Other physical attributes such as plant height, leaf number, total root length, root surface area, root diameter and root branch also increased. The enhanced root morphology of inoculated plants enables them to tolerate drought stress conditions better. Further, the MDA and proline contents were reduced significantly. The antioxidant enzyme activities of SOD and CAT increased. Likewise, the POD and glycyrrhizin contents also increased on inoculation of *A. vagum* under stress conditions.

Likewise, Qiang et al. (2019) isolated a phytohormone-producing endophytic fungi from *Elymus dahuricus*. The isolate LQ1230, identified as *Alternaria alternata*, exhibited 40.12 ± 8.59 μg/mL IAA production, which increased to 319.24 ± 14.88 μg/mL on supplementation of tryptophan in the culture medium. The endophyte was then tested on *Triticum aestivum* L., cultivar zhongmai 175, under mild (55% water-holding capacity) and severe (45% water-holding capacity) stress conditions. The root DW, total DW, photosynthesis and chlorophyll concentration increased under both mild and severe stress conditions. The increased chlorophyll concentration and subsequent photosynthesis indicate the efficient uptake of the available nutrients by the plants.

Developing a well-built root system is indispensable for plants to combat abiotic stress conditions. In one such study, the effect of *Piriformospora indica* inoculation on *Zea mays* seedlings under PEG-6000-induced water stress was observed (Hosseini et al., 2019). Microscopic examination revealed successful colonization of the endophytic fungi in the plant's roots. The inoculated plants exhibited an increase in the maximum axial force (F_{max}), and root growth pressure (σ_{max}). F_{max} and σ_{max} denote the highest root growth pressure at which the elongation rate becomes. The endophyte inoculation also enhanced the time to reach the plant's F_{max} (T_{max}) value. In contrast, no significant change was observed under severe drought stress conditions. This could be attributed to the temporary stress conditions created because of the inoculation of endophyte, which increased the T_{max} value but enhanced the F_{max} and σ_{max}.

To evaluate the efficacy and impact of endophytes over subsequent plant generations, Kumari and Vujanovic (2020) inoculated *Penicillium* (SMCD 2206) and *Paraconiothyrium* sp. (SMCD 2210) in *Pisum sativum* L. plants, irrigated under normal and drought stress conditions. The physiological properties and gene expression of the seeds produced (F2-generation seeds) were evaluated. The germination rate and shoot length of *Penicillium* sp. inoculated F2-generation seedlings increased by ~ 30.61 and 9.37%, respectively, under 5% PEG-8000. Moreover, reduced ROS production was observed. The F2-generation seeds inoculated with *Penicillium* sp. exhibited a 10-fold downregulation, indicating the role of symbiotic association in allocating stress tolerance. Similarly, under drought stress conditions, the SOD and manganese superoxide dismutase (Mn-SOD) gene expression was downregulated. Considering these findings, the endophytic fungi transmit the stress tolerance ability to the second generation and thus represent a plausible solution to mold plants against the changing environment.

In a similar study, the regulatory effect of endophytic *Penicillium chrysogenum*, *Penicillium brevicompactum*, *Eupenicillium osmophilum*, *Alternaria* sp. and *Phaeosphaeria* sp. on *Colobanthus quitensis* was evaluated (Hereme et al., 2020). The inoculated plants exhibited enhanced sugar, proline, rate of photosynthesis, stomatal opening and stomatal conductance. Reduced lipid peroxidation was also observed in inoculated plants. Upregulation of gene expression of *CqNCED1*, *CqABCG25* and *CqRD22* was also observed.

Considering the global water shortage coupled with drought stress conditions, cultivating water-intensive crops such as rice has become problematic in various regions across the globe. Tsai et al. (2020) evaluated the effect of *Piriformospora indica* on *Oryza sativa* L., under 29% PEG-6000-induced drought stress conditions. The inoculated seeds exhibited a survival rate of 82.5%, whereas only 18.4% of the uninoculated seedlings survived. Likewise, the shoot and root weight increased in inoculated plants. The indicator of maximum quantum efficiency, the *Fv*/*F*m ratio, increased.

Colonization of endophytes also enhanced the number of closed stomata under stress conditions. Elevated levels of redox ratio of reduced to oxidized glutathione (GSH/GSSG) were observed in inoculated plants under stress conditions. It indicates the plant's enhanced antioxidant activity, which was also corroborated by the increased activity of CAT and GR.

In another study, endophytic fungi were isolated from *Citrus reticulata* (Sadeghi et al., 2020). The isolates identified as *Penicillium citrinum*, *Aureobasidium pullulans* and *Dothideomycetes* sp. were used as inoculum on a citrus hybrid formed by crossing *Citrus reticulata* × *Citrus limetta*. Under drought stress conditions, inoculation resulted in increases in the shoot DW, root DW, chlorophyll and carotenoid contents. Furthermore, the maximal photochemical efficiency Fv/Fm increased in inoculated plants under stress conditions. On inoculating the three endophytic fungi together, the most effective decrease in MDA content was observed in plants. Similarly, combined inoculation decreased the O_2- and H_2O_2 contents in plants. Combined inoculation exhibited effective results on the antioxidant enzyme systems of plants under stress conditions.

Considering the complexity of drought tolerance, a single gene trait cannot exhume the complete genetic mechanism. Azizi et al. (2021) evaluated various genes and the effect of *Piriformospora indica* inoculation on two cultivars of *Solanum lycopersicum* L. Superluna and Caspian cultivars were tested under normal and drought stress conditions of 50% field capacity. The RWC increased by ~ 3.75 and 11.1% in inoculated Superluna and Caspian cultivars, respectively. Similarly, increases in FW, DW and proline contents were also observed in inoculated plants. An increase in expression levels of late embryogenesis abundant protein, *LEA14* gene, was observed under stress conditions. Likewise, the expression of the *TAS14* gene, a hydrophilic dehydrin of the LEA family, also increased under drought-stress conditions. A much more significant increase in *P. indica* inoculated plants was seen. The expression of the gibberellic acid insensitive *GAI* gene, was also enhanced in inoculated plants under drought conditions.

Zhou et al. (2021) isolated endophytes from *Pinus tabulaeformis*, Chinese red pine. To screen the drought-tolerant endophyte, the plant seedlings were dipped in fermentation broth supernatant (FBS) of the endophytic fungi. Of the tested isolates, plant seedlings treated with FBS of PTD37, PTD56 and PTD78 exhibited promising results on applying 25% PEG-6000 solution. Green and sturdy seedlings were observed, indicating the ability of the isolates to withstand drought conditions. After nine days of PEG-6000-induced stress conditions, the PTD37 FBS-treated seedlings exhibited increases in proline, chlorophyll and leaf water content. Similarly, the antioxidant enzyme activities of POD, CAT and SOD increased. A decrease in MDA concentration was also observed, whereas no significant effect on ABA content was observed.

To evaluate the effect of endophytic fungi on the morphological structure of leaves and the photosynthetic capacity, Liu and Wei (2021) isolated *Acrocalymma vagum*, a DSE. Its effect on *Ormosia hosiei*, under drought stress conditions, was evaluated. The plants were grown under well-watered (75% field capacity), moderately watered (55% field capacity) and low-water (35% field capacity) conditions. In *A. vagum* inoculated plants, the chlorophyll a/b content and carotenoid content increased under all conditions. An increase in Pn value, Gs value and E values was observed under all conditions. Furthermore, Ci values decreased under all conditions. The physical attributes of leaves, such as the leaf length, leaf width, leaf area and leaf DW, were also enhanced under the three conditions. Similar studies reporting the effects of endophytic fungi under drought stress conditions are listed in Table 16.3.

16.2.4 Salinity Tolerance

Salinity stress, which affects agricultural production globally, has increased over time. An estimated 6% of the global land and 19.5% of global arable land are affected by salinity stress conditions. High salinity concentration of the soil creates an osmotic disbalance, thereby reducing plant uptake of minerals and nutrients. Furthermore, the excessive amount of salt injures plant cells, eventually

TABLE 16.3
Effect of Endophytic Fungi on Different Plants Under Drought Stress

Sr. no.	Endophyte	Plant	Response	Stress	References
1.	*Acrocalymma vagum*, *Paraboeremia putaminum* and *Trichoderma viride*	*Astragalus mongholicus* (Mongolian astragalus)	Enhanced plant attributes and growth observed on dual inoculation	70% field capacity (well-watered conditions) 30% field capacity (drought stress conditions)	He et al., 2022
2.	*Piriformospora indica*	*Pimpinella anisum* (anise)	Inoculated plants exhibited increase in DW, seed yield and essential oil yield as compared to uninoculated plants	40–90% field capacity	Abdollahi et al., 2022
3.	*Penicillium chrysogenum* and *P. brevicompactum*	*Colobanthus quitensis* (Antarctic pearlwort)	The expression of expansin multiproteins was modulated in inoculated plants in response to drought stress conditions	Irrigated using 7 mL water (well-watered) and 4 mL water (drought stress conditions)	Morales-Quintana et al., 2021
4.	*Paraconiothyrium estuarinum* and *P. cyclothyrioides*	*Brachiaria brizantha* (palisade grass) and *Megathyrsus maximus* (Guinea grass)	Enhanced shoot length, shoot DW, root DW were observed	Water devoid conditions for 7 and 14 days	Azevedo et al., 2021
5.	*Epicoccum thailandicum*, *Filobasidium magnum*, *Paraphoma rhaphiolepidis* and *Pseudomonas nitroreducens*	*Codonopsis pilosula* (Dangshen)	Inoculated plants exhibited increase in lobetyolin, syringin and atractylolide III content	18–20% moisture content (control) and 10–12% moisture content (drought stress conditions)	Liang et al., 2021
6.	*Talaromyces omanensis*	*Solanum lycopersicum* L. (tomato)	Enhanced reproductive and anatomical characteristics, chlorophyll fluorescence, number of flowers, weight of fruit, shoot DW and root length	15% PEG-6000	Halo et al., 2020

(*continued*)

TABLE 16.3 (Continued)
Effect of Endophytic Fungi on Different Plants Under Drought Stress

Sr. no.	Endophyte	Plant	Response	Stress	References
7.	*Beauveria bassiana*	*Zea mays* L. (maize)	Endophyte inoculated plants exhibited early flowering and drought tolerance as compared to uninoculated plants	Water devoid conditions for 10 days	Kuzhuppillymyal et al., 2020
8.	*Gigaspora margarita*	*Arachis hypogaea* (peanut)	Endophytic fungi synergistically interacted with the arbuscular mycorrhiza community and promoted soil environment and plant's tolerance to drought stress conditions	30–45% water content (well-watered) and 15% water content (drought stress conditions)	Xu et al., 2020
9.	Multiple	*Oryza sativa* L. (rice)	Enhanced shoot length, root and shoot FW, antioxidant, proline and soluble sugar content	0–25% PEG-6000	Pang et al., 2020
10.	*Glycomyces* sp. and *Fusarium* sp.	*Hordeum vulgare* L. (barley)	Plants grown in presence of soil derived endophytes exhibited better tolerance to drought conditions as compared to noninoculated plants	Water devoid conditions for 11 days	Yang et al., 2020
11.	Unidentified	*Oryza sativa* L. (rice)	Enhanced plant height, tiller number, root length, root weight, number of filled grains and production per observed in inoculated plants	1–5 cm of water levels	Syamsia et al., 2020

TABLE 16.3 (Continued)
Effect of Endophytic Fungi on Different Plants Under Drought Stress

Sr. no.	Endophyte	Plant	Response	Stress	References
12.	*Piriformospora indica*	*Hordeum vulgare* L. (barley)	Enhanced activity of PSII, electron transfer chain, accumulation of proteins observed in inoculated plants	25, 50 and 100% field capacity	Ghaffari et al., 2019
13.	*Beauveria bassiana*	*Quercus rubra* L. (Northern red oak)	Enhanced RWC, stomatal conductance and reduced proline accumulation seen	Water devoid conditions for 10–12 days	Ferus et al., 2019
14.	*Paraphoma* sp., *Embellisia chlamydospora*, and *Cladosporium oxysporum*	*Hedysarum scoparium* (sweetvetch), *Glycyrrhiza uralensis* (Chinese liquorice) and *Zea mays* (maize)	Enhanced root length and root biomass	0 to –3.58 MPa induced using PEG-6000	Li et al., 2019b
15.	*Aspergillus niger*	*Helianthus annuus* (sunflower)	Enhanced soil nitrate ions, P and K contents	Drought stress induced using 8% PEG	Seema et al., 2019
16.	*Aspergillus fumigatus*	*Oryza sativa* L. (rice)	Novel antioxidant (Z)-N-(4-hydroxystyryl) formamide from endophytic fungi imparted drought tolerance to the plant	Drought stress	Qin et al., 2019
17.	*Acrocalymma vagum*	*Ormosia hosiei*	Enhanced root FW, root volume, surface area, root tip number in the inoculated plants	35, 55 and 75% field capacity	Liu & Wei, 2019

causing a reduction in plant growth (Parihar et al., 2015; Abdelraheem et al., 2019). The utilization of endophytic fungi to enumerate the salinity conditions in various plants has been reported. For instance, Ben Rhouma et al. (2020) evaluated the effect of isolate B28 under salinity stress. The endophytic fungi B28 was isolated from *Coriandrum sativum*. Identified as *Fusarium oxysporum*, the isolate was optimized to produce a high GA content. The crude GA extract obtained from the endophyte was evaluated for its growth promotion ability in *Solanum lycopersicum*, under salinity stress conditions. The GA-inoculated plants exhibited an increase in plant height, leaf number per plant, fruit number per plant, fresh material and dry material. Furthermore, decreased proline and soluble sugar contents were observed. Accumulation of proline and sugars is an adaptive strategy employed by the plant in response to stress conditions. The decreased content on exogenous application of endophytic GA3 indicates the alleviation of stress conditions.

Another study evaluated the effect of two endophytic fungi, *Penicillium brevicompactum* and *P. chrysogenum*, under salt stress conditions. Isolated from roots of Antarctic plants *Colobanthus quitensis* and *Deschampsia antarctica*, the endophytic fungi were tested on *Lactuca sativa* and *Solanum lycopersicum*. Under stress conditions, the inoculated lettuce and tomato plants increased the final yield. Likewise, the net photosynthetic rate increased by 43% in lettuce and 37% in tomatoes under salt stress conditions. The increased net photosynthetic rate compensates for the enhanced energy demand during stress conditions. The presence of endophytic fungi also enumerated the detrimental effects on water use efficiency. Increased water use efficiency was seen under stress conditions. In contrast, a significant increase in Na^+ accumulation was observed in endophyte-inoculated plants. Though elevated Na^+ concentrations are detrimental for plants, they can also be correlated with enhanced physiological parameters. The enhanced photosynthetic rates contribute to the energy demand to maintain the proton gradient for efficient Na^+/H^+ transport. Thus, endophytic fungi inoculation manages a salt stress scenario by orchestrating a range of physiological factors in concerted action (Molina-Montenegro et al., 2020).

Another group of ascomycetous endophytes known as the DSE were isolated from *Vochysia divergens*. The isolates identified as *Melanconiella elegans* and *Sordariomycetes* sp. were tested on *Vigna unguiculata*. On inducing salt stress conditions, reductions in leaf area, shoot dry mass and root dry mass were observed in uninoculated plants. In *Sordariomycetes* sp. inoculated plants, the leaf area and root dry mass increased by ~ 63% and 40%. However, no significant increase in shoot dry mass was observed. Similarly, the net photosynthetic rate in *M. elegans* inoculated plants was enhanced by ~ 55%. Enhanced nitrogen and phosphorus contents were also observed, indicating better nutritional status of the inoculated plants, as a result of which they possessed a higher capacity for carbon assimilation. The DSEs facilitated the absorption of nitrogen and phosphorus, which helped mitigate the effects of salinity stress (Farias et al., 2020).

Research has also reported that the increased resistance of endophytic fungi to various stress conditions could be attributed to habitat-adapted symbiosis. To corroborate this symbiotically conferred habitat-specific resistance, Morsy et al. (2020) isolated endophytic fungi from different plant samples growing in the harsh habitats of Alabama. *Penicillium chrysogenum* was isolated from a unique salt habitat, and on inoculating *Solanum lycopersicum* L. with *P. chrysogenum* under 300 mM salt stress conditions, enhanced plant health attributes were observed. The inoculated plants exhibited an increase in shoot DW and root DW. Similarly, the average fruit FW increased by ~ 71%, whereas the total fruit weight was enhanced by ~ 160%.

Endophytic fungi were also isolated from four plants growing in saline areas in another study. Two isolates each were isolated from *Avena fatua, Anabasis prostrate and Lolium rigidum* (annual ryegrass), whereas only one endophytic fungus was isolated from *Salsola oppositifolia* (pink saltwort). Under moderate salinity stress conditions, the plants inoculated with *C. coarctatum* and *A. chlamydospore* exhibited a 6.7 and 16.7% increase in the emergence rate. However, under high salinity stress conditions, only the plants inoculated with *A. chlamydospore* exhibited an emergence rate of 16.6%, whereas it was totally inhibited in other treatments. *F. graminearum* inoculated plants exhibited the highest root length, root biomass, shoot biomass and leaf length (Bouzouina et al., 2021).

Similarly, Gonçalves et al. (2021) isolated endophytic fungi from *Salicornia*. The plant can tolerate up to 1 M NaCl stress in the soil by adopting different mechanisms such as salt exclusion, accumulation and excretion. The isolated endophytes DRG94, DRG96 and DRG97, identified as *Phoma* sp., *Alternaria conjuncta* and *Stemphylium* sp., were then inoculated in the host plant to test their plant growth promotion potential under salinity stress conditions. The plants inoculated with *Stemphylium* sp. exhibited the highest aboveground and belowground biomass under 150 mM NaCl stress. Similarly, under stress conditions, the highest root and shoot nitrogen concentration was also observed in plants inoculated with *Stemphylium* sp. Furthermore, the plants inoculated with *Phoma* sp. exhibited the highest chlorophyll fluorescence *Fv*/*Fm* under stress conditions.

Likewise, Ali et al. (2021) evaluated the effect of the endophytic fungi *Aspergillus awamori* on *Vigna radiata*, under salinity stress conditions. Under 150 mM NaCl stress conditions, the inoculated plants exhibited an increase in shoot length, root length, shoot DW, root DW and leaf DW. Similarly, the total chlorophyll, carbohydrate, protein and lipid contents also increased. On the contrary, there was a reduction in total phenolic, flavonoid, proline content and lipid peroxidation. The inoculated plants positively regulated the activity of the antioxidant enzyme system. Reductions in CAT and AAO activities were observed. Furthermore, the application of *A. awamori* enhanced the endogenous IAA content under stress conditions.

A plethora of studies have reported the effects of endophytic fungi in stress mitigation in in vitro conditions. However, affirming these microorganisms' effect "on field trials" is necessary. In a study, Saravi et al. (2021) conducted a field experiment, accessing the growth promotion ability of *Piriformospora indica* and *Trichoderma virens* under different levels of salinity stress. In inoculated plants, a 6.9% reduction in Fo levels was observed. The onset of abiotic stress conditions alters the structure of reaction centers of photosystem II (PSII). This alteration further causes a rise in Fo levels, affecting the optimum fluorescence levels for the PSII system's functioning. Hence, the reduced Fo values indicate the adequate photosynthetic levels maintained in the plant. Furthermore, enhanced *F*m and *F*v values were also observed. *F*v/*F*m indicates the plant's photosynthetic performance; thus, lower values imply stress conditions or photoinhibition.

In addition, evaluating endophytic fungi over generations is another dynamic of paramount importance. To study the effect, Pan et al. (2021) tested 7-year-old endophytic fungi infected *Festuca arundinacea*, under salinity stress conditions. Under stress conditions, the inoculated plants exhibited higher growth than uninoculated plants. The leaf FW, leaf DW and leaf RWC exhibited an increase. Furthermore, better chlorophyll fluorescence was detected. At the same time, lower MDA content, indicating less lipid peroxidation, was seen in inoculated plants. The improved tolerance could be attributed to the relatively stable antioxidant enzyme activity, which helps mitigate ROS accumulation.

A recent study evaluated the plant growth-promoting effect of *Porostereum spadiceum* on *Triticum aestivum*. The endophytic fungi also exhibited plant growth-promoting attributes, such as siderophore production, IAA, phenols and flavonoids. The inoculated plants exhibited higher shoot length, root length, shoot FW, shoot DW, root FW and root DW. The inoculated plants also exhibited reduced ABA activity. On the contrary the salicylic acid activity increased. The endophyte inoculation reduced Na^+ accumulation. On the contrary, the K^+ concentration increased in inoculated plants under stress conditions (Gul et al., 2022).

Likewise, Xu et al. (2022) evaluated the effect of endophytic *Piriformospora indica* and *Azorhizobium caulinodans* on the "sunrise" cultivar of *Solanum lycopersicum*. The inoculated plants exhibited increased shoot length under salinity stress conditions. Under stress conditions, chlorophyll content increased on inoculation of *A. caulinodans*, *P. indica* and dual inoculation. Likewise, an increase in *Fv*/*Fm* by 63.47, 65.41 and 55.45% was seen on inoculation of *A. caulinodans*, *P. indica* and dual inoculation, respectively. Dual inoculation enhanced the SOD activity under stress conditions, whereas no significant difference was seen in the inoculation of *A. caulinodans* and *P. indica*. Similarly, the POD activity increased on inoculation of *P. indica* and dual inoculation, whereas no significant change was seen on inoculation of *A. caulinodans*. However, in the case of CAT, no significant change was observed in inoculated and uninoculated plants. On assessing the fruit quality under stress conditions, the highest activity was observed in *P. indica* inoculated plants based on total sugar and proanthocyanin content. Similar studies reporting the effect of endophytic fungi on salinity stress conditions are listed in Table 16.4.

TABLE 16.4
Effect of Endophytic Fungi on Different Plants Under Salinity Stress

Sr. no.	Endophyte	Plant	Response	Stress	References
1.	*Bipolaris* sp.	*Glycine max* L. (soybean)	Enhanced root and shoot length, root and shoot FW, DW, chlorophyll content, SA, POD. Decrease in ABA, expression level of *GmFDL19*, *GmNARK* and *GmSIN1*	200 mM NaCl stress	Lubna et al., 2022
2.	Unidentified	*Zea mays* (maize)	Enhanced growth parameters, IAA production, antioxidant enzymes and reduced lipid-peroxidation, Na, Cl accumulation	150 mM NaCl stress	Ali et al., 2022
3.	*Piriformospora indica*, *Tricuderma virens*	*Stevia rebaudiana* (candy leaf)	Enhanced antioxidant enzymes, chlorophyll content, Fv/Fm, shoot DW and reduced lipid peroxidation and H_2O_2 content	0–12 dS/m salinity stress	Saravi et al., 2022
4.	*Piriformospora indica*	*Cicer arietinum* L. (chickpea)	Enhanced quantum efficiency, chlorophyll content, RWC, plant height, shoot FW and root volume	6 and 9 dS/m salinity stress	Oskoueian et al., 2022
5.	*Aspergillus terreus*	*Triticum aestivum* (wheat)	Enhanced IAA, GA production, longer roots, shoots, chlorophyll content, GR and reduced activity of CAT and POD	0–180 mM NaCl stress	Khan et al., 2022
6.	*Aspergillus ochraceus*	*Hordeum vulgare* L. (barley)	The endophyte exhibited IAA, antioxidant and antifungal activity and mitigated harmful effects of seawater	0–30% (v/v) seawater	Badawy et al., 2021
7.	*Serendipita indica*	*Hordeum vulgare* L. (barley)	Enhanced shoot DW and reduced accumulation of NaCl	300 mM/L NaCl stress	Sepehri et al., 2021
8.	*Neocamarosporium chichastianum*, *N. goegapense* and *Periconia macrospinosa*	*Cucumis sativus* (cucumber) and *Solanum lycopersicum* (tomato)	Enhanced chlorophyll, proline content and antioxidant activity	2.5–7.5 dS/m salinity stress	Moghaddam et al., 2021
9.	*Epichloë bromicola*	*Hordeum brevisubulatum* (wild barley)	Regulation of some anatomical structures, enhanced leaf xylem, phloem, stem vascular bundles, stem xylem, root meta-xylem, leaf vein thickness, epidermis, cortex and endodermis	0–600 mM NaCl stress	Chen et al., 2021a
10.	*Piriformospora indica*	*Mentha* × *piperita* (peppermint)	Enhanced chlorophyll content, *Fv*/*Fm* ratio, essential oil composition and antioxidant enzymes	0–9 dS/m salinity stress	Khalvandi et al., 2021

TABLE 16.4 (Continued)
Effect of Endophytic Fungi on Different Plants Under Salinity Stress

Sr. no.	Endophyte	Plant	Response	Stress	References
11.	*Epichloë* sp.	*Hordeum brevisubulatum* (wild barley)	Enhanced total amino acid production, DW, potassium, root calcium, nitrogen and phosphorus content and reduced NaCl accumulation	200 mM NaCl stress	Chen et al., 2021b
12.	*Alternaria chlamydosporigena*, *Paraphoma chrysanthemicola* and *Bipolaris sorokiniana*	*Artemisia ordosica* (mugwort)	Enhanced root biomass, IAA, SOD, GSH content	0–3 g NaCl/kg soil	Hou et al., 2021
13.	*Piriformospora indica*	*Stevia rebaudiana* (candy leaf)	Enhanced antioxidant enzymes, phenolic, flavonoid, glucose and proline content	0.85–3 dS/m salinity stress	Forouzi et al., 2020
14.	*Fusarium* sp.	*Oryza sativa* (rice)	Enhanced chlorophyll content, upregulation of genes involved in abiotic stress tolerance, signal perception proteins, transcription factors	150 mM NaCl stress	Sampangi et al., 2020
15.	*Piriformospora indica*	*Artemisia annua* (sweet sagewort)	Enhanced plant height, root and shoot DW, antioxidant enzymes, phenols, flavonoids and reduced MDA content	0–200 mM NaCl stress	Arora et al., 2020

16.3 CONCLUSION AND FUTURE PROSPECTIVE

Considering the global scenario, some of the harsh conditions that plants must endure include a lack of water resources and an increase in soil salinization. A sustainable solution is required to overcome the adverse effects of these abiotic stresses on plant health and productivity. The interaction between endophytic fungi and host plants is a dynamic process that several external factors affect. Based on the vast literature, endophytic fungi are well known for plant growth promotion under various climatic conditions. This endophyte-mediated stress tolerance in plants is an environment-friendly approach that may help sustain crop production. However, various aspects have to be considered before the commercialization of this technology. Future studies should focus on the host specificity of the endophytic fungi so that a single endophyte can be utilized over a wide array of plants. Given the diversified global arable land, the selected endophytic fungi should also be assessed under varying environmental factors. Due to the metabolic complexity and diversity, endophytic fungi possess many undefined genes. However, cutting-edge omics technology must be employed to gain deeper insights into the mechanism and behavior of endophytic fungi. These technologies could help infer the complicated endosymbiosis process by simultaneously investigating the responses at tissue and cellular levels. It is imperative to include metagenomics, metaproteomics, metatranscriptomics and metabolomics data to achieve this. In addition to the molecular and genetic understanding, different delivery methods should be evaluated to enhance the efficacy of the selected endophytic fungi. Moreover, long-term field trials over generations should be carried out to assess the sustainability and effectiveness of the inoculum. Ultimately, establishing a proper network of the end user, i.e., farmers, with the government and other regulatory bodies, would allow the dispensing of the knowledge and standardization of the usage protocols. This inclusive knowledge and additional research could help in understanding and pave the way for the development of suitable inoculum or bio-consortia, which could sustainably improve the health and yield of plants.

REFERENCES

Abdelraheem A, Esmaeili N, O'Connell M, Zhang J. (2019). Progress and perspective on drought and salt stress tolerance in cotton. *Ind Crops Prod* 130:118–129. https://doi.org/10.1016/j.indcrop.2018.12.070.

Abdollahi S, Aliasgharzad N, Zehtabe Salmasi S, Khoshru B. (2022). Effects of endophytic fungus *Piriformospora indica* on growth and essential oil content of *Pimpinella anisum* plant under water deficit conditions. *Water Soil Sci* https://doi.org/10.22034/WS.2021.36333.2298.

Acuña-Rodríguez IS, Ballesteros GI, Atala C, Gundel PE, Molina-Montenegro MA. (2022). Hardening Blueberry plants to face drought and cold events by the application of fungal endophytes. *Agronomy* 12(5):1000. https://doi.org/10.3390/agronomy12051000.

Adhikari P, Pandey A. (2019). Phosphate solubilization potential of endophytic fungi isolated from *Taxus wallichiana* Zucc roots. *Rhizosphere* 9:2–9. https://doi.org/10.1016/j.rhisph.2018.11.002.

Ali AH, Abdelrahman M, Radwan U, El-Zayat S, El-Sayed MA. (2018). Effect of Thermomyces fungal endophyte isolated from extreme hot desert-adapted plant on heat stress tolerance of cucumber. *Appl Soil Ecol* 124:155–162. https://doi.org/10.1016/j.apsoil.2017.11.004.

Ali AH, Radwan U, El-Zayat S, El-Sayed MA. (2019). The role of the endophytic fungus, *Thermomyces lanuginosus*, on mitigation of heat stress to its host desert plant *Cullen plicata*. *Biologia Futura* 70(1):1–7. https://doi.org/10.1556/019.70.2019.01.

Ali R, Gul H, Hamayun M, Rauf M, Iqbal A, Hussain A, Lee IJ. (2022). Endophytic fungi controls the physicochemical status of Maize crop under salt stress. *Pol J Environ Stud* 31(1):561–573. https://doi.org/10.15244/pjoes/134540.

Ali R, Gul H, Hamayun M, Rauf M, Iqbal A, Shah M, Lee IJ. (2021). Aspergillus awamori ameliorates the physicochemical characteristics and mineral profile of mung bean under salt stress. *Chem Biol Technol Agric* 8(1):1–13. https://doi.org/10.1186/s40538-021-00208-9.

Arora M, Saxena P, Abdin MZ, Varma A. (2020). Interaction between *Piriformospora indica* and *Azotobacter chroococcum* diminish the effect of salt stress in *Artemisia annua* L. by enhancing enzymatic and non-enzymatic antioxidants. *Symbiosis* 80(1):61–73. https://doi.org/10.1007/s13199-019-00656-w.

Arora NK. (2019). Impact of climate change on agriculture production and its sustainable solutions. *J Environ Sustain* 2(2):95–96. https://doi.org/10.1007/s42398-019-00078-w.

Azevedo RP, Alves NM, Costa IA, Domingues MIS, Bandória NDA, de Figueiredo UJ, Cardoso PG. (2021). Endophytic fungi assures tropical forage grass growth by water stress tolerances. *Curr Microbiol* 78(12):4060–4071. https://doi.org/10.1007/s00284-021-02672-w.

Aziz L, Hamayun M, Rauf M, Iqbal A, Arif M, Husssin A, Khan SA. (2021a). Endophytic *Aspergillus niger* reprograms the physicochemical traits of tomato under cadmium and chromium stress. *Environ Exp Bot* 186:104456. https://doi.org/10.1016/j.envexpbot.2021.104456.

Aziz L, Hamayun M, Rauf M, Iqbal A, Husssin A, Khan SA, Lee IJ. (2021b). *Aspergillus Flavus* reprogrammed morphological and chemical attributes of *Solanum lycopersicum* through *SlGSH1* and *SlPCS1* genes modulation under heavy metal stress. *J Plant Interact* 16(1):104–115. https://doi.org/10.1080/17429145.2021.1903105.

Azizi M, Fard EM, Ghabooli M. (2021). *Piriformospora indica* affect drought tolerance by regulation of genes expression and some morphophysiological parameters in tomato (*Solanum lycopersicum* L.). *Sci Hortic* 287:110260. https://doi.org/10.1016/j.scienta.2021.110260.

Badawy AA, Alotaibi MO, Abdelaziz AM, Osman MS, Khalil A, Saleh AM, Hashem AH. (2021). Enhancement of seawater stress tolerance in Barley by the endophytic fungus *Aspergillus ochraceus*. *Metabolites* 11(7):428. https://doi.org/10.3390/metabo11070428.

Bang C, Dagan T, Deines P, Dubilier N, Duschl WJ, Fraune S, Bosch TC. (2018). Metaorganisms in extreme environments: do microbes play a role in organismal adaptation? *Zoology* 127:1–19. https://doi.org/10.1016/j.zool.2018.02.004.

Baron NC, Rigobelo EC. (2022). Endophytic fungi: a tool for plant growth promotion and sustainable agriculture. *Mycology* 13(1):39–55. https://doi.org/10.1080/21501203.2021.1945699.

Ben Rhouma M, Kriaa M, Ben Nasr Y, Mellouli L, Kammoun R. (2020). A new endophytic *Fusarium oxysporum* gibberellic acid: optimization of production using combined strategies of experimental designs and potency on tomato growth under stress condition. *BioMed Res Int* 2020: 4587148. https://doi.org/10.1155/2020/4587148.

Bilal L, Asaf S, Hamayun M, Gul H, Iqbal A, Ullah I, Hussain A. (2018). Plant growth promoting endophytic fungi *Asprgillus fumigatus* TS1 and *Fusarium proliferatum* BRL1 produce gibberellins and regulates plant endogenous hormones. *Symbiosis* 76(2):117–127. https://doi.org/10.1007/s13199-018-0545-4.

Bilal S, Shahzad R, Imran M, Jan R, Kim KM, Lee IJ. (2020). Synergistic association of endophytic fungi enhances *Glycine max* L. resilience to combined abiotic stresses: heavy metals, high temperature and drought stress. *Ind Crops Prod* 143:111931. https://doi.org/10.1016/j.indcrop.2019.111931.

Bilal S, Shahzad R, Lee IJ. (2021). Synergistic interaction of fungal endophytes, *Paecilomyces formosus* LHL10 and *Penicillium funiculosum* LHL06, in alleviating multi-metal toxicity stress in *Glycine max* L. *Environ Sci Pollut Res* 28(47):67429–67444. https://doi.org/10.1007/s11356-021-15202-9.

Bita CE, Gerats T. (2013). Plant tolerance to high temperature in a changing environment: scientific fundamentals and production of heat stress-tolerant crops. *Front Plant Sci* 4:273. https://doi.org/10.3389/fpls.2013.00273.

Bodjrenou D, Cheng C, Sun X, Chew H, Liu Y, Li D, Lai Z. (2022). High temperature associated microRNAs and their potential roles in mediating heat tolerance in the leaf of banana inoculated with *Serendipita indica*. *J Hortic Sci Biotechnol* 97(2):171–186. https://doi.org/10.1080/14620316.2021.1971571.

Bouzouina M, Kouadria R, Lotmani B. (2021). Fungal endophytes alleviate salt stress in wheat in terms of growth, ion homeostasis and osmoregulation. *J Appl Microbiol* 130(3):913–925. https://doi.org/10.1111/jam.14804.

Byers EA, Coxon G, Freer J, Hall JW. (2020). Drought and climate change impacts on cooling water shortages and electricity prices in Great Britain. *Nat Commun* 11(1):1–12. https://doi.org/10.1038/s41467-020-16012-2.

Caiyi W, Yuanhua W, Xiuling T. (2004). Recent advances and issues on the endophyte. *Chin J Ecol* 49(3):86–91

Chand K, Shah S, Sharma J, Paudel MR, Pant B. (2020). Isolation, characterization, and plant growth-promoting activities of endophytic fungi from a wild orchid *Vanda cristata*. *Plant Signal Behav* 15(5):1744294. https://doi.org/10.1080/15592324.2020.1744294.

Checker VG, Kushwaha HR, Kumari P, Yadav S. (2018). Role of phytohormones in plant defense: signaling and cross talk. In: Singh A, Singh I, (Eds) *Molecular aspects of plant-pathogen interaction*. Springer Singapore. https://doi.org/10.1007/978-981-10-7371-7_7.

Chen T, White JF, Li C. (2021a). Fungal endophyte *Epichloë bromicola* infection regulates anatomical changes to account for salt stress tolerance in wild barley (*Hordeum brevisubulatum*). *Plant Soil* 461(1):533–546. https://doi.org/10.1007/s11104-021-04828-w.

Chen T, White JF, Li C, Nan Z. (2021b). Exogenous spermidine enhances *Epichloë* endophyte-induced tolerance to NaCl stress in wild barley (*Hordeum brevisubulatum*). *Plant Soil* 468(1):77–95. https://doi.org/10.1007/s11104-021-05109-2.

Del Buono D. (2021). Can biostimulants be used to mitigate the effect of anthropogenic climate change on agriculture? It is time to respond. *Sci Total Environ* 751:141763. https://doi.org/10.1016/j.scitotenv.2020.141763.

Dubey A, Malla MA, Kumar A, Dayanandan S, Khan ML. (2020). Plants endophytes: unveiling hidden agenda for bioprospecting toward sustainable agriculture. *Crit Rev Biotechnol* 40(8):1210–1231. https://doi.org/10.1080/07388551.2020.1808584.

El-Mahdy OM, Mohamed HI, Mogazy AM. (2021). Biosorption effect of *Aspergillus niger* and *Penicillium chrysosporium* for Cd-and Pb-contaminated soil and their physiological effects on *Vicia faba* L. *Environ Sci Pollut Res* 28(47):67608–67631. https://doi.org/10.1007/s11356-021-15382-4.

El-Shafey NM, Marzouk MA, Yasser MM, Shaban SA, Beemster GT, AbdElgawad H. (2021). Harnessing endophytic fungi for enhancing growth, tolerance and quality of Rose-Scented Geranium (*Pelargonium graveolens* (L'Hér) Thunb.) plants under Cadmium stress: a Biochemical Study. *J Fungi* 7(12):1039. https://doi.org/10.3390/jof7121039.

El-Shahir AA, El-Tayeh NA, Ali OM, Abdel Latef AAH, Loutfy N. (2021). The effect of endophytic *Talaromyces pinophilus* on growth, absorption and accumulation of heavy metals of *Triticum aestivum* grown on sandy soil amended by sewage sludge. *Plants* 10(12):2659. https://doi.org/10.3390/plants10122659.

Faeth SH. (2002). Are endophytic fungi defensive plant mutualists? *Oikos* 98(1):25–36. https://doi.org/10.1034/j.1600-0706.2002.980103.x.

Fageria VD. (2001). Nutrient interactions in crop plants. *J Plant Nutr* 24(8):1269–1290. https://doi.org/10.1081/PLN-100106981.

Farias GC, Nunes KG, Soares MA, de Siqueira KA, Lima WC, Neves ALR, de Lacerda CF. (2020). Dark septate endophytic fungi mitigate the effects of salt stress on cowpea plants. *Braz J Microbiol* 51(1):243–253. https://doi.org/10.1007/s42770-019-00173-4.

Ferus P, Barta M, Konôpková J. (2019). Endophytic fungus *Beauveria bassiana* can enhance drought tolerance in red oak seedlings. *Trees* 33(4):1179–1186. https://doi.org/10.1007/s00468-019-01854-1.

Forouzi A, Ghasemnezhad A, Nasrabad RG. (2020). Phytochemical response of Stevia plant to growth promoting microorganisms under salinity stress. *S Afr J Bot* 134:109–118. https://doi.org/10.1016/j.sajb.2020.04.001.

Ghaffari MR, Mirzaei M, Ghabooli M, Khatabi B, Wu Y, Zabet-Moghaddam M, Salekdeh GH. (2019). Root endophytic fungus *Piriformospora indica* improves drought stress adaptation in barley by metabolic and proteomic reprogramming. *Environ Exp Bot* 157:197–210. https://doi.org/10.1016/j.envexpbot.2018.10.002.

Gonçalves DR, Pena R, Zotz G, Albach DC. (2021). Effects of fungal inoculation on the growth of *Salicornia* (Amaranthaceae) under different salinity conditions. *Symbiosis* 84(2):195–208. https://doi.org/10.1007/s13199-021-00783-3.

Gul SL, Moon YS, Hamayun M, Khan SA, Iqbal A, Khan MA, Ali S. (2022). *Porostereum spadiceum*-AGH786 regulates the growth and metabolites production in *Triticum aestivum* L. under salt stress. *Curr Microbiol* 79(6):1–11. https://doi.org/10.1007/s00284-022-02853-1.

Halo BA, Al-Yahyai RA, Al-Sadi AM. (2020). An endophytic *Talaromyces omanensis* enhances reproductive, physiological and anatomical characteristics of drought-stressed tomato. *J Plant Physiol* 249:153163. https://doi.org/10.1016/j.jplph.2020.153163.

Hamayun M, Hussain A, Khan SA, Iqbal A, Lee IJ. (2019). *Aspergillus flavus* promoted the growth of soybean and sunflower seedlings at elevated temperature. *BioMed Res Int* 2019:1295457. https://doi.org/10.1155/2019/1295457.

Hamayun M, Hussain A, Iqbal A, Khan SA, Gul S, Khan H, Lee IJ. (2021a). *Penicillium Glabrum* acted as a heat stress relieving endophyte in soybean and sunflower. *Pol J Environ Stud* 30(4). https://doi.org/10.15244/pjoes/128579.

Hamayun M, Hussain A, Iqbal A, Khan SA, Khan MA, Lee IJ. (2021b). An endophytic fungus *Gliocladium cibotii* regulates metabolic and antioxidant system of *Glycine max* and *Helianthus annuus* under heat stress. *Pol J Environ Stud* 30(2). https://doi.org/10.15244/pjoes/125770.

Hamayun M, Hussain A, Iqbal A, Khan SA, Lee IJ. (2018). Endophytic fungus *Aspergillus japonicus* mediates host plant growth under normal and heat stress conditions. *BioMed Res Int*. https://doi.org/10.1155/2018/7696831.

Hasanuzzaman M, Bhuyan MHM, Zulfiqar F, Raza A, Mohsin SM, Mahmud JA, Fotopoulos V. (2020). Reactive oxygen species and antioxidant defense in plants under abiotic stress: revisiting the crucial role of a universal defense regulator. *Antioxidants* 9(8):681. https://doi.org/10.3390/antiox9080681.

Hassan A, Pariatamby A, Ossai IC, Ahmed A, Muda MA, Barasarathi J, Hamid FS. (2022). Synergistic association of endophytic fungi enhances tolerance, growth, and heavy metal uptake of *Alocasia calidora* in landfill contaminated soil. *Appl Soil Ecol* 170 :104307. https://doi.org/10.1016/j.apsoil.2021.104307.

Hatfield JL, Prueger JH. (2015). Temperature extremes: effect on plant growth and development. *Weather Clim Extrem* 10:4–10. https://doi.org/10.1016/j.wace.2015.08.001.

He C, Liu C, Liu H, Wang W, Hou J, Li X. (2022). Dual inoculation of dark septate endophytes and *Trichoderma viride* drives plant performance and rhizosphere microbiome adaptations of *Astragalus mongholicus* to drought. *Environ Microbiol* 24(1):324–340. https://doi.org/10.1111/1462-2920.15878.

He C, Wang W, Hou J. (2019). Plant growth and soil microbial impacts of enhancing licorice with inoculating dark septate endophytes under drought stress. *Front Microbiol* 10:2277. https://doi.org/10.3389/fmicb.2019.02277.

He YM, Fan XM, Zhang GQ, Li B, Li TG, Zu YQ, Zhan FD. (2020). Effects of arbuscular mycorrhizal fungi and dark septate endophytes on maize performance and root traits under a high cadmium stress. *S Afr J Bot* 134:415–423. https://doi.org/10.1016/j.sajb.2019.09.018.

Hembrom S, Singh B, Gupta SK, Nema AK. (2020). A comprehensive evaluation of heavy metal contamination in foodstuff and associated human health risk: a global perspective. In: Singh P, Singh R, Srivastava V, (Eds) *Contemporary environmental issues and challenges in era of climate change*. Springer, Singapore. pp:33–63. https://doi.org/10.1007/978-981-32-9595-7_2.

Hereme R, Morales-Navarro S, Ballesteros G, Barrera A, Ramos P, Gundel PE, Molina-Montenegro MA. (2020). Fungal endophytes exert positive effects on *Colobanthus quitensis* under water stress but neutral under a projected climate change scenario in Antarctica. *Front Microbiol* 11:264. https://doi.org/10.3389/fmicb.2020.00264.

Hosseini F, Mosaddeghi MR, Dexter AR, Sepehri M. (2019). Effect of endophytic fungus *Piriformospora indica* and PEG-induced water stress on maximum root growth pressure and elongation rate of maize. *Plant Soil* 435(1):423–436. https://doi.org/10.1007/s11104-018-03909-7.

Hou L, Li X, He X, Zuo Y, Zhang D, Zhao L. (2021). Effect of dark septate endophytes on plant performance of *Artemisia ordosica* and associated soil microbial functional group abundance under salt stress. *Appl Soil Ecol* 165:103998. https://doi.org/10.1016/j.apsoil.2021.103998.

Hou L, Yu J, Zhao L, He X. (2020). Dark septate endophytes improve the growth and the tolerance of *Medicago sativa* and *Ammopiptanthus mongolicus* under cadmium stress. *Front Microbiol* 10:3061. https://doi.org/10.3389/fmicb.2019.03061.

Hussain A, Shah M, Hamayun M, Qadir M, Iqbal A. (2022). Heavy metal tolerant endophytic fungi *Aspergillus welwitschiae* improves growth, ceasing metal uptake and strengthening antioxidant system in *Glycine max* L. *Environ Sci Pollut Res* 29(11):15501–15515. https://doi.org/10.1007/s11356-021-16640-1.

Ignatova L, Kistaubayeva A, Brazhnikova Y, Omirbekova A, Mukasheva T, Savitskaya I, Sokolov A. (2021). Characterization of cadmium-tolerant endophytic fungi isolated from soybean (*Glycine max*) and barley (*Hordeum vulgare*). *Heliyon* 7(11):e08240. https://doi.org/10.1016/j.heliyon.2021.e08240.

Imran QM, Falak N, Hussain A, Mun BG, Yun BW. (2021). Abiotic stress in plants; stress perception to molecular response and role of biotechnological tools in stress resistance. *Agronomy* 11(8):1579. https://doi.org/10.3390/agronomy11081579.

Isayenkov SV, Maathuis FJ. (2019). Plant salinity stress: many unanswered questions remain. *Front Plant Sci* 10:80. https://doi.org/10.3389/fpls.2019.00080.

Ismail AH, Mehmood A, Qadir M, Husna AI, Hamayun M, Khan N. (2020). Thermal stress alleviating potential of endophytic fungus *rhizopus oryzae* inoculated to sunflower (*Helianthus annuus* L.) and soybean (*Glycine max* L.). *Pak J Bot* 52(5):1857–1865. http://dx.doi.org/10.30848/PJB2020-5(10).

Ismail Hamayun M, Hussain A, Iqbal A, Khan SA, Lee IJ. (2020a). *Aspergillus niger* boosted heat stress tolerance in sunflower and soybean via regulating their metabolic and antioxidant system. *J Plant Interact* 15(1):223–232. https://doi.org/10.1080/17429145.2020.1771444.

Ismail HM, Hussain A, Khan SA, Iqbal A, Lee IJ. (2020b). An endophytic fungus *Aspergillus violaceofuscus* can be used as heat stress adaptive tool for *Glycine max* L. and *Helianthus annuus* L. *J Appl Bot Food Qual* 93:112. https://doi.org/10.5073/JABFQ.2020.093.014.

Ismail I, Hamayun M, Hussain A, Iqbal A, Khan SA, Ahmad A, Lee IJ. (2022). *Stemphylium Solani* stabilized the physicochemical characteristics of host plant species during stress. *Pol J Environ Stud* 31:(2). https://doi.org/10.15244/pjoes/139300.

Jiang W, Pan R, Buitrago S, Wu C, Abdelaziz ME, Oelmüller R, Zhang W. (2021). Transcriptome analysis of Arabidopsis reveals freezing-tolerance related genes induced by root endophytic fungus *Piriformospora indica*. *Physiol Mol Biol Plants* 27(2):189–201. https://doi.org/10.1007/s12298-020-00922-y.

Jiang W, Pan R, Wu C, Xu L, Abdelaziz ME, Oelmüller R, Zhang W. (2020). *Piriformospora indica* enhances freezing tolerance and post-thaw recovery in Arabidopsis by stimulating the expression of CBF genes. *Plant Signal Behav* 15(4):1745472. https://doi.org/10.1080/15592324.2020.1745472.

Kaddes A, Fauconnier ML, Sassi K, Nasraoui B, Jijakli MH. (2019). Endophytic fungal volatile compounds as solution for sustainable agriculture. *Molecules* 24(6):1065. https://doi.org/10.3390/molecules24061065.

Kapoor D, Singh S, Kumar V, Romero R, Prasad R, Singh J. (2019). Antioxidant enzymes regulation in plants in reference to reactive oxygen species (ROS) and reactive nitrogen species (RNS). *Plant Gene* 19:100182. https://doi.org/10.1016/j.plgene.2019.100182.

Karimi R, Amini H, Ghabooli M. (2022). Root endophytic fungus *Piriformospora indica* and zinc attenuate cold stress in grapevine by influencing leaf phytochemicals and minerals content. *Sci Hortic* 293:110665. https://doi.org/10.1016/j.scienta.2021.110665.

Khalid M, Saeed UR, Haoxin T, Lantian SU, Pei Z, Nan H. (2022). Mutualistic fungus *Piriformospora indica* modulates cadmium phytoremediation properties of host plant via concerted action of enzymatic and non-enzymatic biochemicals. *Pedosphere* 32(2):256–267. https://doi.org/10.1016/S1002-0160(21)60014-0.

Khalid M, Saeed UR, Hassani D, Hayat K, Pei Z, Nan H. (2021). Advances in fungal-assisted phytoremediation of heavy metals: a review. *Pedosphere* 31(3):475–495. https://doi.org/10.1016/S1002-0160(20)60091-1.

Khalvandi M, Amerian M, Pirdashti H, Keramati S. (2021). Does co-inoculation of mycorrhiza and *Piriformospora indica* fungi enhance the efficiency of chlorophyll fluorescence and essential oil composition in peppermint under irrigation with saline water from the Caspian Sea? *PloS ONE* 16(7):e0254076. https://doi.org/10.1371/journal.pone.0254076.

Khan MI, Ali N, Jan G, Hamayun M, Jan FG, Iqbal A, Lee IJ. (2022). Salt stress alleviation in *Triticum aestivum* through primary and secondary metabolites modulation by *Aspergillus terreus* BTK-1. *Front Plant Sci* 13:779623–779623. https://doi.org/10.3389/fpls.2022.779623.

Khan S, Naushad M, Lima EC, Zhang S, Shaheen SM, Rinklebe J. (2021). Global soil pollution by toxic elements: current status and future perspectives on the risk assessment and remediation strategies–A review. *J Hazard Mater* 417:126039. https://doi.org/10.1016/j.jhazmat.2021.126039.

Kumari V, Vujanovic V. (2020). Transgenerational benefits of endophytes on resilience and antioxidant genes expressions in pea (*Pisum sativum* L.) under osmotic stress. *Acta Physiol Plant* 42(4):1–11. https://doi.org/10.1007/s11738-020-03042-y.

Kushwaha AS, Thakur RS, Patel DK, Kumar M. (2022). Impact of arsenic on phosphate solubilization, acquisition and poly-phosphate accumulation in endophytic fungus *Serendipita indica*. *Microbiol Res* 259:127014. https://doi.org/10.1016/j.micres.2022.127014.

Kuzhuppillymyal-Prabhakarankutty L, Tamez-Guerra P, Gomez-Flores R, Rodriguez-Padilla M C, Ek-Ramos MJ. (2020). Endophytic *Beauveria bassiana* promotes drought tolerance and early flowering in corn. *World J Microbiol Biotechnol* 36(3):1–10. https://doi.org/10.1007/s11274-020-02823-4.

Lay-Pruitt KS, Wang W, Prom-U-Thai C, Pandey A, Zheng L, Rouached H. (2022). A tale of two players: the role of phosphate in iron and zinc homeostatic interactions. *Planta* 256(2):1–8. https://doi.org/10.1007/s00425-022-03922-2.

Li C, Zhou K, Qin W, Tian C, Qi M, Yan X, Han W. (2019a). A review on heavy metals contamination in soil: effects, sources, and remediation techniques. *Soil Sediment Contam* 28(4):380–394. https://doi.org/10.1080/15320383.2019.1592108.

Li D, Bodjrenou DM, Zhang S, Wang B, Pan H, Yeh KW, Cheng C. (2021). The endophytic fungus *Piriformospora indica* reprograms banana to cold resistance. *Int J Mol Sci* 22(9):4973. https://doi.org/10.3390/ijms22094973.

Li WK, Hu ZB. (2005). Endophytes and natural medicines. *Chin J Nat Med* 3(4):193–199

Li X, He C, He X, Su F, Hou L, Ren Y, Hou Y. (2019b). Dark septate endophytes improve the growth of host and non-host plants under drought stress through altered root development. *Plant Soil* 439(1):259–272. https://doi.org/10.1007/s11104-019-04057-2.

Li ZG. (2019c). Interplay between hydrogen sulfide and calcium signaling in plant abiotic stress response and adaptation. In: Hasanuzzaman M, Fotopoulos V, Nahar K, Fujita M, (Eds) *Reactive oxygen, nitrogen and sulfur species in plants: production, metabolism, signaling and defense mechanisms*. Wiley. pp:669–683. https://doi.org/10.1002/9781119468677.ch29.

Liang Y, Wei G, Ning K, Zhang G, Liu Y, Dong L, Chen S. (2021). Contents of lobetyolin, syringin, and atractylolide III in *Codonopsis pilosula* are related to dynamic changes of endophytes under drought stress. *Chin Med* 16(1):1–16. https://doi.org/10.1186/s13020-021-00533-z.

Liu Y, Wei X. (2019). Dark septate endophyte improves drought tolerance of *Ormosia hosiei* Hemsley & EH Wilson by modulating root morphology, ultrastructure, and the ratio of root hormones. *Forests* 10(10):830. https://doi.org/10.3390/f10100830.

Liu Y, Wei X. (2021). Dark septate endophyte improves the drought-stress resistance of *Ormosia hosiei* seedlings by altering leaf morphology and photosynthetic characteristics. *Plant Ecol* 222(7):761–771. https://doi.org/10.1007/s11258-021-01135-3.

Llorens E, Sharon O, Camañes G, García-Agustín P, Sharon A. (2019). Endophytes from wild cereals protect wheat plants from drought by alteration of physiological responses of the plants to water stress. *Environ Microbiol* 21(9):3299–3312. https://doi.org/10.1111/1462-2920.14530.

Lubna Khan MA, Asaf S, Jan R, Waqas M, Kim KM, Lee IJ. (2022). Endophytic fungus *Bipolaris* sp. CSL-1 induces salt tolerance in *Glycine max*. L via modulating its endogenous hormones, antioxidative system and gene expression. *J Plant Interact* 17(1):319–332. https://doi.org/10.1080/17429145.2022.2036836.

Lyon C, Saupe EE, Smith CJ, Hill DJ, Beckerman AP, Stringer LC, Aze T. (2022). Climate change research and action must look beyond 2100. *Glob Change Biol* 28(2):349–361. https://doi.org/10.1111/gcb.15871.

Ma Y, Dias MC, Freitas H. (2020). Drought and salinity stress responses and microbe-induced tolerance in plants. *Front Plant Sci* 11 :591911. https://doi.org/10.3389/fpls.2020.591911.

Mahar A, Ali A, Lahori AH, Wahid F, Li R, Azeem M, Zhang Z. (2020). Promising technologies for Cd-contaminated soils: drawbacks and possibilities. In: Roberson E, (Eds) *Environment, climate, plant and vegetation growth*. Springer Cham. pp:63–91. https://doi.org/10.1007/978-3-030-49732-3_3.

Malhi GS, Kaur M, Kaushik P. (2021). Impact of climate change on agriculture and its mitigation strategies: a review. *Sustainability* 13(3):1318. https://doi.org/10.3390/su13031318.

Malik A, Butt TA, Naqvi STA, Yousaf S, Qureshi MK, Zafar MI, Iqbal M. (2020). Lead tolerant endophyte *Trametes hirsuta* improved the growth and lead accumulation in the vegetative parts of *Triticum aestivum* L. *Heliyon* 6(7):e04188. https://doi.org/10.1016/j.heliyon.2020.e04188.

Mishra S, Bharagava RN, More N, Yadav A, Zainith S, Mani S, Chowdhary P. (2019). Heavy metal contamination: an alarming threat to environment and human health. In: Sobti R, Arora N, Kothari R, (Eds) *Environmental biotechnology: for sustainable future*. Springer, Singapore. pp:103–125. https://doi.org/10.1007/978-981-10-7284-0_5.

Moghaddam MSH, Safaie N, Soltani J, Hagh-Doust N. (2021). Desert-adapted fungal endophytes induce salinity and drought stress resistance in model crops. *Plant Physiol Biochem* 160:225–238. https://doi.org/10.1016/j.plaphy.2021.01.022.

Molina-Montenegro MA, Acuña-Rodríguez IS, Torres-Díaz C, Gundel PE, Dreyer I. (2020). Antarctic root endophytes improve physiological performance and yield in crops under salt stress by enhanced energy production and Na^+ sequestration. *Sci Rep* 10(1):1–10. https://doi.org/10.1038/s41598-020-62544-4.

Morales-Quintana L, Barrera A, Hereme R, Jara K, Rivera-Mora C, Valenzuela-Riffo F, Ramos P. (2021). Molecular and structural characterization of expansins modulated by fungal endophytes in the Antarctic *Colobanthus quitensis* (Kunth) Bartl. Exposed to drought stress. *Plant Physiol Biochem* 168:465–476. https://doi.org/10.1016/j.plaphy.2021.10.036.

Morsy M, Cleckler B, Armuelles-Millican H. (2020). Fungal endophytes promote tomato growth and enhance drought and salt tolerance. *Plants* 9(7):877. https://doi.org/10.3390/plants9070877.

Nadarajah KK. (2020). ROS homeostasis in abiotic stress tolerance in plants. *Int J Mol Sci* 21(15):5208. https://doi.org/10.3390/ijms21155208.

Nath R, Sharma GD, Barooah M. (2012). Efficiency of tricalcium phosphate solubilization by two different endophytic *Penicillium* sp. isolated from tea (*Camellia sinensis* L.). *Eur J Exp Biol* 2(4):1354–1358

New M, Liverman D, Schroder H, Anderson K. (2011). Four degrees and beyond: the potential for a global temperature increase of four degrees and its implications. *Philos Trans Royal Soc A* 369(1934):6–19. https://doi.org/10.1098/rsta.2010.0303.

Nguyen HC, Lin KH, Ho SL, Chiang CM, Yang CM. (2018). Enhancing the abiotic stress tolerance of plants: from chemical treatment to biotechnological approaches. *Physiol Plant* 164(4):452–466. https://doi.org/10.1111/ppl.12812.

Oskoueian A, Nezami A, Kafi M, Bagheri A, Lakzian A. (2022). Ability of mycorrhiza arbuscular and endophyte species to improve salinity tolerance in chickpea (*Cicer arietinum* L.). *Env Stresses Crop Sci* 15(1):215–230. https://doi.org/10.22077/ESCS.2020.3572.1877.

Pan L, Cui S, Dinkins RD, Jiang Y. (2021). Plant growth, ion accumulation, and antioxidant enzymes of endophyte-infected and endophyte-free tall fescue to salinity stress. *Acta Physiol Plant* 43(6):1–10. https://doi.org/10.1007/s11738-021-03268-4.

Pang Z, Zhao Y, Xu P, Yu D. (2020). Microbial diversity of upland rice roots and their influence on rice growth and drought tolerance. *Microorganisms* 8(9):1329. https://doi.org/10.3390/microorganisms8091329.

Parihar P, Singh S, Singh R, Singh VP, Prasad SM. (2015). Effect of salinity stress on plants and its tolerance strategies: a review. *Environ Sci Pollut Res* 22(6):4056–4075. https://doi.org/10.1007/s11356-014-3739-1.

Poveda J, Eugui D, Abril-Urías P, Velasco P. (2021). Endophytic fungi as direct plant growth promoters for sustainable agricultural production. *Symbiosis* 85(1):1–19. https://doi.org/10.1007/s13199-021-00789-x.

Qiang X, Ding J, Lin W, Li Q, Xu C, Zheng Q, Li Y. (2019). Alleviation of the detrimental effect of water deficit on wheat (*Triticum aestivum* L.) growth by an indole acetic acid-producing endophytic fungus. *Plant Soil* 439(1):373–391. https://doi.org/10.1007/s11104-019-04028-7.

Qin W, Liu C, Jiang W, Xue Y, Wang G, Liu S. (2019). A coumarin analogue NFA from endophytic *Aspergillus fumigatus* improves drought resistance in rice as an antioxidant. *BMC Microbiol* 19(1):1–11. https://doi.org/10.1186/s12866-019-1419-5.

Raftery AE, Zimmer A, Frierson DM, Startz R, Liu P. (2017). Less than 2 C warming by 2100 unlikely. *Nat Clim Change* 7(9):637–641. https://doi.org/10.1038/nclimate3352.

Ragasa LRP, Joson IV SEA, Bagay WLR, Perez TR, Velarde MC. (2021). Transcriptome analysis reveals involvement of oxidative stress response in a copper-tolerant *Fusarium oxysporum* strain. *Fungal Biol* 125(6):435–446. https://doi.org/10.1016/j.funbio.2021.01.001.

Rastogi A, Yadav S, Hussain S, Kataria S, Hajihashemi S, Kumari P, Brestic M. (2021). Does silicon really matter for the photosynthetic machinery in plant? *Plant Physiol Biochem* 169:40–48. https://doi.org/10.1016/j.plaphy.2021.11.004.

Rauf M, Awais M, Ud-Din A, Ali K, Gul H, Rahman MM, Arif M. (2021). Molecular mechanisms of the 1-aminocyclopropane-1-carboxylic acid (ACC) deaminase producing *Trichoderma asperellum* MAP1 in enhancing wheat tolerance to waterlogging stress. *Front Plant Sci* 11:614971. https://doi.org/10.3389/fpls.2020.614971.

Raza H, Bilal H, Raza A, Shoukat MB, Tariq R, Rasheed A, Ramzan M. (2021). Impact of climate change on agriculture and food security: a review. *Ind J Pure App Biosci* 9(3):110–118. http://dx.doi.org/10.18782/2582-2845.8699.

Rhaman MS, Imran S, Rauf F, Khatun M, Baskin CC, Murata Y, Hasanuzzaman M. (2020). Seed priming with phytohormones: an effective approach for the mitigation of abiotic stress. *Plants* 10(1):37. https://doi.org/10.3390/plants10010037.

Ripa FA, Cao WD, Tong S, Sun JG. (2019). Assessment of plant growth promoting and abiotic stress tolerance properties of wheat endophytic fungi. *BioMed Res Int*. https://doi.org/10.1155/2019/6105865.

Ritonga FN, Chen S. (2020). Physiological and molecular mechanism involved in cold stress tolerance in plants. *Plants* 9(5):560. https://doi.org/10.3390/plants9050560.

Rizwan M, Ali S, Maqbool A. (2019). A critical review on the effects of zinc at toxic levels of cadmium in plants. *Environ Sci Pollut Res* 26(7):6279–6289. https://doi.org/10.1007/s11356-019-04174-6.

Sachdev S, Ansari SA, Ansari MI, Fujita M, Hasanuzzaman M. (2021). Abiotic stress and reactive oxygen species: generation, signaling, and defense mechanisms. *Antioxidants* 10(2):277. https://doi.org/10.3390/antiox10020277.

Sadeghi F, Samsampour D, Seyahooei MA, Bagheri A, Soltani J. (2020). Fungal endophytes alleviate drought-induced oxidative stress in mandarin (*Citrus reticulata* L.): toward regulating the ascorbate–glutathione cycle. *Sci Hortic* 261:108991. https://doi.org/10.1016/j.scienta.2019.108991.

Saijo Y, Loo EPI. (2020). Plant immunity in signal integration between biotic and abiotic stress responses. *New Phytol* 225(1):87–104. https://doi.org/10.1111/nph.15989.

Saikkonen K, Wäli P, Helander M, Faeth SH. (2004). Evolution of endophyte–plant symbioses. *Trends Plant Sci* 9(6):275–280 https://doi.org/10.1016/j.tplants.2004.04.005.

Sampangi-Ramaiah MH, Dey P, Jambagi S, Vasantha Kumari MM, Oelmüller R, Nataraja KN, Uma Shaanker R. (2020). An endophyte from salt-adapted Pokkali rice confers salt-tolerance to a salt-sensitive rice variety and targets a unique pattern of genes in its new host. *Sci Rep* 10(1):1–14. https://doi.org/10.1038/s41598-020-59998-x.

Sangamesh MB, Jambagi S, Vasanthakumari MM, Shetty NJ, Kolte H, Ravikanth G, Uma Shaanker R. (2018). Thermotolerance of fungal endophytes isolated from plants adapted to the Thar Desert, India. *Symbiosis* 75(2):135–147. https://doi.org/10.1007/s13199-017-0527-y.

Saravi HB, Gholami A, Pirdashti H, Firouzabadi MB, Asghari H. (2021). The response of stevia (*Stevia rebaudiana* Bertoni) photosystem II photochemistry to fungi symbiosis and spermidine application under saline water irrigation. *Russ Agric Sci* 47(1):32–36. https://doi.org/10.3103/S106836742101016X.

Saravi HB, Gholami A, Pirdashti H, Firouzabadi MB, Asghari H, Yaghoubian Y. (2022). Improvement of salt tolerance in *Stevia rebaudiana* by co-application of endophytic fungi and exogenous spermidine. *Ind Crops Prod* 177:114443. https://doi.org/10.1016/j.indcrop.2021.114443.

Schulz B, Boyle C, Draeger S, Römmert AK, Krohn K. (2002). Endophytic fungi: a source of novel biologically active secondary metabolites. *Mycol Res* 106(9):996–1004. https://doi.org/10.1017/S0953756202006342.

Schulz B, Römmert AK, Dammann U, Aust HJ, Strack D. (1999). The endophyte-host interaction: a balanced antagonism? *Mycol Res* 103(10):1275–1283. https://doi.org/10.1017/S0953756299008540.

Seema N, Hamayun M, Ara H, Khan RS. (2019). Inoculation of sunflower with endophytic fungi alters soil physio-chemical properties to nullify drought stress. *Sarhad Journal of Agriculture* 35(1). http://dx.doi.org/10.17582/journal.sja/2019/35.1.109.115.

Sepehri M, Ghaffari MR, Khayam Nekoui M, Sarhadi E, Moghadam A, Khatabi B, Hosseini Salekdeh G. (2021). Root endophytic fungus *Serendipita indica* modulates barley leaf blade proteome by increasing the abundance of photosynthetic proteins in response to salinity. *J Appl Microbiol* 131(4):1870–1889. https://doi.org/10.1111/jam.15063.

Shadmani L, Jamali S, Fatemi A. (2021a). Effects of root endophytic fungus, *Microdochium bolleyi* on cadmium uptake, translocation and tolerance by *Hordeum vulgare* L. *Biologia* 76(2):711–719. https://doi.org/10.2478/s11756-020-00598-5.

Shadmani L, Jamali S, Fatemi A. (2021b). Isolation, identification, and characterization of cadmium-tolerant endophytic fungi isolated from barley (*Hordeum vulgare* L.) roots and their role in enhancing phytoremediation. *Braz J Microbiol* 52(3):1097–1106. https://doi.org/10.1007/s42770-021-00493-4.

Shahabivand S, Parvaneh A, Aliloo AA. (2020). Different response of *Alyssum montanum* and *Helianthus annuus* to cadmium bioaccumulation mediated by the endophyte fungus *Serendipita indica*. *Acta Ecol Sin* 40(4):315–322. https://doi.org/10.1016/j.chnaes.2019.09.002.

Shahzad A, Ullah S, Dar AA, Sardar MF, Mehmood T, Tufail MA, Haris M. (2021). Nexus on climate change: agriculture and possible solution to cope future climate change stresses. *Environ Sci Pollut Res* 28(12):14211–14232. https://doi.org/10.1007/s11356-021-12649-8.

Shukla J, Mohd S, Kushwaha AS, Narayan S, Saxena PN, Bahadur L, Kumar M. (2022). Endophytic fungus *Serendipita indica* reduces arsenic mobilization from root to fruit in colonized tomato plant. *Environ Pollut* 298:118830. https://doi.org/10.1016/j.envpol.2022.118830.

Sodhi GK, Saxena S. (2023a). Plant growth promotion and abiotic stress mitigation in rice using endophytic fungi: advances made in the last decade. *Environ Exp Bot* 209:105312. https://doi.org/10.1016/j.envexpbot.2023.105312.

Sodhi GK, Saxena S. (2023b). Role of endophytic fungi in promoting plant health. In: Singh I, Rajpal VR, Navi SS, (Eds) *Fungal resources for sustainable economy: current status and future perspectives*. Springer Nature Singapore. pp:319–345. https://doi.org/10.1007/978-981-19-9103-5_12.

Sodhi GK, Saxena S. (2023c). Promising drought and salinity tolerance features of *Nigrospora* species existing as endophytes in *Oryza sativa*. *3 Biotech* 13(8):262. https://doi.org/10.1007/s13205-023-03679-9.

Sodhi GK, Saxena S. (2023d). Plant growth-promoting endophyte *Nigrospora oryzae* mitigates abiotic stress in rice (*Oryza sativa* L.). *FEMS Microbiol Ecol* 99(9):fiad094. https://doi.org/10.1093/femsec/fiad094.

Stone JK, Polishook JD, White JF. (2004). Endophytic fungi. In: Foster M, Bills G, (Eds) *Biodiversity of fungi*. Elsevier Academic Press, Burlington. pp:241–270. https://doi.org/10.1016/B978-012509551-8/50015-5.

Su Z, Zeng Y, Li X, Perumal AB, Zhu J, Lu X, Lin F. (2021a). The Endophytic Fungus *Piriformospora Indica*-Assisted Alleviation of Cadmium in Tobacco. *J Fungi* 7(8):675. https://doi.org/10.3390/jof7080675.

Su ZZ, Dai MD, Zhu JN, Liu XH, Li L, Zhu XM, Lin FC. (2021b). Dark septate endophyte *Falciphora oryzae*-assisted alleviation of cadmium in rice. *J Hazard Mater* 419:126435. https://doi.org/10.1016/j.jhazmat.2021.126435.

Syamsia S, Kadir M, Idham A, Noerfitryani N. (2020). Utilization of local aromatic rice endophytic fungi to promote the growth and yield of rice plant in drought stress conditions. In: *IOP Conference Series: Earth and Environmental Science*. Vol: 486. IOP Publishing. https://doi.org/10.1088/1755-1315/486/1/012134.

Tabande L, Sepehri M, Yasrebi J, Zarei M, Ghasemi-Fasaei R, Khatabi B. (2022). A comparison between the function of *Serendipita indica* and *Sinorhizobium meliloti* in modulating the toxicity of zinc oxide nanoparticles in alfalfa (*Medicago sativa* L.). *Environ Sci Pollut* 29(6):8790–8803. https://doi.org/10.1007/s11356-021-16287-y.

Thuiller W. (2007). Climate change and the ecologist. *Nature* 448(7153):550–552. https://doi.org/10.1038/448550a.

Tiwari P, Bae H. (2020). Horizontal gene transfer and endophytes: an implication for the acquisition of novel traits. *Plants* 9(3):305. https://doi.org/10.3390/plants9030305.

Tsai HJ, Shao KH, Chan MT, Cheng CP, Yeh KW, Oelmüller R, Wang SJ. (2020). *Piriformospora indica* symbiosis improves water stress tolerance of rice through regulating stomata behavior and ROS scavenging systems. *Plant Signal Behav* 15(2):1722447. https://doi.org/10.1080/15592324.2020.1722447.

Wang M, Tian P, Gao M, Li M. (2021). The Promotion of Festuca sinensis under Heavy Metal Treatment Mediated by Epichloë Endophyte. *Agronomy* 11(10):2049. https://doi.org/10.3390/agronomy11102049.

Wang S, Dai H, Wei S, Skuza L, Chen Y. (2022). Effects of Cd-resistant fungi on uptake and translocation of Cd by soybean seedlings. *Chemosphere* 291:132908. https://doi.org/10.1016/j.chemosphere.2021.132908.

Wang X, Tian Z, Xi Y, Guo Y. (2022). Identification of endophytic fungi with ACC deaminase-producing isolated from halophyte *Kosteletzkya Virginica*. *Plant Signal Behav* 17(1):2152224. https://doi.org/10.1080/15592324.2022.2152224.

Xu FJ, Song SL, Ma CY, Zhang W, Sun K, Tang MJ, Dai CC. (2020). Endophytic fungus improves peanut drought resistance by reassembling the root-dwelling community of arbuscular mycorrhizal fungi. *Fungal Ecol* 48:100993. https://doi.org/10.1016/j.funeco.2020.100993.

Xu Z, Pehlivan N, Ghorbani A, Wu C. (2022). Effects of *Azorhizobium caulinodans* and *Piriformospora indica* co-inoculation on growth and fruit quality of tomato (*Solanum lycopersicum* L.) under salt stress. *Horticulturae* 8(4):302. https://doi.org/10.3390/horticulturae8040302.

Yan L, Zhu J, Zhao X, Shi J, Jiang C, Shao D. (2019). Beneficial effects of endophytic fungi colonization on plants. *Appl Microbiol Biotechnol* 103(8):3327–3340. https://doi.org/10.1007/s00253-019-09713-2.

Yang L, Schröder P, Vestergaard G, Schloter M, Radl V. (2020). Response of barley plants to drought might be associated with the recruiting of soil-borne endophytes. *Microorganisms* 8(9):1414. https://doi.org/10.3390/microorganisms8091414.

Yu Y, Teng Z, Mou Z, Lv Y, Li T, Chen S, Zhao Z. (2021). Melatonin confers heavy metal-induced tolerance by alleviating oxidative stress and reducing the heavy metal accumulation in *Exophiala pisciphila*, a dark septate endophyte (DSE). *BMC Microbiol* 21(1):1–11. https://doi.org/10.1186/s12866-021-02098-1.

Zhai Y, Chen Z, Malik K, Wei X, Li C. (2022). Effect of fungal endophyte *Epichloë bromicola* infection on Cd tolerance in Wild Barley (*Hordeum brevisubulatum*). *J Fungi* 8(4):366. https://doi.org/10.3390/jof8040366.

Zhang H, Zhu J, Gong Z, Zhu JK. (2022). Abiotic stress responses in plants. *Nat Rev Genet* 23(2):104–119. https://doi.org/10.1038/s41576-021-00413-0.

Zhou L, Li C, White JF, Johnson RD. (2021). Synergism between calcium nitrate applications and fungal endophytes to increase sugar concentration in Festuca sinensis under cold stress. *PeerJ* 9:e10568. http://doi.org/10.7717/peerj.10568.

Zhou XR, Dai L, Xu GF, Wang HS. (2021). A strain of Phoma species improves drought tolerance of *Pinus tabulaeformis*. *Sci Rep* 11(1):1–11. https://doi.org/10.1038/s41598-021-87105-1.

Żurek G, Wiewióra B, Rybka K, Prokopiuk K. (2022). Different response of perennial ryegrass—*Epichloë* endophyte symbiota to the elevated concentration of heavy metals in soil. *J Appl Genet* 63(1):47–59. https://doi.org/10.1007/s13353-021-00661-0.

17 Plant-Associated Fungi
Regulation of Temperature and Drought Stress Conditions

*Meghna Dhar, Kajal Sorout, Deepansh Sharma and Sunny Dholpuria**
*Corresponding author.

17.1 INTRODUCTION

Human lives are dependent upon a variety of factors for their existence on Earth. These factors include abiotic components like oxygen, light, water, temperature and humidity, and biotic components like plants, microbes and decomposers (Satrovic et al., 2024). Among these, plants are one of the major components, which control the flow of energy across all ecosystems via photosynthesis. One can find plants in almost every part of the world with different climatic conditions related to temperature, humidity and water availability (Leisner et al., 2023). Many plants have evolved themselves to withstand these conditions, especially those related to extreme temperature and drought conditions. The presence of long roots, succulents to store water, spines for protection and short leaves are some of the external modifications found in arid-region plants. Along with these, the presence of certain genes in their DNA (Xia et al., 2023) and the production of secondary metabolites also helps them to survive under extreme conditions (Khayri et al., 2023). In the present condition, a number of human activities have created tremendous pressures on the environment, leading to a rise in global temperatures. This increase in temperatures is not an isolated event but extends to the alteration of weather patterns, resulting in the intensification of drought conditions in numerous regions across the globe. Moreover, these anthropogenic stresses caused by climate change, manifested in increasing global temperatures and more frequent and severe droughts, present a substantial threat to plant performance and crop production (Cotrina Cabello et al., 2023). The plant-associated microbiome, a complex community of microorganisms living in and on plant tissues, plays a significant role in plant health and productivity. This microbial community, which comprises bacteria, fungi, archaea, viruses and other microscopic entities, interacts with plants in symbiotic, mutualistic, commensal or sometimes pathogenic ways, thereby influencing plant growth, development and survival (Muhammad et al., 2023). The plant-associated microbiome has emerged as a critical modulator of biotic and abiotic stress impacts on plant survival. For instance, certain plant-associated microbes can induce systemic resistance in plants against pests and pathogens, reducing the detrimental effects of biotic stresses. They can also improve plant tolerance to abiotic stresses by altering plant hormone levels, enhancing nutrient uptake or mitigating the effects of environmental extremes (Chen et al., 2023). In the context of biotic stresses, some microbes can outcompete or inhibit pathogens, induce systemic resistance in plants or modulate plant immune responses. For example, certain strains of *Pseudomonas* and *Bacillus* bacteria have been shown to protect plants against fungal pathogens by producing antimicrobial compounds or by triggering plant defense mechanisms (Trivedi et al., 2022; Ayaz et al., 2023). Regarding abiotic stresses, microbes can enhance plant

DOI: 10.1201/9781003486275-20

resilience by improving nutrient uptake and utilization, modifying root architecture or producing plant growth-promoting hormones. For instance, arbuscular mycorrhizal fungi form mutualistic associations with plant roots, enhancing nutrient absorption and helping plants cope with nutrient-poor conditions. Similarly, certain bacteria can produce hormones that promote plant growth under drought or salinity stress (Thangavel et al., 2022). Temperature and drought are two crucial environmental factors that significantly affect plant growth and development, both directly and indirectly. Enzyme-catalyzed processes within plants are highly temperature-dependent, and their rates can double with every 10°C increase within a certain range. However, once the temperature exceeds the optimal range, these processes can slow down or even stop, leading to thermal stress. This stress can cause damage to proteins and membranes, disrupt cellular homeostasis and lead to the production of reactive oxygen species (Suzuki, 2023). In extreme cases, high temperatures can cause cell death and reduce crop yield. On the other hand, low temperatures can also be detrimental to plants. Some plants can't survive freezing temperatures, while others can only tolerate them for a short time. Cold stress can lead to a reduction in photosynthesis, as the enzymes involved in this process are temperature sensitive. This can reduce growth and productivity. Additionally, cold temperatures can lead to ice formation within plant tissues, damaging cells (Ding & Yang, 2022). Drought, or water scarcity, can also significantly impact plants. When water supply is inadequate, stomata close to prevent water loss, which in turn limits the plant's ability to take in carbon dioxide for photosynthesis (Castro et al., 2023). Drought can cause the plant to wilt and increase susceptibility to pests and diseases. Despite these challenges, many plants have developed strategies to survive in high temperatures and drought conditions. Some plants, known as xerophytes, have evolved various adaptations such as small leaves, thick cuticles and deep root systems to minimize water loss and maximize water uptake. Other plants produce heat-shock proteins in response to high temperatures to protect their cells from thermal stress (Zhang et al., 2022). Harnessing the beneficial interactions between plants and their microbiome could provide innovative strategies for sustainable agriculture, aiding in stress management and enhancing crop productivity under changing environmental conditions.

17.2 NATURAL RESISTANCE OF PLANTS AGAINST ABIOTIC STRESS

Environmental alterations, whether precipitated by human activities or sudden seasonal weather shifts, present a formidable survival challenge to plants. As stationary organisms, plants are compelled to withstand these adverse environmental conditions. Among the myriad environmental changes, factors such as water shortages or droughts, water logging, salinity extremes, temperature variations, heavy metals and solar radiation pose significant threats to the optimal growth and development of plants (Hossain et al., 2021). Consequently, these stressors lead to global reductions in both biomass and grain yields. In harsh environments, various types of abiotic stresses tend to co-exist. For instance, salt stress often coincides with drought conditions, a situation that can be further exacerbated by extreme temperatures (Sabagh et al., 2021). These interconnected stressors compound the survival challenges faced by plants, underscoring the relentless struggle they endure in adapting to continually changing environmental conditions. To survive, they have developed a remarkable array of defense mechanisms. These mechanisms involve a complex interplay of sensing, signaling and response mechanisms.

The five general plant defense metabolites are:

- Cuticle: This waxy layer acts as the first line of defense, protecting plants from water loss, pathogens and UV radiation.
- Unsaturated fatty acids (UFAs): These molecules act as membrane modulators and precursors to oxylipins, which play important roles in signaling and defense.

- Reactive species scavengers: These molecules protect cells from damage caused by reactive oxygen species (ROS). Examples include antioxidants like glutathione and ascorbic acid.
- Molecular chaperones: These proteins help stabilize other proteins and subcellular structures under stress conditions.
- Compatible solutes: These small organic molecules accumulate in cells under stress conditions, protecting them from dehydration and other damage. Examples include proline, glycine betaine and sugars (Bhandari et al., 2023).

These defensive metabolites are coordinated by a complex network of signaling molecules and regulatory factors; including stress phytohormones (PHs) which act as chemical messengers, regulating responses to various stresses (Singh & Roychoudhury, 2023). An example is abscisic acid (ABA), which plays a key role in drought tolerance, and another is salicylic acid (SA), which is involved in disease resistance. Polyamines helps in regulating plant growth via modulating gene transcription, and posttranslational modifications. These molecules also act as important antioxidants with a high ROS-scavenging capacity in both free and conjugated forms (Singh & Roychoudhury, 2023). Hydrogen sulfide (H_2S) also plays an important role in stress tolerance by regulating antioxidant defenses and gene expression; likewise, nitric oxide (NO) is also involved in various physiological processes, including stress signaling and defense responses. Phytochromes are light-sensitive proteins that regulate plant growth and development, and also play a role in stress responses (Yadav et al., 2021). Strigolactones and brassinosteroids, which are relatively newly discovered plant hormones, are increasingly recognized for their role in stress tolerance, particularly against extreme temperatures and drought (Nolan et al., 2020). By understanding these defense mechanisms, scientists can develop new strategies to help plants cope with the ever-increasing challenges of a changing environment. Apart from these and many other biomolecules and compounds, the DNA of different plant species also plays an important role in mitigating stress conditions. DNA harbors many genes which are expressed as per the environmental need, such as heat shock protein genes, drought resistance and salinity resistance genes (dos Santos et al., 2022; Zhang et al., 2022).

17.3 REGULATION OF TEMPERATURE VIA PLANT-ASSOCIATED FUNGI

Heat stress has a profound and multifaceted impact on plant growth and development, ranging from physiological disruptions to visible damage and economic losses. High temperatures can impair seed germination and early seedling growth. This can manifest as delayed emergence, reduced seedling number and weaker initial establishment. Plant growth and development are significantly slowed down under heat stress. This can affect shoot and root elongation, leaf expansion and overall biomass production (Singh & Roychoudhury, 2023). High temperatures can cause excessive water loss, leading to wilting and drooping of leaves. In severe cases, foliage and reproductive organs can scorch and burn, compromising photosynthesis and fruit production. The cumulative effects of all these physiological and visible disturbances ultimately result in substantial yield losses. Additionally, heat stress can trigger cellular damage and death; further compromising plant function (Wahab et al., 2023). Arbuscular mycorrhiza fungi (AMF) fungi form symbiotic relationships with approximately 70–90% of all terrestrial plant species. This extensive association highlights their vital role in plant health and ecosystem functioning. AM fungi extend their filamentous network (hyphae) beyond the plant's root zone, enabling access to nutrients like phosphorus, nitrogen and water that are often unavailable to the plant directly. AM fungi can help plants cope with various environmental stresses such as drought, salinity and heavy metal contamination. This is achieved through various mechanisms, including improved nutrient and water uptake, enhanced root growth and increased production of stress-protective hormones (Shi et al., 2023). The Earth's climate is changing at an alarming rate, with rising temperatures being one of the most significant consequences. This warming trend is particularly pronounced in high-latitude ecosystems, such as the Arctic and alpine regions. These regions are experiencing a faster rate of warming than the global average, leading to dramatic changes in the composition and functioning of their ecological communities. Examples include predicted alteration of nitrogen-fixing taxa assembly in sub-Arctic tundra. Moss-associated cyanobacteria are important nitrogen fixers while the exact contribution varies. Cyanobacteria play a significant role in fixing nitrogen in high-latitude ecosystems. Their ability to convert atmospheric nitrogen (N_2) into usable forms (e.g., ammonium) helps support moss growth and contributes to overall ecosystem productivity (Wu et al., 2023). It has been found that the warming temperatures associated with climate change negatively impact the diversity and abundance of diazotrophic bacteria (bacteria capable of nitrogen fixation) within the peat moss community. This decline in diversity could potentially lead to reduced overall nitrogen fixation rates, resulting in changes in ecosystem structure and function (Kilner et al., 2023). Also, mosses play a crucial role in carbon sequestration in high-latitude ecosystems. If warming temperatures negatively impact moss growth due to reduced nitrogen availability, this could impact the ability of these ecosystems to store carbon (Kilner et al., 2023). Furthermore, alterations in the bacterial community composition of peat moss could have cascading effects on other ecosystem processes, such as decomposition and nutrient cycling (Klarenberg et al., 2022). Another such example where warming is reported to affect the rate at which mycorrhizal fungal mycelium (the network of fungal filaments that live in symbiosis with plant roots) decomposes is in peatland. Peatlands are wetlands characterized by the accumulation of partially decomposed organic matter called peat. Peatlands are important carbon sinks that store a significant amount of the world's terrestrial carbon. This change is likely caused by shifts in the composition of the microbial decomposer community, which is the group of microorganisms that break down dead organic matter. These shifts could involve an increase in the abundance of microbes that are more efficient at decomposing fungal material (Maillard et al., 2022). AMF inoculated plants show increased tolerance against temperature stress in a number of plant species. Inoculating wheat plants with *Glomus intraradices*, an AMF strain, significantly enhances their performance under drought conditions. This association causes an increase in the growth of the wheat plant as well as its overall production. The fungi resulted in greater photosynthetic rate in the plant which increased the biomass production (Sun et al., 2023). Perennial ryegrass (*Lolium perenne* L.) is a valuable and versatile grass species cultivated across temperate regions for biofuel and biochemical production, highlighting its economic and environmental significance. Inoculating AM plants with *Rhizophagus irregularis* significantly improved their growth and physiological parameters under both low-temperature and high-temperature stress compared to non-mycorrhizal (NM) plants as was evident by increased plant height, shoot and root dry weight, higher leaf chlorophyll and carotenoid content, increased net photosynthesis rate and increased photochemical efficiency (Wei et al., 2023). AMF-inoculated soybeans also exhibited greater growth compared to NM plants under high-temperature stress. Specifically, increased leaf area, shoot and root dry weights, and shoot and root length were observed in AMF-colonized plants, suggesting enhanced biomass production and resource utilization. Beyond increased growth, AMF symbiosis also influences other plant characteristics, such as leaf number, area and delayed senescence (Jumrani et al., 2022). In cucumber, an AMF, *Diversispora versiformis*, provides positive benefits against heat stress conditions. A significant variation can be observed between fungus-treated and nontreated cucumber plants. The treatment caused increased chlorophyll content, osmolytes (sucrose, fructose and betaine) as well as enhanced *Hsp70* gene expression in the leaves (Liu et al., 2024). A direct impact of AMF on plant photosystem (PS) II is also one of the ways by which these fungi makes plants heat tolerant. Maize grown in AMF-enriched soil can withstand heat temperature stress, and this happens as a result of the conversion of inactive β and γ centers to active α centers, and QB nonreducing centers to reducing centers of PS II (Mathur & Jajoo, 2020). Ectomycorrhizal fungi can also help to alleviate the temperature stress conditions in plants, and these can be a potential candidate as a biofertilizer to overcome heat stress in seedlings. Oak seedlings inoculated with ectomycorrhiza can withstand heat stress conditions. This association causes increased phenolics and polyamine content, especially spermidine and spermine, in the plant (Kebert et al., 2022).

It is not only the fungus which can make plants heat tolerant, sometimes the products secreted by them are also sufficient to give the same results. The incubation of Lanzhou lily (an edible crop in China) with trichokonins can give heat resistance properties to the plants. Trichokonins are antimicrobial peptaibols which are produced by *Trichoderma longibrachiatum* strain SMF2 and their application causes several modifications in plant species. These enhance the antioxidant enzyme system, which increases the level of heat-resistance-associated phytohormones and the relative water content as well as chlorophyll content of the plant (Cao et al., 2023). Similarly, Zhinengcong, a crude extract from the endophytic fungus *Paecilomyces variotii*, can help *Solanum lycopersicum* to overcome heat stress-induced reproductive defects. The extract causes repression in the genes related to the production of reactive oxygen species and also does not allow their accumulation in anther, pollen grain or pollen tube (Cui et al., 2022). Apart from this, the thermophilic fungi obtained from the plants of hot desert regions can help other plants to have some thermotolerance properties. Endophytic *Thermomyces* sp., obtained from the roots of *Cullen plicata*, are able to protect cucumber plants against heat stress conditions. The positive effects are due to an increase in a number of factors that include photosynthetic rate, root length, flavonoids, total sugar and antioxidant enzyme system (Ali et al., 2018). A number of other examples have been listed in Table 17.1,

TABLE 17.1
Effects of Different Fungal Species on Host Plants in Alleviating Drought Stress

S. no.	Fungal species	Plant species	Mechanism	References
1.	*Rhizopus oryzae*	*Glycine max L.* and *Helianthus annuus L.*	Increases in the activity of antioxidant enzymes, ascorbic acid oxidase and catalase. Also enhances the concentration of proline and chlorophyll	Ismail et al., 2020 ,
2.	*Rhizophagus irregularis*	*Solanum lycopersicum* L.; *Capiscum annuum* L.; *Cucumis sativus* L.	Increases strength of plant as well as productivity and quality of fruit	Reva et al., 2021
3.	*Microdochium majus*, *Meyerozyma guilliermondi* and *Aspergillus aculeatu*	*Moringa oleifera*	Increase in primary and secondary metabolites, enhancement in antioxidant enzyme system	Javed et al., 2022
4.	*Funneliformis mosseae*	*Lactuca sativa* L.	Increases net photosynthetic rate by enhancing the concentration of chlorophyll a and chlorophyll b	Yan et al., 2021
5.	*Glomus intraradices*	*Asparagus officinalis* L.	Enhancement in nutrient uptake by the plant	Yeasmin et al., 2019
6.	AMF consortia (mainly comprised of *Rhizophagus irregulariae*, *Funneliformis mosseae* and *Glomus* species)	*Zea mays* L.	Enhancement in photosynthetic system	Mathur et al., 2021

(continued)

TABLE 17.1 (Continued)
Effects of Different Fungal Species on Host Plants in Alleviating Drought Stress

S. no.	Fungal species	Plant species	Mechanism	References
7	*Talaromyces purpureogenus*	*Triticum aestivum* L.	Improves antioxidant system and enhances phenolic content	Kaur and Saxena, 2023
8.	*Penicillium oxalicum*	*Zea mays* L. (maize)	Improves water use efficiency and promotes root development	Zhang et al., 2023
9.	*Piriformospora indica*	*Arabidopsis thaliana*	Regulates stress-responsive genes and enhances stomatal closure	Li et al., 2023
10	*Trichoderma virens*	*Solanum lycopersicum* L. (tomato)	Promotes root growth and increases production of osmoprotectants	Adedayo & Babalola, 2023
11	*Serendipita indica*	*Oryza sativa* L. (rice)	Enhances drought tolerance by modulating plant hormone signaling	Zheng et al., 2023.
12	Endophytic fungi (diverse species)	Various plant species	Mechanisms can vary depending on the fungal species and plant host. May improve water uptake and nutrient acquisition, enhance stress tolerance through antioxidant activity and production of compatible solutes, or promote root growth	Bakhshi et al., 2023

which shows the positive effects of plant–fungi association in overcoming drought stress conditions in different types of plants.

The fungi not only help to overcome the heat stress condition but in certain situations they also help in overcoming the stress conditions related to low temperature. The AMF effect on snapdragon plants under low temperature suggests that AMF promote growth and resilience in snapdragon through enhanced nutrient absorption, stimulated root activity, improved photosynthesis and increased water use efficiency (Wei et al., 2023). Another AMF, *Glomus mosseae*, can enhance low temperature stress in *Vaccinium ashei* (rabbiteye blueberry) and *Vaccinium corymbosum* (highbush blueberry). The fungi cause an increase in antioxidant molecules in leaves along with increased sugar and proline concentrations (Liu et al., 2017). AMF also modulate gene expression in plants by upregulating and downregulating certain genes which helps to encounter cold stress conditions as observed in maize seedlings. AMF stimulate cyclic electron flow in chloroplast, and phosphoric acid production in mitochondria, glyoxylic acid cycle and tricarboxylic acid cycle. All these effects induce low-temperature tolerance in maize seedlings (Li et al., 2020). The incubation of banana plant, *Musa acuminate*, with an endophytic fungi, *Piriformospora indica*, can make this plant species resistant to cold stress conditions. The interaction makes a number of changes in the plant to provide significant tolerance ability. In leaves, catalase activity and soluble sugar content are enhanced, and along with these changes other parts of the plants undergo many other effects. Malondialdehyde (MDA) and hydrogen peroxide (H_2O_2) are reduced and the superoxide dismutase (SOD) activity

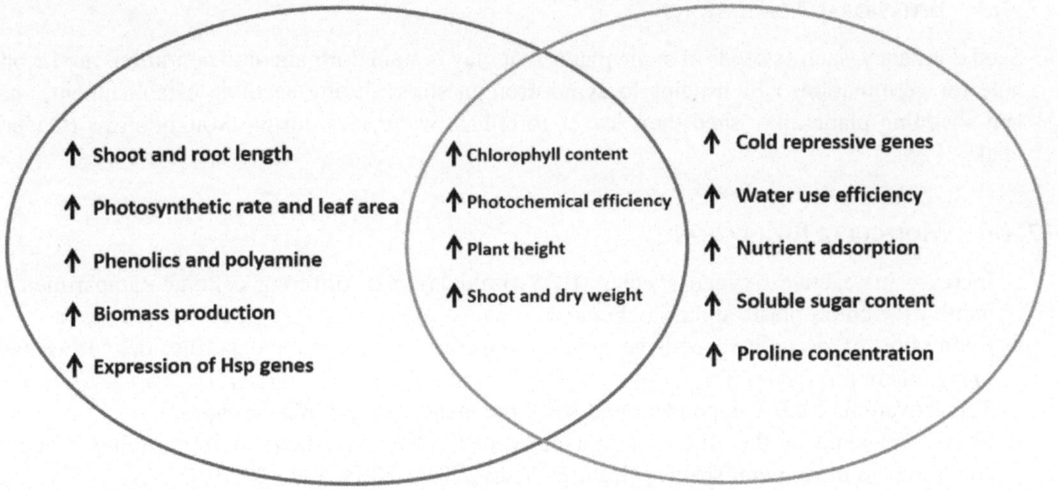

FIGURE 17.1 Fungi–plant association: Different parameters mitigating temperature stress in plants. Red circle depicts heat stress conditions; blue circle depicts cold stress conditions. The overlapping region depicts the parameters affected in both conditions. Upright arrow indicates an increment.

increases. Apart from these, the fungi also induce a significant increase in the expression of cold-responsive genes copper/zinc superoxide dismutase 1C (CSD1C), transcription factor WHIRLY 1 (Why 1), high expression of osmotically responsive gene 1 (HOS1) and CRT/DRE-binding factor 7-1 (CBF7-1) (Li et al., 2021). Therefore, a number of factors are involved in regulating heat and cold stress conditions in plants which are regulated by fungal association (Figure 17.1).

17.4 PLANTS AND DROUGHT RESISTANCE

Plants have evolved remarkable properties of adaptation to cope with drought stress. These adaptations can be categorized into four main groups.

17.4.1 Morphological Modulation

a. Modification in root systems: plants can develop deeper and more extensive root systems to access water sources deeper in the soil.
b. Reduction in leaf surface area: smaller leaves or a thicker cuticle help in minimizing water loss through transpiration.
c. Waxy coatings: some plants have waxy coatings on their leaves to reduce water loss.
d. Furthermore, there is stomata closure where plants can close their stomata, which are pores on the underside of their leaves, to conserve water (Naikwade, 2023).

17.4.2 Physiological Modulation

a. Osmotic adjustment: plants can accumulate solutes, such as sugars and amino acids, in their cells. This helps to maintain cell turgor and prevents dehydration.
b. Production of compatible solutes: plants can produce compatible solutes, such as proline and glycine betaine, which help to protect cells from damage caused by drought stress (Naikwade, 2023).

17.4.3 Behavioral Adaptations

a. Seed dormancy, such as seeds of some plants that may remain dormant until conditions are favorable for germination, which helps to avoid drought stress during seedling establishment, and leaf-shedding plants may shed their leaves to reduce water loss during drought stress (Qaderi et al., 2023).

17.4.4 Molecular Regulation

a. Increase in reactive oxygen species (ROS) production in different cellular compartments, namely the chloroplasts and mitochondria.
b. Expression of drought-responsive genes, transcription factors, aquaporins, dehydrins and many others.
c. The activation of ABA-dependent and ABA-independent signaling cascades.
d. These are some of the different molecular methods which help in overcoming drought conditions in many plant species (Kaur & Asthir, 2017; Hura et al., 2023).

17.5 REGULATION OF DROUGHT VIA PLANT-ASSOCIATED FUNGI

Fungi and plants engage in a trade-off, providing each other with essential resources to thrive in challenging environments. There are a number of ways by which different fungi species help plants to overcome drought stress conditions (Madouh & Quoreshi, 2023). One of the direct influences is via regulation of plant hormone production responsible for mitigating abiotic stress conditions. Out of the many different plant hormones, abscisic acid (ABA) is the main one whose overproduction helps in overcoming drought stress conditions. This hormone regulates the closing of stomata and directly influences plant growth during drought stress by modulating primary metabolism, secondary metabolite production, root structure and its secretion (Muhammad Aslam et al., 2022). Many fungi are directly or indirectly involved in the synthesis of ABA. In *Solanum lycopersicum*, AM fungi have been found to upregulate the expression of key genes involved in the ABA signaling pathway. These genes translate into 14-3-3 proteins which regulate the activity of many other proteins via protein–protein interactions. The plant–AM association significantly upregulates the expression of *TFT5*, *TFT7*, *TFT9* and *TFT10* (14-3-3) genes under drought stress, which causes an alteration in the transpiration rate as well as stomatal behavior (Xu et al., 2018). In tea plant roots AM fungi also cause an increase in the concentration of ABA to overcome drought stress conditions (Liu et al., 2024). Sometimes, fungi act as a direct source of ABA to help the host plant. *Pinus tabulaeformis*, a tree found in arid regions, receives ABA directly from different types of endophytic fungi found in them (Zhou et al., 2021).

During drought conditions, plants lose water, which lowers their internal water pressure and causes a number of other problems in the plant. Osmolytes include a number of low-molecular-weight molecules like glycine betaine, trehalose, proline, mannitol, amino acids and others, which helps plants to overcome drought stress conditions. Osmolytes have proved to be a driving force for water absorption in roots, helping in maintaining turgor pressure, lowering ROS levels and stabilizing cellular proteins (Ghosh et al., 2021). AM fungi can contribute to maintain osmoregulation by helping plants to accumulate compatible solutes, such as sugars and amino acids, in their cells. Trehalose, a sugar produced by AMF, assists plants to resist drought stress by maintaining biological nitrogen fixation (Zulfiqar et al., 2020). In the case of pedunculate oak (*Quercus robur* L.) plant seedlings, the association with ectomycorrhizal (ECM) fungi can greatly enhance the drought resistance properties. This association increases the production of polyamines and osmolytes like glycine betaine, spermine and spermidine (Kebert et al., 2023). Among many osmolytes, proline is the main one that has been found to help plants in overcoming drought stress, and many fungi directly affect its concentration inside plant cells. The inoculation of many plant species with endophytic fungi, *Serendipita indica*, alleviates drought stress by modulating a number of processes and the accumulation of proline is the major one. The other effects include improving water uptake, increasing chlorophyll content and the synthesis of antioxidant enzymes (Boorboori & Zhang, 2022). In plants like wheat and tomato, an endophytic fungi strain, *Zopfiella erostrata*, enhances the proline accumulation in the leaves to overcome drought stress (Miranda et al., 2023). A plant, *Araucaria araucana*, which grows in extreme drought conditions due to volcanism, fires and low rainfall, is another good example of AMF and endophytic fungi effects. These fungi increase the proline concentration in plant leaves along with other effects (Chávez et al., 2023).

It has been found that both plant hormones and osmolytes help in scavenging ROS species to overcome this stress. Fungal symbionts cause great diversity in differential gene expression of the host plant, which enables them to overcome drought stress (Aimone et al., 2023). Among many different genes they also enhance the expression of genes involved in the synthesis of antioxidant enzymes, which helps in reducing ROS. The association of endophytic fungi *Piriformospora indiac* and trifoliate orange (*Poncirus trifoliata*) improves drought resistance ability in the plant by activating an antioxidant defense system. The fungi not only enhance the activities of SOD, peroxidase and glutathione reductase but also upregulate the expression of PtFe-SOD and PtCu/Zn-SOD genes (Cao et al., 2023). *Solanum tuberosum*, one of the most common crops across the world, is also sensitive toward drought stress and can be modulated with synergetic effects of fungi. Inoculating this plant with AMF strain, *Claroideoglomus lamellosum* HMC26, improves drought resistance properties which occurs due to an increase in the activities of antioxidant enzyme system (Nahuelcura et al., 2023). The improvement in the antioxidant system has also been observed in the case of inoculating summer savory (*Satureja hortensis* L.) with AMF, *Funneliformis mosseae*. This association causes an enhancement in the activity of SOD, ascorbate peroxidase and guaiacol peroxidase up to 133.8%, 71.4% and 61%, respectively (Rasouli et al., 2023).

Apart from these parameters there are a number of other factors which are altered during the fungi–plant relationship in overcoming drought stress. These factors include modifications in root architecture, improving chlorophyll content, increasing nutrient absorption and improving soil architecture (Fales, 2023). The growth of grapevine (*Vitis vinifera*) under a drought stress condition has been favored by the presence of AMF, which results in an increase in the chlorophyll content in the leaves that causes an enhancement of the photosynthetic rate of the plant. AMF incubation can cause nearly a two-fold increase in chlorophyll a content and about a one and a half increase in chlorophyll b content in leaves (Ye et al., 2022). A similar AMF effect provides drought resistance properties in tomato plant also (*Lycopersicon esculentum* Mill. cv. Campbell 33). This association causes a number of effects on the host plant that include a reduction in ROS, improving leaf water potential and osmolyte concentration, but the compositions of chlorophyll a and chlorophyll b are greatly affected, which makes the plant drought resistant (Soussani et al., 2023). The fungi–plant association induces not only a single effect to overcome stress conditions, hence sometimes there are two or more major factors contributing to stress resistance properties. In the case of maize plant, AMF and dark septate endophytic incubation produce major effects that include an increase in chlorophyl contents but a decrease in the expression of aquaporin genes (*ZmPIP1* and *ZmPIP2*). The downregulation of aquaporin genes prevents water loss through cell membranes and is a positive response against water deficit soil (Gong et al., 2023). However, there is also an upregulation of aquaporin genes on inoculating the plant with AMF and this helps in enabling more efficient water absorption. This has been found in the case of the symbiotic association of wheat (*Triticum aestivum*) with AMF, *Funneliformis mosseae* (Asadollahi et al., 2023).

17.6 CONCLUSION

A number of exogenous and endogenous fungi, along with their products, help plants to combat temperature and drought stress conditions. However, there is still a need for improvements in this association, and many plant species do not show a positive correlation with fungi. This enables the fungi to confer abiotic resistance properties to certain plant species. Thus, new fungal species still need to be explored for the betterment of plants, and this can be achieved via modern biotechnological tools like gene cloning and tissue engineering.

REFERENCES

Adedayo, A.A. and Babalola, O.O. 2023. Fungi that promote plant growth in the rhizosphere boost crop growth. *Journal of Fungi* 9:239.

Aimone, C.D., Giauque, H. and Hawkes, C.V. 2023. Fungal symbionts generate water-saver and water-spender plant drought strategies via diverse effects on host gene expression. *Phytobiomes Journal* 7:172–183.

Ali, A.H., Abdelrahman, M., Radwan, U., El-Zayat, S. and El-Sayed, M.A. 2018. Effect of thermomyces fungal endophyte isolated from extreme hot desert-adapted plant on heat stress tolerance of cucumber. *Applied Soil Ecology* 124:155–162.

Al-Khayri, J.M., Rashmi, R., Toppo, V., Chole, P.B., Banadka, A., Sudheer, W.N., Nagella, P., Shehata, W.F., Al-Mssallem, M.Q., Alessa, F.M. and Almaghasla, M.I. 2023. Plant secondary metabolites: The weapons for biotic stress management. *Metabolites* 13:716.

Asadollahi, M. Iranbakhsh, A Ahmadvand, R., Ebadi, M. and Mehregan I. 2023. Synergetic effect of water deficit and arbuscular mycorrhizal symbiosis on the expression of aquaporins in wheat (Triticum aestivum L.) roots: Insights from NGS RNA-sequencing. *Physiology and Molecular Biology of Plants* 29:195–208.

Ayaz, M., Li, C.H., Ali, Q., Zhao, W., Chi, Y.K., Shafiq, M. Ali, F., Yu, X.Y., Yu, Q., Zhao, J.T. and Yu, J.W. 2023. Bacterial and fungal biocontrol agents for plant disease protection: Journey from lab to field, current status, challenges, and global perspectives. *Molecules* 28:6735.

Bakhshi, S., Eshghi, S. and Banihashemi, Z. 2023. Application of candidate endophytic fungi isolated from extreme desert adapted trees to mitigate the adverse effects of drought stress on maize (Zea mays L.). *Plant Physiology and Biochemistry* 202:107961.

Bhandari, U., Gajurel, A., Khadka, B., Thapa, I., Chand, I., Bhatta, D., Poudel, A., Pandey, M., Shrestha, S. and Shrestha, J. 2023. Morpho-physiological and biochemical response of rice (Oryza sativa L.) to drought stress: A review. *Heliyon* 9(3).

Boorboori, M.R. and Zhang, H.Y. 2022. The role of Serendipita indica (Piriformospora indica) in improving plant resistance to drought and salinity stresses. *Biology* 11:952.

Cao, J.L., W.X., Zou, Y.N. and Wu, Q.S. 2023. An endophytic fungus, Piriformospora indica, enhances drought tolerance of trifoliate orange by modulating the antioxidant defense system and composition of fatty acids. *Tree Physiology* 43:452–466.

Cao, X., Sui, J., Li, H., Yue, W., Liu, T., Hou, D., Liang, J. and Wu, Z. 2023. Enhancing heat stress tolerance in Lanzhou lily (Lilium davidii var. unicolor) with Trichokonins isolated from Trichoderma longibrachiatum SMF2. *Frontiers in Plant Science* 14:1182977.

Castro, H., Dias, M.C., Sousa, J.P. and Freitas, H. 2023. Functional groups response to water deficit in mediterranean ecosystems. *Plants* 12:1471.

Chávez, D., Rivas, G., Machuca, Á., Santos, C., Deramond, C., Aroca, R. and Cornejo, P. 2023. Contribution of arbuscular mycorrhizal and endophytic fungi to drought tolerance in Araucaria araucana seedlings. *Plants* 12:2116.

Chen, L., Liu, Y., Xu, Z. and Song, Y. 2023. Beneficial microbe-plant interactions under biotic/abiotic stress conditions. *Frontiers in Microbiology* 14:1294042.

Cotrina Cabello, G.G., Ruiz Rodriguez, A., Husnain Gondal, A., Areche, F.O., Flores, D.D.C., Astete, J.A.Q., Camayo-Lapa, B.F., Yapias, R.J.M., Jabbar, A., Yovera Saldarriaga, J. and Salas-Contreras, W.H. 2023. Plant adaptability to climate change and drought stress for crop growth and production. *CABI Reviews*. https://doi.org/10.1079/cabireviews.2023.0004

Cui, X., Liu, S., Zhang, L., Guo, X., Li, T., Zhang, X., Wang, Q., Zeng, W., Huang, J., Duan, Q. and Cao, Y. 2022. Endophytic extract Zhinengcong alleviates heat stress-induced reproductive defect in Solanum lycopersicum. *Frontiers in Plant Science* 13:977881.

Ding, Y. and Yang, S. 2022. Surviving and thriving: How plants perceive and respond to temperature stress. *Developmental Cell* 57:947–958.

dos Santos, T.B., Ribas, A.F., de Souza, S.G.H., Budzinski, I.G.F. and Domingues, D.S. 2022. Physiological responses to drought, salinity, and heat stress in plants: A review. *Stresses* 2:113–135.

Fales, R.J. 2023. *Climate change impacts on kelp: Physiological responses across habitats, species, and populations* (Doctoral dissertation, University of Washington).

Ghosh, U.K., Islam, M.N., Siddiqui, M.N. and Khan, M.A.R. 2021. Understanding the roles of osmolytes for acclimatizing plants to changing environment: A review of potential mechanism. *Plant Signaling & Behavior* 16:1913306.

Gong, M., Bai, N., Wang, P., Su, J., Chang, Q. and Zhang, Q. 2023. Co-Inoculation with Arbuscular Mycorrhizal fungi and Dark Septate endophytes under drought stress: Synergistic or competitive effects on maize growth, photosynthesis, root hydraulic properties and Aquaporin? *Plants* 12:2596.

Hossain, A., Skalicky, M., Brestic, M., Maitra, S., Ashraful Alam, M., Syed, M.A., Hossain, J., Sarkar, S., Saha, S., Bhadra, P. and Shankar, T., 2021. Consequences and mitigation strategies of abiotic stresses in wheat (Triticum aestivum L.) under the changing climate. *Agronomy*, 11(2):241.

Hura, T., Hura, K. and Ostrowska, A. 2023. Drought-stress induced physiological and molecular changes in plants 2.0. *International Journal of Molecular Sciences* 24:1773.

Ismail, A.H., Mehmood, A.S.I.F., Qadir, M.U.H.A.M.M.A.D., Husna, A.I., Hamayun, M.U.H.A.M.M.A.D. and Khan, N.A.E.E.M. 2020. Thermal stress alleviating potential of endophytic fungus rhizopus oryzae inoculated to sunflower (Helianthus annuus L.) and soybean (Glycine max L.). *Pakistan Journal of Botany* 52:1857–1865.

Javed, J., Rauf, M., Arif, M., Hamayun, M., Gul, H., Ud-Din, A., Ud-Din, J., Sohail, M., Rahman, M.M. and Lee, I.J. 2022. Endophytic fungal consortia enhance basal drought-tolerance in Moringa oleifera by upregulating the antioxidant enzyme (APX) through Heat shock factors. *Antioxidants* 11:1669.

Jumrani, K., Bhatia, V.S., Kataria, S., Alamri, S.A., Siddiqui, M.H. and Rastogi, A. 2022. Inoculation with arbuscular mycorrhizal fungi alleviates the adverse effects of high temperature in soybean. *Plants* 11: 2210.

Kaur, G. and Asthir, B. 2017. Molecular responses to drought stress in plants. *Biologia plantarum* 61:201–209.

Kaur, R. and Saxena, S. 2023. Evaluation of drought-tolerant endophytic fungus Talaromyces purpureogenus as a bioinoculant for wheat seedlings under normal and drought-stressed circumstances. *Folia Microbiologica* 68:781–799.

Kebert, M., Kostić, S., Čapelja, E., Vuksanović, V., Stojnić, S., Markić, A.G., Zlatković, M., Milović, M., Galović, V. and Orlović, S. 2022. Ectomycorrhizal fungi modulate Pedunculate Oak's heat stress responses through the alternation of polyamines, phenolics, and osmotica content. *Plants* 11:3360.

Kebert, M., Kostić, S., Stojnić, S., Čapelja, E., Markić, A.G., Zorić, M., Kesić, L. and Flors, V. 2023. A fine-tuning of the plant hormones, polyamines and osmolytes by Ectomycorrhizal fungi enhances drought tolerance in Pedunculate Oak. *International Journal of Molecular Sciences* 24:7510.

Kilner, C.L., Carrell, A.A., Wieczynski, D.J., Votzke, S., DeWitt, K., Yammine, A., Shaw, J., Pelletier, D.A., Weston, D.J. and Gibert, J.P. 2023. Climate change factors interactively shift Peatland functional microbial composition in a whole-ecosystem warming experiment. *bioRxiv* 2023-03.

Klarenberg, I.J., Keuschnig, C., Russi Colmenares, A.J., Warshan, D., Jungblut, A.D., Jónsdóttir, I.S. and Vilhelmsson, O. 2022. Long-term warming effects on the microbiome and nifH gene abundance of a common moss species in sub-Arctic tundra. *New Phytologist* 234:2044–2056.

Leisner, C.P., Potnis, N. and Sanz-Saez, A. 2023. Crosstalk and trade-offs: Plant responses to climate change-associated abiotic and biotic stresses. *Plant, Cell & Environment 46*:2946–2963.

Li, D., Bodjrenou, D.M., Zhang, S., Wang, B., Pan, H., Yeh, K.W., Lai, Z. and Cheng, C. 2021. The endophytic fungus Piriformospora indica reprograms banana to cold resistance. *International Journal of Molecular Sciences* 22:4973.

Li, L., Feng, Y., Qi, F. and Hao, R. 2023. Research progress of Piriformospora indica in improving plant growth and stress resistance to plant. *Journal of Fungi* 9:965.

Li, S., Yang, W., Guo, J., Li, X., Lin, J. and Zhu, X. 2020. Changes in photosynthesis and respiratory metabolism of maize seedlings growing under low temperature stress may be regulated by arbuscular mycorrhizal fungi. *Plant Physiology and Biochemistry* 154:1–10.

Liu, C.Y., Hao, Y., Wu, X.L., Dai, F.J., Abd-Allah, E.F., Wu, Q.S. and Liu, S.R. 2024. Arbuscular mycorrhizal fungi improve drought tolerance of tea plants via modulating root architecture and hormones. *Plant Growth Regulation* 102:13–22.

Liu, X.M., Xu, Q.L., Li, Q.Q., Zhang, H. and Xiao, J.X. 2017. Physiological responses of the two blueberry cultivars to inoculation with an arbuscular mycorrhizal fungus under low-temperature stress. *Journal of Plant Nutrition* 40:2562–2570.

Liu, X.R., Rong, Z.Y., Tian, X., Hashem, A., Abd_Allah, E.F., Zou, Y.N. and Wu, Q.S. 2023. Mycorrhizal fungal effects on plant growth, Osmolytes, and CsHsp70s and CsPIPs expression in leaves of cucumber under a short-term heat stress. *Plants* 12:2917.

Madouh, T.A. and Quoreshi, A.M. 2023. The function of arbuscular mycorrhizal fungi associated with drought stress resistance in native plants of arid desert ecosystems: A review. *Diversity* 15:391.

Maillard, F., Fernandez, C.W., Mundra, S., Heckman, K.A., Kolka, R.K., Kauserud, H. and Kennedy, P.G. 2022. Warming drives a 'hummockification' of microbial communities associated with decomposing mycorrhizal fungal necromass in peatlands. *New Phytologist* 234:2032–2043.

Mathur, S. and Jajoo, A. 2020. Arbuscular mycorrhizal fungi protects maize plants from high temperature stress by regulating photosystem II heterogeneity. *Industrial Crops and Products* 143:111934.

Mathur, S., Agnihotri, R., Sharma, M.P., Reddy, V.R. and Jajoo, A. 2021. Effect of high-temperature stress on plant physiological traits and mycorrhizal symbiosis in maize plants. *Journal of Fungi* 7:867.

Miranda, V., Silva-Castro, G.A., Ruiz-Lozano, J.M., Fracchia, S. and García-Romera, I. 2023. Fungal endophytes enhance wheat and tomato drought tolerance in terms of plant growth and biochemical parameters. *Journal of Fungi* 9:384.

Muhammad Aslam, M., Waseem, M., Jakada, B.H., Okal, E.J., Lei, Z., Saqib, H.S.A., Yuan, W., Xu, W. and Zhang, Q. 2022. Mechanisms of abscisic acid-mediated drought stress responses in plants. *International Journal of Molecular Sciences* 23:1084.

Muhammad, M., Waheed, A., Wahab, A., Majeed, M., Nazim, M., Liu, Y.H., Li, L. and Li, W.J. 2023. Soil salinity and drought tolerance: An evaluation of plant growth, productivity, microbial diversity, and amelioration strategies. *Plant Stress* 11:100319.

Nahuelcura, J., Ortega, T., Peña, F., Berríos, D., Valdebenito, A., Contreras, B., Santander, C., Cornejo, P. and Ruiz, A. 2023. Antioxidant response, phenolic compounds and yield of Solanum tuberosum tubers inoculated with Arbuscular Mycorrhizal fungi and growing under water Stress. *Plants* 12:4171.

Naikwade, P.V. 2023. Plant responses to drought stress: Morphological, physiological, molecular approaches, and drought resistance. In *Plant Metabolites under Environmental Stress* 15:149–183. Apple Academic Press.

Nolan, T.M., Vukašinović, N., Liu, D., Russinova, E. and Yin, Y. 2020. Brassinosteroids: Multidimensional regulators of plant growth, development, and stress responses. *The Plant Cell* 32:295–318.

Qaderi, M.M. 2023. Environmental regulation of weed seed dormancy and germination. *Seeds* 2:259–277.

Rasouli, F., Amini, T., Skrovankova, S., Asadi, M., Hassanpouraghdam, M.B., Ercisli, S., Buckova, M., Mrazkova, M. and Mlcek, J. 2023. Influence of drought stress and mycorrhizal (Funneliformis mosseae) symbiosis on growth parameters, chlorophyll fluorescence, antioxidant activity, and essential oil composition of summer savory (Satureja hortensis L.) plants. *Frontiers in Plant Science* 14:1151467.

Reva, M., Cano, C., Herrera, M.A. and Bago, A. 2021. Arbuscular mycorrhizal inoculation enhances endurance to severe heat stress in three horticultural crops. *HortScience* 56:396–406.

Sabagh, A.E., Hossain, A., Islam, M.S., Iqbal, M.A., Amanet, K., Mubeen, M., Nasim, W., Wasaya, A., Llanes, A., Ratnasekera, D. and Singhal, R.K. 2021. Prospective role of plant growth regulators for tolerance to abiotic stresses. In Aftab, T., Hakeem, K.R. (eds) *Plant growth regulators: Signalling under stress condition* :1–38. Springer, Cham.

Satrovic, E., Cetindas, A., Akben, I. and Damrah, S. 2024. Do natural resource dependence, economic growth and transport energy consumption accelerate ecological footprint in the most innovative countries? The moderating role of technological innovation. *Gondwana Research* 127:116–130.

Shi, J., Wang, X. and Wang, E. 2023. Mycorrhizal symbiosis in plant growth and stress adaptation: From genes to ecosystems. *Annual Review of Plant Biology* 74:569–607.

Singh, A. and Roychoudhury, A. 2023. Abscisic acid in plants under abiotic stress: Crosstalk with major phytohormones. *Plant Cell Reports* 42:961–974.

Soussani, F.E., Boutasknit, A., Ben-Laouane, R., Benkirane, R., Baslam, M. and Meddich, A. 2023. Arbuscular Mycorrhizal fungi and compost-based biostimulants enhance fitness, physiological responses, yield, and quality traits of drought-stressed tomato plants. *Plants* 12:1856.

Sun, W. and Shahrajabian, M.H., 2023. The application of arbuscular mycorrhizal fungi as microbial biostimulant, sustainable approaches in modern agriculture. *Plants*, 12(17):3101.

Suzuki, N. 2023. Fine tuning of ROS, redox and energy regulatory systems associated with the functions of chloroplasts and mitochondria in plants under heat stress. *International Journal of Molecular Sciences* 24:1356.

Thangavel, P., Anjum, N.A., Muthukumar, T., Sridevi, G., Vasudhevan, P. and Maruthupandian, A. 2022. Arbuscular mycorrhizae: Natural modulators of plant–nutrient relation and growth in stressful environments. *Archives of Microbiology* 204:264.

Trivedi, P., Batista, B.D., Bazany, K.E. and Singh, B.K. 2022. Plant–microbiome interactions under a changing world: Responses, consequences and perspectives. *New Phytologist* 234:1951–1959.

Wahab, A., Muhammad, M., Munir, A., Abdi, G., Zaman, W., Ayaz, A., Khizar, C. and Reddy, S.P.P. 2023. Role of arbuscular mycorrhizal fungi in regulating growth, enhancing productivity, and potentially influencing ecosystems under abiotic and biotic stresses. *Plants* 12:3102.

Wei, H., Li, X., He, W., Kuang, Y., Wang, Z., Hu, W., Tang, M. and Chen, H. 2023. Arbuscular mycorrhizal symbiosis enhances perennial ryegrass growth during temperature stress through the modulation of antioxidant defense and hormone levels. *Industrial Crops and Products* 195:116412.

Wei, L.I., Yan-Lin, Z.H.A.I., Hong-Shuang, X.I.N.G., Li-Jun, X.I.N.G. and Shao-Xia, G.U.O. 2023. Arbuscular mycorrhizal fungi promote photosynthesis in Antirrhinum majus L. under low-temperature and weak-light conditions. *Notulae Botanicae Horti Agrobotanici Cluj-Napoca* 51:13012–13012.

Wu, J., Guo, G., Chen, D., Zhao, X., Long, M., Wang, D. and Li, X. 2023. Review of diversity and nitrogen fixation potential of bryophyte-cyanobacteria associations. *Biodiversity Science* 31:23081.

Xia, L., Sun, S., Han, B. and Yang, X. 2023. NAC domain transcription factor gene GhNAC3 confers drought tolerance in plants. *Plant Physiology and Biochemistry* 195:114–123.

Xu, L., Li, T., Wu, Z., Feng, H., Yu, M., Zhang, X. and Chen, B. 2018. Arbuscular mycorrhiza enhances drought tolerance of tomato plants by regulating the 14-3-3 genes in the ABA signaling pathway. *Applied Ssoil Ecology* 125:213–221.

Yadav, B., Jogawat, A., Gnanasekaran, P., Kumari, P., Lakra, N., Lal, S.K., Pawar, J. and Narayan, O.P. 2021. An overview of recent advancement in phytohormones-mediated stress management and drought tolerance in crop plants. *Plant Gene 25*:100264.

Yan, Z., Ma, T., Guo, S., Liu, R. and Li, M. 2021. Leaf anatomy, photosynthesis and chlorophyll fluorescence of lettuce as influenced by arbuscular mycorrhizal fungi under high temperature stress. *Scientia Horticulturae* 280:109933.

Ye, Q., Wang, H. and Li, H. 2022. Arbuscular mycorrhizal fungi improve growth, photosynthetic activity, and chlorophyll fluorescence of vitis vinifera l. cv. ecolly under drought stress. *Agronomy* 12:1563.

Yeasmin, R., Bonser, S.P., Motoki, S. and Nishihara, E. 2019. Arbuscular mycorrhiza influences growth and nutrient uptake of asparagus (Asparagus officinalis L.) under heat stress. *HortScience* 54:846–850.

Zhang, H., Zhu, J., Gong, Z. and Zhu, J.K. 2022. Abiotic stress responses in plants. *Nature Reviews Genetics* 23:104–119.

Zhang, K., Gao, N., Li, Y., Dou, S., Liu, Z., Chen, Y., Ma, C. and Zhang, H., 2023. Responses of maize (Zea mays L.) seedlings growth and physiological traits triggered by polyvinyl chloride microplastics is dominated by soil available nitrogen. *Ecotoxicology and Environmental Safety*, 252:114618.

Zheng, M., Zhong, S., Wang, W., Tang, Z., Bu, T. and Li, Q. 2023. Serendipita indica promotes the growth of Tartary Buckwheat by stimulating hormone synthesis, metabolite production, and increasing systemic resistance. *Journal of Fungi* 9:1114.

Zhou, X.R., Dai, L., Xu, G.F. and Wang, H.S. 2021. A strain of Phoma species improves drought tolerance of Pinus tabulaeformis. *Scientific Reports* 11:7637.

Zulfiqar, F., Akram, N.A. and Ashraf, M. 2020. Osmoprotection in plants under abiotic stresses: New insights into a classical phenomenon. *Planta* 251:1–17.

18 Endophytes for Sustainable Agricultural Practices

Boon for Mirzapur District, U.P., India

*Sneha Dwivedi and Alok Kumar Singh**
*Corresponding author.

18.1 INTRODUCTION

The term "sustainable development" has many different meanings and interpretations. From all these angles, it seems clear that the key to human civilization's survival and ability to meet its various needs without jeopardizing the ability of future generations to meet their own is through sustainable development. The goal of sustainable development is to achieve human development objectives while balancing natural systems' capacity to provide the ecosystem service and the natural attributes that underpin society and the economy. It strives for a society in which the resources and living conditions are used to meet requirements without jeopardizing the stability and integrity of natural resources. Constraints facing the global community have a negative effect on the production of environmentally friendly and sustainable agriculture. Many limiting variables, such as insufficient or inconsistent rainfall, insufficient water supplies and water-use efficiency, and declining soil fertility coupled with a drop in soil nutrients, impede improvements in crop production and productivity. Abiotic stressors, urbanization, land degradation, overuse of synthetic agrochemicals, harmful pests and phytopathogens, and overexploitation of agricultural soil are some of the causes of these limitations. By 2030, there will be more than 9 billion people on the planet, which will put further demand on these already limited resources (Hannah et al., 2017). This enormous undertaking is further complicated by the problem of climate change and its detrimental consequences on crop output. Variations in the local climate have a significant impact on plant growth and productivity, even at a smaller scale. In addition to the effects of variations in the elements of the climate – such as temperature, humidity, radiation and precipitation – anthropogenic activities are also changing the climate, which has detrimental effects on the sustainable productivity of crops in the local ecological system and the agricultural industry as a whole, even globally.

Due to the cornerstone of a revived "green revolution in agriculture production," reports stressing the significance and advantageous qualities of microbial-based formulations in boosting crop development and productivity have made the headlines (Baez et al., 2017; Singh et al., 2016). To achieve environmental sustainability and balance, other options are desperately needed. Furthermore, it is essential to proceed in this direction because imbalances in the microbiome of the various plant–host relationships and a decline in beneficial microbial diversity frequently contribute to plant infestation by pests and disease-causing pathogens, as well as plants' susceptibility to various environmental stresses. To synthesize enzymes or metabolites that can suppress or inhibit the detrimental effects of plant pathogens, as well as to produce crucial phytohormones and guarantee the plant's resistance to environmental stress, plant microbiomes are an essential partner in defending plants from stress (Le et al., 2017).

DOI: 10.1201/9781003486275-21

This review chapter is aimed at showcasing specific research articles from the Web of Science that have explored the roles of endophytic bacteria and fungi as phytopathogens that hinder bio-inoculants and as possible promoters of plant growth. This chapter clarifies the methods utilized to accomplish these goals. With regard to clever biotechnological techniques, the current level of knowledge regarding the potential of endophytic bacteria and fungi has also been emphasized, as has the potential application of these significant partners of the plant microbiomes as biological inoculants in sustainable crop management. In addition, the necessity of employing high-throughput molecular approaches to the bioprospects for novel endophytes in various ecological niche, define their roles, and screen and also optimize their fitness for environmental stress in order to effectively survive and compete in the field has been discussed, in order to emphasize the significant roles that microbial resources play and their potential to improve agro-food crop production in the sustainable and environmentally friendly ways.

18.2 ENDOPHYTES

The world of plants makes the relationships between all the living things on Earth more clear. In their different ecological niches, microbes and plants coexist in intimate harmony. A symbiotic relationship frequently exists between the plant host and its diverse community of microbes. In the many environmental niches of these plant hosts, microorganisms can be classified as endophytic, rhizospheric or epiphytic. Endophytes are the most closely related and intimate group of microorganisms that live in a symbiotic connection with their plant host among all of these other species. To put it simply, endophytes are microbial forms that help colonize plant tissues without having any negative effects. Without exception, every live plant is colonized by a variety of endophytes; in fact, plants serve as the mutualistic, beneficial host to a wide range of endophytic microorganisms, which is essential to the growth and health of plants. Through horizontal transfer from soil to plants or vertical planting, endophytic bacteria can infiltrate and colonize plants. In either case, interactions between microbes and plants still result in mutualistic advantages (Verma & White, 2018; Huang et al., 2016). Through this mutualistic relationship, which is highly advantageous to both parties, endophytic microbes are able to provide their plant host with a variety of benefits, including increased nutrient availability, defense against pathogens and destructive pests, enhanced plant resistance to environmental stressors, assistance in regulating the host plant's development and assistance in managing weed growth. Because endophytic bacteria employ various techniques to bestow these crucial tasks on their host plant, the features they acquire are seen to be advantageous (Santoyo et al., 2016).

18.3 BACTERIAL ENDOPHYTES

Plants must establish mutualistic relationships with other living organisms in the ecological system to successfully flourish in their ecological niche. The relationship that exists between various microbes and plants is a prime example of a beneficial association (Miliute et al., 2015; Mishra et al., 2013). Certain tissue-colonizing bacteria, such as *Bradyrhizobium* sp. SUTNa-2, *Pseudomonas granadensis* T6 and *Rhizobium larrymoorei* E2 (Greetatorn et al., 2019; Shen et al., 2019), develop a tight relationship with their host plant and even benefit it in favorable and unfavorable circumstances. One benefit that endophytic bacteria provide to their host plants is that they aid in mitigating different growth-limiting biotic and abiotic impacts. Endophytic bacteria give their host plant the ability to withstand stress. They enhance the growth of the host plant and cause allelopathic effects in it (Kushwaha et al., 2020). The bioactivities of selected endophytic bacteria which help host plants flourish and survive in their specific ecological niches are summarized in Table 18.1.

Plants tissues such as the root tissues, stems, leaves, various types of seeds, fruits, and also tubers, ovules and nodules, have been found to harbor bacterial endophytes (Nair et al., 2014). On the

TABLE 18.1
Biological Activities of Bacterial Endophytes to Increase Plant Productivity

Bacterial endophytes	Host plants	Bioactivity impacts	References
Curtobacterium sp. SAK 1	*Glycine max*	PGP and salinity stress tolerance	Khan et al., 2019
Bacillus tequilensis (PBE1)	*Solanum lycopersicum*	PGP and biocontrol of pathogens	Bhattacharya et al., 2019
Bradyrhizobium sp. SUTNa-2	*Oryza sativa*	Plant-growth-promoting	Greetatorn et al., 2019
Pseudomonas granadensis T6; *Rhizobium larrymoorei* E2	*Oryza sativa*	Plant-growth-promoting and pesticide tolerance	Shen et al., 2019
Bacillus amyloliquefaciens EPP90; *Bacillus subtilis*; *Bacillus pumilus*	*Pennicetum glaucum*	PGP and abiotic stress tolerance	Kushwaha et al., 2020

other hand, bacterial endophytes are more common in the root tissues than in aerial parts of plants. Numerous exploratory research projects have suggested the enormous agro-biotechnological potential of using endophytic bacteria as bio-inoculants to create a long-lasting, environmentally friendly and sustainable agricultural production system.

18.4 FUNGAL ENDOPHYTES

In their intimate, mutualistic beneficial relationship with their plant's host, fungal endophytes enhance their host plant's ecology by helping them to withstand harmful biotic and abiotic pressures. Endophytes, in turn, gain nourishment and defense from the plant (Suryanarayanan, 2017). Plant host tissues like stems, fruits, flowers, roots, leaves and branches are colonized by fungi endophytes; this process occurs asymptomatically and has no negative effects (Saikkonen et al., 2006). They play a major role in the enormous biodiversity of the kingdom of fungus. It is well known that fungi endophytes benefit their plant hosts. Among them are the prevention or reduction of harm from pests or harmful insects. Furthermore, reports suggest that these fungal endophytes penetrate plants, making them less vulnerable to the damaging effects of pests. By halting the pest's growth and the different development phases as well as its eating habits and reproductive stages, endophytes provide advantages to their host plants and ultimately impact the pest's overall survival. The example of the maize plant has been used to emphasize the containment of pest damage ascribed to fungal endophytes (Cherry et al., 2004). Additionally, reports of insect damage control have been made for banana, faba bean, common bean plants, as well as tomato, various types of cotton and coffee plants. Fungal endophytes have been shown by researchers to be successful at suppressing and containing pest damage to plants (Jaber & Araj, 2018; Rondot & Reineke, 2018). Because fungal endophytes produce secondary mycotoxigenic compounds in their host plant that are poisonous to pests, the damage caused by the pests is lessened (Gurulingappa et al., 2010).

Given the numerous benefits that endophytic bacteria and fungi have been shown to have for agrobiotechnology, it is not unexpected that the use of endophytic microbial formulations has gained popularity as a secure, sustainable and ecologically friendly method of enhancing agricultural outputs. Because fungal endophytes can produce essential compounds that support plant growth, inhibit pathogens and pests that damage plants, and confer immunity, fitness and abiotic stress-tolerant traits on plants, they have been shown to be a viable option for achieving sustainable crops for intensification (Table 18.2).

TABLE 18.2
Use of Fungal Endophytes to Increase Plant Productivity

Fungal endophytes	Host plants	Bioactive impacts	References
Paecilomyces variotii, *Penicillium purpurogenum*	*Caralluma acutangula*	Plant-growth-promoting	Ali et al., 2019
Aspergillus oryzae	*Raphanus sativus*	Plant-growth-promoting and biocontrol	Sun et al., 2018
Yarrowia lipolytica	*Euphorbia milii* L.	Plant-growth-promoting and salinity stress tolerance	Gul et al., 2019
Piriformospora indica	*Cymbidium aloifolium*	Plant-growth-promoting and abiotic stress tolerance	Shah et al., 2019
Mucor sp.	*Arabidopsis arenosa*	Metal toxicity tolerance	Domka et al., 2019

18.5 THE PROCESSES OF PLANT TISSUE COLONIZATION OF ENDOPHYTES INTO HOST PLANTS

"How do the endophytic microbes enter and colonize the tissues of their plant hosts?" (Gul et al., 2019; Shah et al., 2019). The process by which microbes enter and eventually colonize their plant host is fraught with complexity. Typically, the process of colonization starts with an early communication through the plant root exudates and the related microbes' tentative entry. Important organic compounds that function as chemo-attractants are produced by the plant's rhizospheric zones and root exudates. These chemicals aid in the transmission of signals or communication between microorganisms and plant roots, which in turn helps to attract more germs. This subsequently starts the endophytic entry process, which allows the plant host's tissues to become colonized.

18.6 THE POSSIBILITY OF INCREASING AGRICULTURAL PRODUCTIVITY WITH ENDOPHYTIC BACTERIA AND FUNGI

In order to maintain the environmental balance and achieve sustained productivity from intensified crop productions, it is imperative that we look beyond these obstacles in a modern, agriculturally sustainable manner. Multiple constraints currently negatively affect improvements in the productivity of the agriculture food system. The utilization of endophytic bacteria and fungus as a sustainable alternative to increase agricultural productivity has garnered significant interest in recent times. The potential of endophytic fungi is demonstrated by their ability to stimulate plant growth (Figure 18.1), induce tolerance characteristics in response to abiotic challenges and increase immunological fitness in several plant species (Kumar et al., 2019). They can also be used to biologically control phytopathogens, damaging pests and insects. Endophytic microorganisms exhibit a variety of multifaceted functional features that have a positive impact on plant productivity due to their capacity to penetrate, settle in and permanently colonize plant tissue. Therefore, the question that needs to be answered is: How do endophytes actually manage to provide plants these advantageous functional traits? One characteristic that sets endophytes apart is their multifunctionality, since they can operate as both agents for biocontrol and for enhancing plant development and yields. According to previous studies, endophytic bacteria play a crucial role in the interaction they have with their plant hosts in the ecological environment because they provide the host plant with advantageous functional traits. These advantages include a greater capacity to withstand abiotic stress, the ability to fight off or inhibit pathogens that cause disease and the ability to encourage the plant's uptake of nutrients and growth.

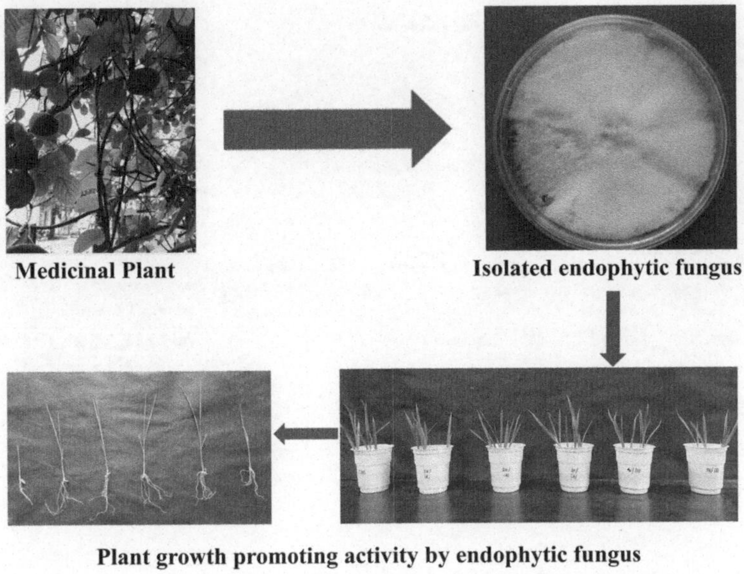

FIGURE 18.1 The functional activity of endophytic fungi that benefit plants.

Studies have demonstrated the functional characteristics of endophytes. They can, as previously said, boost the amount of nutrients that plants receive, inhibit plant diseases, pests, insects and nematodes, cause a plant to create phytohormones and make a plant more resilient to abiotic pressures. Moreover, they are capable of fixing, solubilizing and mobilizing necessary components for the plant's use (Khan et al., 2016; Yadav & Yadav, 2019). Through several ways, endophytic bacteria also contribute to increases in plant productivity. These mechanisms include the synthesis of enzymes and other necessary bioactive metabolites, as well as the breakdown of inorganic nutritional materials from the soil to allow them to reach the roots of their host plant (Chhipa & Deshmukh, 2019). In addition to these advantages, endophytes also increase production by shielding their host plant from diseases by controlling the synthesis of crucial phytohormones, which in turn affects the host plant's physiological response. The advantageous relationship endophytes have with plants in terms of abiotic stress tolerance is another important way in which they contribute to increased agricultural productivity (Ripa et al., 2019; Etminani & Harighi, 2018). According to reports in the literature, endophytes are beneficial in mitigating abiotic stressors because of a variety of processes that have been emphasized in various research works. These processes endow their plant host with the ability to withstand abiotic stress. They achieve this by creating metabolites that function against stress, activating and expressing the genes responsive to the stress and creating scavengers like the reactive oxygen species to deal with free radicals. Researchers' reports from the literature support the theory that endophytes can assist plants in mitigating environmentally induced stresses, such as drought, higher temperatures, excess salinity and frequent flooding, all of which are classified as abiotic stressors. Climate change is known to have adverse effects on the growth and survival of vulnerable plants. The capacity of endophytes to withstand biotic stressors by inhibiting phytopathogens through antagonistic activities is another crucial functional role they play in raising agricultural output. Through the production of physiological and defensive pathways in the host plant that are genetically connected to disease-causing pathogens and pests, endophytes have also been demonstrated to suppress phytopathogens.

Endophytes generate and secrete bioactive metabolites such as salicylic acid, jasmonic acid, siderophores, phytoalexins and volatile organic carbon to efficiently suppress phytopathogens and

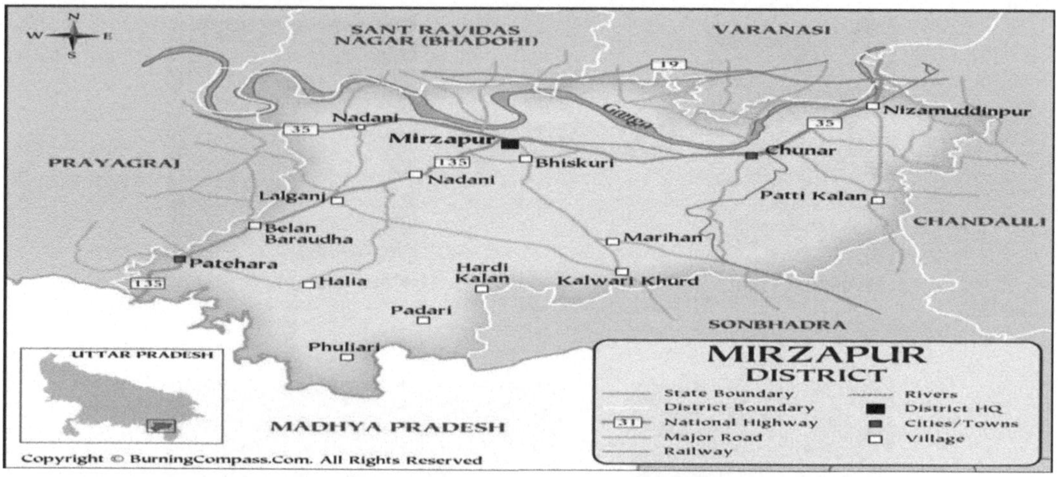

FIGURE 18.2 Map of the study area.

pests. It is well known that each of these bioactive metabolites has strong inhibitory effects on pests and phytopathogens that could endanger the plant (Etminani & Harighi, 2018).

18.7 THE RESEARCH AREA

Mirzapur-Vindhyachal is geographically located in the southeastern area of the state of Uttar Pradesh. It is located approximately 30 miles (48 km) southwest of Varanasi on the Ganga River (Figure 18.2). Mirzapur was most likely established in the 1700s. West-central India's Vindhya Range, sometimes called Vindhyachal, is a complicated, irregular network of plateau escarpments, hill ranges, highlands and mountain ridges. In terms of geology, the Vindhyas are not technically a single mountain range. The term Vindhyas refers to a group of hill ranges in central India, including the Satpura Range, although its exact boundaries are not well defined. Today, the phrase mostly refers to the Madhya Pradesh escarpment and its hilly outcrops that run north of and roughly parallel to the Narmada river. Depending on the definition, the range includes places like Gujarat in the west, Chhattisgarh in the east and Uttar Pradesh and Bihar in the north.

In Indian mythology and history, the Vindhya regions plays a major role. The southern border of the Āryāvarta, the ancient Indo-Aryan peoples' homeland, is mentioned in several ancient writings as being the Vindhyas. Despite the fact that Indo-Aryan languages are now spoken south of the Vindhyas, the range is still regarded as the historical divide separating north and south India. The Vindhya Range served as the inspiration for the old Vindhya Pradesh.

Endophytes isolated from Mirzapur district include *Aspergillus niger*, *Alternaria* spp., *Curvularia* spp., *Fusarium* spp., *Rhizopus* spp., *Mucor* sp. and *Penicillium* sp. from the *Buchanania* species and the prevalence of endophytic fungal genus isolated from *Celastrus* species include *Aspergillus fumigatus*, *Penicillium* spp., *Alternaria* sp., *Curvularia* sp., *Fusarium* sp., etc. (Dwivedi et al., 2023).

18.8 ENDOPHYTES IN SUSTAINABLE AGRICULTURAL INTENSIFICATION: A CONCLUSION AND PROSPECTS FOR THE FUTURE

It has been demonstrated that bacterial and fungal endophytes have a great deal of promise as environmentally friendly natural resources that may be used in agricultural food production to enable crop cultivation to be intensified, increasing agriculture's productivity and environmental sustainability. These could be accomplished by using more affordable, easily accessible natural bioresources in a sustainable manner. The various uses of endophytes in agricultural crop production may be beneficial, according to current research on these organisms' potential uses, as bioeffectors, biofertilizers, biocontrol agents that reduce biopesticidal risks and biostimulants.

In order to reduce the usage of traditional agrochemicals and to ensure the sustainable activity of the ecological balance, endophytic organisms, including synergistic consortia of both bacterial and fungal endophytes, should also be developed and supported in their role as bioformulated inoculants for use in agriculture. Now is the time to save the ecosystem. Increasing the intensity of sustainable agriculture techniques is one way to achieve this. In the next green revolution, endophytic microbes will be crucial to our efforts to preserve the ecosystem. In the near future, these different organisms could also be used more effectively in the field of applications as bio-inoculants for the intensification of sustainable agriculture, as more is known about the genetics, metabolics and physiological processes of endophytic microorganisms and the different symbiotic interrelationships among plants through their advanced biotechnological screening and the investigative processes.

However, there are still gaps that need to be filled. Future studies should focus on the following important areas:

1. It is advised to look into the community and functions of endophytic microbes using cutting-edge biotechnological techniques (omics). It is possible that additional exploratory research into the endomicrobiome of plant tissue will reveal the novel endophytic bacteria with important functional properties that can also be further utilized in the ongoing efforts to improve crops in a sustainable manner. To get the best results, the next generation is to use molecular techniques. Utilizing more advanced biotechnological instruments will be crucial for researching the endomicrobiome's functional features related to transcriptomics, proteomics and genomes.
2. A thorough bioprospecting investigation of the endophytic microorganisms from various ecological niches – such as the marine environment, harsh environments, etc. – is required to identify and characterize novel endophytes with particular characteristics that may be even advantageous to crop production.
3. In order to get over established environmental obstacles, there is a need for efficient screening for the significant and vital metabolites, so that they may be used correctly in the field.

REFERENCES

Ali, Sajid, Sumera Afzal Khan, Muhammad Hamayun, Amjad Iqbal, Abdul Latif Khan, Anwar Hussain, and Mohib Shah. "Endophytic fungi from Caralluma acutangula can secrete plant growth promoting enzymes." *Fresenius Environment Bulletin* 28 (2019): 2688–2696.

Baez-Rogelio, Antonino, Yolanda Elizabeth Morales-García, Verónica Quintero-Hernández, and Jesús Muñoz-Rojas. "Next generation of microbial inoculants for agriculture and bioremediation." *Microbial Biotechnology* 10, no. 1 (2017): 19–21.

Bhattacharya, Arpita, Ved Prakash Giri, Satyendra Pratap Singh, Shipra Pandey, Priyanka Chauhan, Sumit Kumar Soni, Suchi Srivastava, Poonam C. Singh, and Aradhana Mishra. "Intervention of bio-protective endophyte Bacillus tequilensis enhance physiological strength of tomato during Fusarium wilt infection." *Biological Control* 139 (2019): 104074.

Cherry, Andy J., Agnassim Banito, Denis Djegui, and Chris Lomer. "Suppression of the stem-borer Sesamia calamistis (Lepidoptera; Noctuidae) in maize following seed dressing, topical application and stem injection with African isolates of Beauveria bassiana." *International Journal of Pest Management* 50, no. 1 (2004): 67–73.

Chhipa, H., and S. K. Deshmukh. "Fungal endophytes: rising tools in sustainable agriculture production." In Sumita Jha (Ed.)*Endophytes and secondary metabolites*. Springer International Publishing AG, Cham (2019): 1–24.

Domka, Agnieszka, Piotr Rozpądek, Rafał Ważny, and Katarzyna Turnau. "Mucor sp.—An endophyte of Brassicaceae capable of surviving in toxic metal-rich sites." *Journal of Basic Microbiology* 59, no. 1 (2019): 24–37.

Dwivedi, Sneha, Ankita Rai, Alok Kumar Singh, Manju Srivastava, and Anand Prakash Singh. "Isolation of fungal endophytes at diferent pH from two wild plants of Mirzapur district (UP) India." *Flora and Fauna* 29, no. 2 (2023): 263–268

Etminani, Faegheh, and Behrouz Harighi. "Isolation and identification of endophytic bacteria with plant growth promoting activity and biocontrol potential from wild pistachio trees." *The Plant Pathology Journal* 34, no. 3 (2018): 208.

Greetatorn, Teerana, Shun Hashimoto, Sukanlaya Sarapat, Panlada Tittabutr, Nantakorn Boonkerd, Toshiki Uchiumi, and Neung Teaumroong. "Empowering rice seedling growth by endophytic Bradyrhizobium sp. SUTN9-2." *Letters in Aapplied Microbiology* 68, no. 3 (2019): 258–266.

Gul Jan, Farzana, Muhammad Hamayun, Anwar Hussain, Gul Jan, Amjad Iqbal, Aman Khan, and In-Jung Lee. "An endophytic isolate of the fungus Yarrowia lipolytica produces metabolites that ameliorate the negative impact of salt stress on the physiology of maize." *BMC Microbiology* 19 (2019): 1–10.

Gurulingappa, Pampapathy, Gregory A. Sword, Gregory Murdoch, and Peter A. McGee. "Colonization of crop plants by fungal entomopathogens and their effects on two insect pests when in planta." *Biological Control* 55, no. 1 (2010): 34–41.

Hannah, Lee, Marc Steele, Emily Fung, Pablo Imbach, Lorriane Flint, and Alan Flint. "Climate change influences on pollinator, forest, and farm interactions across a climate gradient." *Climatic Change* 141 (2017): 63-75.

Huang, Yali, Zaoyuan Kuang, Wenfeng Wang, and Lixiang Cao. "Exploring potential bacterial and fungal biocontrol agents transmitted from seeds to sprouts of wheat." *Biological Control* 98 (2016): 27–33.

Jaber, Lara R., and Salah-Eddin Araj. "Interactions among endophytic fungal entomopathogens (Ascomycota: Hypocreales), the green peach aphid Myzus persicae Sulzer (Homoptera: Aphididae), and the aphid endoparasitoid Aphidius colemani Viereck (Hymenoptera: Braconidae)." *Biological Control* 116 (2018): 53–61.

Khan, Abdul Latif, Ahmed Al-Harrasi, Ahmed Al-Rawahi, Zainab Al-Farsi, Aza Al-Mamari, Muhammad Waqas, Sajjad Asaf et al. "Endophytic fungi from frankincense tree improves host growth and produces extracellular enzymes and indole acetic acid." *PloS one* 11, no. 6 (2016): e0158207.

Khan, Muhammad Aaqil, Sajjad Asaf, Abdul Latif Khan, Ihsan Ullah, Sajid Ali, Sang-Mo Kang, and In-Jung Lee. "Alleviation of salt stress response in soybean plants with the endophytic bacterial isolate Curtobacterium sp. SAK1." *Annals of Microbiology* 69 (2019): 797–808.

Kumar, Vinay, Ravindra Soni, Lata Jain, Biplab Dash, and Reeta Goel. "Endophytic fungi: recent advances in identification and explorations." In *Advances in endophytic fungal research: present status and future challenges* (2019): 267–281.

Kushwaha, Prity, Prem Lal Kashyap, Pandiyan Kuppusamy, Alok Kumar Srivastava, and Rajesh Kumar Tiwari. "Functional characterization of endophytic bacilli from pearl millet (Pennisetum glaucum) and their possible role in multiple stress tolerance." *Plant Biosystems-An International Journal Dealing with all Aspects of Plant Biology* 154, no. 4 (2020): 503–514.

Le Cocq, Kate, Sarah J. Gurr, Penny R. Hirsch, and Tim H. Mauchline. "Exploitation of endophytes for sustainable agricultural intensification." *Molecular Plant Pathology* 18, no. 3 (2017): 469–473.

Miliute, Inga, Odeta Buzaite, Danas Baniulis, and Vidmantas Stanys. "Bacterial endophytes in agricultural crops and their role in stress tolerance: a review." *Zemdirbyste-Agriculture* 102, no. 4 (2015): 465–478.

Mishra, Sandhya, Ram Sanmukh Upadhyay, and Chandra Shekhar Nautiyal. "Unravelling the beneficial role of microbial contributors in reducing the allelopathic effects of weeds." *Applied Microbiology and Biotechnology* 97 (2013): 5659–5668.

Nair, Dhanya N., and S. J. T. S. W. J. Padmavathy. "Impact of endophytic microorganisms on plants, environment and humans." *The Scientific World Journal* 2014 (2014): 250693.

Rondot, Yvonne, and Annette Reineke. "Endophytic Beauveria bassiana in grapevine Vitis vinifera (L.) reduces infestation with piercing-sucking insects." *Biological Control* 116 (2018): 82–89.

Ripa, Farhana Alam, Wei-dong Cao, Shuai Tong, and Jian-guang Sun. "Assessment of plant growth promoting and abiotic stress tolerance properties of wheat endophytic fungi." *BioMed Research International* 2019 (2019): 6105865.

Saikkonen, Kari, Päivi Lehtonen, Marjo Helander, Julia Koricheva, and Stanley H. Faeth. "Model systems in ecology: dissecting the endophyte–grass literature." *Trends in Plant Science* 11, no. 9 (2006): 428–433.

Santoyo, Gustavo, Gabriel Moreno-Hagelsieb, Ma del Carmen Orozco-Mosqueda, and Bernard R. Glick. "Plant growth-promoting bacterial endophytes." *Microbiological Research* 183 (2016): 92–99.

Shah, Sujit, Bir Bhadur Thapa, Krishna Chand, Shreeti Pradhan, Anjana Singh, Ajit Varma, Laxmi Sen Thakuri, Puspraj Joshi, and Bijaya Pant. "Piriformospora indica promotes the growth of the in-vitro-raised Cymbidium aloifolium plantlet and their acclimatization." *Plant Signaling & Behavior* 14, no. 6 (2019): 1596716.

Shen, Fo-Ting, Jui-Hung Yen, Chien-Sen Liao, Wen-Ching Chen, and Yi-Ting Chao. "Screening of rice endophytic biofertilizers with fungicide tolerance and plant growth-promoting characteristics." *Sustainability* 11, no. 4 (2019): 1133.

Singh, Mahendra, M. L. Dotaniya, Amit Mishra, C. K. Dotaniya, K. L. Regar, and Manju Lata. "Role of biofertilizers in conservation agriculture." In Jaideep Kumar Bisht, Vijay Singh Meena, Pankaj Kumar Mishra, Arunava Pattanayak (Eds.) *Conservation agriculture: an approach to combat climate change in Indian Himalaya* (2016): 113–134.

Sun, Bo-Tong, Komivi Senyo Akutse, Xiao-Feng Xia, Jun-Hui Chen, Xin Ai, Yang Tang, Qian Wang, Bo-Wen Feng, Mark S. Goettel, and Min-Sheng You. "Endophytic effects of Aspergillus oryzae on radish (Raphanus sativus) and its herbivore, Plutella xylostella." *Planta* 248 (2018): 705–714.

Suryanarayanan, T. S. "Fungal endophytes: an eclectic review." *Kavaka* 48, no. 1 (2017): 1–9.

Verma, Satish K., and James F. White. "Indigenous endophytic seed bacteria promote seedling development and defend against fungal disease in browntop millet (Urochloa ramosa L.)." *Journal of Applied Microbiology* 124, no. 3 (2018): 764–778.

Yadav, Anurag, and Kusum Yadav. "Plant growth-promoting endophytic bacteria and their potential to improve Agricultural crop yields." In Dhananjaya Pratap Singh, Ratna Prabha (Eds.) *Microbial interventions in agriculture and environment*, Volume 3, Soil and Crop Health Management (2019): 143–169.

19 Revolutionizing Sustainable Agriculture and Medicinal Plant Cultivation in China through Endophyte Innovation

*Mohsin Khan, Nizakat Bibi, Zahoor Ul Haq, Zixuan Wang, Maria Kalsoom, Fazal ur Rehman and Junqiang Zheng**
*Corresponding author.

19.1 INTRODUCTION

Endophytes are microorganisms, predominantly fungi or bacteria, that inhabit plant tissues without causing obvious or negative harm to the host (Harrison & Griffin, 2020). These symbiotic partnerships can be mutualistic, in which both the endophyte and the host plant benefit, or commensalist, in which the endophyte benefits while having no effect on the plant (El-Metwally et al., 2023). Depending on the kind of endophyte and the type of host plant, different endophytes can be found in different places within plants (Compant et al., 2021). Plant cells can have endophytes living in them and, as a result, they develop a strong bond with the plant without endangering the host's cells. The tissues that make up vascular bundles include xylem and phloem, and endophytes are known to live there. The gaps between vascular cells serve as an excellent location for them for colonizing (Gangwar et al., 2017). Hydathodes, also known as water pores, are microscopic holes found in plant tissues that permit water to escape. Endophytes colonize these pores and benefit from the abundant nutrients and moisture available at these sites (Ilahi et al., 2021). Gas exchange is conducted via stomata in leaves. The presence of endophytes near stomata also affects the host plant's reaction to external stimuli. Endophytes also frequently settle in the tissues of the roots, where they may engage with the root cells of the host plant and promote the uptake of nutrients. Furthermore, through seeds, endophytes get transmitted vertically. They live in the seed tissues and are inherited by the plant's progeny (Weyens et al., 2009).

De Bary (1866) was the first to identify the existence of other (nonpathogenic) organisms within plants in addition to disease-causing microorganisms. He found microbial cells in the tissues of plants that were examined under a microscope. However, this observation remained unexamined until the end of the 20th century when the term "endophytes" was defined. The term endophyte was first defined by De Bary in 1866, who stated that "any organisms that grows within plant tissues are termed as endophytes." Nevertheless, different scholars have continued to define endophytes differently (Wilson, 1995). Furthermore, the endophytic communities have been split into many subgroups, such as "obligate" or "facultative," which are connected to all kinds of plants, as a result of in-depth research on these groups of microorganisms (Schouten, 2019). In recent years, numerous attempts have been made to discover the origin of endophytic organisms in various species (Samreen et al., 2021). Endophyte genome sizes and origins have been evaluated by researchers by linking

 DOI: 10.1201/9781003486275-22

genome size with bacterial lifestyle. Many experts believe that the definition of endophytes is appropriate for the concept that they dwell inside plant species, where the environment is more stable than the soil where the plant develops (Gouda et al., 2016).

19.2 SIGNIFICANCE OF ENDOPHYTES IN PLANT–MICROBE INTERACTIONS

Endophytic microorganisms are most commonly associated with economically important plants such as maize, wheat, sorghum, wheat and soybean. Endosphere culturable microbial endophytes occupy distinct zones (Lambers et al., 2009). Gram-negative and Gram-positive bacteria such as rhizobacteria dominate the plant endosphere, whereas some endophytic mycorrhizal fungi found in the soil also form tight associations with plant roots (Mitter et al., 2013). Endophytes are economically important, as they help in crop protection. It is well recognized that some endophytes can impart disease resistance (Compant et al., 2021). They can assist in defending crops against harmful fungi, bacteria and other pathogens by colonizing plant tissues, thereby lowering the need for chemical pesticides. Plant pathogen growth is inhibited by endophytes, which can function as biocontrol agents. The economics of sustainable and ecologically friendly agriculture practices are affected by this biocontrol capability (Fontana et al., 2021). Microorganisms linked to plants benefit from the diversity that endophytes provide as they affect the dynamics of plant communities, nutrient cycling and overall ecosystem stability, while this variety is critical to the health of the ecosystem. Endophytes may affect the makeup of plant communities, which could contribute to ecological succession. The establishment and survival of plant species at various phases of succession can be impacted by their presence and actions (Van Der Heijden et al., 2008). Ecosystems can be stabilized by endophytes, particularly in regions that are vulnerable to environmental stressors like salt, drought or nutrient shortage. They might strengthen plant communities' ability to adapt to shifting environmental circumstances (Table 19.1) (Govindasamy et al., 2018).

TABLE 19.1
Endophytic Bacterial and Fungal Species Associated with Plants

Plants	Endophytes	Functions	References
Maize (*Zea mays*)	*Bacillus, Rhizobium, Pseudomonas, Xanthomonas, Azospirillum*	Improved agricultural yield, biocontrol of plant diseases, nitrogen fixation and improved plant nutrition uptake	(Fisher et al., 1992)
Wheat (*Triticum aestivum*)	*Azotobacter chroococcum, Cladosporium herbarum, Bacillus circulans*	Producing phytohormones, fixing nitrogen, phosphate solubilization, encouraging plant growth, and secreting lytic enzymes, as well as the creation of agents of biocontrol	(Afridi et al., 2019)
Cowpea (*Vigna unguiculata*)	*Azotobacter, Azospirillum, Rhizobium*	Boost crop yields, nitrogen fixation, plant development and phytohormone synthesis	(Swarnalakshmi et al., 2020)
Soybean (*Glycine max*)	*Rhizobium, Methylobacterium*	Facilitate the plant to fix nitrogen	(Minami et al., 2016)
Rice (*Oryza sativa*)	*Azotobacter*	Synthesis of siderophores, phosphate solubilization and nitrogen fixation	(Sahoo et al., 2014)
Tomato (*Solanum lycopersicum*)	*Azospirillum, Pseudomonas*	Fixing nitrogen, promoting plant development and solubilizing phosphate	(Moustaine et al., 2017)

(continued)

TABLE 19.1 (Continued)
Endophytic Bacterial and Fungal Species Associated with Plants

Plants	Endophytes	Functions	References
Black pepper (*Piper nigrum*)	*Pseudomonas*	Growth hormone production, antibiotics, siderophores, hydrogen cyanide and phosphate solubilization	(Lau et al., 2020)
Gastrodia elata Blume (*G. elata*)	Mushroom, *Armillaria*	It can treat headache, dizziness, numbness of limbs, children's convulsions, seizures, tetanus and other diseases	(Gong et al., 2023)
Pinellia ternata (Araceae)	*Clavispora, Penicillium, Fusarium, Penicillium, Longispora, Fungicillium, Cefomyces, Aspergillus, Fungicillium, Microspora*	It has the effect of eliminating phlegm and dampness, reducing regurgitation and stopping vomiting, eliminating fluffiness and dispersing. It mainly treats phlegm and cough, vomiting and nausea	(Liu et al., 2009)
Panax quinquefolius L. (Araliaceae)	*Fusarium* spp., *Sphaerococcus* spp., *Clonocladium* spp., *Sclerospora* spp., *Chitospora, Schizophyllum, Photopelia, Penicillium, Streptospora, Polynospora, Sitospora, Gibberellum, Gibberellum, Acanthophora, Leptocladosporium, Pseudotrichospora, Botrispora*	It has the effects of hemostasis, activating blood circulation and removing blood stasis, antifatigue, antiaging, resisting oxygen deficiency, lowering blood sugar and improving immune function	(Zhang et al., 2011)
Salvia miltiorrhiza Bunge (labiate)	*Gliocladium, Aspergillus, Penicillium, Chrysospora, Dendrospora, Botrytis, Fusarium, Trichospora, Pleurospora, Humus, Viriomyces, Trichospora, Trichospora*	It has the effect of promoting blood circulation and removing blood stasis, reducing heat and calming nerves, reducing swelling and relieving pain, treating coronary heart disease, myocardial infarction, angina pectoris and so on	(Wang et al., 2013)

Most endospheric microorganisms may have capable unique genes for boosting plant growth. However, standard isolation techniques have only isolated and identified 1% of the entire microbial community in the endosphere. As a result, additional research into their separation could turn them into valuable assets in future agricultural productivity (Adeleke & Babalola, 2022). Microbial endophytes have the ability to directly affect plant growth through a variety of processes, including nitrogen fixation, the synthesis of biological control agents to combat soil and plant pathogens, the production of secondary metabolites like antibiotics, siderophores, ammonia and hydrogen cyanide, and the phosphate solubilization process (Liu et al., 2017; Zhang & Guo, 2020). In order to ensure sustainable agriculture, it is appropriate to utilize plant-microbe resources. It is suggested that microbial endophytes play a pivotal role in solving environmental problems caused by stressed soil (Eid et al., 2021). Actinobacteria, Bacteroidetes, Firmicutes, Proteobacteria, and endophytic bacterial phyla have been found in a number of investigations on *Arabidopsis thaliana*, *Brassica chinensis*,

G. max, *V. vinifera* and *Zea mays* (Chen et al., 2019). The genetic makeup of plant-associated microbial endophytes has been revealed by the development of molecular technologies and bioinformatics applied to microbial genomic investigations (Slama et al., 2021).

19.3 DIVERSITY OF ENDOPHYTES IN CHINESE FLORA

Researchers have dedicated considerable time to the study of endophytes, and over the past two decades, they have unearthed substantial findings. One noteworthy discovery is the ability of endophytes to replicate the secondary metabolites of their host plants (Chaudhuri et al., 2017).

Mycologists have studied many fungal endophytes in plants which grow in the heavy-metal-polluted areas. For example, in a Pb-Zn mine in China, about 495 endophytic fungi from 20 taxa were isolated from six major plant species. The most prevalent isolates were *Phoma* spp., *Alternaria* spp. and *Peyronellaea* spp. Similarly, 186 endophytic fungi in a Pb-Zn mine located in Southwest China were isolated from various plant species. The rate of isolation of endophytic fungi varied from 43% to 85% (Zheng et al. 2013). Li et al. (2012) studied five distinct plant species from the Baima Snow Mountain (4,000–4,300 m altitude), southwest China, and isolated over 600 endophytic fungi from them. Internal transcribed spacer (ITS) sequencing analysis and morphological features identified 43 distinct taxa, with *Cephalosporium*, *Sirococcus*, *Penicillium* and *Aspergillus* being the main genera that were widely spread throughout all five plant species.

19.4 BENEFICIAL EFFECTS OF ENDOPHYTES ON PLANT GROWTH

Endophytic microorganisms exhibit various beneficial effects on the growth and development of plants (Adeleke et al., 2021). Presented below some important endophytes harnessed for the growth and development of plants are described.

19.4.1 Uptake and Improvement of Nutrients

Some bacterial and fungal endophytes have a role in nutrient cycling to make them available for plants. They can absorb important nutrients like nitrogen and phosphorus by solubilization and accessibility increment. Endophytes help in the production of enzymes and organic acids, which leads to solubilization of nitrogen and phosphorus, making them available for plants (Yuan et al., 2010). For example, *Rhizobia* and *Azospirillum* spp. are nitrogen-fixing bacteria which form symbiotic relationships with leguminous plants and convert atmospheric nitrogen into a soluble form which can be utilized by plants (Widnyana & Javandira, 2016). Arbuscular mycorrhizal fungi and ectomycorrhizal fungi have mutualistic relationships with plants and are helpful in enhancing nitrogen and phosphorus availability to plants (Soumare et al., 2015).

19.4.2 Increased Resistance to Abiotic Stress

Some endophytes can help plants to survive in harsh conditions like heavy metal stress, salinity, high temperature and drought stress. These endophytes may produce proteins, enzymes and other substances which help plants to survive in harsh conditions. For example, *Pseudomonas aeruginosa* and *Pseudomonas putida* produce ACC deaminase enzymes, which are helpful in reducing ethylene hormones and increase the ability of plants to tolerate saline stress (Bi & Xue, 2023). Endophytes have many genes that can induce the production of different metabolites, which help plants survive under stress conditions. They may contain genes that produce antioxidants, stress coping proteins and other enzymes. *Enterobacter* spp. have the ability to encode heat shock protein and antioxidant enzymes that help in the mitigation of drought stress. Similarly, *Rhizobium* spp. are responsible for mitigating heavy metal stress in leguminous plants (Govindasamy et al., 2018). Endophytes have the ability to strengthen the cell wall by inducing some structural molecules like lignin and callose. Therefore, a strong cell wall enhances resistance to abiotic stress (Dora et al., 2022).

19.4.3 Enhanced Herbivore Resistance

Endophytes have the ability to produce secondary metabolites that prevent grazing, therefore endophytes can increase resistance to herbivores (Zhang et al., 2023). Some endophytes have the ability to generate secondary metabolites which cause toxicity to plant tissues and prevent herbivores from feeding on them, for example, terpenoids and phenolic chemicals (Wink, 2015). Jasmonic acid (JA) is also produced which has the ability to defend the plant against herbivores (Wu & Baldwin, 2010). For example, *Fusarium oxysporum* and *Claviceps purpurea* are endophytic fungi that produce various secondary metabolites like terpenoid and alkaloids which may have toxic effects against herbivores (Fite et al., 2023). *Altertnia cerealis* produces phenolic compounds and terpenoid which prevent herbivores from feeding on host plants (Mahmoud et al., 2022). Similarly, various species of *Pseudomonas* can produce jasmonic acid which is a signaling compound that protects plants from herbivore attacks.

19.4.4 Increase in Heavy Metal Tolerance

Endophytes have the ability to mitigate heavy metal stress and help in phytoremediation in polluted environments (Ma et al., 2016). In the presence of endophytes, plants produce proteins like phytochelatins and metallothioneins, and in the presence of these proteins heavy metals are chelated (Jan et al., 2019). For example, *Pseudomonas putida*, *Bacillus subtilis*, *Rhuzobium* spp. and arbuscular mycorrhizal fungi are some endophytes which can reduce metal toxicity in plants (Sharma & Kumar, 2021).

19.5 DIRECT PLANT GROWTH PROMOTION (DPGP) MECHANISMS BY FUNGI AND BACTERIA

19.5.1 Phytohormone Synthesis

Phytohormones like auxins, cytokinin and gibberellin are produced by some bacterial and fungal endophytes. These hormones enable the division, differentiation and elongation of plant cells. Endophytes possess enzymes in their body, such as terpene for the syntheses of gibberellins, isopentenyl transferase for cytokinin and tryptophan aminotransferase for auxins (Peng et al., 2023). For example, *Fusarium fujikuroi* is an endopytic fungus which can produce gibberellin (Wang et al., 2022). Various strains of *Rhizobium* produce auxins and cytokines (Parthiban et al., 2016).

19.5.2 Solubilization of Phosphorus

Organic acid helps in the solubilization of phosphorus and makes it available for use by plants; the types of organic acid which are produce by endophytes are gluconic acids and oxalic acids (Lü et al., 2023). *Bacillus megaterium* (Kang et al., 2014), *Enterobacter cloacae* (Pande et al., 2017) and *Azospirillum brasilense* (Tahir et al., 2013) are some of the endophytes responsible for solubilization of phosphorus.

19.5.3 Drought Resistance

Some endophytes help plants to survive in a drought environment, as they have the ability to promote water intake. These types of endophytes have the ability to increase the length and number of roots, which may help plants to get water from deeper soil (Harman & Uphoff, 2019). *Bacillus amyloliquefaciens* (Poudel et al., 2021), *Pseudomonas fluorescens* (Aslam et al., 2022) and *Enterobacter* spp. are some of the endophytes which can mitigate drought stress in plants.

19.5.4 Antagonistic Effects

Endophytes have the ability to produce compounds which protect plants from pathogenic attacks and act as a shield for host plants (Di et al., 2023). Antimicrobial substances like antifungals and antibiotics are produced by endophytes which prevent the growth of plant pathogens (Morales-Cedeño et al., 2021). For example, endophytic *Streptomyces* spp. and *Bacillus subtilis* are responsible for an antagonistic effect (Kaur et al., 2023)

19.6 INDIRECT PLANT GROWTH PROMOTION (IPGP) MECHANISMS AND THEIR ROLE IN DISEASE RESISTANCE

19.6.1 Manufacturing of Antimicrobial Substances

19.6.1.1 Secondary Metabolites and Antibiotics

Endophytic bacteria and fungi produce secondary metabolites and antibiotics. These metabolites and antibiotics protect plants from the attacks of microorganisms and from illness (Li et al., 2023). Tetracycline, penicillin and streptomycin are examples of antibiotics (Caruso et al., 2022). The endophytes responsible for the production of these antibiotics are *Bacillus* spp., such as bacitracin and polymyxin, and *Xylaria* fungi (Caruso et al., 2022).

19.6.1.2 Immune System Priming

Systematic resistance can be produced by different types of endophytes which improve the immunology of plants. As a results, plants become more resistant to disease attacks. Plants express more pathogenesis-related (PR) protein when endophytes are present. The PR proteins are important to plants in protecting them from pathogen attacks (Enebe & Babalola, 2019).

19.6.1.3 Induced Plant Systemic Resistance (ISR)

The signaling chemicals are produced by endophytes which enable plants to gain systematic resistance and make the plants less susceptible to the attacks of pathogens in the future (Zhu et al., 2022). Endophytes increase the ability of plants to absorb more nutrients and decrease the amount of available nutrients to pathogens (Johnston-Monje & Raizada, 2011), for example, Trichoderma spp., *Bacillus subtilis* and *Pseudomonas fluorescens.*

19.7 ROLES OF ENDOPHYTES IN SUSTAINABLE AGRICULTURE

Sustainable agriculture is boosted by endophytes through the production of bioenergy and biofuel plants, and also by reducing the dependence of plants on chemical fertilizers (Lugtenberg et al., 2016). For example, endophytes like *Pseudomonas*, *Bacillus* and *Enterobacter* enhance photosynthesis and stress tolerance and increase nutrient efficiency in plants, thus helping to create sustainable agriculture (Figure 19.1) (Eid et al., 2021).

19.7.1 Diminishing Reliance on Chemical Sources

The dependency on chemical fertilizers and synthetic chemicals like pesticides and herbicides is decreased by endophytes. Therefore, they lead toward more sustainable and environmentally friendly agricultural practices by increasing information about the environmental, ecological and

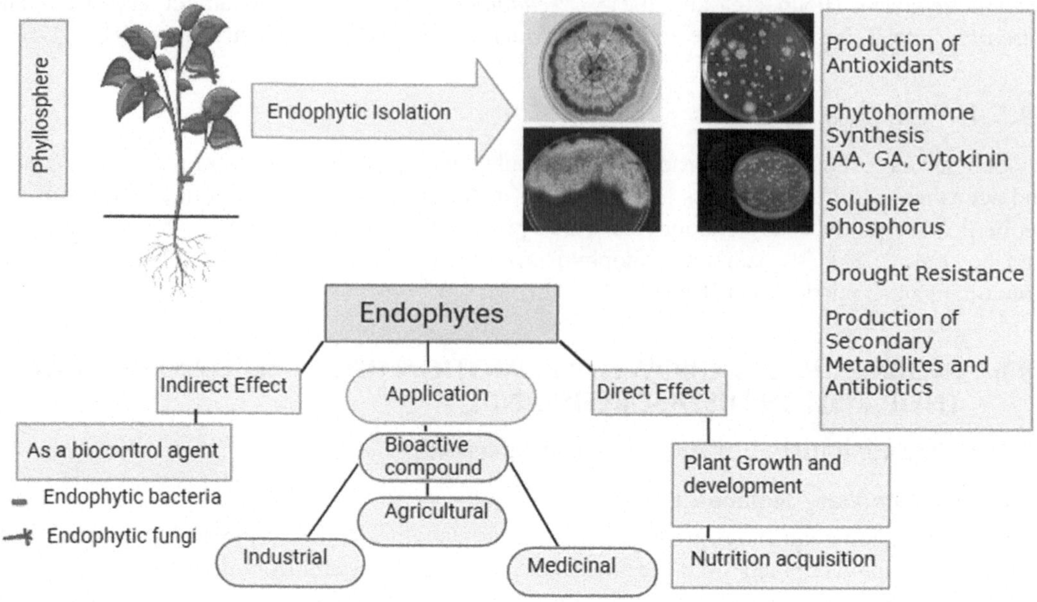

FIGURE 19.1 Isolation of endophytes and their role in sustainable agriculture.

health issues of the old agricultural practices' reliance on chemical fertilizers (Bamisile et al., 2021). Examples of some of these endophytes are *Glomus intraradices* (Lanfranco et al., 2016), *Trichoderma harizianum* and *Bacillus subtilis* (El Enshasy et al., 2020).

19.7.1.1 Nitrogen Fixation

Certain types of endophytes have the ability to carry out nitrogen fixation and make it available for use by plants. As a result, less artificial nitrogen fertilizers are required (Harindintwali et al., 2020). Examples of such endophytes are *Acetobacter diazotrophicus* (Zhang et al., 2022), Frankia spp., and *Herbaspirillum* bacteria (Ramos et al., 2020).

19.7.1.2 Solubilization of Phosphorus

The production of organic acids like oxalic acid and gluconic acids is carried out by endophytic bacteria. This help in phosphorus solubilization and makes it available for plants. As a result, the runoff of phosphorus to the environment can be decreased (Etesami, 2020). Among the endophytes that solubilize phosphorus are *Bacillus megaterium* (Kang et al., 2014), *Enterobacter cloacae* (Pande et al., 2017) and *Azospirillum brasilense* (Tahir et al., 2013).

19.7.1.3 Abiotic Stress Resistance

Abiotic stresses like salinity and drought stress can be resisted by plants due to the presence of endophytes. As a result, sustainable agriculture can be promoted as the endophytes enable plants to survive in harsh environmental conditions (Shi et al., 2023). Some of these endophytes are *Azospirillum brasilense* (Cao et al., 2023), *Pseudomonas fluorescens* (Ali et al., 2023), arbuscular mycorrhizal fungi (Ujvári et al., 2021) and *Bacillus subtilis* (Lachu et al., 2022).

19.7.1.4 Encouraging Plant Growth

Some endophytes are able to produce phytohormones. These endophytes help in the growth and development of plants. As a result, less growth regulators are needed (Shi et al., 2023), for example,

Enterobacter spp., *Gluconacetobacter diazotrophicus* (Meneses et al., 2011) and *Phyllobacterium* bacteria produce auxin (Ali et al., 2017), while *Streptomyces* and *Actinoplanes* produce auxins, cytokinesis and gibberellin (Olanrewaju & Babalola, 2019),

19.7.2 Supporting Crops for Biofuel and Bioenergy

19.7.2.1 Production of Biomass

The production of bioenergy plants like switchgrass and *Miscanthus* are supported by some endophytes. The increment of biomass produced for biofuel is supported by these plants. For example, mycorrhizal fungi and the bacterial endophytes *Azospirillum* and *Burkholderia* have the ability to promote the growth and production of bioenergy crops (Shi et al., 2023).

19.7.2.2 Enhanced Nutrient Uptake and Better Nutrient Use Efficiency

Some endophytes have the ability to increase the uptake of nutrients. As a result, they can decrease overfertilization. For example, the bacterial endophytes such as the genera *Rhizobium* and *Bradyrhizobium* have a symbiotic relationship with bioenergy leguminous crops and as result increase the uptake of nutrients in bioenergy plants (Xu & Geelen, 2018).

19.7.2.3 Soil Microbial Diversity

Total microbial diversity in the soil can be increased by some endophytes (Jiang et al., 2021), for example, *Bacillus* spp. and *Pseudomonas* spp. (Mahapatra et al., 2022).

19.8 CASE STUDY AND RESEARCH IN CHINA

19.8.1 Chinese Medicinal Herbs

Research on the endophytes which are connected to Chinese traditional medicinal herbs has been carried out by researchers. Research has showed how some endophytes like *Astragalus membranaceus*, *Panax ginseng* and *Scutellaria baicalensis* take part in the production of bioactive chemicals which have important medicinal applications (Yang et al., 2016).

19.8.2 Cultivation of Rice and Wheat

Chinese scientists have carried out research on the endophytes in staple crops like wheat and rice. They have showed that they can enhance disease resistance, the use of nutrient efficiency and total crop output, which are the main objectives of their research. For example, *Bacillus* and *Pseudomonas* are bacterial endophytes found in rice and wheat (Tian et al., 2020).

19.8.3 Tea Plants (*Camellia sinensis*)

Due to the importance of tea growing in China, research on endophytes have also been done in these plants. Endophytes play an important role in the stress tolerance, health and leaf quality of tea plants. The research shows that bacterial endophytes like *Pseudomonas* and *Bacillus* and endophyte from fungi like *Phomopsis* and *Colletotrichum* have beneficial effects on *Camellia sinensis* (Xie et al., 2020).

19.8.4 Biofuel Crops

Certain endophytes have been studied for an increase in the growth and yield of bioenergy crops for the production of biofuel (Mei et al., 2021). Such endophytes include *Herbaspirillum seropedicae*

(Schlemper et al., 2018), *Azospirillum brasilense* (Santos et al., 2021) and *Pseudomonas fluorescens* (Farrar et al., 2014).

19.8.5 Bioprospecting for Novel Compounds

Scientists have carried out research on endophytes which may be helpful in the production of novel chemicals and compounds which can be used in agriculture, medicine and other industries (Raimi & Adeleke, 2021). Examples of such types of endophytes are *Streptomyces* (Matsumoto & Takahashi, 2017), *Aspergillus, Fusarium* and *Xylaria* (Devi et al., 2020).

19.9 CHALLENGES IN HARNESSING ENDOPHYTE POTENTIAL

It is very challenging to fully understand the complex relationship between plants and endophytes because of the interaction between these systems. Large-scale research is required to understand the variety of endophytes and their importance in different plant species and habitats.

Endophytes are unpredictable in nature, so it is challenging to gain consistent results across different agricultural situations (Johnson et al., 2013). It is challenging to apply the results of small-scale laboratory studies to large-scale agricultural landscapes. Therefore, understanding the properties of plant–endophyte interactions is the need of the hour (Hanani et al., 2021). Still, a vast amount of research is required on how to apply endophytes on different cropping systems. Many endophytes show host-specific properties which show that they affects different plants. Broad research is required to study the use of endophytes for specific environments and specific plants. Comprehensive measures and precautions are required to test the efficacy of endophytes because it is difficult to obtain regulatory licenses for the commercial use of endophytes (Bailey, 2014).

19.10 FUTURE PROSPECTS FOR UTILIZING ENDOPHYTES IN ADVANCING SUSTAINABLE AGRICULTURE

19.10.1 Accurate Farming

Improvements in the knowledge about plant–microbe interactions, such as knowledge about endophytes that have the ability to develop sustainable agriculture practices, are needed. An understanding of endophytes' impacts on specific plants and environmental conditions is also required (Nair et al., 2014). For example, *Epichloe* is a fungal endophyte which has an association with tall fescue (*Festuca arundinacea*). This endophyte also helps plants to tolerate environmental stresses like drought and heat stress (Rojas et al., 2016). *Glycine max* has an association with nitrogen-fixing bacteria like *Bradyrhizobium japonicum.* These bacteria are able to make nodules on the roots of *Glycine max* and convert atmospheric nitrogen into a form which can be used by plants (Meena et al., 2018).

19.10.2 Microbial Groups

Microbial consortia investigation, including endophytic bacteria, can show complete results for improvements in the growth and physiology of crops. It can increase the ability of crops to mitigate stress and maintain the health of crops (Ren et al., 2023). Examples of such endophytic microorganisms are *Rhizobium japonicum* (Dwivedi et al., 2015), *Rhizobium leguminosarum* (Sánchez-Cañizares et al., 2018), *Pseudomonas fluorescens* and *Bacillus subtilis* (Hol et al., 2013).

19.10.3 Climate-Adaptive Vegetables

Climate change and harsh environmental conditions adversely affect the growth of vegetables. Therefore, to mitigate harsh environmental conditions, the use of endophytes is required to protect against environmental stresses like drought, heat and cold temperatures (Ren et al., 2023). A few of these endophytes are *Bacillus subtilis* (Lachu et al., 2022), arbuscular mycorrhizal fungi (Ujvári et al., 2021), *Pseudomonas fluorescens* (Ali et al., 2023) and *Azospirillum brasilense* (Zeffa et al., 2019).

19.10.4 Biological Remedies for Agriculture Illnesses

Endophytes can be used as a biocontrol and to protect against plant infection, which can decrease the reliance of plants on chemical fertilizers and pesticides (Rong et al., 2020). Examples of such endophytes are *Aspergillus harzianum, Aspergillus fluorescens, Bacillus subtilis, Gloeosporium colletotrichum, Fusarium oxysporum, Alternaria alternata* and *Gryphonostoma griseus*.

19.11 CONCLUSION

Research on endophytes in China has paved the way for sustainable agriculture, allowing a convergence of traditional farming methods with scientific advancements. This research not only sheds light on the potential benefits of endophytes but also underscores the importance of integrating age-old agricultural practices with modern scientific approaches for enhanced sustainability. The research on endophytes has showed that it can decrease the use of chemical fertilizers, pesticides and herbicides in China and other countries. Endophytes may be sustainable and environmental friendly and may be use as biocontrol agents, as they are used to protect plants from diseases.

REFERENCES

Adeleke, B. S., & Babalola, O. O. (2022). Roles of plant endosphere microbes in agriculture-a review. *Journal of Plant Growth Regulation*, *41*(4), 1411–1428.

Adeleke, B. S., Babalola, O. O., & Glick, B. R. (2021). Plant growth-promoting root-colonizing bacterial endophytes. *Rhizosphere*, *20*, 100433.

Afridi, M. S., Mahmood, T., Salam, A., Mukhtar, T., Mehmood, S., Ali, J., Khatoon, Z., Bibi, M., Javed, M. T., & Sultan, T. (2019). Induction of tolerance to salinity in wheat genotypes by plant growth promoting endophytes: Involvement of ACC deaminase and antioxidant enzymes. *Plant Physiology and Biochemistry*, *139*, 569–577.

Ali, A., Tabbasum, I., Azeem, H., Ölmez, F., Deveci, G., Khalid, B., & Mehtab, M. (2023). Bacterial endophytes, a resilient way toward sustainable agriculture: Provide plant growth promotion and biocontrol of plant pathogens. *Journal of Global Innovations in Agricultural Sciences*, *11*(2), 153–174.

Ali, S., Charles, T. C., & Glick, B. R. (2017). Endophytic phytohormones and their role in plant growth promotion. In Doty, S. (eds) *Functional importance of the plant microbiome: Implications for agriculture, forestry and bioenergy* (pp. 89–105) Springer, Cham. https://doi.org/10.1007/978-3-319-65897-1_6.

Aslam, M. M., IDRIS, A. L., Zhang, Q., Weifeng, X., KARANJA, J. K., & Wei, Y. (2022). Rhizosphere microbiomes can regulate plant drought tolerance. *Pedosphere*, *32*(1), 61–74.

Bailey, K. L. (2014). The bioherbicide approach to weed control using plant pathogens. In D. P. Abrol (Ed.), *Integrated pest management* (pp. 245–266). Elsevier. Academic Press. https://doi.org/https://doi.org/10.1016/B978-0-12-398529-3.00014-2

Bamisile, B. S., Siddiqui, J. A., Akutse, K. S., Ramos Aguila, L. C., & Xu, Y. (2021). General limitations to endophytic entomopathogenic fungi use as plant growth promoters, pests and pathogens biocontrol agents. *Plants*, *10*(10), 2119.

Bi, Y., & Xue, Z. (2023). Dark septate endophyte inoculation enhances antioxidant activity in Astragalus membranaceus var. mongholicus under heat stress. *Physiologia Plantarum*, *175*(6), e14054.

Cao, M., Narayanan, M., Shi, X., Chen, X., Li, Z., & Ma, Y. (2023). Optimistic contributions of plant growth-promoting bacteria for sustainable agriculture and climate stress alleviation. *Environmental Research*, *217*, 114924.

Caruso, D. J., Palombo, E. A., Moulton, S. E., & Zaferanloo, B. (2022). Exploring the promise of endophytic fungi: A review of novel antimicrobial compounds. *Microorganisms*, *10*(10), 1990.

Chen, Q.-L., Cui, H.-L., Su, J.-Q., Penuelas, J., & Zhu, Y.-G. (2019). Antibiotic resistomes in plant microbiomes. *Trends in Plant Science*, *24*(6), 530–541.

Choudhury, F. K., Rivero, R. M., Blumwald, E., & Mittler, R. (2017). Reactive oxygen species, abiotic stress and stress combination. *The Plant Journal*, *90*(5), 856–867.

Compant, S., Cambon, M. C., Vacher, C., Mitter, B., Samad, A., & Sessitsch, A. (2021). The plant endosphere world–bacterial life within plants. *Environmental Microbiology*, *23*(4), 1812–1829.

de Bary, A. (1866). *Morphologie und physiologie der pilze, flechten und myxomyceten* (Vol. 1). Engelmann.

Devi, R., Kaur, T., Guleria, G., Rana, K. L., Kour, D., Yadav, N., Yadav, A. N., & Saxena, A. K. (2020). Fungal secondary metabolites and their biotechnological applications for human health. In A. A. Rastegari, A. N. Yadav, & N. Yadav (Eds.), *New and future developments in microbial biotechnology and bioengineering* (pp. 147–161). Elsevier. https://doi.org/https://doi.org/10.1016/B978-0-12-820528-0.00010-7

Di, R., Liu, L., Shoaib, N., Xi, B., Zhou, Q., & Yu, G. (2023). Sheath Blight of Maize: An overview and prospects for future research directions. *Agriculture*, *13*(10), 2006.

Dora, S., Terrett, O. M., & Sánchez-Rodríguez, C. (2022). Plant–microbe interactions in the apoplast: Communication at the plant cell wall. *The Plant Cell*, *34*(5), 1532–1550.

Dwivedi, S. L., Sahrawat, K. L., Upadhyaya, H. D., Mengoni, A., Galardini, M., Bazzicalupo, M., Biondi, E. G., Hungria, M., Kaschuk, G., & Blair, M. W. (2015). Advances in host plant and rhizobium genomics to enhance symbiotic nitrogen fixation in grain legumes. *Advances in Agronomy*, *129*, 1–116.

Eid, A. M., Fouda, A., Abdel-Rahman, M. A., Salem, S. S., Elsaied, A., Oelmüller, R., Hijri, M., Bhowmik, A., Elkelish, A., & Hassan, S. E.-D. (2021). Harnessing bacterial endophytes for promotion of plant growth and biotechnological applications: An overview. *Plants*, *10*(5), 935.

El-Metwally, M. M., Mekawey, A. A. I., El-Halmouch, Y., & Naga, N. G. (2023). Symbiotic relationships with fungi: From mutualism to parasitism. In Rashad, Y.M., Baka, Z.A.M., Moussa, T.A.A. (Eds) *Plant mycobiome: Diversity, interactions and uses* (pp. 375–413). Springer, Cham. https://doi.org/10.1007/978-3-031-28307-9_15.

El Enshasy, H. A., Ambehabati, K. K., El Baz, A. F., Ramchuran, S., Sayyed, R., Amalin, D., Dailin, D. J., & Hanapi, S. Z. (2020). Trichoderma: Biocontrol agents for promoting plant growth and soil health. In Yadav, A., Mishra, S., Kour, D., Yadav, N., Kumar, A. (Eds) *Agriculturally important fungi for sustainable agriculture: Volume 2: Functional annotation for crop protection* (pp. 239–259) Fungal Biology. Springer, Cham. https://doi.org/10.1007/978-3-030-48474-3_8.

Enebe, M. C., & Babalola, O. O. (2019). The impact of microbes in the orchestration of plants' resistance to biotic stress: A disease management approach. *Applied Microbiology and Biotechnology*, *103*, 9–25.

Etesami, H. (2020). Enhanced phosphorus fertilizer use efficiency with microorganisms. In *Nutrient dynamics for sustainable crop production* (pp. 215–245).

Farrar, K., Bryant, D., & Cope-Selby, N. (2014). Understanding and engineering beneficial plant–microbe interactions: Plant growth promotion in energy crops. *Plant Biotechnology Journal*, *12*(9), 1193–1206.

Fisher, P., Petrini, O., & Scott, H. L. (1992). The distribution of some fungal and bacterial endophytes in maize (Zea mays L.). *New Phytologist*, *122*(2), 299–305.

Fite, T., Kebede, E., Tefera, T., & Bekeko, Z. (2023). Endophytic fungi: versatile.

Fontana, D. C., de Paula, S., Torres, A. G., de Souza, V. H. M., Pascholati, S. F., Schmidt, D., & Dourado Neto, D. (2021). Endophytic fungi: Biological control and induced resistance to phytopathogens and abiotic stresses. *Pathogens*, *10*(5), 570.

Gangwar, M., Saini, P., Nikhanj, P., & Kaur, S. (2017). Plant growth-promoting microbes (PGPM) as potential microbial bio-agents for eco-friendly agriculture. In Adhya, T., Mishra, B., Annapurna, K., Verma, D., Kumar, U. (eds) Advances in Soil Microbiology: Recent Trends and Future Prospects. Microorganisms for Sustainability, vol 4. Springer, Singapore. https://doi.org/10.1007/978-981-10-7380-9_3.

Gong, M.-q., Lai, F.-f., Chen, J.-z., Li, X.-h., Chen, Y.-j., & He, Y. (2023). Traditional uses, phytochemistry, pharmacology, applications, and quality control of Gastrodia elata Blume: A comprehensive review. *Journal of Ethnopharmacology*, 319, 117128.

Gouda, S., Das, G., Sen, S. K., Shin, H.-S., & Patra, J. K. (2016). Endophytes: A treasure house of bioactive compounds of medicinal importance. *Frontiers in Microbiology*, *7*, 1538.

Govindasamy, V., George, P., Raina, S. K., Kumar, M., Rane, J., & Annapurna, K. (2018). Plant-associated microbial interactions in the soil environment: Role of endophytes in imparting abiotic stress tolerance to crops. In Bal, S., Mukherjee, J., Choudhury, B., Dhawan, A. (eds) *Advances in crop environment interaction* (pp. 245–284). Springer, Singapore. https://doi.org/10.1007/978-981-13-1861-0_10

Hanani, A., Bagarello, V., Davino, S., & Valentini, D. F. (2021). Apulian olive sapwood endophytes to cope with Xylella fastidiosa: community analysis, antagonists screening, colonization efficiency, and genes expression.

Harindintwali, J. D., Zhou, J., Yang, W., Gu, Q., & Yu, X. (2020). Biochar-bacteria-plant partnerships: Eco-solutions for tackling heavy metal pollution. *Ecotoxicology and Environmental Safety, 204*, 111020.

Harman, G. E., & Uphoff, N. (2019). Symbiotic root-endophytic soil microbes improve crop productivity and provide environmental benefits. *Scientifica, 2019,* 9106395.

Hol, W., Bezemer, T. M., & Biere, A. (2013). Getting the ecology into interactions between plants and the plant growth-promoting bacterium Pseudomonas fluorescens. *Frontiers in Plant Science, 4*, 43491.

Ilahi, H., Hsouna, J., Ellouze, W., Gritli, T., Chihaoui, S. A., Barhoumi, F., ... & Mnasri, B. (2021). Phylogenetic study of rhizobia nodulating pea (Pisum sativum) isolated from different geographic locations in Tunisia. *Systematic and Applied Microbiology, 44*(4), 126221.

Jan, R., Khan, M. A., Asaf, S., Lee, I.-J., & Kim, K. M. (2019). Metal resistant endophytic bacteria reduces cadmium, nickel toxicity, and enhances expression of metal stress related genes with improved growth of Oryza sativa, via regulating its antioxidant machinery and endogenous hormones. *Plants, 8*(10), 363.

Jiang, Z., Yang, S., Pang, Q., Xu, Y., Chen, X., Sun, X., Qi, S., & Yu, W. (2021). Biochar improved soil health and mitigated greenhouse gas emission from controlled irrigation paddy field: Insights into microbial diversity. *Journal of Cleaner Production, 318*, 128595.

Johnson, L. J., de Bonth, A. C., Briggs, L. R., Caradus, J. R., Finch, S. C., Fleetwood, D. J., Fletcher, L. R., Hume, D. E., Johnson, R. D., & Popay, A. J. (2013). The exploitation of epichloae endophytes for agricultural benefit. *Fungal Diversity, 60*, 171–188.

Johnston-Monje, D., & Raizada, M. (2011). 4.58 Plant and Endophyte relationships: nutrient management. In Murray Moo-Young (ed.), *Comprehensive biotechnology,* 2nd edn. Academic, Burlington, 4, 713–727. Elsevier.

Kang, S.-M., Radhakrishnan, R., You, Y.-H., Joo, G.-J., Lee, I.-J., Lee, K.-E., & Kim, J.-H. (2014). Phosphate solubilizing Bacillus megaterium mj1212 regulates endogenous plant carbohydrates and amino acids contents to promote mustard plant growth. *Indian Journal of Microbiology, 54*, 427–433.

Kaur, T., Khanna, K., Sharma, S., & Manhas, R. K. (2023). Mechanistic insights into the role of actinobacteria as potential biocontrol candidates against fungal phytopathogens. *Journal of Basic Microbiology, 63*(11), 1196–1218.

Lachu, K., Kamle, M., Borah, R., Tiwari, B., & Kumar, P. (2022). Endophytic bacteria: Application against biotic and abiotic stresses and plant health improvements for sustainable agriculture. In Singh, A.K., Tripathi, V., Shukla, A.K., Kumar, P. (eds) *Bacterial endophytes for sustainable agriculture and environmental management* (pp. 1–21). Springer. https://doi.org/10.1007/978-981-16-4497-9_1

Lambers, H., Mougel, C., Jaillard, B., & Hinsinger, P. (2009). *Plant-microbe-soil interactions in the rhizosphere: An evolutionary perspective*. Springer.

Lanfranco, L., Bonfante, P., & Genre, A. (2016). The mutualistic interaction between plants and arbuscular mycorrhizal fungi. *Microbiology Spectrum, 4*(6), 10–1128. doi: 10.1128/microbiolspec.funk-0012-2016.

Lau, E. T., Tani, A., Khew, C. Y., Chua, Y. Q., & San Hwang, S. (2020). Plant growth-promoting bacteria as potential bio-inoculants and biocontrol agents to promote black pepper plant cultivation. *Microbiological Research, 240*, 126549.

Li, H. Y., Shen, M., Zhou, Z. P., Li, T., Wei, Y. L., & Lin, L. B. (2012). Diversity and cold adaptation of endophytic fungi from five dominant plant species collected from the Baima Snow Mountain, Southwest China. *Fungal Diversity, 54*, 79–86.

Li, Z., Xiong, K., Wen, W., Li, L., & Xu, D. (2023). Functional endophytes regulating plant secondary metabolism: Current status, prospects and applications. *International Journal of Molecular Sciences, 24*(2), 1153.

Liu, H., Carvalhais, L. C., Crawford, M., Singh, E., Dennis, P. G., Pieterse, C. M., & Schenk, P. M. (2017). Inner plant values: Diversity, colonization and benefits from endophytic bacteria. *Frontiers in Microbiology, 8*, 2552.

Liu, L., Yu, L., Pearl, D. K., & Edwards, S. V. (2009). Estimating species phylogenies using coalescence times among sequences. *Systematic Biology, 58*(5), 468–477.

Lü, Z.-W., Liu, H.-Y., Wang, C.-L., Chen, X., Huang, Y.-X., Zhang, M.-M., Huang, Q.-L., & Zhang, G.-F. (2023). Isolation of endophytic fungi from Cotoneaster multiflorus and screening of drought-tolerant fungi and evaluation of their growth-promoting effects. *Frontiers in Microbiology*, *14*, 1267404.

Lugtenberg, B. J., Caradus, J. R., & Johnson, L. J. (2016). Fungal endophytes for sustainable crop production. *FEMS Mmicrobiology Ecology*, *92*(12), fiw194.

Ma, Y., Rajkumar, M., Zhang, C., & Freitas, H. (2016). Beneficial role of bacterial endophytes in heavy metal phytoremediation. *Journal of Environmental Management*, *174*, 14–25.

Mahapatra, S., Yadav, R., & Ramakrishna, W. (2022). Bacillus subtilis impact on plant growth, soil health and environment: Dr. Jekyll and Mr. Hyde. *Journal of applied microbiology*, *132*(5), 3543–3562.

Mahmoud, G. A., Abdel-Sater, M. A., Al-Amery, E., & Hussein, N. A. (2022). Production of terpenoid phytoalexins as a plant response against Alternaria cerealis AUMC 14484 controlled by Trichoderma harzianum. *13*, 12–33.

Matsumoto, A., & Takahashi, Y. (2017). Endophytic actinomycetes: Promising source of novel bioactive compounds. *The Journal of Antibiotics*, *70*(5), 514–519.

Meena, R. S., Vijayakumar, V., Yadav, G. S., & Mitran, T. (2018). Response and interaction of Bradyrhizobium japonicum and arbuscular mycorrhizal fungi in the soybean rhizosphere. *Plant Growth Regulation*, *84*, 207–223.

Mei, C., Amaradasa, S., Sikaroodi, M., Zhang, X., Gillevet, P., Nowak, J., & Lowman, S. (2021). Potential application of plant growth promoting bacteria in bioenergy crop production. In J. White, A. Kumar, & S. Droby (Eds.), *Microbiome stimulants for crops* (pp. 109–123). Elsevier. https://doi.org/https://doi.org/10.1016/B978-0-12-822122-8.00014-5

Meneses, C. H., Rouws, L. F., Simões-Araújo, J. L., Vidal, M. S., & Baldani, J. I. (2011). Exopolysaccharide production is required for biofilm formation and plant colonization by the nitrogen-fixing endophyte Gluconacetobacter diazotrophicus. *Molecular Plant-Microbe Interactions*, *24*(12), 1448–1458.

Minami, T., Anda, M., Mitsui, H., Sugawara, M., Kaneko, T., Sato, S., Ikeda, S., Okubo, T., Tsurumaru, H., & Minamisawa, K. (2016). Metagenomic analysis revealed methylamine and ureide utilization of soybean-associated Methylobacterium. *Microbes and Environments*, *31*(3), 268–278.

Mitter, B., Brader, G., Afzal, M., Compant, S., Naveed, M., Trognitz, F., & Sessitsch, A. (2013). Advances in elucidating beneficial interactions between plants, soil, and bacteria. *Advances in Agronomy*, *121*, 381–445.

Morales-Cedeño, L. R., del Carmen Orozco-Mosqueda, M., Loeza-Lara, P. D., Parra-Cota, F. I., de Los Santos-Villalobos, S., & Santoyo, G. (2021). Plant growth-promoting bacterial endophytes as biocontrol agents of pre-and post-harvest diseases: Fundamentals, methods of application and future perspectives. *Microbiological Research*, *242*, 126612.

Moustaine, M., Elkahkahi, R., Benbouazza, A., Benkirane, R., & Achbani, E. (2017). Effect of plant growth promoting rhizobacterial (PGPR) inoculation on growth in tomato (Solanum lycopersicum L.) and characterization for direct PGP abilities in Morocco. *International Journal of Environment, Agriculture and Biotechnology*, *2*(2), 238708.

Nair, D. N., & Padmavathy, S. J. T. S. W. J. (2014). Impact of endophytic microorganisms on plants, environment and humans. *The Scientific World Journal*, *2014*(1), 250693.

Olanrewaju, O. S., & Babalola, O. O. (2019). Streptomyces: Implications and interactions in plant growth promotion. *Applied Microbiology and Biotechnology*, *103*, 1179–1188.

Pande, A., Pandey, P., Mehra, S., Singh, M., & Kaushik, S. (2017). Phenotypic and genotypic characterization of phosphate solubilizing bacteria and their efficiency on the growth of maize. *Journal of Genetic Engineering and Biotechnology*, *15*(2), 379–391.

Parthiban, P., Shijila Rani, A., Mahesh, V., & Ambikapathy, V. (2016). Studies on biosynthesis of auxin in rhizobium and their impact on growth of Vigna mungo L. *Pharmaceutical and Biological Evaluation*, *3*(3), 371–376.

Peng, M., Jiang, Z., Zhou, F., & Wang, Z. (2023). From salty to thriving: Plant growth promoting bacteria as nature's allies in overcoming salinity stress in plants. *Frontiers in Microbiology*, *14*, 1169809.

Poudel, M., Mendes, R., Costa, L. A., Bueno, C. G., Meng, Y., Folimonova, S. Y., Garrett, K. A., & Martins, S. J. (2021). The role of plant-associated bacteria, fungi, and viruses in drought stress mitigation. *Frontiers in Microbiology*, *12*, 743512.

Raimi, A., & Adeleke, R. (2021). Bioprospecting of endophytic microorganisms for bioactive compounds of therapeutic importance. *Archives of Microbiology*, *203*(5), 1917–1942.

Ramos, A. C., Melo, J., de Souza, S. B., Bertolazi, A. A., Silva, R. A., Rodrigues, W. P., Campostrini, E., Olivares, F. L., Eutrópio, F. J., & Cruz, C. (2020). Inoculation with the endophytic bacterium Herbaspirillum seropedicae promotes growth, nutrient uptake and photosynthetic efficiency in rice. *Planta*, *252*, 1–8.

Ren, F., Ji, N., & Zhu, Y. (2023). Research progress of α-Glucosidase inhibitors produced by microorganisms and their applications. *Foods*, *12*(18), 3344.

Rojas, X., Guo, J., Leff, J. W., McNear, D. H., Fierer, N., & McCulley, R. L. (2016). Infection with a shoot-specific fungal endophyte (Epichloë) alters tall fescue soil microbial communities. *Microbial Ecology*, *72*, 197–206.

Rong, S., Xu, H., Li, L., Chen, R., Gao, X., & Xu, Z. (2020). Antifungal activity of endophytic Bacillus safensis B21 and its potential application as a biopesticide to control rice blast. *Pesticide Biochemistry and Physiology*, *162*, 69–77.

Sahoo, R. K., Ansari, M. W., Dangar, T. K., Mohanty, S., & Tuteja, N. (2014). Phenotypic and molecular characterisation of efficient nitrogen-fixing Azotobacter strains from rice fields for crop improvement. *Protoplasma*, *251*, 511–523.

Samreen, T., Naveed, M., Nazir, M. Z., Asghar, H. N., Khan, M. I., Zahir, Z. A., Kanwal, S., Jeevan, B., Sharma, D., & Meena, V. S. (2021). Seed associated bacterial and fungal endophytes: Diversity, life cycle, transmission, and application potential. *Applied Soil Ecology*, *168*, 104191.

Sánchez-Cañizares, C., Jorrín, B., Durán, D., Nadendla, S., Albareda, M., Rubio-Sanz, L., Lanza, M., González-Guerrero, M., Prieto, R. I., & Brito, B. (2018). Genomic diversity in the endosymbiotic bacterium Rhizobium leguminosarum. *Genes*, *9*(2), 60.

Santos, M. S., Nogueira, M. A., & Hungria, M. (2021). Outstanding impact of Azospirillum brasilense strains Ab-V5 and Ab-V6 on the Brazilian agriculture: Lessons that farmers are receptive to adopt new microbial inoculants. *Revista Brasileira de Ciência do Solo*, *45*, e0200128.

Schlemper, T. R., Dimitrov, M. R., Gutierrez, F. A. S., van Veen, J. A., Silveira, A. P., & Kuramae, E. E. (2018). Effect of Burkholderia tropica and Herbaspirillum frisingense strains on sorghum growth is plant genotype dependent. *PeerJ*, *6*, e5346.

Schouten, A. (2019). Endophytic fungi: definitions, diversity, distribution and their significance in plant life. In A. Schouten (Ed.), *Endophyte biotechnology: Potential for agriculture and pharmacology* (pp. 6–31). CABI Wallingford UK. https://doi.org/10.1079/9781786399427.0006

Sharma, P., & Kumar, S. (2021). Bioremediation of heavy metals from industrial effluents by endophytes and their metabolic activity: Recent advances. *Bioresource Technology*, *339*, 125589.

Shi, J., Wang, X., & Wang, E. (2023). Mycorrhizal symbiosis in plant growth and stress adaptation: From genes to ecosystems. *Annual Review of Plant Biology*, *74*, 569–607.

Slama, H. B., Cherif-Silini, H., Bouket, A. C., Silini, A., Alenezi, F. N., Luptakova, L., Vallat, A., & Belbahri, L. (2021). Biotechnology and bioinformatics of endophytes in biocontrol, bioremediation, and plant growth promotion. *Endophytes: Mineral Nutrient Management, 3*, 181–205.

Soumare, A., Diop, T., Manga, A., & Ndoye, I. (2015). Role of arbuscular mycorrhizal fungi and nitrogen fixing bacteria on legume growth under various environmental stresses. *International Journal of Bioscience*, *7*(4), 31–46.

Swarnalakshmi, K., Yadav, V., Tyagi, D., Dhar, D. W., Kannepalli, A., & Kumar, S. (2020). Significance of plant growth promoting rhizobacteria in grain legumes: Growth promotion and crop production. *Plants*, *9*(11), 1596.

Tahir, M., Mirza, M. S., Zaheer, A., Dimitrov, M. R., Smidt, H., & Hameed, S. (2013). Isolation and identification of phosphate solubilizer 'Azospirillum, Bacillus' and 'Enterobacter' strains by 16SrRNA sequence analysis and their effect on growth of wheat ('Triticum aestivum'L.). *Australian Journal of Crop Science*, *7*(9), 1284–1292.

Tian, B., Xie, J., Fu, Y., Cheng, J., Li, B., Chen, T., Zhao, Y., Gao, Z., Yang, P., & Barbetti, M. J. (2020). A cosmopolitan fungal pathogen of dicots adopts an endophytic lifestyle on cereal crops and protects them from major fungal diseases. *The ISME Journal*, *14*(12), 3120–3135.

Ujvári, G., Turrini, A., Avio, L., & Agnolucci, M. (2021). Possible role of arbuscular mycorrhizal fungi and associated bacteria in the recruitment of endophytic bacterial communities by plant roots. *Mycorrhiza*, *31*(5), 527–544.

Van Der Heijden, M. G., Bardgett, R. D., & Van Straalen, N. M. (2008). The unseen majority: Soil microbes as drivers of plant diversity and productivity in terrestrial ecosystems. *Ecology Letters*, *11*(3), 296–310.

Wang, H.-N., Ke, X., Zhou, J.-P., Liu, Z.-Q., & Zheng, Y.-G. (2022). Recent advances in metabolic regulation and bioengineering of gibberellic acid biosynthesis in Fusarium fujikuroi. *World Journal of Microbiology and Biotechnology*, *38*(8), 131.

Wang, Q., Kuang, H., Su, Y., Sun, Y., Feng, J., Guo, R., & Chan, K. (2013). Naturally derived anti-inflammatory compounds from Chinese medicinal plants. *Journal of Ethnopharmacology*, *146*(1), 9–39.

Weyens, N., van der Lelie, D., Taghavi, S., & Vangronsveld, J. (2009). Phytoremediation: Plant–endophyte partnerships take the challenge. *Current Opinion in Biotechnology*, *20*(2), 248–254.

Widnyana, I. K., & Javandira, C. (2016). Activities Pseudomonas spp. and Bacillus sp. to stimulate germination and seedling growth of tomato plants. *Agriculture and Agricultural Science Procedia*, *9*, 419–423.

Wilson, D. (1995). Endophyte: The evolution of a term, and clarification of its use and definition. *Oikos*, *73*(2), 274–276. https://doi.org/10.2307/3545919

Wink, M. (2015). Modes of action of herbal medicines and plant secondary metabolites. *Medicines*, *2*(3), 251–286.

Wu, J., & Baldwin, I. T. (2010). New insights into plant responses to the attack from insect herbivores. *Annual Review of Genetics*, *44*, 1–24.

Xie, H., Feng, X., Wang, M., Wang, Y., Kumar Awasthi, M., & Xu, P. (2020). Implications of endophytic microbiota in Camellia sinensis: A review on current understanding and future insights. *Bioengineered*, *11*(1), 1001–1015.

Xu, L., & Geelen, D. (2018). Developing biostimulants from agro-food and industrial by-products. *Frontiers in Plant Science*, *9*, 1567.

Yang, L., Yang, C., Li, C., Zhao, Q., Liu, L., Fang, X., & Chen, X.-Y. (2016). Recent advances in biosynthesis of bioactive compounds in traditional Chinese medicinal plants. *Science Bulletin*, *61*(1), 3–17.

Yuan, Z.-l., Zhang, C.-l., & Lin, F.-c. (2010). Role of diverse non-systemic fungal endophytes in plant performance and response to stress: Progress and approaches. *Journal of Plant Growth Regulation*, *29*, 116–126.

Zeffa, D. M., Perini, L. J., Silva, M. B., de Sousa, N. V., Scapim, C. A., Oliveira, A. L. M. D., ... & Azeredo Gonçalves, L. S. (2019). Azospirillum brasilense promotes increases in growth and nitrogen use efficiency of maize genotypes. *Plos One*, *14*(4), e0215332.

Zhang, L., Ravipati, A. S., Koyyalamudi, S. R., Jeong, S. C., Reddy, N., Smith, P. T., Bartlett, J., Shanmugam, K., Münch, G., & Wu, M. J. (2011). Antioxidant and anti-inflammatory activities of selected medicinal plants containing phenolic and flavonoid compounds. *Journal of Agricultural and Food Chemistry*, *59*(23), 12361–12367.

Zhang, X.-G., & Guo, S.-J. (2020). Diversity and bioactivity of endophytes from Angelica sinensis in China. *Frontiers in Microbiology*, *11*, 522688.

Zhang, X., Tong, J., Dong, M., Akhtar, K., & He, B. (2022). Isolation, identification and characterization of nitrogen fixing endophytic bacteria and their effects on cassava production. *PeerJ*, *10*, e12677.

Zhang, Y., Zhang, S., & Xu, L. (2023). The pivotal roles of gut microbiota in insect plant interactions for sustainable pest management. *NPJ Biofilms and Microbiomes*, *9*(1), 66.

Zheng, Y. K., Chen, C., Ren, D., & Gu, Y. F. (2013). Genetic diversity of the plant endophytic fungi in a Pb-Zn mine area in Hanyuan, Sichuan Province. *China Journal of Sichuan Agricultural University*, *31*(3), 308–313.

Zhu, L., Huang, J., Lu, X., & Zhou, C. (2022). Development of plant systemic resistance by beneficial rhizobacteria: Recognition, initiation, elicitation and regulation. *Frontiers in Plant Science*, *13*, 952397.

20 Endophytes in Bioremediation of Heavy Metals

*Nizakat Bibi, Mohsin Khan, Maria Kalsoom, Fazal ur Rehman, Mahrukh, Najeeba Parre Paker, Muhammad Farooq Hussain Munis and Hassan Javed Chaudhary**
*Corresponding author.

20.1 INTRODUCTION

The Earth's crust is naturally composed of metals, and the composition of these metals varies depending on the location, creating structural differences between nearby concentrations (Dung et al., 2013). Regarding the metabolic processes of living things, some heavy metals are crucial in small quantities. Because many industrial operations release heavy metals into the environment, they become exceedingly hazardous and dangerous contaminants of soil and water due to their high solubility (Jaishankar et al., 2014). When these metals are discharged into the environment, they may flow off into surface waters and soils, contaminating both, or they may leach into subterranean waters and deposit in groundwater (Mishra et al., 2019). It has been difficult for humanity to deal with the manmade, unavoidable issue of heavy metal contamination. Heavy metals thus constitute a possible environmental pollutant that may participate in trophic transmission in food systems. The physiological bio-toxic effects of heavy metals are mostly caused by their relative oxidation state, which also determines their toxicity (Mitra et al., 2022). These metals interfere with biological organisms' ability to function normally and keep them from taking part in bioreactions when they interact with DNA, proteins and enzymes to generate extremely stable bio-toxic chemicals (Dimian et al., 2022). Lead, cadmium, arsenic and chromium are examples of heavy metals, which are defined as having an atomic number greater than 20 and a density greater than 5 grams per cubic centimeter, which are extremely hazardous and have mutagenic, carcinogenic and genotoxic effects (Han et al., 2020). These metals have strong inhibitory effects on biodegradation activities, which makes them directly linked to concerns about biological toxicity and environmental contamination (Olaniran et al., 2013). The primary origins of hazardous metal emissions into the environment are a variety of activities and processes, such as the burning and extraction of toxic waste into the atmosphere, as well as excessive wastewater and effluent emissions that contaminate soils and surface waters with dangerous chemicals (Kumar et al., 2022). The ongoing use of dangerous metals in manmade processes from the mining, mining, tanning, painting, and textile and dye industries, among others, has greatly increased and is now destructive to the diversity of life on Earth (Kahlon et al., 2018). There is a significant concern about environmental contamination due to the careless release of dangerous metals from these industrial operations, frequently at high quantities (beyond allowable limits). Therefore, for the sake of environmental safety, their removal or remediation is now required (Kahlon et al., 2018). The efficacy of various techniques, including coagulation, flocculation, biosorption, chemical precipitation, ion exchange, filtration, reverse osmosis, membrane process and aerobic and anaerobic microbial degradation, in removing metals from various environmental media has been thoroughly studied in the past (Singh et al., 2022). Because of its low cost and environmentally benign character, biosorption is regarded as one of these processes'

DOI: 10.1201/9781003486275-23

novel technologies and remediation approaches (Torres, 2020). In general, bioremediation entails the utilization of microorganisms to reduce the bioavailability of metals or the adsorption of soil contaminants on organic matter. However, the majority of studies advise using microbes rather than organic matter in order to eliminate or lower the quantity of heavy metals in polluted areas (Sayqal & Ahmed, 2021). Intrinsic characteristics of microbial communities have been successfully applied to clean up contaminated locations and decrease, decompose or immobilize harmful substances (Jo et al., 2020). Phytoremediation is an inexpensive and green technique of removing heavy metals from soil and water by using plants that are specifically chosen to accumulate metals (Ashraf et al., 2019). Although it has limitations, preliminary research suggests that phytoremediation is a potential cleanup option for a range of polluted locations (Ashraf et al., 2019). The biological feature of this kind of treatment method directly contributes to many of phytoremediation's drawbacks and benefits (Dubchak & Bondar, 2019).

20.2 PHYTOREMEDIATION: A SUSTAINABLE APPROACH

The amount of anthropogenic and natural activities contaminating the soil, water and food chain is increasing daily, contributing to environmental contamination. Phytoremediation is a relatively harmless, green cleaning technique that uses plants as a means of extraction, destruction or containment as well as to degrade, eliminate, confine or immobilize contaminants such as metals, insecticides, hydrocarbons and chlorinated solvents, that are present in soil and water. The rhizosphere may be effectively stimulated to break down organic compounds by a variety of plant species (Hassan et al., 2023; Raza et al., 2023). These plants often have large, fibrous roots that form a broad rhizosphere. They include maize, wheat, soybean, peas, beans and some common grasses. Furthermore, a variety of tree species are capable of absorbing and decomposing some organic pollutants. One of many approaches that can be used for phytoremediation of metals and other inorganic compounds is phytoextraction, the process by which metals from the soil are absorbed and concentrated in the roots and shoots of plants; using plant roots, a process known as rhizofiltration can be used to extract metals from wastewater (Tangahu et al., 2011). Application of plants to prevent the spread of metals in the environment is known as phytostabilization; releasing volatile materials like compounds containing arsenic or mercury into the atmosphere is known as phytovolatilization (Ekta & Modi, 2018). Plants that may naturally collect high concentrations of metal have been identified, and these are being researched for the purpose of phytoremediation of metal-containing environments. Unfortunately, even hyper-accumulating plants only grow to a modest size when exceptionally high metal concentrations are present. In other words, even those plants that have the capacity to hyper-accumulate metals are inhibited from growing when exposed to elevated metal concentrations (Tovar-Sánchez et al., 2023). Over 400 plant species from at least 45 plant families have been identified as having metal hyper-accumulation. These plants, known as hyper-accumulators, are often observed growing in regions where the level of metal content in the soil is higher. Fabaceae, Scrophulariaceae, Asterraceae, Euphorbiaceae, Brassicaceae and Lamiaceae are a few of these families. Crops that are extremely promising for bioaccumulation of Cu, Co, Se, Cd/Zn and Ni include alpine pennycress (*Thlaspi caerulescens*), *Haumaniastrum*, *Astragalus racemosus*, *Ipomea alpine*, *Haumaniastrum robertii* and *Sebertia acuminate* (Tovar-Sánchez et al., 2023). For instance, plants with phytotransformation activity may have laccases that may break down anilines like triaminotoluene, nitroreductases that are helpful in breaking down nitroaromatics, and dehalogenases that break down chlorinated solvents and pesticides. Complex aromatic chemicals like flavonoids and coumarins are an essential class of plant components in root colonization. Interestingly, these chemicals do not accumulate in soil much because soil microflora consume them and use the decreased carbon and nitrogen to break them down. The structural similarity of these aromatic plant molecules to various organic contaminants, such as polycyclic aromatic hydrocarbons (PAHs) and polychlorinated biphenyls (PCBs), and per chloroethylene

(PHC), is fortunate and offers a way to use rhizosphere activities naturally for the remediation of contaminants (Ilyas et al., 2022). Water and soil contaminated with heavy metals can result in a shortage of wholesome food and clean water, which is one of the world's most urgent concerns (Hayyat et al., 2020). The scientific community suggests using biological solutions based in the lab and field to eliminate heavy metals from contaminated water and soil. While phytoremediation is a promising method for removing metals from polluted areas, there are several drawbacks, such as:

1. It takes several years to clean up polluted areas.
2. The majority of species that are hyper-accumulators grow slowly, have poor biomass and are difficult to culture.
3. Plant bioaccumulation capacity may be lowered as a result of pests. The biodiversity of the area may be impacted by the introduction of new species in polluted areas.
4. It is of concern how contaminated biomass should be disposed of.
5. Climate conditions are the primary determinant of the sustainability of plant-based metal remediation.
6. Sites with low to moderate concentrations of heavy metals may benefit from the use of phytoremediation technology.
7. If polluted biomass is not managed properly, metals may find their way into the food chain.
8. Only soils, sediments and shallow groundwater can be treated by phytoremediation.
9. Soil phytoremediation is limited to soil on the surface.
10. A 100% decrease is not possible using phytoremediation.

20.3 ROLE OF ENDOPHYTIC MICROORGANISMS IN PHYTOREMEDIATION

The small area of soil where soil microorganisms and secretions from roots directly affect it is known as the rhizosphere (Lei et al., 2024). An intricate and dynamic ecology is created in this region by the different symbiotic connections that plants have with microbes. Most plant species have symbiotic interactions with these fungi, known as arbuscular mycorrhizae. The plant provides food and shelter to the fungi and the fungi help the plant absorb more water and nutrients, particularly phosphorus. The fungus known as ectomycorrhiza is associated symbiotically with the roots of several trees, including oaks and pines (Adnan et al., 2022). In order to facilitate the passage of nutrients between the fungus and the plant, the fungi create a sheath around the terminals of the roots. Nodules are formed by certain bacteria, such as *Rhizobium* species, on the roots of leguminous plants. The plant gives the bacteria sugars and other organic compounds if the bacteria fix atmospheric nitrogen and convert it into a form that plants can use for growth. Certain plants have symbiotic associations with actinomycetes, a kind of filamentous bacteria (Shanthi, 2021).

In the rhizosphere, these bacteria support the cycling of nutrients and the control of disease. Through a variety of processes, such as the synthesis of plant hormones, nutrient solubilization and pathogen defense, some bacteria in the rhizosphere support plant growth. The general health and vitality of plants are enhanced by these microorganisms (Shanthi, 2021). Endophytic fungi are able to dwell inside plant tissues without endangering their host. Certain endophytes provide advantages to the host plant, including better nutrition absorption, heightened stress tolerance and greater resistance to infections. Other nitrogen-fixing bacteria, such as *Azospirillum* and *Azotobacter*, can associate with the roots of nonleguminous plants in addition to rhizobium–legume symbiosis. These microorganisms transform air nitrogen into forms that are useful to plants. The rhizosphere's symbiotic connections are essential to the sustainability of ecosystems, plant health and the cycling of nutrients (Saha & Bauddh, 2020). Both the environment and living things are poisoned by pollutants. There have been attempts to remove these contaminants using biological agents, and a range of fungi and bacteria have been effective in many of these removals (Gnanasalomi et al., 2013). The potential for bioremediation of a class of bacteria called endophytes has been investigated recently. Endophytes have traditionally been studied for their plant growth-promoting properties, biocontrol activities and production of bioactive compounds (Sim et al., 2019). Therefore, endophytes are promising microorganisms in search for novel instruments for pollution bioremediation. Over the past 20 years, bioremediation has become a more viable option, particularly for the cleanup of large, contaminated areas (Griffin, 2013). Endophytic bacteria have the ability to decrease phytotoxicity and enhance the recruitment and heavy metal build-up in the biomass of aboveground plants during the phytoremediation of inorganic contaminants (Cherian et al., 2012). Many endophytes have the ability to break down organic contaminants and their resistance to heavy metals most likely stems from their exposure to these substances when they are found in the plant/soil niche (Prakash, 2021). Due to their involvement in the process of bioaugmentation, bacterial endophytes may be more beneficial than bacteria that are introduced to the soil (Graj et al., 2013). Numerous bacterial strains that were isolated from grapevine (*Vitis vinifera* L.) plants demonstrated resistance to lead, mercury, nickel, manganese and zinc. When commercially available metabolic inhibitors like 2,4-dinitrophenol (DNP) or N,N′dicyclohexylcarbodiimide (a specific ATPase inhibitor, DCC) are present, the endophytic bacteria *Bacillus* sp. reduced cadmium to about 94%. Likewise, the introduction of *Serratia nematodiphila* LRE07 endophytic bacteria to *Solanum nigrum* L. in the presence of cadmium eased growth inhibition (Yadav et al., 2021). It has been reported that heavy metal bioremediation can be accomplished by the use of endophytic bacteria (Table 20.1) For example, L14 (EB L14), which were isolated from the cadmium hyperaccumulator *Solanum nigrum* L.

TABLE 20.1
Types of Microbes Which Are Used Against Heavy Metal Stress

Microbes or mixtures	Types/names	Pollution source	Efficiency	Mechanisms	References
Mixtures of microbes and sludge	Water treatment sludge	Electroplating wastewater	Cu^{2+} and Cr^{6+} adsorption capacities are about 3.5 mg g^{-1} and 1.7	Biosorption	Singh et al., 2022
	Sludge was enhanced by 43.3% of Desulfovibiro's sulfate-reducing bacteria	Electroplating wastewater	6.666 mg g^{-1}	Biosorption and bioreduction	Sharma & Kumar, 2021
	Combination of sewage sludge with *Candida lipolytica*	Electroplating effluent	100% in 3 days	Precipitation by biological means	Ma et al., 2016
Main bacteria strains	*Acidithiobacillus ferrooxidans*	Dewatered metal plating sludge	In 20 days, 97% of Zn, 96% of Cu, 93% of Ni, 84% of Pb, 67% of Cd and 34% of Cr were soluble in metals	Bioleaching	Lopez et al., 2022

TABLE 20.1 (Continued)
Types of Microbes Which Are Used Against Heavy Metal Stress

Microbes or mixtures	Types/names	Pollution source	Efficiency	Mechanisms	References
	Acidithiobacillus ferrooxidans	Electroplating sludge	With acid leaching, Cr is 80.9% and Ni is 65.8%; moreover, 6.0% of Cr and 11.7% of Ni are extracted further using bioleaching	Combined bioleaching and acid leaching	Mei & Flinn, 2010
	Arthrobacter sp. X34 *Bacillus* sp. Y9-2 *Rahnella* sp. Y9602	Simulated groundwater	0 95% precipitated in 5 days 73% precipitated in 5 days	Biomineralization	Bilal et al., 2019
Combinations of bioresource and microalgae	Calcined eggshells and *Chlorella vulgaris*	Acid mine drainage	100% of Cd, 100% of As, 99.5% of Cu, 99.7% of Fu, 99.8% of Mn and 99.9% of Zn in 6 days	Biosorption	Domka et al., 2019
Mixed plants and endophytic bacteria	Combination of *Microbacterium arborescens* HU33, *Pantoea stewartii* ASI11, *Leptochloa fusca* and *Enterobacter* sp. U38	Tannery effluent	135 ± 4.2, 19 ± 1.3, 8550 ± 42, 450 ± 9.5, 174 ± 3.1, 15 ± 0.72, 11 ± 0.21, 5 ± 0.29, 7 ± 0.65 (mg kg^{-1} root)	Bioaccumulation and phytoextraction	Jan et al., 2019
Single culture or mixtures	*Bacillus cereus*, *Pseudomonas putida*	Tannery effluent	51.9% of Cr (VI)	Bioreduction	Pecoraro et al., 2021

20.4 MECHANISMS OF METAL SEQUESTRATION BY ENDOPHYTES

Endophytes have the ability to mitigate heavy metal stress and help in phytoremediation in polluted environments. When plants harbor endophytes, they stimulate the production of proteins such as phytochelatins and metallothioneins. These proteins play a crucial role in binding and immobilizing heavy metals through a chelation process within the plant tissues (Ma et al., 2016)

20.4.1 Production of Extracellular Polysaccharides

Endophytes produce extracellular polysaccharides that can bind to heavy metals and build complex structures with them. This structure may inhibit the metal from being accessible to or absorbed by the plant roots (Ma et al., 2011). For example, *Pseudomonas putida*, *Pseudomonas aeruginosa*, *Bacillus cereus*, *Bacillus licheniformis* and *Bacillus thuringiensis* are some endophytes that have the ability to produce extracellular polysaccharides which can be HM (Fakhar et al., 2022).

20.4.2 Production of Metallothionein

The metallothionein protein is enhanced by some endophytes. These proteins have the ability to bind to heavy metals, shielding them from toxicity and preventing them from participating in crucial biological processes (Lopez et al., 2022). For example, *Enterobacter cloacae*, *Pseudomonas putida*, *Bacillus subtilis*, *Trichoderma harzianum* and arbuscular mycorrhizal fungi (AMF) (various species) are the endophytes which are able to enhance the metallothionein protein (Zulfiqar et al., 2023).

20.4.3 Production of Indole Acetic Acid (IAA)

Certain endophytes have the capacity to grow roots and may generate IAA, or hormones that promote plant growth. An increase in root biomass can improve heavy metal storage in the root and stop it from moving to the plant's upper sections.(Mei & Flinn, 2010). Some examples of these endophytes are *Pseudomonas fluorescens*, *Bacillus amyloliquefaciens*, *Azospirillum brasilense* and *Rhizobium leguminosarum* (Saranraj et al., 2022).

20.4.4 Production of Siderophores

Certain endophytes generate siderophores, which bind to metal ions, particularly iron, which might be able to withstand heavy metals and lessen their availability to plants (Pecoraro et al., 2021). *Pseudomonas aeruginosa*, *Bacillus subtilis*, *Rhizobium* spp., *Enterobacter cloacae* and *Azotobacter vinelandii* are examples of such endophytes (Ma et al., 2011; Pahari et al., 2017).

20.4.5 Phosphate Solubilization

Endophytic bacteria produce organic acids like gluconic acids, oxalic acids and organic acids, which then help in the solubilization of phosphorus and make it available for plants (Ma et al., 2011). Examples include *Pseudomonas fluorescens*, *Bacillus megaterium*, *Rhizobium leguminosarum*, *Azospirillum, brasilense* and *Enterobacter cloacae.*

20.4.6 Enhanced Enzyme Production of Antioxidants

When plants come into contact with endophytes, they may release antioxidant enzymes. These antioxidants may aid in lowering the oxidative stress brought on by metals. (Jan et al., 2019). For example, *Trichoderma harzianum*, *Bacillus subtilis*, *Pseudomonas putida*, *Azospirillum brasilense* and *Enterobacter cloacae* are some such types of endophytes (Bilal et al., 2020).

20.4.7 Intracellular Accumulation

Endophytes function as "biofilters," which may aid in the buildup of heavy metals within their cells. Thus, they might aid in preventing heavy metals from getting to plants' essential organs (Padhan et al., 2021). *Pseudomonas putida*, *Bacillus subtilis*, *Enterobacter cloacae*, AMF and *Rhizobium* spp. are examples of such endophytes.

20.4.8 Improved Absorption of Nutrients

Endophytes indirectly aid in metal sequestration by facilitating the plant's access to vital nutrients. Because heavy metals compete for the same absorption pathways, a healthy diet can reduce the amount absorbed (Kumar et al., 2016). Examples include *Azospirillum* spp., *Bacillus* spp., *Pseudomonas* spp., *Rhizobium* spp. and AMF.

20.5 CONTRIBUTION TO PLANT HEALTH AND RHIZOSPHERE ENVIRONMENT

Research shows that different endophytes are suitable for detoxification and stress mitigation in plants. The roles that endophytes perform in numerous processes are highlighted in the following (Domka et al., 2019).

20.5.1 Production of Substances that Promote Plant Growth

Endophytes generate phytohormones, which enhance the growth and development of plants. Consequently, they supports the defense mechanism against various stressors (Eid et al., 2021). Examples of such endophytes are *Pseudomonas fluorescens*, *Bacillus subtilis*, *Azospirillum brasilense*, *Rhizobium leguminosarum* and *Trichoderma harzianum* (Santoyo et al., 2021).

20.5.2 Induced Systemic Resistance

Endophytes have the ability to synthesize signaling compounds that enable plants to establish systemic resistances and withstand additional infections in the future. Various endophytes improve plants' capacity to absorb nutrients. Additionally, endophytes reduce the amount of nutrients that diseases can access, while simultaneously increasing the nutrients that plants can access to combat these pathogens (Lamalakshmi Devi et al., 2017). *Pseudomonas fluorescens*, *Bacillus subtilis*, *Trichoderma harzianum*, *Azospirillum brasilense* and *Rhizobium leguminosarum* are such types of endophytes (Xia et al., 2022).

20.5.3 Production of Stress-Specific Metabolites

Various types of endophytes can produce metabolites which are specific to heavy metal tolerance that may help to prevent heavy metal build up in plant tissues (Phurailatpam & Mishra, 2020). Some examples of these endophytes are *Pseudomonas putida*, *Bacillus subtilis*, AMF, *Rhizobium leguminosarum* and *Trichoderma harzianum* (Kumar & Verma, 2018).

20.5.4 Transformation of Metal

It is possible for various endophytes to change harmful metals into safe versions. Processes like immobilization, precipitation or conversion to a less hazardous oxidation state may be involved in this shift (Sharma & Kumar, 2021). Examples include *Pseudomonas putida*, *Bacillus subtilis*, AMF, *Rhizobium leguminosarum* and *Trichoderma harzianum.*

20.5.5 Increased Uptake of Metals

Certain endophytes can assist plants in absorbing necessary metals and preventing or reducing the intake of harmful chemicals. Hence, plants are capable of selective uptake (Lodewyckx, 2001). Examples of such types of endophytes are *Pseudomonas fluorescens*, *Bacillus subtilis*, AMF, *Rhizobium* spp. and *Trichoderma* spp. (Doty, 2008).

20.5.6 Plant Defense Mechanism Stimulation

Certain endophytes possess the ability to trigger the plant's defense system in response to heavy metal stress. This procedure could improve detoxification and tolerance to heavy metals (Bilal et al., 2019). Examples include *Pseudomonas putida*, *Bacillus subtilis*, AMF, *Rhizobium leguminosarum* and *Trichoderma harzianum* (Shahzad et al., 2019).

20.6 POTENTIAL APPLICATIONS IN BIOTECHNOLOGY-ASSISTED PHYTOREMEDIATION

Phytoremediation is the process by which plants can mitigate the heavy metals from polluted land using biological agents like bacteria or genetically modified plants. This process can be applied to different environmental problems involving the poisoning of land and water (Bilal et al., 2019). Some of the biotechnological uses for phytoremediation supported by biotechnology are given below.

20.6.1 Improved Metal Uptake and Tolerance

The use of biotechnological process and tools can enhance the ability of plants to absorb metals. For this purpose, genes are inserted into plants which may help the plants to live in soil having high concentrations of heavy metals (Mani & Kumar, 2014). Examples include *Enterobacte*r spp., *Azospirillum* spp., *Actinomycetes*, fungal endophytes (e.g., *Fusarium* spp., *Penicillium* spp.), and Cyanobacteria (e.g., *Nostoc* spp.) (Shahzad et al., 2019).

20.6.2 Enhanced Organic Pollutant Degradation

Engineered microbes are applied to the rhizosphere of plants which can break down organic pollutants. These microbes can carry specific genes that carry a code for specific enzymes that breaks down specific types of contamination (Koźmińska et al., 2018). *Pseudomonas putida*, *Escherichia coli*, *Bacillus subtilis*, *Stenotrophomonas maltophilia* and *Burkholderia spp*. are some examples of such microbes (Kumar et al., 2021).

20.6.3 Using Phytostabilization to Manage Erosion

The root structure of plants can be modified by biotechnological treatments which help in the increment of the soil solubilization. Therefore, it may help in soil erosion (Gayathiri et al., 2022).

20.6.4 Enhanced Assimilation of Nutrients for the Restoration of Sites

With the help of biotechnology, plants can be altered to increase their ability for nutrient uptake. In polluted soils, this could help to reset soils which were depleted, and overall ecosystem regeneration could be increased (El-Ramady, 2014).

20.7 CONCLUSION

Endophytes are capable of decreasing the amount of heavy metals in contaminated soil. In this chapter we have highlighted the role of endophytes in the phytoremediation of heavy metals. The phytoremediation technique is successful due to the symbiotic relationship between host plants and endophytes. Different mechanisms are adapted by endophytes such as metal sequestration, detoxification and increased plant growth promotion to decrease the negative effects of heavy metals. Therefore, endophytes are among the best available options to mitigate heavy metals and increase plant growth in contaminated soil.

REFERENCES

Adnan, M., Wahid, F., Fahad, S., Arif, M., Shi, S. M., & He, X. H. (2022). Mycorrhizae and soil health.

Ashraf, S., Ali, Q., Zahir, Z. A., Ashraf, S., & Asghar, H. N. (2019). Phytoremediation: Environmentally sustainable way for reclamation of heavy metal polluted soils. *Ecotoxicology and Environmental Safety*, *174*, 714–727.

Bilal, S., Shahzad, R., Imran, M., Jan, R., Kim, K. M., & Lee, I.-J. (2020). Synergistic association of endophytic fungi enhances Glycine max L. resilience to combined abiotic stresses: Heavy metals, high temperature and drought stress. *Industrial Crops and Products*, *143*, 111931.

Bilal, S., Shahzad, R., Khan, A. L., Al-Harrasi, A., Kim, C. K., & Lee, I.-J. (2019). Phytohormones enabled endophytic Penicillium funiculosum LHL06 protects Glycine max L. from synergistic toxicity of heavy metals by hormonal and stress-responsive proteins modulation. *Journal of Hazardous Materials*, *379*, 120824.

Cherian, S., Weyens, N., Lindberg, S., & Vangronsveld, J. (2012). Phytoremediation of trace element–contaminated environments and the potential of endophytic bacteria for improving this process. *Critical Reviews in Environmental Science and Technology*, *42*(21), 2215–2260.

Dimian, R., Sheta, A. A. S., Alia, R., & Abd-Elwahed, M. S. (2022). Assessment of Pollution by Heavy Metals at Landforms of Selected Areas in the Nile Valley and Desert Fringes.

Domka, A. M., Rozpądek, P., & Turnau, K. (2019). Are fungal endophytes merely mycorrhizal copycats? The role of fungal endophytes in the adaptation of plants to metal toxicity. *Frontiers in Microbiology*, *10*, 371.

Doty, S. L. (2008). Enhancing phytoremediation through the use of transgenics and endophytes. *New Phytologist*, *179*(2), 318–333.

Dubchak, S., & Bondar, O. (2019). Bioremediation and phytoremediation: Best approach for rehabilitation of soils for future use. In Gupta. D.K and Voronina. A. (Eds), *Remediation Measures for Radioactively Contaminated Areas* (pp. 201–221), Springer.

Dung, T. T. T., Cappuyns, V., Swennen, R., & Phung, N. K. (2013). From geochemical background determination to pollution assessment of heavy metals in sediments and soils. *Reviews in Environmental Science and Bio/Technology*, *12*, 335–353.

Eid, A. M., Fouda, A., Abdel-Rahman, M. A., Salem, S. S., Elsaied, A., Oelmüller, R., Hijri, M., Bhowmik, A., Elkelish, A., & Hassan, S. E.-D. (2021). Harnessing bacterial endophytes for promotion of plant growth and biotechnological applications: An overview. *Plants*, *10*(5), 935.

Ekta, P., & Modi, N. R. (2018). A review of phytoremediation. *Journal of Pharmacognosy and Phytochemistry*, *7*(4), 1485–1489.

El-Ramady, H. R. (2014). Integrated nutrient management and postharvest of crops. *Sustainable Agriculture Reviews*, *13*, 163–274.

Fakhar, A., Gul, B., Gurmani, A. R., Khan, S. M., Ali, S., Sultan, T., Chaudhary, H. J., Rafique, M., & Rizwan, M. (2022). Heavy metal remediation and resistance mechanism of Aeromonas, Bacillus, and Pseudomonas: A review. *Critical Reviews in Environmental Science and Technology*, *52*(11), 1868–1914.

Gayathiri, E., Gobinath, R., Ganapathy, G., Salunkhe, A. A., Jayanthi, J., Ragunathan, M., & Pourghasemi, H. R. (2022). Studies on plant selection framework for soil bioengineering application. In H. R. Pourghasemi (Ed.), *Computers in Earth and Environmental Sciences* (pp. 299–317). Elsevier.

Gnanasalomi, V. D. V., Jebapriya, G. R., & Gnanadoss, J. J. (2013). Bioremediation of hazardous pollutants using fungi. *International Journal of Computing Algorithm*, *2*, 273–278.

Graj, W., Lisiecki, P., Szulc, A., Chrzanowski, Ł., & Wojtera-Kwiczor, J. (2013). Bioaugmentation with petroleum-degrading consortia has a selective growth-promoting impact on crop plants germinated in diesel oil-contaminated soil. *Water, Air, & Soil Pollution*, *224*, 1–15.

Griffin, M. R. (2013). Biocontrol and bioremediation: Two areas of endophytic research which hold great promise. In V. C. Verma & A. C. Gange (Eds.), *Advances in Endophytic Research* (pp. 257–282). Springer.

Han, R., Zhou, B., Huang, Y., Lu, X., Li, S., & Li, N. (2020). Bibliometric overview of research trends on heavy metal health risks and impacts in 1989–2018. *Journal of Cleaner Production*, *276*, 123249.

Hassan, A., Hamid, F. S., Pariatamby, A., Suhaimi, N. S. M., Ling, K. N., & Mohan, P. (2023). Bioaugmentation-assisted bioremediation and biodegradation mechanisms for PCB in contaminated environments: A review on sustainable clean-up technologies. *Journal of Environmental Chemical Engineering*, 11(3), 110055.

Hayyat, M. S., Adnan, M., Awais, M., Bilal, H. M., Khan, B., & Rahman, H. A. (2020). Effect of heavy metal (Ni) on plants and soil: A review. *Int. J. Appl.* Res, 6(7), 313–318.

Ilyas, H., Mal, J., Masih, I., & van Hullebusch, E. D. (2022). Bioremediation Technologies for the Treatment of water contaminated by organic and inorganic contaminants. In R. Selvasembian, E. D. van Hullebusch, & J. Mal (Eds.), *Biotechnology for Environmental Protection* (pp. 61–129). Springer Nature Singapore.

Jaishankar, M., Tseten, T., Anbalagan, N., Mathew, B. B., & Beeregowda, K. N. (2014). Toxicity, mechanism and health effects of some heavy metals. *Interdisciplinary Toxicology*, *7*(2), 60.

Jan, R., Khan, M. A., Asaf, S., Lee, I.-J., & Kim, K. M. (2019). Metal resistant endophytic bacteria reduces cadmium, nickel toxicity, and enhances expression of metal stress related genes with improved growth of Oryza sativa, via regulating its antioxidant machinery and endogenous hormones. *Plants*, *8*(10), 363.

Jo, J., Oh, J., & Park, C. (2020). Microbial community analysis using high-throughput sequencing technology: A beginner's guide for microbiologists. *Journal of Microbiology*, *58*, 176–192.

Kahlon, S. K., Sharma, G., Julka, J., Kumar, A., Sharma, S., & Stadler, F. J. (2018). Impact of heavy metals and nanoparticles on aquatic biota. *Environmental Chemistry Letters*, *16*, 919–946.

Koźmińska, A., Wiszniewska, A., Hanus-Fajerska, E., & Muszyńska, E. (2018). Recent strategies of increasing metal tolerance and phytoremediation potential using genetic transformation of plants. *Plant Biotechnology Reports*, *12*, 1–14.

Kumar, A., & Verma, J. P. (2018). Does plant—microbe interaction confer stress tolerance in plants: A review? *Microbiological Research*, *207*, 41–52.

Kumar, D., Singh, D., Barman, S., & Kumar, N. (2016). Heavy metal and their regulation in plant system: An overview. In A. Singh, S. M. Prasad, & R. P. Singh (Eds.), *Plant Responses to Xenobiotics* (pp. 19–38). Springer.

Kumar, J., Malik, N. A., & Atri, N. S. (2021). Aromatic and medicinal plants for phytoremediation: A sustainable approach. In T. Aftab & K. R. Hakeem (Eds.), *Medicinal and Aromatic Plants: Healthcare and Industrial Applications* (pp. 485–543). Springer.

Kumar, P., Gacem, A., Ahmad, M. T., Yadav, V. K., Singh, S., Yadav, K. K., Alam, M. M., Dawane, V., Piplode, S., & Maurya, P. (2022). Environmental and human health implications of metal (loid) s: Source identification, contamination, toxicity, and sustainable clean-up technologies. *Frontiers in Environmental Science*, *10*, 949581.

Lamalakshmi Devi, E., Kumar, S., Basanta Singh, T., Sharma, S. K., Beemrote, A., Devi, C. P., Chongtham, S., Singh, C. H., Yumlembam, R. A., & Haribhushan, A. (2017). Adaptation strategies and defence mechanisms of plants during environmental stress. In M. Ghorbanpour & A. Varma (Eds.), *Medicinal Plants and Environmental Challenges* (pp. 359–413). Springer.

Lei, Y., Bi, Y., Dong, X., Li, H., Gao, X., Li, X., & Yan, Z. (2024). Effects of salinity on iron-organic carbon binding in the rhizosphere of Kandelia obovata: Insights from root exudate analysis. *Science of the Total Environment*, 955, 177214.

Lodewyckx, C. (2001). A potential role for bacterial endophytes in phytoremediation of heavy metal-contaminated soils. Limburgs Universitair Centrum (Belgium).

Lopez, A. M. Q., Silva, A. L. D. S., Maranhão, F. C. D. A., & Ferreira, L. F. R. (2022). Plant growth promoting bacteria: Aspects in metal bioremediation and phytopathogen management. In A. Kumar (Ed.), *Microbial Biocontrol: Sustainable Agriculture and Phytopathogen Management: Volume 1* (pp. 51–78). Springer International Publishing.

Ma, Y., Prasad, M., Rajkumar, M., & Freitas, H. (2011). Plant growth promoting rhizobacteria and endophytes accelerate phytoremediation of metalliferous soils. *Biotechnology Advances*, *29*(2), 248–258.

Ma, Y., Rajkumar, M., Zhang, C., & Freitas, H. (2016). Beneficial role of bacterial endophytes in heavy metal phytoremediation. *Journal of Environmental Management*, *174*, 14–25.

Mani, D., & Kumar, C. (2014). Biotechnological advances in bioremediation of heavy metals contaminated ecosystems: An overview with special reference to phytoremediation. *International Journal of Environmental Science and Technology*, *11*, 843–872.

Mei, C., & Flinn, B. S. (2010). The use of beneficial microbial endophytes for plant biomass and stress tolerance improvement. *Recent Patents on Biotechnology*, *4*(1), 81–95.

Mishra, S., Bharagava, R. N., More, N., Yadav, A., Zainith, S., Mani, S., & Chowdhary, P. (2019). Heavy metal contamination: An alarming threat to environment and human health. In *Environmental Biotechnology: For Sustainable Future* (pp. 103–125).

Mitra, S., Chakraborty, A. J., Tareq, A. M., Emran, T. B., Nainu, F., Khusro, A., Idris, A. M., Khandaker, M. U., Osman, H., & Alhumaydhi, F. A. (2022). Impact of heavy metals on the environment and human health: Novel therapeutic insights to counter the toxicity. *Journal of King Saud University-Science*, *34*(3), 101865.

Olaniran, A. O., Balgobind, A., & Pillay, B. (2013). Bioavailability of heavy metals in soil: Impact on microbial biodegradation of organic compounds and possible improvement strategies. *International Journal of Molecular Sciences*, *14*(5), 10197–10228.

Padhan, D., Rout, P. P., Kundu, R., Adhikary, S., & Padhi, P. P. (2021). Bioremediation of heavy metals and other toxic substances by microorganisms. In J. A. Parray, A. H. A. E. Mahmoud, & R. Sayyed (Eds.), *Soil Bioremediation: An Approach Towards Sustainable Technology* (pp. 285–329). Wiley Online Library.

Pahari, A., Pradhan, A., Nayak, S., & Mishra, B. B. (2017). Bacterial siderophore as a plant growth promoter. In J. K. Patra, C. N. Vishnuprasad, & G. Das (Eds.), *Microbial Biotechnology: Volume 1. Applications in Agriculture and Environment* (pp. 163–180). Springer.

Pecoraro, L., Wang, X., Shah, D., Song, X., Kumar, V., Shakoor, A., Tripathi, K., Ramteke, P. W., & Rani, R. (2021). Biosynthesis pathways, transport mechanisms and biotechnological applications of fungal siderophores. *Journal of Fungi*, *8*(1), 21.

Phurailatpam, L., & Mishra, S. (2020). Role of plant endophytes in conferring abiotic stress tolerance. In M. Hasanuzzaman (Ed.), *Plant Ecophysiology and Adaptation under Climate Change: Mechanisms and Perspectives II: Mechanisms of Adaptation and Stress Amelioration* (pp. 603–628). Springer.

Prakash, J. (2021). Potential application of endophytes in bioremediation of heavy metals and organic pollutants and growth promotion: Mechanism, challenges, and future prospects. In G. Saxena, V. Kumar, & M. P. Shah (Eds.), *Bioremediation for Environmental Sustainability* (pp. 91–121). Elsevier.

Raza, T., Qadir, M. F., Khan, K. S., Eash, N. S., Yousuf, M., Chatterjee, S., ... & Oetting, J. N. (2023). Unrevealing the potential of microbes in decomposition of organic matter and release of carbon in the ecosystem. *Journal of Environmental Management*, 344, 118529.

Saha, L., & Bauddh, K. (2020). Sustainable agricultural approaches for enhanced crop productivity, better soil health, and improved ecosystem services. In K. Bauddh, S. Kumar, R. P. Singh, & J. Korstad (Eds.), *Ecological and Practical Applications for Sustainable Agriculture* (pp. 1–23). Springer.

Santoyo, G., Guzmán-Guzmán, P., Parra-Cota, F. I., Santos-Villalobos, S. d. l., Orozco-Mosqueda, M. d. C., & Glick, B. R. (2021). Plant growth stimulation by microbial consortia. *Agronomy*, *11*(2), 219.

Saranraj, P., Sayyed, R., Kokila, M., Sudha, A., Sivasakthivelan, P., Devi, M. D., Naz, R., & Yasmin, H. (2022). Plant growth-promoting and biocontrol metabolites produced by endophytic pseudomonas fluorescence. In R. Z. Sayyed & V. G. Uarrota (Eds.), *Secondary Metabolites and Volatiles of PGPR in Plant-Growth Promotion* (pp. 349–381). Springer International Publishing.

Sayqal, A., & Ahmed, O. B. (2021). Advances in heavy metal bioremediation: An overview. *Applied Bionics and Biomechanics*, *2021*, 1609149.

Shahzad, R., Bilal, S., Imran, M., Khan, A. L., Alosaimi, A. A., Al-Shwyeh, H. A., Almahasheer, H., Rehman, S., & Lee, I.-J. (2019). Amelioration of heavy metal stress by endophytic Bacillus amyloliquefaciens RWL-1 in rice by regulating metabolic changes: Potential for bacterial bioremediation. *Biochemical Journal*, *476*(21), 3385–3400.

Shanthi, V. (2021). Actinomycetes: Implications and prospects in sustainable agriculture. In I. Inamuddin, M. I. Ahamed, R. Boddula, & M. Rezakazemi (Eds.), *Biofertilizers: Study and Impact* (pp. 335–370). Wiley Online Library.

Sharma, P., & Kumar, S. (2021). Bioremediation of heavy metals from industrial effluents by endophytes and their metabolic activity: Recent advances. *Bioresource Technology*, *339*, 125589.

Sim, C. S. F., Chen, S. H., & Ting, A. S. Y. (2019). Endophytes: Emerging tools for the bioremediation of pollutants. In I. Inamuddin, M. I. Ahamed, R. Boddula, & M. Rezakazemi (Eds.), *Emerging and Eco-Friendly Approaches for Waste Management* (pp. 189–217). Springer.

Singh, A., Pal, D. B., Mohammad, A., Alhazmi, A., Haque, S., Yoon, T., Srivastava, N., & Gupta, V. K. (2022). Biological remediation technologies for dyes and heavy metals in wastewater treatment: New insight. *Bioresource Technology*, *343*, 126154.

Tangahu, B. V., Sheikh Abdullah, S. R., Basri, H., Idris, M., Anuar, N., & Mukhlisin, M. (2011). A review on heavy metals (As, Pb, and Hg) uptake by plants through phytoremediation. *International Journal of Chemical Engineering*, *2011*, 939161.

Torres, E. (2020). Biosorption: A review of the latest advances. *Processes*, *8*(12), 1584.

Tovar-Sánchez, E., Concepción-Acosta, C. M., Sánchez-Reyes, A., Sánchez-Cruz, R., Folch-Mallol, J. L., & Mussali-Galante, P. (2023). Aspergillus luchuensis, an endophyte fungus from the metal hyperaccumulator plant Prosopis laevigata, promotes its growth and increases metal translocation. *Plants*, *12*(6), 1338.

Xia, Y., Liu, J., Chen, C., Mo, X., Tan, Q., He, Y., Wang, Z., Yin, J., & Zhou, G. (2022). The multifunctions and future prospects of endophytes and their metabolites in plant disease management. *Microorganisms*, *10*(5), 1072.

Yadav, A. N., Kour, D., Kaur, T., Devi, R., Yadav, A., Dikilitas, M., Abdel-Azeem, A. M., Ahluwalia, A. S., & Saxena, A. K. (2021). Biodiversity, and biotechnological contribution of beneficial soil microbiomes for nutrient cycling, plant growth improvement and nutrient uptake. *Biocatalysis and Agricultural Biotechnology*, *33*, 102009.

Zulfiqar, U., Haider, F. U., Maqsood, M. F., Mohy-Ud-Din, W., Shabaan, M., Ahmad, M., Kaleem, M., Ishfaq, M., Aslam, Z., & Shahzad, B. (2023). Recent advances in microbial-assisted remediation of cadmium-contaminated soil. *Plants*, *12*(17), 3147.

21 Mycorrhizae Importance in Future Agriculture and Mycorrhizal Culturing Methods

Bashir Uddin, Najeeba Parre Paker, Mazhar Rafique and Zahoor ul Haq**
*Corresponding authors.

21.1 MYCORRHIZAL FUNGI

In nature, plants live in association with fungi around their root zone, which benefits the plant but can cause harm depending on the nature of association. Some promote plant development under particular circumstances, enabling plants to endure in harsh conditions (Hiruma et al., 2018). AM fungi belong to the fungal taxon of Glomeromycotina, comprising the mutualistic fungi which promote plant growth under nutrient-limiting conditions by way of transporting nutrients to host plants (Bruns et al., 2018). AM fungi are believed to be monophyletic of the Ordovician period of about 480 million years ago (Delaux, 2017). Eighty percent of terrestrial plant species contain AM fungus (Neuenkamp et al., 2018). Frank first used the term "mycorrhizae" in 1885 to refer to the symbiotic relationship between plant roots and fungi. Mycorrhiza means "fungus root" and the most common type is the AM fungi type which produces a specialized structure such as arbuscules and vesicles in the cortical region of the root (Powell & Bagyaraj, 2018). This symbiotic association is considered about 450 million years old and enhances plant growth in nutrient-deficient soil (Hoysted et al., 2018). From an evolutionary point of view, AM fungi symbiosis is an ancient mutualistic relationship, and an important adaptation in the terrestrialization of plants (Humphreys et al., 2010).

21.2 IMPORTANCE OF MYCORRHIZAE

AM fungi are a key component of soil biota, with a positive role in the production of crops and a sustainable ecosystem in new plant production strategies (Gianinazzi et al., 2010). AM fungi developed along with vascular plants and are widespread in the terrestrial zone. AM fungi are common in agricultural and natural environments because they improve the availability of plant nutrients, especially phosphate, and provide a number of other advantages to the environment and host plants, both directly and indirectly (Cardoso et al., 2017). AM fungi benefit their host plant with up to 80% of the nitrogen (N) and 100% of the phosphate required for the proliferation and growth of plants (Luginbuehl et al., 2017). AM fungi effectively scavenge P_i from the large volume of soil to root cortical cells and hence bypass direct root uptake of P_i (Ferlian et al., 2018).

Mycorrhizal associations play a significant role in shaping ecosystem structure and function by influencing the distribution and recycling of phosphorus (P) among different species, facilitating its movement between aboveground and belowground parts of the ecosystem (Cavagnaro et al., 2015). In order to improve plant development mechanisms, AM fungi are more concerned with ensuring that soil P and trace nutrients are fully available and absorbed (Ceballos et al., 2013). Generally

DOI: 10.1201/9781003486275-24

speaking, AM fungi that colonize plant roots improve nutritional status, hormone activity, growth rate and yield, to further advance plant development (Hart et al., 2015). Plant parasitic nematodes have the ability to either promote or inhibit AM fungal colonization of roots and spore development. Consequently, nematode penetration, development and reproduction may be influenced by AM fungi (Chabi et al., 2016). AM fungi absorb essential nutrients that plants need, such as N and P, and transfer them to the host plants, where they assimilate carbon (Augé et al., 2016) enabling plants to cope with stress phases (Jung et al., 2012).

AM fungi assist host plants in several ways, including improved nutrient and water uptake, defense against biotic and abiotic stressors and improved soil structure (Smith & Smith, 2012). In both natural and agricultural situations, the primary function of AM fungi is to modify the soil structure in a way that benefits plants, increases production and protects against stress (Leifheit et al., 2014). Without compacting the soil, the highly branched AM fungus mycelium's dense hyphae form a three-dimensional matrix that entangles and cross-links soil particles.

Additionally, it has been shown that glycoprotein is a crucial component in the stabilization of soil masses (Singh et al., 2013). In addition to strengthening the structure of the soil, AM fungi aid in lowering the amount of nutrients that drain from it, especially P and N, though it is possible that other mineral nutrients are also involved. When combined, AM fungi create closed nutrient cycles that integrate the soil's nutrient fluxes and support long-term soil fertility (Cavagnaro et al., 2015). Numerous advantages of AM fungi have created numerous prospects for commercial application. As a result, over the past few decades, the markets for AM fungi have expanded significantly, with an increase in the number of participants, goods and market size (Vosátka et al., 2008).

21.3 MYCORRHIZAL ROLE IN AGRICULTURAL PRODUCTIVITY

The bulk of people on the planet primarily depend on agriculture for their livelihood. However, this may encounter major difficulties in the 21st century, mostly as a result of rising global food consumption combined with depleting natural resources, including water and land availability, the effects of chemicals on the environment and the depletion of mineral phosphate fertilizer supplies (*Crossay et al.*, 2017). Over the past few decades, it has been discovered that agricultural methods have a negative impact on the ecosystem services (soil nutrients supply, water retention and filtration) (Jónsson & Davíðsdóttir, 2016). The main obstacle to achieving sustainable food security on a global scale is determining a substitute for inorganic fertilizers. This is due to the fact that the manufacturing of inorganic fertilizer uses a significant portion of the global energy budget and the availability of essential raw materials, especially phosphorus (Damar, 2016). Soil organisms have been employed as a bio-fertilizer for biotechnological applications in agronomy and the environment in an effort to produce sustainable agriculture. Consequently, AM fungi are among these species and have been successfully used in both domains (Berruti et al., 2016). AM fungi are known as essential components of the agro-ecosystem (Thirkell et al., 2017). Over 400 million years ago, a symbiotic relationship between AM fungi and their host plants existed (Selosse et al., 2015). Therefore, such an association is mainly based on a succession of biological actions that can lead to beneficial impacts in both natural and agro-ecosystems (Van Der Heijden et al., 2015). In an agricultural system, the mycorrhizal group of fungi proliferates, which allows the roots to operate in a state of good equilibrium with their environment. In a symbiotic connection with plants, mycorrhizal fungi dwell with the roots and help the plants absorb nutrients in exchange for carbohydrates. As a result, the improved plant performance has been demonstrated; consequently, mycorrhizal plants are typically more competitive and able to withstand a wider range of environmental challenges than nonmycorrhizal plants (Kostenko et al., 2018). AM fungi have the potential to assist plants in overcoming stressful conditions and generally improve their performance in challenging circumstances (Balestrini et al., 2017). The plant–AM symbiosis is primarily responsible for the plant's increased capacity to absorb nutrients and to withstand a variety of biotic and abiotic stressors, such as drought, salinity, flooding, cold and heavy metals (Lenoir et al., 2016).

AM fungi improve plant performance in normal and stress conditions along with soil aggregate stability (de León et al., 2018). AM fungi are in high demand in agriculture where modern practices can harness the potential benefit of mycorrhization; they are usually formed with the roots of the majority of plant species of *Poaceae* family and many other crop plants. There are a number of evidences which favor the role of AM fungi in phosphorus and zinc nutrition of plants (Ortas & Bykova, 2018). Since the production of phosphorus (P) chemical fertilizers is expensive and there are fewer sources available, plant adaptation techniques like mycorrhizal association may be a useful tactic as an affordable and simple agricultural management technique (Azcón-Aguilar & Barea, 2015). Additionally, AM fungi help plants defend themselves against infections, which is crucial to agriculture (Martín-Robles et al., 2018). AM fungi are thought to be a typical mechanism for agricultural soil because of their high inclination for reproduction, which may occur at the expense of giving the plant host inputs (Chagnon et al., 2013). AM fungi provide essential nutrients from the soil to the plant through their spreading network of hyphae in exchange for carbon from the host plant, which helps to reduce the dependency on inorganic fertilizers (Fester & Sawers, 2011). Although many woody perennial plants, including grapes, raspberries, currants and many fruit and nut trees, can also form a symbiotic association with AM fungi, the majority of research on the interaction between AM fungi and plants has focused on arable crops. The direct benefits of AM fungi to crops include increased resistance to disease (Jung et al., 2012), improved capacity relative to nonmycorrhizal plants, increased resistance to drought and other harsh soil conditions and indirect assistance to the crop through improved soil structure and soil nutrient retention (Berdeni et al., 2018). The agricultural forestry and environmental sectors utilize AM fungi to improve soil health by fixing carbon (Selvakumar et al., 2016).

21.4 MYCORRHIZAL CULTURE ON A COMMERCIAL SCALE

In recent years, the generation of mycorrhizal inocula for commercial use has advanced significantly from fungal proliferation in on-site nursery plots (Estrada et al., 2013). Up until 2001, a range of manure products produced by AM fungal inocula was used in the agricultural zone (Gianinazzi & Vosátka, 2004). The AM fungal inocula are preserved for use as plant and soil improvement materials. These materials have a wide range of applications, such as in horticulture, forestry, erosion management and sustainable agriculture. Even though the AM fungal inocula may have a favorable, neutral or unfavorable effect on plant growth, quality control must be ensured by examining the performance and contents of the inocula (Yao et al., 2018).

The large-scale production of AM fungi is limited due to the obligate biotrophic nature of these fungi. However, conventional techniques such as trap culture can be a successful tool for the propagation in plant growth and recovery of land in a small-scale and cost-effective way. Presently, AM fungal inocula are produced by utilizing various methods such as an in vitro, greenhouse or field-based methods (Gianinazzi & Vosátka, 2004). A pure culture of AM fungi propagules is achieved in in vitro methods, but the preparation of AM fungal inocula for large-scale application is still in its infancy (Selvakumar et al., 2016).

Utilizing host plants injected with a starter inoculum of infected root spores, AM fungi can be propagated in a pot culture. As a bio-fertilizer, media containing hyphae, infected roots and spores can be employed. The generation of AM fungi is dependent on the host type, environmental factors and species of the fungus (Asrianti et al., 2016). Due to the obligate biotrophic nature of AM fungi – that is, the requirement that they grow in symbiosis with host plant roots in order to complete their life cycle and produce propagules – the generation of AM fungal inocula on a wide scale is restricted. Commercially accessible AM fungal inocula range in concentration from potting media with a low inoculum concentration to a high concentration of AM fungi propagules in carrier materials (Douds Jr et al., 2010). Maximum agricultural benefits are achieved by using a suitable type of mycorrhizal fungi inoculation in the soil. Therefore, for the best way to culture and maintain

AM fungi production, the utilization of VAM inocula should become commercially viable (Bholay et al., 2018; Douds et al., 1993).

21.5 ISSUES IN CULTURING

The abundance and diversity of particular AM fungi species includes those with large hyphal networks such as *Gigaspoaceae* against other species that have increased resistance against pathogens or even parasites (Tkacz & Poole, 2015, Berruti et al., 2016). The effective storage of bio-inoculants is a major hindrance in AM fungal inocula production (Arora et al., 2010). Generally, there are three procedures to preserve fungi: continuous growth method, drying with silica gel or by air and reduction of available water in the cells (Rodrigues & Rodrigues, 2018). The obligatory biotrophic characteristic of AM fungi limits their large-scale production, as it makes them reliant on their host plant for survival. There are several ways to cultivate AM fungi in large quantities while maintaining host roots, such as using soil-substrate techniques like pot cultures and free methods like hydroponics and aeroponics (Liu & Yang, 2008). Regarding their design and commercialization quality, these techniques suffer from specified advantages and disadvantages (Ijdo et al., 2011). One of the major disadvantages is the high-quality production and uncontaminated inoculum. Furthermore, long-term storage and propagule maintenance in the artificial medium need skillful workers and are costly (Douds Jr et al., 2010). The primary problem in AM fungi research is the use of molecular markers for the intra- and interspecific quantification and identification of AM fungi (Crossay et al., 2017).

21.6 ESTABLISHED METHODS OF MYCORRHIZAL CULTURING

AM fungal inocula are now made using in vitro, greenhouse and field-based techniques. One in vitro technique used is the monoxenic cultivation of sterile AM fungal spores using Ri T DNA. Compared to traditional techniques (pot/trap culture), the in vitro root organ culture system is preferable since it produces clean, viable propagules free of contamination in a small area (Rodrigues & Rodrigues, 2017). Native soil inoculum (NSOIL) was shown by Urgiles et al. (2014) to enhance seedling development and growth in pioneer tree species in the Andean region, highlighting its potential benefits for reforestation and ecological restoration efforts. In order to construct a standard culture system, a dual-compartment plate with a nylon mesh or membrane was used. One of the best reported techniques for producing AM fungal spores devoid of plant roots is the culture method that was initially presented by St-Arnaud et al. (1996).

21.6.1 Culturing Methods (Substrate-Based Production Systems)

The traditional method of producing AM fungi involves growing plants and their symbionts in a soil or sand substrate, though a variety of substrate replacements and amendments are also frequently employed. This includes nonsterile hydroponic systems (physical soil substrates, such as sand and perlite) because the existence of a carrier medium affects the proliferation of AM fungi. Large-scale AM mushroom cultivation can be carried out in single pots made of various materials, such as plastic or earthenware, and in different sizes (Millner & Kitt, 1992) or from medium-sized bags and pots to highly raised or grounded beds (Douds Jr et al., 2006). The cultivation of AM fungi is typically carried out in growth chambers or greenhouses under regulated or semicontrolled conditions, making it easier to manage and regulate the temperature and humidity levels. On the other hand, large-scale cultivation of AM fungi may be carried out in an open environment, such as on farms (Douds Jr et al., 2006), and sporadically on field plots (Dodd et al., 1990), depending on the climate and host plant.

21.6.2 Production Parameters (AM Fungi)

Large-scale mass production of AM fungi is occasionally started with single species that are specific to the site of application but that are not always known to the species level. Small-scale mass production of AM fungi, for example in sand/soil or another substrate, is typically initiated with individual identified species or a consortium of selected AM fungi species (Gaur & Adholeya, 2002). To start AM fungus development, isolated spores or a combination of spores and mycorrhizal root species are typically used as the starting inoculum (Gaur & Adholeya, 2002). To prepare the mixed inoculum, the roots are first dried and then finely chopped, while the spores are often obtained through wet sieving and pouring. Additionally, the hyphal network of AM fungi found in soil is utilized in a mixed inoculum (Gaur & Adholeya, 2002), and the mixed inoculum is of particular concern only for those AM fungi species having intraradical spores and vesicles (Klironomos & Hart, 2002). Furthermore, plants can be directly injected with single or mixed spores; plantlets can also be precolonized prior to being moved into beds or other containers (Douds Jr et al., 2006). Reputable global culture collections, including INVAM, BEG and GINCO, can frequently ensure the availability of accurately identified individual species and offer a transparent picture of the traceability of organisms using their repository identification codes. Therefore, the culture collection serves as a repository for inocula deposited by corporations or offers material while adhering to regulations (material transfer agreement). The spore vouchers play a major role in determining traceability, which ensures the isolate's identity and place of origin. Techniques to verify that the injected isolate is the only source of intraradical colonization in the field are still being developed.

21.6.3 AM Host Plants

Plants like maize (*Zea mays*), Bahiagrass (*Paspalum notatum*) and onion and leek (*Allium* spp.) are frequently employed for the large-scale production of AM fungi. These plants are employed because of a number of benefits, including their short lifespan, adequate root system growth, maximum level of colonization by a wide variety of AM fungi and low P tolerance. In addition, they exhibit minimal pathogen susceptibility, yellow symptoms of colonized roots against white uncolonized roots (like maize and leek) and maximal temperature tolerance (Millner & Kitt, 1992). The C4 grass and stressed tropical plants killed during fall in temperate regions could favor sporulation (Douds Jr et al., 2006). These authors claim that the *Bahia* grass species they used in their example is unlikely to transmit crop viruses during inoculum treatment, in contrast to most crop plants. Host-dependent sporulation of AM fungi is thought to be a significant factor in the formation of inoculum. Gaur and Adholeya (2002) showed that after they infected five different fodder crop plants with a consortium of native AM fungus species, the quantity of infectious propagule generation was strongly dependent on the host plant species. The type of AM fungal inoculum – which may include both spore and mixed inoculum – markedly determines the host plant, or fungus linked with it. High intraradical colonization levels, for instance, are crucial for mixed spore–root inocula, however they may not be required to achieve spore inoculum. The report of the INVAM website (http://invam.caf.wvu.edu) showed that the number of spores in some of the pot cultures in the collection declines after a period of successive propagation cycles. In response to this issue, they recommended switching up the hosts and recommending that C4 Sudan grass (*Sorghum sudanese*), which is frequently employed as a host by INVAM, be replaced with C3 red clover (*Trifolium pratense*), which has no connection to the earlier host.

Furthermore, several *Glomus* species speed up their sporulation in conjunction with a C4 host after receiving N fertilizers; yet, other *Gigasporaceae* spp. after N fertilization increase the amount of spores they produce in conjunction with a C3 host plant (Egerton-Warburton et al. 2007). Such a reduction in sporulation may not occur in producing fields or in beds where AM fungal species are connected with a variety of hosts.

21.6.4 Substrates and Amendments

Propagation of AM fungi species on a large-scale with various substrate types, either pure, mixed or both, have been used. For this purpose, soil type recommended most often was sandy, as well as pure sand, to a range of substitutes such as peat, glass beads, vermiculite, perlite, compost and calcinated clay (Sylvia & Schenck, 1983; Millner & Kitt, 1992; Neumann & George, 2004; Douds Jr et al., 2006; Ma et al., 2007). For a variety of uses, several replacements for soil, sands and additives have been explored. For instance, comparatively inert materials used to dilute nutrient-rich soil and compost include vermiculite and perlite (Douds Jr et al., 2006). On the contrary, nutrient-deficient soils are amended with compost or other organic substrates (Ma et al., 2007). The root colonization of AM fungi is influenced by a variety of organic materials. Inert substrates can be used as a carrier medium to assess the growth of fungi and roots in situations where plant nourishment was predominantly supplied by nutrient solution (Lee & George, 2005). The particle size of the substrate has been demonstrated to have a significant impact on the sporulation of AM fungi and is necessary for proper drainage, humidity and aeration (Gaur & Adholeya, 2000). Sand particle sizes ranging between 250–850 µm are required for the development of several *Glomus* spp. using *Z. mays* as host plants in soil-free culture, according to Gaur and Adholeya (2000).

21.6.5 Substrate-Free Cultivation Systems

There are several different substrate-free growing methods available, generally referred to as "solution culture methods," which mostly differ in how the nutrient-rich solution is applied and how it is aerated. However, in static systems, an aeration pump is required to aerate the nutrient solution in order to prevent roots from experiencing oxygen constraint. However, vigorous nutrient solution flow and air bubble bursts may harm the fragile extraradical hyphal network's growth. The pumps may be periodically turned on to address this issue and reduce the influence on the growth and development mechanism of AM fungi (Hawkins & George, 1997). A variation to the system described above is called nutrient flow techniques (NFTs), in which a thin layer of nutrient solution enters the area where the host roots and AM fungus species grow. In order to prevent the issue of inadequate aeration and to enhance the relative area for gaseous exchange, this nutrient film was employed in the early 1980s to culture AM fungi (Elmes & Mosse, 1984). More recently, this method was used as a substrate-based system by Lee and George (2005). There is a form of hydroponics called aeroponics, in which host plant root and AM fungi are suspended in a mist of nutrient solution (Zobel et al., 1976). Gaseous exchange is enabled by the liquid coating surrounding the root and the microdroplets that are sprayed into the culture medium, increasing its aeration. The size of the tiny droplets produced by the different mist application procedures differs. In substrate-free production, such as hydroponics (also known as aeroponics), precolonized host plants are grown before being included into the system. For pre-inoculation, surface-sterilized host plant seedlings and AM fungus propagules are needed. These are typically precultures in pots on a substrate that includes a sand and perlite mixture for a few weeks. The container where the host roots and AM fungus grow is typically protected to prevent the growth of algae.

21.6.6 Production Parameters (AM Fungi)

Glomus intraradices, *mosseae*, *fasciculatum* and *caledonium* are among the species of *Glomus* that have been effectively employed to set up substrate-free inoculum production (Dugassa et al., 1995; Tajini et al., 2009). While *G. intraradices* was successfully grown in a tripartite hydroaeroponics system with beans and rhizobia, *G. rosea* was unable to grow in this system (Tajini et al., 2009). However, the cause of the failure to develop is unknown, because AM fungus can spread in a damp environment if there is sufficient aeration.

21.6.7 AM Host Plants

Similar to the substrate-based systems, a variety of plant species have been employed in substrate-free cultivation. Solution culture methods have been used to cultivate the roots of at least 21 genera of host plants (Jarstfer & Sylvia, 1997). The following plant species are recommended for static hydroponics: linseed (*Linum usitatissimum* L.) and wheat (*Triticum aestivum* L.). It has also been claimed that sweet potatoes (*Ipomoea batatas* L.), Bahiagrass (*P. notatum* Flügge), and Sudan grass (*S. sudanese* Staph.) are suitable for aeroponic cultivation of AM fungus (Jarstfer & Sylvia, 1997; Muhammad et al., 2000). The choice of host plant species typically affects the level of colonization. In addition, it is important to remember that nutritional requirements may vary throughout host plant species.

21.6.8 Problems of Producing Mycorrhizae on a Large Scale

Use of AM fungi on a commercial scale in agricultural productivity is problematic due to difficulties in the production of sterile, economical and sufficient amounts of inoculum. Moreover, its transportation to the field with cheap transportation cost is also desirable but needs to be achieved practically. It is impossible to check that the test soil is completely devoid of other dangerous bacteria, and the weight and volume of the inoculum substrate might result in outrageously high transit costs. As a result, ensuring consistent quality of inoculum is a very tough undertaking. For the AM fungal species, *Rhizophagus irregularis*, biotechnology is one of the most effective large-scale productions of AM fungi that can be highly concentrated in a sterile carrier, making it a great candidate for application in cassava cropping (Beballos et al., 2013). *R. irregularis* has been used as a model AM fungi organism by many researchers due to its easy growth in a laboratory in vitro culture system that is similar to that used in a large-scale commercial in vitro AM fungi culture. This is another compelling reason to use this species to increase cassava production. If in vitro produced *R. irregularis* is able to actually increase cassava yield in the field over uninoculated cassava, then this fungus could be used as a model species for future molecular-based AM fungi breeding and improvement programs due to its capacity to produce large numbers of genetically new *R. irregularis* that have an influence on plant growth and then their rapid production in commercial quantities. Thus, plant growth promotion by following AM fungi inoculation has been widely observed in many plant species and in several scientific research articles (Smith & Read, 2008). However, the majority of these have not been introduced in realistic agricultural sectors with unsterilized soil that already contains AM fungi. Additionally, some examples showed the efficacy of AM fungi inoculation over conventional fertilizers regarding significant increases in a realistic agricultural management system. Therefore, cassava is considered to be one of the only crop plants shown to benefits from AM fungi inoculation and that nourishes a huge population (approximately 1 billion people in 105 countries).

However, previously application of AM fungi inocula on a large scale was not successful because it needs a huge amount of inocula consolidated with the soil. For instance, an experiment was conducted on cassava plants where 500 g of soil having AM fungi inocula was required and added to each plant root (Ceballos et al., 2013). Besides that, primary knowledge of the local people about AM fungi and their application is also considered mandatory for promising research.

We contend that because these systems necessitate a very different management approach than the farmer's conventional management approach, they are vulnerable to low-quality inoculum production and are unlikely to last. Furthermore, considering the quantity of material that must be carried, stored and administered under actual farming conditions, the quantity of inocula needed for a farming unit – that is, per hectare – would make the technique costly. On the other hand, farmers are more inclined to use an inoculum that is of guaranteed quality, portable and simple to apply during planting. A fungus that was grown in vitro and utilized in these tests was supplied at a concentration that required 0.25 milliliters of undiluted liquid product per plant. Additionally, there was very little labor expense involved in spraying the AM fungus during planting.

21.6.9 Environmental Impact of Adding a Nonlocal AM Fungus

The addition of an AM fungi inoculum on a commercial scale in any non-indigenous area could pose a danger and this area has not yet been well studied (Schwartz et al., 2006). In this chapter, we suggest two important tests that should be made before commercial use of such inocula. First, the indigenous AM fungi community should be addressed seriously, as the addition of commercial inocula could potentially induce changes in the structure and diversity of indigenous AM fungi. Second, the addition of commercial inocula may introduce significant changes in alleles of the local AM fungal inocula. An environmental impact assessment should be performed before the introduction of commercial inocula in the field.

21.7 CONCLUSION

Mycorrhizal fungi are a key component of the symbiotic association for plant growth and stress tolerance. Their impact on carbon fixation is also of key interest in the changing environment. To harness the wide-ranging benefits of mycorrhizal fungi in agro-ecosystems, the commercial scale of mycorrhizal fungi with proper production methods and quality checks are important. This chapter described the benefits of mycorrhizal fungi for plant growth and stress tolerance. Moreover, various pilot-scale studies conducted in lab and field conditions for mycorrhizal culturing have been studied which show potential opportunities of mycorrhizae production on a commercial scale.

REFERENCES

Asrianti A, Tuheteru FD, Kandari AH, IS Mekuo M. (2016). Status culture of Arbuscular Mycorrhizal Fungi isolated from rhizosphere of Endemic Endangered Species of Kalapi (Kalappia celebica Kosterm). *Eur J Sustain Dev* 5(4), 395–402.

Arora NK, Khare E, Maheshwari DK. (2010). Plant growth promoting rhizobacteria: constraints in bioformulation, commercialization, future strategies. In DK Maheshwari (Ed.), "*Plant growth health promoting bacteria*." pp. 97–116. Springer.

Augé RM, Toler HD, Saxton AM. (2016). Mycorrhizal stimulation of leaf gas exchange in relation to root colonization, shoot size, leaf phosphorus nitrogen: a quantitative analysis of the literature using meta-regression. *Front Plant Sci* 7(1), 1084.

Azcón-Aguilar C, Barea JM. (2015). Nutrient cycling in the mycorrhizosphere. *J Soil Sci Plant Nutr* 15(2), 372–396.

Balestrini R, Chitarra W, Fotopoulos V, Ruocco M. (2017). Potential role of beneficial soil microorganisms in plant tolerance to abiotic stress factors. In M. Lukac, P. Grenni, & M. Gamboni (Eds.), "*Soil biological communities ecosystem resilience*." pp. 191–207. Springer, Cham.

Beballos I, Ruiz M, Fernandez C, Peña R, Rodríguez A, Sanders IR. (2013). The *in vitro* mass-produced model mycorrhizal fungus, *Rhizophagus irregularis*, significantly increases yields of the globally important food security crop cassava. *Plos One* 8(8), e70633.

Berdeni D, Cotton TE, Daniell TJ, Bidartondo M, Cameron DD, Evans KL. (2018). The effects of arbuscular mycorrhizal fungal colonisation on nutrient status, growth, productivity canker resistance of apple (*Malus pumila*). *Front Microbiol* 9, 1461.

Berruti A, Lumini E, Balestrini R, Bianciotto V. (2016). Arbuscular mycorrhizal fungi as natural biofertilizers: let's benefit from past successes. *Front Microbiol* 6, 1559.

Bholay AD, Barbosa S, Jadhav K. (2018). *In vitro* cultivation of Vesicular-Arbuscular Mycorrhizal fungi its biological efficacy. *J Curr Microbiol App Sci* 7(3), 931–941.

Bruns TD, Corradi N, Redecker D, Taylor JW, Öpik M. (2018). Glomeromycotina: what is a species why should we care. *New Phytol* 220(4), 963–967.

Cardoso E, Nogueira MA, Zangaro W. (2017). Importance of Mycorrhizae in tropical soils. In JL de Azevedo & MC Quecine (Eds.), "*Diversity benefits of microorganisms from the tropics*." pp. 245–267. Springer.

Cavagnaro TR, Bender SF, Asghari HR, van der Heijden MG. (2015). The role of arbuscular mycorrhizas in reducing soil nutrient loss. *Trends Plant Sci* 20(5), 283–290.

Ceballos I, Ruiz M, Fernandez C, Peña R, Rodríguez A, Sanders IB. (2013). The *in vitro* Mass-Produced Model Mycorrhizal Fungus, *Rhizophagus irregularis*, significantly increases yields of the globally important food security crop Cassava. *Plos One* 8(8), e70633.

Chabi A, Hountondji FC, Ogunsola B, Lawouin L, Coyne D, Wiemken A, Oehl F. (2016). Effect of two species of arbuscular mycorrhizal fungi inoculation on development of micro-propagated yam plantlets suppression of Scutellonema bradys (Tylenchideae). *J Entomol Nematol* 8(1), 1–10.

Chagnon PL, Bradley RL, Maherali H, Klironomos JN. (2013). A trait-based framework to underst life history of mycorrhizal fungi. *Trends Plant Sci* 18(9), 484–491.

Crossay T, Antheaume C, Redecker D, Bon L, Chedri N, Richert C, Guentas L, Cavaloc Y, Amir H. (2017). New method for the identification of arbuscular mycorrhizal fungi by proteomic-based biotyping of spores using MALDI-TOF-MS. *Sci Report* 7(1), 14306.

Damar HE. (2016). Phosphorus: the underrated element for feeding the world. *Trends Plant Sci* 21, 461–463.

de León DG, Neuenkamp L, Moora M, Öpik M, Davison J, Peña-Venegas CP, Vasar M, Jairus T, Zobel M. (2018). Arbuscular mycorrhizal fungal communities in tropical rain forest are resilient to slash- -burn agriculture. *J Trop Ecol* 34(3), 186–199.

Delaux PM. (2017). Comparative phylogenomics of symbiotic associations. *New Phytol* 213(1), 89–94.

Dodd J, Arias I, Koomen I, Hayman D. (1990). Agriculture, the management of populations of vesicular-arbuscular mycorrhizal fungi in acid-infertile soils of a savanna ecosystem. *Plant Soil* 122(2), 229 240.

Douds DD, Janke RR, Peters SE. (1993). VAM fungus spore populations colonization of roots of maize soybean under conventional low-input sustainable. *Agr Ecosyst Environ* 43(4), 325–335.

Douds Jr DD, Nagahashi G, Hepperly PR. (2010). On-farm production of inoculum of indigenous arbuscular mycorrhizal fungi assessment of diluents of compost for inoculum production. *Bioresour Technol* 101(7), 2326–2330.

Douds Jr DD, Nagahashi G, Pfeffer PE, Reider C, Kayser WM. (2006). On-farm production of AM fungus inoculum in mixtures of compost vermiculite. *Bioresour Technol* 97(6), 809–818.

Dugassa DG, Grunewaldt SG, Schönbeck F. (1995). Growth of *Glomus intraradices* its effect on linseed in hydroponic culture. *Mycorrhiza* 5(4), 279–282.

Egerton-Warburton LM, Johnson NC, Allen EB. (2007). Mycorrhizal community dynamics following nitrogen fertilization: a cross-site test in five grassl. *Ecol Monogr* 77(4), 527–544.

Elmes RP, Mosse B. (1984). Vesicular–arbuscular endomycorrhizal inoculum production II experiments with maize (*Zea mays*) other hosts in nutrient flow culture. *Can J Bot* 62(7), 1531–1536.

Estrada B, Aroca R, Azcón-Aguilar C, Barea JM, Ruiz-Lozano JM. (2013). Importance of native arbuscular mycorrhizal inoculation in the halophyte *Asteriscus maritimus* for successful establishment growth under saline conditions. *Plant Soil* 370(1–2), 175–185.

Ferlian O, Biere A, Bonfante P, Buscot F, Eisenhauer N, Fernandez I, Hause B, Herrmann S, Krajinski-Barth F, Meier IC, Pozo MJ. (2018). Growing research networks on mycorrhizae for mutual benefits. *Trends Plant Sci* 23, 975–984.

Fester T, Sawers R. (2011). Progress challenges in agricultural applications of arbuscular mycorrhizal fungi. *Crit Rev Plant Sci* 30(5), 459–470.

Gaur A, Adholeya A. (2000). Effects of the particle size of soil-less substrates upon AM fungus inoculum production. *Mycorrhiza* 10(1), 43–48.

Gaur A, Adholeya A. (2002). Arbuscular-mycorrhizal inoculation of five tropical fodder crops inoculum production in marginal soil amended with organic matter. *Biol Fert Soils* 35(3), 214–218.

Gianinazzi S, Gollotte A, Binet MN, van Tuinen D, Redecker D, Wipf D. (2010). Agroecology: the key role of arbuscular mycorrhizas in ecosystem services. *Mycorrhiza* 20(8), 519–530.

Gianinazzi S, Vosátka M. (2004). Inoculum of arbuscular mycorrhizal fungi for production systems: science meets business. *Can J Bot* 82(8), 1264–1271.

Hart M, Ehret DL, Krumbein A, Leung C, Murch S, Turi C, Franken P. (2015). Inoculation with arbuscular mycorrhizal fungi improves the nutritional value of tomatoes. *Mycorrhiza* 25(5), 359–376.

Hawkins HJ, George E. (1997). Hydroponic culture of the mycorrhizal fungus *Glomus mosseae* with *Linum usitatissimum* L, *Sorghum bicolor* L. *Triticum aestivum* L. *Plant Soil* 196(1), 143–149.

Hiruma K, Kobae Y, Toju H. (2018). Beneficial associations between Brassicaceae plants fungal endophytes under nutrient-limiting conditions: evolutionary origins host–symbiont molecular mechanisms. *Curr Opin Plant Biol* 44, 145–154.

Hoysted GA, Kowal J, Jacob A, Rimington WR, Duckett JG, Pressel S, Orchard S, Ryan MH, Field KJ, Bidartondo MI. (2018). A mycorrhizal revolution. *Curr Opin Plant Biol* 44, 1–6.

Humphreys CP, Franks PJ, Rees M, Bidartondo MI, Leake JR, Beerling DJ (2010). Mutualistic mycorrhiza-like symbiosis in the most ancient group of land plants. *Nat Commun*, 1(1), 103.

Ijdo M, Cranenbrouck S, Declerck, S. (2011). Methods for large-scale production of AM fungi: past, present, and future. *Mycorrhiza*, 21, 1–16.

Jarstfer AG, Sylvia, DM. (1997). Isolation, culture detection of arbuscular mycorrhizal fungi. In CJ Hurst (Ed.), *Manual of Environmental Microbiology* (pp. 406–412). American Society of Microbiology, Washington, DC..

Jónsson JÖ, Davíðsdóttir B. (2016). Classification valuation of soil ecosystem services. *Agr Syst* 145, 24–38.

Jung SC, Martinez-Medina A, Lopez-Raez JA, Pozo MJ. (2012). Mycorrhiza-induced resistance priming of plant defenses. *J Chem Ecol* 38(6), 651–664.

Klironomos JN, Hart MM. (2002). Colonization of roots by arbuscular mycorrhizal fungi using different sources of inoculum. *Mycorrhiza* 12(4), 181–184.

Kostenko V, Pechko V, Ivanova O. (2018). Impact of mycorrhizal fungi on walnuts grapes resistance to pathogens in Ukrainian orchards–a review. *Ukr J Ecol* 8(1), 533–541.

Lee YJ, George E. (2005). Development of a nutrient film technique culture system for arbuscular mycorrhizal plants. *Hort Sci* 40(2), 378–380.

Leifheit EF, Veresoglou SD, Lehmann A, Morris EK, Rillig MC. (2014). Multiple factors influence the role of arbuscular mycorrhizal fungi in soil aggregation—a meta-analysis. *Plant Soil* 374(1–2), 523–537.

Lenoir I, Fontaine J, Sahraoui AL. (2016). Arbuscular mycorrhizal fungal responses to abiotic stresses: a review. *Phytochemistry* 123, 4–15.

Liu WK, Yang QC. (2008). Integration of mycorrhization photoautotrophic micropropagation *in vitro*: feasibility analysis for mass production of mycorrhizal transplants inoculants of arbuscular mycorrhizal fungi. *Plant Cell Tiss Org* 95(2), 131–139.

Luginbuehl LH, Menard GN, Kurup S, Van Erp H, Radhakrishnan GV, Breakspear A, Oldroyd GE, Eastmond PJ. (2017). Fatty acids in arbuscular mycorrhizal fungi are synthesized by the host plant. *Science* 356, eaan0081.

Ma N, Yokoyama K, Marumoto T. (2007). Effect of peat on mycorrhizal colonization effectiveness of the arbuscular mycorrhizal fungus *Gigaspora margarita*. *Soil Sci Plant Nutr* 53(6), 744–752.

Martín-Robles N, Lehmann A, Seco E, Aroca R, Rillig MC, Milla R. (2018). Impacts of domestication on the arbuscular mycorrhizal symbiosis of 27 crop species. *New Phytol* 218(1), 322–334.

Millner P, Kitt D. (1992). The Beltsville method for soilless production of vesicular-arbuscular mycorrhizal fungi. *Mycorrhiza* 2(1), 9–15.

Mohammad A, Khan AG, Kuek C. (2000). Improved aeroponic culture of inocula of arbuscular mycorrhizal fungi. *Mycorrhiza* 9(6), 337–339.

Neuenkamp L, Moora M, Öpik M, Davison J, Gerz M, Männistö M, Jairus T, Vasar M, Zobel M. (2018). The role of plant mycorrhizal type status in modulating the relationship between plant arbuscular mycorrhizal fungal communities. *New Phytol* 220, 1236–1247.

Neumann E, George E. (2004). Colonisation with the arbuscular mycorrhizal fungus *Glomus mosseae* (Nicol & GerD) enhanced phosphorus uptake from dry soil in *Sorghum bicolor*. *Plant Soil* 261(1–2), 645–255.

Ortas I, Bykova A. (2018). The effect of Mycorrhiza inoculation Phosphorus application on Phosphorus efficiency of wheat plants. *Commun Soil Sci Plant Anal* 49(10), 1199–1207.

Powell CL Bagyaraj DJ. (2018). VA mycorrhizae: why all the interest? In CL Powell & DJ Bagyaraj (Eds.), "*VA mycorrhiza*." pp.1–3. CRC press.

Rodrigues K, Rodrigues B. (2018). Maximizing the shelf life of monoxenically produced carrier-based AM fungal bio-inocula for maintaining its long-term viability.

Rodrigues KM, Rodrigues BF. (2017). Development of carrier based *in vitro* produced arbuscular mycorrhizal (AM) fungal inocula for organic agriculture.

Schwartz MW, Hoeksema JD, Gehring CA, Johnson NC, Klironomos JN, Abbott LK, Pringle A. (2006). The promise the potential consequences of the global transport of mycorrhizal fungal inoculum. *Ecol Lett* 9(5), 501–515.

Selosse MA, Strullu-Derrien C, Martin FM, Kamoun S, Kenrick P. (2015). Plants, fungi oomycetes: a 400-million-year affair that shapes the biosphere. *New Phytol* 206(2), 501–506.

Selvakumar G, Kim K, Walitang D, Chanratana M, Kang Y, Chung B, Sa T. (2016). Trap culture technique for propagation of Arbuscular Mycorrhizal Fungi using different host plants. 한국토양비료학회지 49(5), 608–613.

Singh PK, Singh M, Tripathi BN. (2013). Glomalin: an arbuscular mycorrhizal fungal soil protein. *Protoplasma* 250(3), 663–669.

Smith S, Read D. (2008). "*Mycorrhizal symbiosis edn*". Cambridge: Academic Press, Elsevier.

Smith SE, Smith FA. (2012). Fresh perspectives on the roles of arbuscular mycorrhizal fungi in plant nutrition growth. *Mycologia* 104(1), 1–13.

St-Arnaud M, Hamel C, Vimard B, Caron M, Fortin JA. (1996). Enhanced hyphal growth spore production of the arbuscular mycorrhizal fungus *Glomus intraradices* in an in vitro system in the absence of host roots. *Mycol Res* 100(3), 328–332.

Sylvia DM, Schenck NC. (1983). Application of superphosphate to Mycorrhizal plants sitimulates sporulation of phosphorus-tolerant versicular-Arbuscular my Corrhizal Fungi. *New Phytol* 95(4), 655–661.

Tajini F, Suriyakup P, Vailhe H, Jansa J, Drevon JJ. (2009). Assess suitability of hydroaeroponic culture to establish tripartite symbiosis between different AMF species, beans, rhizobia. *BMC Plant Biol* 9(1), 73.

Thirkell TJ, Charters MD, Elliott AJ, Sait SM, Field KJ. (2017). Are mycorrhizal fungi our sustainable saviours? Considerations for achieving food security. *J Ecol* 105(4), 921–929.

Tkacz A, Poole P. (2015). Role of root microbiota in plant productivity. *J Exp Bot*, 66(8), 2167–2175.

Urgiles N, Strauß A, Loján P, Schüßler A. (2014). Cultured arbuscular mycorrhizal fungi and native soil inocula improve seedling development of two pioneer trees in the Andean region. *New For*, 45(6), 859–874.

Van Der Heijden MG, Martin FM, Selosse MA, Sanders IR. (2015). Mycorrhizal ecology evolution: the past, the present, the future. *New Phytol* 205(4), 1406–1423.

Vosátka M, Albrechtová J, Patten R. (2008). The international market development for mycorrhizal technology. In A. Varma & A.C. Kharkwal (Eds.), "*Mycorrhiza*." pp. 419–438. Springer.

Yao Q, Yan W, Wei J. (2018). Establishment of a quality test for Arbuscular Mycorrhizal Inoculum. *Int J Agri Biol* 20(4), 926–932.

Zobel RW, Del Tredici P, Torrey JG. (1976). Method for growing plants aeroponically. *Plant Physiol* 57(3), 344–346.

22 Nonthermal Decontamination Techniques for Extending the Shelf-Life of Date Fruit

A Review

*Abdulhakeem Alzahrani**
*Corresponding author.

22.1 INTRODUCTION

The preservation of date fruit has been a practice dating back thousands of years, particularly in regions where this fruit is a dietary staple. Dates, known for their delightful sweetness and rich nutritional profile, are harvested in abundance during the harvesting season. They are laden with vital nutrients such as dietary fiber, vitamins and minerals (Tang et al., 2013). The cultivation of date palms has historically supported many communities' livelihoods. Dates are crucial to the nutrition and economy of regions like the Middle East and North Africa (MENA). They are a staple food with cultural and economic importance. Globally, dates are recognized as a healthy snack and ingredient in various culinary applications. Therefore, ensuring the safety and extended shelf-life of date fruit has broader implications for international trade and the food industry (Vignali et al., 2022). However, one of the significant challenges associated with date fruit is its relatively short shelf-life due to microbial contamination. Dates are highly susceptible to post-harvest spoilage, which not only results in economic losses but also hinders their distribution to distant markets (Younis et al., 2023). Therefore, finding effective preservation techniques is vital to harnessing the full potential of this fruit.

Traditional date preservation methods, like sun drying and chemical treatments, have been used for centuries. Though somewhat effective, they often fail to completely eradicate microbial contaminants and ensure extended shelf-life. Furthermore, consumers are becoming increasingly aware of the potential health hazards linked to chemical treatments (Wilson et al., 2019). These techniques reduce the need for chemical preservatives and excessive heat, minimizing nutrient loss in date fruit. Nonthermal decontamination methods, such as high-pressure processing (HPP), pulsed electric fields (PEF), ultraviolet (UV) irradiation, cold plasma treatment and ozone treatment, offer innovative preservation solutions (Maqsood et al., 2020; Singh & Golding, 2020; Sarraf et al., 2021). These methods reduce microbial loads while preserving the fruit's nutritional quality and sensory attributes, addressing the longstanding challenge of extending date fruit's shelf-life. Here, we review the multifaceted significance of nonthermal decontamination techniques for the preservation of dates. It addresses a longstanding challenge in regions heavily reliant on date palm cultivation while also catering to the growing international demand for this nutritious fruit. By enhancing shelf-life, ensuring food safety and preserving nutritional quality, these techniques hold the potential to revolutionize date fruit preservation practices, benefiting both local economies and global consumers. We strive to explore diverse nonthermal techniques, evaluating their effectiveness and implications, thereby highlighting a subject of utmost significance within the realms of food science and agriculture.

 DOI: 10.1201/9781003486275-25

22.2 DATE FRUIT: AN OVERVIEW

22.2.1 A Brief Description of Date Fruit

Date fruits are cultivated from *Phoenix dactylifera*, known as date palm, a cherished delicacy enjoyed for millennia. Recently, global production has soared to 9.45 million metric tons (FAO, 2020). With a heritage from ancient civilizations, it symbolizes sustenance, culture and economic significance (Sarraf et al., 2021). Known for its sweetness and nutritional profile, date fruit is emblematic of the MENA. As a dietary staple, dates provide dietary fiber, essential vitamins and minerals, enhancing its status as a nutritional powerhouse (Maqsood et al., 2020). Beyond nutrition, date palm cultivation supports community livelihoods, intertwined with cultural and economic identities. The economic contribution of date fruit, however, depends on its shelf-life.

22.2.2 Nutritional Value and Importance

Date fruit, revered for its cultural, nutritional and economic significance, offers a rich profile of essential nutrients. It is packed with sweetness and contains dietary fibers, B-complex vitamins (B6 and B5) and vital minerals like potassium, magnesium and iron (Tang et al., 2013; Darré et al., 2022; Table 22.1). Date fruit also contains antioxidants such as flavonoids and carotenoids, enhancing its nutritional value. Its global recognition as a natural, healthy snack and versatile culinary ingredient highlights the importance of ensuring its safety and extended shelf-life for international trade and the food industry.

TABLE 22.1
Nutritional Value and Health Benefits of Date Fruit

Macronutrient	Percentage content	Health benefits	References
Carbohydrate	50–88	Energy source	Tang et al., 2013; Djaoud et al., 2020
Protein	1–7	Source of amino acids, aid in metabolism	Hussain et al., 2020; Mahomoodally et al., 2023
Fatty acids	0.1–0.5	Aid in metabolism, normal body function, hormones, etc.	Ibourki et al., 2021; Darré et al., 2022
Dietary fiber	6.4–11.5	Useful as roughage in food, as prebiotics for gut bacteria	Mrabet et al., 2019; Kamal-Eldin et al., 2020
Minerals (magnesium, manganese, phosphorus, iron, calcium, potassium, sodium and zinc)	<0.001–0.9	Aid in metabolism, normal body function, etc.	Benmeziane-Derradji, 2019; Perveen & Bokahri, 2020
Vitamins (thiamin, riboflavin, niacin, ascorbic acid, pyridoxine and vitamin A)	<0.001	As enzyme cofactors, aid in metabolism, normal body function, etc.	Bouhali et al., 2017; Alharbi et al., 2021
Carotenoids, phenolics, cinnamic acids, flavonoid glycosides, flavones, flavonols, etc.	<0.001	As antioxidants	Al-Shwyeh, 2019; Alam et al., 2021

22.2.3 Microbial Contamination as a Challenge

Despite its rich history and nutritional value, date fruit faces a significant threat: microbial contamination. This contamination causes spoilage, foodborne illness, economic losses and public health concerns. Spoilage due to microbes reduces the shelf-life of dates. Studies have isolated yeasts, molds, lactic acid bacteria, *Staphylococcus aureus* and *E. coli* from date fruits. These microorganisms can grow on the surface or inside the fruits, leading to spoilage when their levels reach or exceed 6.0 log CFU per gram (Zamir et al., 2018). Contamination can occur during harvesting and postharvest stages (Jdaini et al., 2022). This challenge threatens the postharvest integrity of date fruit and requires innovative solutions.

22.3 TRADITIONAL METHODS FOR THE PRESERVATION OF DATES AND THEIR LIMITATIONS

22.3.1 Traditional Methods of Date Fruit Preservation

Traditional methods of date fruit preservation have been used for centuries to extend the shelf-life of this fruit. Some of these methods (Homayouni et al., 2015; Hazbavi et al., 2015; Sarraf et al., 2021; Baroyi et al., 2023) include the following:

1) **Sun drying**: Sun drying involves laying out freshly harvested date fruit under the sun to naturally dehydrate them. The process harnesses the sun's energy to evaporate moisture from the fruit, effectively reducing the risk of spoilage. Sun drying is also popular due to its role in enhancing the natural sweetness of date fruit by concentrating sugars.
2) **Open-air drying**: Open-air drying is a similar technique but often involves the use of elevated platforms or drying racks. This method allows for better air circulation, which can aid in preventing mold growth and ensuring more uniform drying.
3) **Chemical treatment** with sulfur dioxide, potassium sorbate and sodium benzoate to prevent the growth of yeast and molds on dates. This also helps in inhibiting enzymatic browning, thus conserving the fruit's natural color.
4) **Date canning**: This involves sealing dates in jars and pasteurizing them in a water bath to extend their shelf-life
5) **Preservation in syrup**: Dates can be preserved in syrup to extend their shelf-life. This method has been practiced for a long time and has undergone additional development work
6) **Filling with almond mixture**: This involves pitting and filling the cavity of dates with a suitable mixture of almonds that has been developed to create filled dates. The resulting filled date can be coated with chocolate/biscuit/wafer, if desired.

22.3.2 Challenges and Limitations of Traditional Methods

Traditional techniques have been practiced for centuries and offer some advantages in terms of simplicity and low cost. While these methods have been instrumental in preserving date fruit historically, they come with certain limitations (Palumbo et al., 2022; Muhialdin et al., 2022) such as:

1) **Microbial contamination**: One of the primary challenges of traditional methods is their limited effectiveness in eliminating microbial contaminants. Date fruit, being naturally high in sugars and moisture, is prone to yeast and mold growth, even after undergoing sun drying or chemical treatments.
2) **Nutrient loss**: Traditional methods, especially sun drying, can result in nutrient loss. Prolonged exposure to heat and sunlight may lead to the degradation of some essential vitamins and antioxidants in date fruit.

3) **Chemical residues**: The use of chemical treatments raises concerns about the presence of chemical residues in date fruit. Consumers are becoming increasingly health-conscious, seeking natural and chemical-free food options.
4) **Inconsistent quality**: Traditional methods may lead to variations in the quality and safety of preserved date fruit. Factors like weather conditions during drying can affect the final product.
5) **Limited shelf-life extension**: While traditional methods can extend the shelf-life of date fruit to some extent, they may not provide the desired level of long-term stability, especially for international trade or extended storage.

As a result, there is a growing interest in exploring modern, nonthermal decontamination techniques that can offer more effective and sustainable solutions to extend the shelf-life of date fruit while preserving its nutritional value, texture and sensory attributes.

22.4 NONTHERMAL DECONTAMINATION TECHNIQUES

Nonthermal decontamination techniques offer effective alternatives to thermal processing, preserving food without high temperatures, which can cause nutrient loss and alter the taste and texture. Methods such as high-pressure processing, pulsed electric fields, ultraviolet irradiation, cold plasma treatment, ionizing radiation and ozone treatment mitigate microbial risks and maintain nutritional quality. They also address concerns over chemical residues and energy consumption in food preservation. The following sections describe the significance of nonthermal techniques, their principles, applications and recent research findings, highlighting their promising role in the future of food preservation (Figure 22.1, Table 22.2).

22.4.1 High-Pressure Processing (HPP)

High-pressure processing (HPP), a nonthermal food preservation method, shows promise in enhancing safety and extending the shelf-life of date fruit. HPP involves subjecting food to increased pressures, typically 100–800 megapascals (MPa). This technique can eradicate or significantly

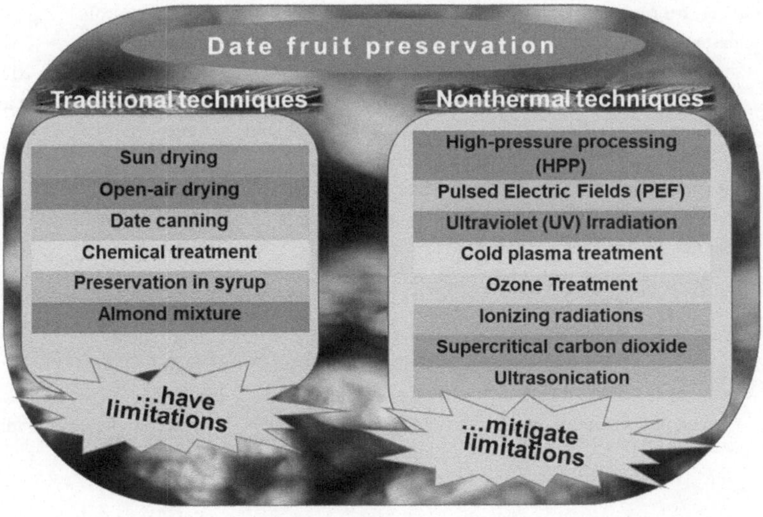

FIGURE 22.1 Traditional and nonthermal technologies for the preservation of date fruit.

TABLE 22.2
Nonthermal Decontamination Methods in the Preservation of Fruits/Vegetables

Technique	Applicability	Benefits	Limitations	References
High-pressure processing (HPP)	Fruit juices and vegetable smoothies, fruits, dates, vegetables, rice, milk, cheese, chicken, beef, pork, dips, sauces	Enhanced microbial safety; minimal nutrient loss; extended shelf-life	High equipment cost; texture changes; limited applications	Pérez-Lamela et al., 2021; Houška et al., 2022; Koutsoumanis et al., 2022
Pulsed electric fields (PEF)	Pine nuts, apples, cinnamon, fruit juices, dates, carrots, coconut water, tomatoes, etc.	Enhanced food safety; nutrient preservation; prolonged shelf-life	High energy consumption; limited penetration; equipment complexity	Jadhav et al., 2021; Naliyadhara et al., 2022; Younis et al., 2023
Ultraviolet (UV) irradiation	Fresh-cut mangoes, fruit juice, dates, mushrooms, tomatoes, almonds, cheese, pork, beef, etc.	Effective microbial control; minimal chemical residue; eco-friendly preservation	Surface treatment; energy intensity; limited microbial spectrum	Moreno et al., 2017; Bhatnagar et al. 2022; Darré et al., 2022
Cold plasma treatment	Rice, butter, fruit juice, dates, apples, strawberries, lettuces, almonds, milk, pork, bacon, etc.	Microbial inactivation; minimal nutrient loss; sustainable approach	Equipment complexity; limited penetration; safety concerns	Zhang et al., 2021; Maghsoudi et al. 2023; Amir-Mojahedi et al. 2023
Ozone treatment	Bananas, dates, pineapples, parsley, cakes, cut fruits like pineapples, guavas, bananas, strawberries, grapes, chilli, chicken, etc.	Broad-spectrum microbial control; no residual chemicals; environmentally friendly	Variable efficacy; equipment maintenance; regulatory constraints	Mahmoud and Abdel-Khalik, 2022; Piechowiak et al., 2022
Ionizing radiation	Grapes, fruit juices, dates, garlic, pasta, etc.	Comprehensive microbial inactivation; prolonged shelf-life; preservation without chemicals	Limited control; regulatory compliance; equipment costs	Zarbakhsh & Rastegar, 2019; Bisht et al., 2021; Bhatnagar et al., 2022
Supercritical carbon dioxide	Fruit juices, dates, carrots, apples, basil leaves, soya flour, cheese, fish/fish products, etc.	Microbial control; minimal nutrient loss; reduced chemical usage	Equipment complexity; limited microbial control; equipment costs	Silva et al., 2020; Bushnaq et al., 2022; AlYammahi et al., 2023
Ultrasonication	Apples, dates, olives, flaxseed, carrots, peppers, pork, etc.	Enhanced safety; nutrient preservation; sustainable approach	Limited penetration; equipment complexity; energy intensity	Khademi et al., 2018; Jadhav et al., 2021; Arya et al., 2023

reduce harmful microorganisms, including bacteria, yeasts, molds and pathogens, without high temperatures (Houška et al., 2022; Koutsoumanis et al., 2022). The mechanisms of HPP are complex.

22.4.1.1 Principles and Mechanisms

The application of high pressure leads to several key effects:

1) **Microbial inactivation**: High pressure damages the cellular structure of microorganisms. The increased pressure disrupts the integrity of cell membranes and cell walls, leading to microbial inactivation. This effect is especially beneficial in reducing spoilage microorganisms and pathogens that can compromise the quality and safety of date fruit.
2) **Enzyme inhibition**: HPP can also deactivate the enzymes responsible for ripening and deterioration processes in date fruit. This enzyme inhibition helps to maintain the product's sensory attributes and nutritional content over an extended shelf-life.
3) **Homogenization**: HPP can homogenize the texture and structure of date fruit, leading to improved product uniformity and consistency. This is particularly relevant for date-based products such as pastes and spreads.
4) **Minimal impact on quality**: Importantly, HPP is a nonthermal process, which means it does not involve the application of heat. This lack of thermal exposure helps to retain the natural color, flavor and nutritional content of date fruit, offering a significant advantage over traditional thermal processing methods.

22.4.1.2 Application of High-Pressure Processing in Date Fruit Preservation

High-pressure processing has been found to have diverse applications in date fruit preservation. The technique is especially suitable for date products that require extended shelf-life while preserving their quality. Some notable applications (Balakrishna et al., 2020; Houška et al., 2022) include:

1) **Date paste**: Date paste, a common ingredient in various culinary applications, benefits from HPP. The process helps eliminate microorganisms that can cause spoilage and extend the paste's shelf-life.
2) **Date-based snacks**: HPP is used to enhance the safety and shelf-life of date-based snacks, such as date bars and energy bites. These snacks are popular among health-conscious consumers and require effective preservation methods.
3) **Ready-to-eat dates**: For convenience, ready-to-eat dates can be subjected to HPP. This not only ensures food safety but also extends the product's storage life.
4) **Date juices and syrups**: Date juices and syrups, prized for their natural sweetness, can be effectively preserved using HPP without the need for chemical preservatives.

22.4.1.3 Recent Research Findings

Recent research into the application of HPP in date fruit preservation has yielded valuable insights into the technique's effectiveness and potential areas for improvement (Pérez-Lamela et al., 2021; Aganovic et al., 2021)

1) **Microbial reduction**: Studies have consistently demonstrated that HPP is highly effective in reducing the microbial load in date products. Recent research has focused on optimizing pressure and processing time parameters to achieve maximum microbial inactivation while minimizing the impact on sensory quality.
2) **Quality maintenance:** Researchers have explored the impact of HPP on the sensory attributes of date fruit, including color, flavor and texture. Findings indicate that, when properly applied, HPP can maintain the quality of date products, making them more appealing to consumers.

3) **Nutritional preservation**: One of the key advantages of HPP is its ability to preserve the nutritional content of date fruit. Recent studies have examined changes in nutrient profiles during HPP and have highlighted the minimal nutrient degradation associated with this nonthermal technique.
4) **Packaging considerations**: Research has also delved into the role of packaging in HPP for date products. The choice of packaging materials and design can influence the efficacy of HPP and the overall quality of preserved date fruit.
5) **Consumer acceptance**: Understanding consumer preferences and perceptions regarding HPP-treated date products is an emerging area of research. Recent studies have sought to gauge consumer acceptance of HPP as a preservation method for date fruit and its impact on product trust and choice.

22.4.2 Pulsed Electric Fields (PEF) in Date Fruit Preservation

Pulsed electric fields (PEF) are a nonthermal decontamination technique with potential for enhancing the safety and shelf-life of date fruit. PEF applies short, high-intensity electric pulses to food, creating changes in the electrical potential within microorganism cells and, to a lesser extent, plant cells (Nowosad et al., 2021). This disrupts cellular membranes, leading to microbial and enzyme inactivation while maintaining the nutritional and sensory attributes of date fruit.

22.4.2.1 Principles and Mechanisms

PEF's mechanisms of action are characterized by several key effects (Mtaoua et al., 2017):

1) **Microbial inactivation**: The high-intensity electric pulses cause irreversible damage to microorganisms' cell membranes. This disruption leads to the leakage of cellular contents and eventual cell death. PEF is effective against bacteria, yeasts, molds and pathogens, making it an exceptional choice for date fruit preservation.
2) **Enzyme inhibition**: PEF can also target enzymes responsible for date fruit ripening and degradation. By deactivating these enzymes, PEF helps extend the shelf-life of date fruit and preserves its sensory qualities.
3) **Minimal thermal exposure**: A distinct advantage of PEF is its minimal thermal impact. Unlike conventional thermal processing methods, PEF operates at ambient or mildly elevated temperatures, avoiding the thermal degradation of heat-sensitive nutrients and sensory attributes in date fruit.

22.4.2.2 Application of PEF in Date Fruit Preservation

PEF has found diverse applications in the preservation of date fruit, offering effective solutions for extending shelf-life and ensuring food safety. Some prominent applications include the following (Naliyadhara et al., 2022; Younis et al., 2023):

1) **Date paste**: Date paste, a versatile ingredient in a wide range of food products, can be preserved using PEF. This technique eliminates microorganisms that can lead to spoilage, extending the paste's shelf-life while maintaining its natural sweetness and nutritional profile.
2) **Date snacks**: Ready-to-eat date-based snacks like bars, energy bites and coated dates can benefit significantly from PEF treatment. It provides an effective means of microbial decontamination without compromising product quality.
3) **Date-based beverages**: Date juices, concentrates and syrups, prized for their natural sweetness, can be effectively preserved using PEF. This nonthermal technique allows date beverages to maintain their fresh flavor and nutritional content.

22.4.2.3 Recent Research Findings

Recent research on PEF application in date fruit preservation has yielded noteworthy findings (Mtaoua et al., 2017; Naliyadhara et al., 2022):

1) **Microbial reduction**: Studies have consistently demonstrated PEF's ability to reduce microbial loads in date products effectively. Recent research has focused on optimizing pulse parameters such as intensity, duration and frequency to achieve maximum microbial inactivation while preserving product quality.
2) **Quality maintenance**: Researchers have examined the impact of PEF on the sensory attributes of date fruit, including color, flavor and texture. Recent findings indicate that PEF, when properly applied, can maintain the sensory qualities that consumers value.
3) **Nutritional preservation**: An essential advantage of PEF is its ability to preserve the nutritional content of date fruit. Recent studies have investigated changes in nutrient profiles during PEF treatment and have consistently demonstrated minimal nutrient degradation.
4) **Packaging considerations**: Research has delved into the influence of packaging materials and design on PEF treatment. The choice of packaging can affect treatment efficacy and the overall quality of preserved date fruit products.
5) **Consumer acceptance**: Emerging research areas include consumer perceptions and preferences regarding PEF-treated date products. Understanding consumer acceptance and its impact on product trust and choice is vital for the successful adoption of PEF in date fruit preservation.

22.4.3 Ultraviolet (UV) Irradiation in Date Fruit Preservation

Ultraviolet (UV) irradiation is a nonthermal decontamination method that uses ultraviolet light to improve the safety and shelf-life of date fruit. UV-C light, with wavelengths from 200 to 280 nanometers, has germicidal properties that damage the DNA and RNA of microorganisms (Araque et al., 2018; Sonntag et al., 2023). It targets nucleic acids within microbial cells, causing breaks in DNA and RNA strands, preventing microorganisms from replicating and functioning, effectively rendering them harmless.

22.4.3.1 Principles and Mechanisms

The mechanisms of UV irradiation for date fruit preservation can be summarized as follows (Moreno et al., 2017; Darré et al., 2022)

1) **Microbial inactivation**: UV-C light penetrates the cell walls of microorganisms and damages their genetic material, leading to microbial inactivation. It is particularly effective against bacteria, viruses, molds and yeasts that are common contaminants in date fruit.
2) **Enzyme inhibition**: UV irradiation can also inhibit the activity of enzymes responsible for the degradation and ripening of date fruit. By targeting these enzymes, UV treatment helps extend the shelf-life of date products.
3) **Minimal thermal impact**: UV irradiation is a nonthermal technique, meaning it does not introduce heat that could compromise the sensory and nutritional quality of date fruit. This characteristic is especially valuable in preserving the natural attributes of date products.

22.4.3.2 Application of UV Irradiation in Date Fruit Preservation

UV irradiation has a wide range of applications in date fruit preservation, offering effective solutions for extending shelf-life while ensuring food safety (Darré et al., 2022; Sonntag et al., 2023)

1) **Whole date fruit**: Date fruits can be exposed to UV-C light for microbial decontamination. This treatment can be applied before or after packaging to eliminate surface contaminants and extend the date's shelf-life.
2) **Date syrups and concentrates**: Date syrups and concentrates, commonly used in beverages and confectionery, can benefit from UV irradiation to ensure microbial safety and prolong product stability.
3) **Date-based snacks**: UV treatment is suitable for date-based snacks like bars and coated dates. It helps maintain microbial quality without altering the sensory attributes that consumers value.

22.4.3.3 Recent Research Findings

Recent research on UV irradiation in date fruit preservation has yielded significant findings (Moreno et al., 2017; Bhatnagar et al., 2022; Darré et al., 2022)

1) **Microbial reduction**: Studies have consistently demonstrated UV-C light's effectiveness in reducing microbial loads in date fruit. Recent research has focused on optimizing exposure times and intensities to achieve maximum microbial inactivation while preserving product quality.
2) **Quality maintenance**: Researchers have investigated the impact of UV irradiation on the sensory attributes of date fruit, including color, flavor and texture. Recent findings indicate that when appropriately applied, UV treatment minimally affects these sensory qualities.
3) **Nutritional preservation**: One of the strengths of UV irradiation is its minimal thermal impact. Recent studies have examined changes in nutrient profiles during UV treatment and have consistently shown that this technique preserves the nutritional content of date fruit effectively.
4) **Packaging considerations**: Emerging research areas include the influence of packaging materials and design on UV treatment efficacy. Understanding how packaging can affect treatment outcomes is vital for optimizing the preservation process.
5) **Consumer acceptance**: Research has delved into consumer perceptions and preferences regarding UV-treated date products. Understanding consumer acceptance is essential for the successful adoption of UV irradiation as a preservation method for date fruit.

22.4.4 Cold Plasma Treatment for the Preservation of Dates

Cold plasma treatment, also known as nonthermal atmospheric plasma, is an innovative technique used to enhance the safety and shelf-life of date fruit. This method generates a cold plasma by applying electrical energy to a gas, typically air or nitrogen, at atmospheric pressure (Pankaj et al., 2014; Sarangapani et al., 2018) The resulting plasma consists of a mix of charged particles, including electrons, ions, free radicals and UV photons.

22.4.4.1 Principles and Mechanisms

The mechanisms of cold plasma treatment for date fruit preservation can be summarized as follows (Sarangapani et al., 2018; Zhang et al., 2021):

1) **Microbial inactivation**: Cold plasma generates reactive oxygen and nitrogen species (RONS) and UV radiation. These species disrupt the cellular structures and genetic material of microorganisms, effectively inactivating them. This method is particularly effective against bacteria, molds and yeasts that commonly contaminate date fruit.
2) **Surface decontamination**: Cold plasma treatment is especially suitable for decontaminating the surface of date fruit. It can be applied directly to the fruit's skin, reducing the risk of surface contamination during postharvest handling and processing.

3) **Minimal thermal impact**: Cold plasma operates at or near room temperature, making it a nonthermal technique. This characteristic minimizes the risk of heat-induced damage to the sensory and nutritional qualities of date fruit.

22.4.4.2 Application of Cold Plasma in Date Fruit Preservation

Cold plasma treatment offers a versatile range of applications in date fruit preservation (Mai-Prochnow et al., 2016; Maghsoudi et al., 2023):

1) **Surface decontamination**: Cold plasma is effective at decontaminating the surface of date fruit. It can be used as a prepackaging or postharvest treatment to reduce the microbial load on the fruit's skin.
2) **Date syrups and concentrates**: Date syrups and concentrates, commonly used in various food products, can be treated with cold plasma to enhance microbial safety and extend product stability.
3) **Date-based snacks**: Date-based snacks, such as bars and confections, can benefit from cold plasma treatment to ensure microbial quality while preserving the sensory attributes that consumers appreciate.

22.4.4.3 Recent Research Findings

Recent research on cold plasma treatment in date fruit preservation has yielded noteworthy findings (Maghsoudi et al., 2023; Amir-Mojahedi et al., 2023):

1) **Microbial reduction**: Studies have consistently demonstrated the effectiveness of cold plasma in reducing microbial counts on date fruit surfaces. Recent research has explored optimizing treatment parameters to achieve maximum microbial inactivation while preserving product quality.
2) **Quality maintenance**: Researchers have investigated the impact of cold plasma treatment on the sensory attributes of date fruit, including color, flavor and texture. Recent findings indicate that when applied correctly, cold plasma minimally affects these sensory qualities.
3) **Nutritional preservation**: Cold plasma's nonthermal nature ensures minimal impact on the nutritional content of date fruit. Recent studies have confirmed that this technique effectively preserves the nutritional profile of date fruit.
4) **Packaging considerations**: Emerging research areas include evaluating the influence of packaging materials and design on cold plasma treatment outcomes. Understanding packaging's role in the preservation process is crucial for optimizing treatment efficacy.
5) **Consumer acceptance**: Research has explored consumer perceptions and preferences regarding cold plasma-treated date products. Understanding consumer acceptance is vital for the successful adoption of this innovative preservation method.

22.4.5 Ozone Treatment for the Preservation of Dates

Ozone treatment is a nonthermal decontamination technique that relies on the powerful oxidizing properties of ozone (O_3) gas to enhance the safety and shelf-life of date fruit. Ozone is an unstable molecule, readily breaking down into oxygen (O_2) and a highly reactive oxygen atom (O·). It is this reactivity that makes ozone a potent antimicrobial agent (Homayouni et al., 2015; Pandiselvam et al., 2017).

22.4.5.1 Principles and Mechanisms

The mechanisms of ozone treatment for date fruit preservation can be summarized as follows (Khalil, 2016; Mahmoud and Abdel-Khalik, 2022):

1) **Microbial inactivation**: Ozone gas reacts with microbial cell membranes and intracellular components, disrupting their structure and function. This oxidative stress leads to microbial inactivation. Ozone exhibits effectiveness against a wide range of microorganisms, encompassing bacteria, molds, yeasts and viruses.
2) **Chemical degradation**: Ozone further engages with organic compounds, encompassing enzymes and mycotoxins produced by certain molds, leading to chemical transformations. These chemical reactions can further enhance the safety and quality of date fruit.
3) **Minimal residue**: Ozone naturally decomposes into oxygen, leaving no chemical residues on the treated date fruit. This characteristic is advantageous for food safety and consumer acceptance.

22.4.5.2 Application of Ozone Treatment in Date Fruit Preservation

Ozone treatment offers several valuable applications in date fruit preservation (Benkeblia, 2013):

1) **Surface decontamination**: Ozone can be applied as a gas or in solution to sanitize the surface of date fruit. It effectively reduces microbial contamination on the fruit's skin, minimizing the risk of postharvest spoilage.
2) **Washing and rinse waters**: Ozone-treated water can be used in washing and rinsing processes during date fruit processing. This ensures that the fruit remains free from microbial contaminants acquired during handling and transport.
3) **Storage and packaging**: Ozone can be employed in storage facilities and packaging to maintain the microbial quality of date fruit throughout the supply chain. It helps extend the shelf-life of packaged date products.

22.4.5.3 Recent Research Findings

Recent research on ozone treatment in date fruit preservation has yielded notable findings (Piechowiak et al., 2022):

1) **Microbial reduction**: Studies have consistently demonstrated the effectiveness of ozone in reducing microbial counts on date fruit surfaces. Recent research has focused on optimizing ozone concentrations, exposure times and treatment conditions to achieve maximum microbial inactivation.
2) **Quality maintenance**: Ozone treatment has been shown to minimally affect the sensory attributes of date fruit, including color, flavor and texture. Recent findings indicate that when applied correctly, ozone preserves the sensory qualities consumers expect.
3) **Nutritional preservation**: Ozone's nonthermal nature ensures minimal impact on the nutritional content of date fruit. Recent studies have confirmed that this technique effectively preserves the nutritional profile of date fruit.
4) **Environmental considerations**: Emerging research areas include evaluating the environmental impact of ozone treatment and exploring sustainable ozone generation methods. Understanding the environmental aspects of this technology is essential for its widespread adoption.

22.4.6 Decontamination of Date Fruits by Ionizing Radiation

22.4.6.1 Principles and Mechanisms

Ionizing radiation, including gamma rays, X-rays and electron beams, is a cutting-edge nonthermal technique for enhancing the safety and extending the shelf-life of date fruits. Gamma rays come from isotopes like cobalt-60 or cesium-137, X-rays from up to 5 MeV and electron beams from accelerators up to 10 MeV (Bhatnagar et al., 2022). The underlying mechanisms can be summarized as:

1) **Microbial inactivation**: These methods destroy microbes in the fruit by ionizing water molecules, creating reactive nitrogen species and free radicals. These radicals damage microbial DNA and oxidize cellular components like lipids and carbohydrates. This causes fatal genetic mutations, breaks and cross-linking, preventing replication or repair. Consequently, microorganisms lose their ability to grow and reproduce, significantly reducing their population.
2) **Minimal residue**: Ionizing radiation does not deter nutrient content, neither does it leave any harmful residues and it can penetrate food material even through the packaging material. These helps to provide safety without compromising the nutrient content.

22.4.6.2 Application of Ionizing Radiation in Date Fruit Preservation

Decontamination by ionizing radiation is typically applied in specialized facilities equipped with gamma irradiators, X-ray generators and electron accelerators. The date fruits are carefully loaded into the irradiation chamber, where they are exposed to controlled doses of ionizing radiation, thereby having the following applications (Zarbakhsh & Rastegar, 2019; Al-Bachir et al., 2019):

1) **Deep penetration and decontamination:** This nonthermal process ensures deep penetration that keeps the fruit free from harmful pathogens, including bacteria, molds and yeasts, without the use of heat or chemicals.
2) **Quality attributes retention:** The exposure duration and intensity are precisely calibrated to achieve the desired microbial reduction while preserving the quality, nutritional value and sensory attributes of the date fruits.

22.4.6.3 Recent Research Findings

Recent research into ionizing radiation has shown its promise in safe, fast, easy and effective decontamination of date fruits and has been outlined as follows (Bisht et al., 2021; Bhatnagar et al., 2022):

1) **Microbial reduction**: These types of radiation are utilized to reduce microbial contamination, including bacteria, molds and yeasts, thereby minimizing spoilage and increasing the fruit's postharvest storage duration.
2) **Quality maintenance**: Ionizing radiation, such as gamma rays, has been applied effectively in date fruit preservation to extend shelf-life and enhance safety.
3) **Nutritional preservation** The controlled application of ionizing radiation ensures that date fruits maintain their sensory and nutritional attributes while meeting stringent food safety standards.
4) **Regulatory regulations**: Researchers have investigated the potential formation of radiolytic products, which could impact the nutritional quality of date fruits. These studies have aimed to establish safe dose limits and provide scientific evidence for regulatory agencies.

22.4.7 Supercritical Carbon Dioxide-Based Decontamination of Date Fruits

22.4.7.1 Principles and Mechanisms

Supercritical carbon dioxide (SC-CO_2) technology has garnered considerable attention as a nonthermal decontamination technique for preserving date fruits. The principle behind SC-CO_2 decontamination lies in the unique properties of carbon dioxide when it reaches its supercritical state. This can be outlined as follows (Chen et al., 2022):

1) **Solubility and diffusion:** Supercritical carbon dioxide (SC-CO_2) penetrates date fruit tissues due to its low viscosity and high diffusivity in its supercritical state, reaching microbial cells.

2) **Microbial inactivation**: SC-CO_2 disrupts microbial cell membranes and denatures cellular proteins, leading to microbial inactivation.
3) **Minimal thermal impact**: SC-CO_2 operates at relatively low temperatures, minimizing heat-induced nutrient loss and preserving the sensory attributes of date fruits.
4) **Selective compounds extraction:** SC-CO_2 can selectively extract undesirable compounds, such as mycotoxins, from date fruits, contributing to their safety and quality.

22.4.7.2 Application in Date Fruit Preservation

The application of SC-CO_2 in date fruit preservation involves placing the fruit in a sealed chamber where carbon dioxide is subjected to controlled pressure and temperature conditions and has the following applications (Silva et al., 2020; Arumugham et al., 2022):

1) **Microbial reduction**: Supercritical carbon dioxide (SC-CO_2) demonstrates efficacy in reducing microbial contaminants on date fruits.
2) **Preservation of nutritional quality:** SC-CO_2 preserves the nutritional quality of date fruits by minimizing heat exposure and nutrient degradation during the decontamination process.
3) **Environmental sustainability:** The application of SC-CO_2 aligns with sustainable food preservation practices, reducing the need for chemical treatments and minimizing environmental impact.

22.4.7.3 Recent Research Findings

In conclusion, SC-CO_2-based decontamination is a promising nonthermal technique for enhancing the safety and quality of date fruits (Bushnaq et al., 2022; AlYammahi et al., 2023):

1) **Microbial reduction**: Recent research in SC-CO_2-based decontamination of date fruits has focused on optimizing process parameters, such as pressure, temperature and treatment time, to achieve maximum microbial reduction while maintaining fruit quality. Studies have also explored the use of SC-CO_2 in combination with other decontamination techniques to enhance its efficacy further.
2) **Quality and nutrient value maintenance**: Moreover, investigations have addressed the impact of SC-CO_2 treatment on the physicochemical and sensory properties of date fruits, ensuring that treated fruits meet consumer expectations. Researchers have also studied the mechanism of microbial inactivation and the role of SC-CO_2 in preserving the nutritional value and bioactive compounds of date fruits.

22.4.8 Ultrasonication-Based Decontamination of Date Fruits

22.4.8.1 Principles and Mechanisms

Ultrasonication, a nonthermal decontamination technique, has emerged as a promising method to enhance the safety and quality of date fruits. Ultrasonication relies on high-frequency sound waves (ultrasound) beyond the range of human hearing, typically above 20 kHz, to induce physical and chemical effects and can be summarized as follows (Wang et al., 2017):

1) **Cavitation-induced microbial inactivation**: Ultrasonication generates high-frequency sound waves that create microbubbles in the liquid surrounding date fruits. The rapid expansion and collapse of these microbubbles generate shockwaves, leading to a phenomenon called cavitation. Cavitation can damage microbial cells, leading to their inactivation.
2) **Shear forces and disruption:** Ultrasonication creates intense shear forces and turbulence in the liquid, which can disrupt microbial cell membranes, reducing cell viability.

3) **Enhanced mass transfer:** Ultrasonication enhances mass transfer processes, improving the penetration of decontaminating agents or heat into date fruits, which aids in microbial inactivation.

22.4.8.2 Application in Date Fruit Preservation

Ultrasonication decontaminates date fruits by immersing them in an ultrasonic bath or using ultrasound probes. The frequency and power are controlled for effective decontamination without compromising fruit quality. This brief treatment lasts only a few minutes, minimizing negative impacts on the fruit (Kang et al., 2017).

1) **Microbial reduction:** Ultrasonication has been applied to date fruits for the reduction of microbial contamination. Studies have demonstrated its effectiveness in reducing pathogenic microorganisms and extending the shelf-life of date fruits.
2) **Enzyme inactivation**: Ultrasonication can be used to inactivate enzymes that contribute to the deterioration of date fruit quality. This process helps maintain the texture and color of the fruit during storage.
3) **Enhanced mass transfer**: Ultrasonication facilitates the penetration of antimicrobial agents or sanitizers into date fruits, improving the efficacy of decontamination processes.
4) **Nutritional preservation**: The non-thermal nature of ultrasonication minimizes heat-related nutrient loss, preserving the nutritional quality of date fruits.

22.4.8.3 Recent Research Findings

Recent research on ultrasonication for decontaminating date fruits focuses on optimizing parameters like frequency, power, and treatment time for maximum microbial reduction while minimizing adverse effects. Studies also explore combining ultrasonication with other techniques to enhance efficacy (Khademi et al., 2018; Arya et al., 2023).

1) **Enhanced microbial safety**: Recent research has demonstrated that ultrasonication effectively reduces microbial contamination in date fruits, enhancing their safety and extending shelf-life.
2) **Improved nutritional quality**: Studies have shown that ultrasonication preserves the nutritional quality of date fruits by minimizing nutrient degradation during the decontamination process.
3) **Optimized process parameters**: Ongoing research focuses on optimizing ultrasonication parameters, such as frequency and intensity, to achieve the best results in terms of microbial reduction and quality preservation.
4) **Sustainable decontamination:** Recent findings emphasize the sustainability of ultrasonication-based decontamination as it reduces the reliance on chemical treatments, aligning with modern food safety and sustainability trends.

22.5 COMPARATIVE ANALYSIS OF NONTHERMAL TECHNIQUES

Nonthermal decontamination techniques have emerged as a crucial aspect of date fruit preservation, aiming to extend shelf-life, ensure food safety and preserve nutritional quality. In this section, we conduct a comparative analysis of these techniques, focusing on their effectiveness in microbial decontamination, impact on nutritional quality and practical considerations such as cost and scalability (Trinetta et al., 2011; Buniowska et al., 2017).

22.5.1 Effectiveness in Microbial Decontamination (Al-Juhaimi et al., 2018; Roobab et al., 2022)

1) **High-pressure processing (HPP)**: HPP has proven highly effective in reducing microbial contamination in date fruit. The application of high pressure disrupts the cell membranes of microorganisms, rendering them nonviable. Recent research has optimized HPP parameters to achieve maximal microbial inactivation. However, the effectiveness can vary depending on the specific microorganism and the pressure and time conditions used.
2) **Pulsed electric fields (PEF)**: PEF is another technique that shows promise in microbial decontamination. Short bursts of electrical pulses cause irreversible damage to microbial cell membranes. Recent studies indicate significant reductions in microbial loads on date fruit. However, the efficacy may depend on the microbial species and the electrical field strength applied.
3) **Ultraviolet (UV) irradiation**: UV irradiation is effective in reducing surface microbial contamination. It damages microbial DNA, preventing replication. Recent research has focused on optimizing UV exposure times and intensities to enhance effectiveness. However, its efficiency is limited to surface decontamination and it may not penetrate deep into the fruit.
4) **Cold plasma treatment**: Cold plasma treatment offers effective microbial inactivation on date fruit surfaces. Plasma generates reactive species that disrupt microbial membranes and biomolecules. Recent findings suggest that this technique can achieve high levels of microbial reduction. However, plasma treatment requires careful control of parameters to avoid damaging the fruit's sensory attributes.
5) **Ozone treatment**: Ozone is a potent antimicrobial agent, effective against a wide range of microorganisms. Recent research has demonstrated its effectiveness in reducing microbial contamination on date fruit surfaces. However, achieving uniform ozone distribution throughout a batch can be challenging.

22.5.2 Impact on Nutritional Quality (Asl et al., 2022; Lotfy et al., 2022)

1) **High-pressure processing (HPP)**: HPP generally has a minimal impact on the nutritional quality of date fruit. It is a nonthermal technique, which helps preserve the fruit's essential nutrients, including vitamins and minerals. Recent studies have confirmed that HPP maintains the nutritional profile of date fruit effectively.
2) **Pulsed electric fields (PEF)**: PEF is a nonthermal technique that exerts minimal thermal stress on date fruit. Recent research indicates that PEF preserves nutritional quality, including vitamins and antioxidants, while achieving microbial decontamination.
3) **Ultraviolet (UV) irradiation**: UV irradiation does not introduce heat during treatment, which is beneficial for preserving the nutritional quality of date fruit. However, excessive UV exposure can lead to nutrient degradation, particularly of vitamins. Recent studies aim to optimize UV conditions to minimize nutrient loss.
4) **Cold plasma treatment**: Cold plasma treatment is a nonthermal technique that offers advantages in maintaining nutritional quality. Recent findings suggest that it preserves the nutritional value of date fruit effectively, including vitamins, antioxidants and minerals.
5) **Ozone treatment**: Ozone treatment is nonthermal and minimally affects the nutritional quality of date fruit. Recent research indicates that it preserves the nutritional profile, making it a suitable choice for maintaining essential nutrients.

22.5.3 Practical Considerations (Vignali et al., 2022; Lopes et al., 2023)

1) **High-pressure processing (HPP)**: HPP equipment can be costly to purchase and maintain. Scalability depends on the availability of industrial-scale machines. However, it offers the advantage of batch processing and minimal energy consumption.

2) **Pulsed electric fields (PEF)**: PEF systems can be expensive to install and maintain. Scalability can be achieved through modular systems. Energy consumption varies with the scale and configuration of the equipment.
3) **Ultraviolet (UV) irradiation**: UV systems are relatively affordable, and small-scale units are readily available. Scalability depends on the number of lamps or the treatment duration. Energy consumption is generally low.
4) **Cold plasma treatment**: Cold plasma systems can be complex and may require specialized equipment. Scalability depends on the design of the treatment chamber. Energy consumption can vary depending on the plasma generation method.
5) **Ozone treatment**: Ozone generators can vary in cost, with smaller units being more affordable. Scalability depends on the ozone production capacity. Energy consumption is generally low.

In conclusion, comparing nonthermal decontamination techniques for date fruit preservation shows varying effectiveness in microbial decontamination, preserving nutritional quality and practicality. Each technique has strengths and limitations, suitable for different applications and production scales. The choice should align with goals, resource availability and desired product quality. Ongoing research refines these techniques, improving their effectiveness and feasibility in date fruit preservation.

22.6 COMBINATION APPROACHES FOR THE PRESERVATION OF DATES

Preserving date fruit involves challenges like microbial contamination, shelf-life extension and nutritional retention. Traditional and nonthermal techniques address these issues individually. Combining nonthermal techniques is promising, leveraging synergies, successful examples and future prospects for enhancing date fruit preservation (Pinela & Ferreira, 2017; Ghidelli & Pérez-Gago, 2018).

22.6.1 Synergistic Effects of Combining Nonthermal Techniques

Combining nonthermal techniques enhances microbial decontamination while preserving nutritional quality. Techniques like high-pressure processing (HPP) and pulsed electric fields (PEF) can synergize, leading to enhanced microbial inactivation. HPP disrupts microbial membranes, increasing susceptibility to PEF treatment and resulting in higher microbial reduction with minimal exposure time.

Additionally, combining nonthermal techniques can overcome individual limitations. For example, while ultraviolet (UV) irradiation is effective for surface decontamination, it may not penetrate deeply into date fruit. Cold plasma treatment following UV irradiation can complement it by reaching microbial contaminants within porous structures, ensuring comprehensive decontamination (Giannakourou & Tsironi, 2021).

22.6.2 Examples of Successful Combinations

1) **HPP and PEF**: The combination of HPP and PEF has been applied to date fruit preservation with promising results. Research has shown that this approach achieves greater microbial reduction compared to single treatments (Putnik et al., 2020; Bigi et al., 2022). Additionally, it helps maintain the fruit's sensory attributes and nutritional quality.
2) **UV irradiation and cold plasma**: UV irradiation followed by cold plasma treatment has been explored for surface decontamination. UV irradiation effectively reduces surface contaminants, while cold plasma complements this by addressing contaminants within the fruit's microstructure. This combination ensures thorough decontamination without compromising nutritional quality (Ouf et al., 2015; Misra et al., 2019; Chen et al., 2020; Maghsoudi et al., 2023).

3) **Ozone treatment and modified atmosphere packaging (MAP)**: Ozone treatment can be combined with MAP to extend the shelf-life of date fruit. Ozone effectively reduces surface microbes, while MAP creates a modified environment that inhibits microbial growth and enzymatic reactions. Together, they enhance preservation and maintain fruit quality during storage (Jemni et al., 2016; Fang & Wakisaka, 2021).

22.6.3 Challenges and Future Prospects

While combination approaches hold promise, several challenges and future prospects must be considered (Sonne et al., 2012; Cano-Lamadrid & Artés-Hernández, 2021; Aaliya et al., 2021):

1) **Optimization**: Determining the optimal sequence and parameters for combined treatments requires extensive research. Factors like treatment order, duration and intensity must be fine-tuned to maximize benefits.
2) **Cost**: Implementing combination approaches can involve higher equipment and operational costs. Economical solutions and scalability need to be explored to make these techniques accessible to the industry.
3) **Sensory attributes**: Maintaining the sensory attributes of date fruit, such as texture and flavor, while achieving microbial decontamination, remains a challenge. Research should focus on minimizing any potential sensory changes.
4) **Nutritional quality**: Combination approaches should prioritize the retention of nutritional quality. Continuous research into the impact on vitamins, antioxidants and other essential nutrients is essential.
5) **Consumer acceptance**: Assessing consumer acceptance of date fruit treated with combination approaches is crucial. Consumer perceptions regarding safety, quality and taste should be considered.
6) **Regulatory compliance**: Ensuring that combination treatments comply with regulatory standards is imperative for market acceptance. Safety and labeling regulations need to be met.
7) **Future innovations**: As technology evolves, new nonthermal techniques and combinations thereof may emerge. Researchers should stay at the forefront of innovation to develop novel approaches for date fruit preservation.

22.7 CONCLUSIONS AND FUTURE DIRECTIONS

Nonthermal techniques are reshaping food preservation, but significant research challenges remain. Understanding the mechanisms of these methods, optimizing processing parameters and exploring synergies between techniques are key challenges. Long-term effects on food quality and safety also need more study. Identifying these gaps will improve the understanding of each technique's risks and benefits, advancing the field. Scaling up nonthermal processes for widespread use in the food industry presents various prospects and challenges, including cost-effectiveness, equipment design and regulatory compliance. Understanding these factors will guide stakeholders in moving from innovation to real-world impact. In conclusion, the future of nonthermal techniques in food preservation is promising but complex. This review offers insights into emerging technologies and research areas for further exploration, contributing to the ongoing development of nonthermal methods in food preservation.

REFERENCES

Aaliya, B., Valiyapeediyekkal Sunooj, K., Navaf, M., Parambil Akhila, P., Sudheesh, C., Ahmad Mir, S., Sabu, S., Sasidharan, A., Theingi Hlaing, M., & George, J. (2021). Recent trends in bacterial decontamination of food products by hurdle technology: A synergistic approach using thermal and non-thermal processing techniques. *Food Research International (Ottawa, Ont.)*, *147*, 110514. https://doi.org/10.1016/j.foodres.2021.110514

Aganovic, K., Hertel, C., Vogel, R. F., Johne, R., Schlüter, O., Schwarzenbolz, U., Jäger, H., Holzhauser, T., Bergmair, J., Roth, A., Sevenich, R., Bandick, N., Kulling, S. E., Knorr, D., Engel, K. H., & Heinz, V. (2021). Aspects of high hydrostatic pressure food processing: Perspectives on technology and food safety. *Comprehensive Reviews in Food Science and Food Safety*, *20*(4), 3225–3266. https://doi.org/10.1111/1541-4337.12763

Alam, M. Z., Alhebsi, M. S., Ghnimi, S., & Kamal-Eldin, A. (2021). Inability of total antioxidant activity assays to accurately assess the phenolic compounds of date palm fruit (Phoenix dactylifera L.). *NFS Journal*, *22*, 32–40 https://doi.org/10.1016/j.nfs.2021.01.001

Al-Bachir, M., Al-Ghzawi, A., & Al-Ghzawi, A. L. (2019). Microbial load reduction in packaged and non-packaged dates (Phoenix dactylifera L.) using gamma irradiation. *Food Microbiology*, *84*, 103246. https://doi.org/10.1016/j.scienta.2018.12.035

Alharbi, K. L., Raman, J., & Shin, H. J. (2021). Date fruit and seed in nutricosmetics. *Cosmetics*, *8*(3), 59. https://doi.org/10.3390/cosmetics8030059

Al-Juhaimi, F., Ghafoor, K., Özcan, M. M., Jahurul, M. H. A., Babiker, E. E., Jinap, S., ... & Zaidul, I. S. M. (2018). Effect of various food processing and handling methods on preservation of natural antioxidants in fruits and vegetables. *Journal of Food Science and Technology*, *55*, 3872–3880. https://doi.org/10.1007/s13197-018-3370-0

Al-Shwyeh, H. A. (2019). Date palm (Phoenix dactylifera L.) fruit as potential antioxidant and antimicrobial agents. *Journal of Pharmacy & Bioallied Sciences*, *11*(1), 1. https://doi.org/10.4103/jpbs.JPBS_168_18

AlYammahi, J., Darwish, A. S., Lemaoui, T., AlNashef, I. M., Hasan, S. W., Taher, H., & Banat, F. (2023). Parametric analysis and machine learning for enhanced recovery of high-value sugar from date fruits using supercritical CO2 with co-solvents. *Journal of CO2 Utilization*, *72*, 102511. https://doi.org/10.1016/j.jcou.2023.102511

Amir-Mojahedi, M. S., Maghsoudi, H., Balvardi, M., Ganjovi, A., & Taraz, M. (2023). The effect of atmospheric cold plasma processing on microbial and organoleptic properties of Mazafati dates. *Iranian Journal of Biosystems Engineering*, *53*(4), 327–340. https://doi.org/10.22059/IJBSE.2023.349229.665499

Araque, L. C. O., Rodoni, L. M., Darré, M., Ortiz, C. M., Civello, P. M., & Vicente, A. R. (2018). Cyclic low dose UV-C treatments retain strawberry fruit quality more effectively than conventional pre-storage single high fluence applications. *LWT*, *92*, 304–311. https://doi.org/10.1016/j.lwt.2018.02.050

Arumugham, T., AlYammahi, J., Rambabu, K., Hassan, S. W., & Banat, F. (2022). Supercritical CO2 pretreatment of date fruit biomass for enhanced recovery of fruit sugars. *Sustainable Energy Technologies and Assessments*, *52*, 102231. https://doi.org/10.1016/j.seta.2022.102231

Arya, S. S., More, P. R., Ladole, M. R., Pegu, K., & Pandit, A. B. (2023). Non-thermal, energy efficient hydrodynamic cavitation for food processing, process intensification and extraction of natural bioactives: A review. *Ultrasonics Sonochemistry*, 106504. https://doi.org/10.1016/j.ultsonch.2023.106504

Asl, P. J., Rajulapati, V., Gavahian, M., Kapusta, I., Putnik, P., Khaneghah, A. M., & Marszałek, K. (2022). Non-thermal plasma technique for preservation of fresh foods: A review. *Food Control*, *134*, 108560. https://doi.org/10.1016/j.foodcont.2021.108560

Balakrishna, A. K., Wazed, M. A., & Farid, M. (2020). A review on the effect of high pressure processing (HPP) on gelatinization and infusion of nutrients. *Molecules (Basel, Switzerland)*, *25*(10), 2369. https://doi.org/10.3390/molecules25102369

Baroyi, S. A. H. M., Yusof, Y. A., Ghazali, N. S. M., Al-Awaadh, A. M., Kadota, K., Mustafa, S., Abu Saad, H., Shah, N. N. A. K., & Fikry, M. (2023). Determination of physicochemical, textural, and sensory properties of date-based sports energy gel. *Gels (Basel, Switzerland)*, *9*(6), 487. https://doi.org/10.3390/gels9060487

Benkeblia, N. (2013, November). Storage technologies for date palm fruits (Phoenix dactylifera L.): Current and potential techniques. . Acta Hortic. 994, 151–154. https://doi.org/10.17660/ActaHortic.2013.994.14

Benmeziane-Derradji, F. (2019). Nutritional value, phytochemical composition, and biological activities of Middle Eastern and North African date fruit: An overview. *Euro-Mediterranean Journal for Environmental Integration*, *4*(1), 39. https://doi.org/10.1007/s41207-019-0132-y

Bhatnagar, P., Gururani, P., Bisht, B., Kumar, V., Kumar, N., Joshi, R., & Vlaskin, M. S. (2022). Impact of irradiation on physico-chemical and nutritional properties of fruits and vegetables: A mini review. *Heliyon, 8*. https://doi.org/10.1016/j.heliyon.2022.e10918

Bigi, F., Maurizzi, E., Quartieri, A., De Leo, R., Gullo, M., & Pulvirenti, A. (2022). Non-thermal techniques and the "hurdle" approach: How is food technology evolving? *Trends in Food Science & Technology, 132*, 11–39. https://doi.org/10.1016/j.tifs.2022.12.015

Bisht, B., Bhatnagar, P., Gururani, P., Kumar, V., Tomar, M. S., Sinhmar, R., ... & Kumar, S. (2021). Food irradiation: Effect of ionizing and non-ionizing radiations on preservation of fruits and vegetables–a review. *Trends in Food Science & Technology, 114*, 372–385. https://doi.org/10.1016/j.tifs.2021.06.002

Bouhlali E. T., Ramchoun, M., Alem, C., Ghafoor, K., Ennassir, J., & Zegzouti, Y. F. (2017). Functional composition and antioxidant activities of eight Moroccan date fruit varieties (Phoenix dactylifera L.). *Journal of the Saudi Society of Agricultural Sciences, 16*(3), 257–264. https://doi.org/10.1016/j.jssas.2015.08.005

Buniowska, M., Carbonell-Capella, J. M., Frigola, A., & Esteve, M. J. (2017). Bioaccessibility of bioactive compounds after non-thermal processing of an exotic fruit juice blend sweetened with Stevia rebaudiana. *Food Chemistry, 221*, 1834–1842. https://doi.org/10.1016/j.foodchem.2016.10.093

Bushnaq, H., Krishnamoorthy, R., Abu-Zahra, M., Hasan, S. W., Taher, H., Alomar, S. Y., ... & Banat, F. (2022). Supercritical technology-based date sugar powder production: Process modeling and simulation. *Processes, 10*(2), 257. https://doi.org/10.3390/pr10020257

Cano-Lamadrid, M., & Artés-Hernández, F. (2021). By-products revalorization with non-thermal treatments to enhance phytochemical compounds of fruit and vegetables derived products: A review. *Foods, 11*(1), 59. https://doi.org/10.3390/foods11010059

Chen, Y. Q., Cheng, J. H., & Sun, D. W. (2020). Chemical, physical and physiological quality attributes of fruit and vegetables induced by cold plasma treatment: Mechanisms and application advances. *Critical Reviews in Food Science and Nutrition, 60*(16), 2676–2690. https://doi.org/10.1080/10408398.2019.1654429

Chen, Z., Spilimbergo, S., Mousavi Khaneghah, A., Zhu, Z., & Marszałek, K. (2022). The effect of supercritical carbon dioxide on the physiochemistry, endogenous enzymes, and nutritional composition of fruit and vegetables and its prospects for industrial application: A overview. *Critical Reviews in Food Science and Nutrition*, 1–15. Advance online publication. https://doi.org/10.1080/10408398.2022.2157370

Darré, M., Vicente, A. R., Cisneros-Zevallos, L., & Artés-Hernández, F. (2022). Postharvest ultraviolet radiation in fruit and vegetables: Applications and factors modulating its efficacy on bioactive compounds and microbial growth. *Foods (Basel, Switzerland), 11*(5), 653. https://doi.org/10.3390/foods11050653

Djaoud, K., Arkoub-Djermoune, L., Remini, H., Sait, S., Tazarourte, M., Hadjal, S.,.. & Boulekbache-Makhlouf, L. (2020). Syrup from common date variety (Phoenix dactylifera L.): Optimization of sugars extraction and their quantification by high performance liquid chromatography. *Current Nutrition & Food Science, 16*(4), 530–542. https://doi.org/10.2174/1573401315666190115160950

Fang, Y., & Wakisaka, M. (2021). A review on the modified atmosphere preservation of fruits and vegetables with cutting-edge technologies. *Agriculture, 11*(10), 992. https://doi.org/10.3390/agriculture11100992

FAOSTAT. 2020. Agro-Statistics. Database. Rome: Food & Agri. Organiz. United Nations. www.Fao.Org/Faostat. Assessed 05 September, 2023.

Ghidelli, C., & Pérez-Gago, M. B. (2018). Recent advances in modified atmosphere packaging and edible coatings to maintain quality of fresh-cut fruits and vegetables. *Critical Reviews in Food Science and Nutrition, 58*(4), 662–679. https://doi.org/10.1080/10408398.2016.1211087

Giannakourou, M. C., & Tsironi, T. N. (2021). Application of processing and packaging hurdles for fresh-cut fruits and vegetables preservation. *Foods (Basel, Switzerland), 10*(4), 830. https://doi.org/10.3390/foods10040830

Hazbavi, I., Khoshtaghaza, M., Mostaan, A., Banakar, A., 2015. Effect of postharvest hot-water and heat treatment on quality of date palm (cv. Stamaran). *Journal Saudi Society Agriculture Science, 14*, 153–159. https://doi.org/10.1016/j.jssas.2013.10.003

Homayouni, A., Azizi, A., Keshtiban, A. K., Amini, A., & Eslami, A. (2015). Date canning: A new approach for the long time preservation of date. *Journal of Food Science and Technology, 52*(4), 1872–1880. https://doi.org/10.1007/s13197-014-1291-0

Houška, M., Silva, F. V. M., Evelyn, Buckow, R., Terefe, N. S., & Tonello, C. (2022). High pressure processing applications in plant foods. *Foods (Basel, Switzerland), 11*(2), 223. https://doi.org/10.3390/foods11020223

Hussain, M. I., Farooq, M., & Syed, Q. A. (2020). Nutritional and biological characteristics of the date palm fruit (Phoenix dactylifera L.)–A review. *Food Bioscience*, *34*, 100509. https://doi.org/10.1016/j.fbio.2019.100509

Ibourki, M., Azouguigh, F., Jadouali, S. M., Sakar, E. H., Bijla, L., Majourhat, K., ... & Laknifli, A. (2021). Physical fruit traits, nutritional composition, and seed oil fatty acids profiling in the main date palm (Phoenix dactylifera L.) varieties grown in Morocco. *Journal of Food Quality*, *2021*, 1–12. https://doi.org/10.1155/2021/5138043

Jadhav, H. B., Annapure, U. S., & Deshmukh, R. R. (2021). Non-thermal technologies for food processing. *Frontiers in Nutrition*, *8*, 657090. https://doi.org/10.3389/fnut.2021.657090

Jdaini, K., Alla, F., M'hamdi, H., Guerrouj, K., Parmar, A., Elhoumaizi, M. A. (2022). Effect of harvesting and post-harvest practices on the microbiological quality of dates fruits (Phoenix dactylifera L.). *Journal Saudi Society Agriculture Science*, *21*, 552–559. https://doi.org/10.1016/j.jssas.2022.04.001

Jemni, M., Chniti, S., Harbaoui, K., Ferchichi, A., & Artés, F. (2016). Partial vacuum and active modified atmosphere packaging for keeping overall quality of dates. *Journal New Science Agriculture Biotechnology*, *29*, 1656–1665.

Kamal-Eldin, A., George, N., Sobti, B., AlRashidi, N., Ghnimi, S., Ali, A. A., ... & Hamed, F. (2020). Dietary fiber components, microstructure, and texture of date fruits (Phoenix dactylifera, L.). *Scientific Reports*, *10*(1), 21767. https://doi.org/10.1038/s41598-020-78713-4

Kang, D., Jiang, Y., Xing, L., Zhou, G., & Zhang, W. (2017). Inactivation of Escherichia coli O157:H7 and Bacillus cereus by power ultrasound during the curing processing in brining liquid and beef. *Food Research International (Ottawa, Ont.)*, *102*, 717–727. https://doi.org/10.1016/j.foodres.2017.09.062

Khademi, S., Madadlou, A., & Saboury, A. A. (2018). Ultrasound-assisted extraction of date (Phoenix dactylifera L.) seed protein isolate: Influence of operating parameters and process optimization. *Food Biophysics*, *13*(4), 400–409. https://doi.org/10.1016/j.lwt.2020.110816

Khalil, H. A. (2016). Effect of ozone application on postharvest quality and microbiological state of "Zaghloul" date palm fruits. *Journal of Plant Production*, *7*(1), 43–51.

Koutsoumanis, K., Alvarez-Ordóñez, A., Bolton, D., Bover-Cid, S., Chemaly, M., Davies, R., De Cesare, A., Herman, L., Hilbert, F., Lindqvist, R., Nauta, M., Peixe, L., Ru, G., Simmons, M., Skandamis, P., Suffredini, E., Castle, L., Crotta, M., Grob, K., … Allende, A. (2022). The efficacy and safety of high-pressure processing of food. *EFSA Journal. European Food Safety Authority*, *20*(3), e07128. https://doi.org/10.2903/j.efsa.2022.7128

Lopes, S. J., de Souza Sant'Ana, A., & Freire, L. (2023). Non-thermal emerging processing Technologies: Mitigation of microorganisms and mycotoxins and sensory and nutritional properties maintenance in clean label fruit juices. *Food Research International*, *168*, 112727. https://doi.org/10.1016/j.foodres.2023.112727

Lotfy, K., Al-Qahtani, S. M., Al-Harbi, N. A., El-Absy, K. M., Bu Shulaybi, F. A., Alali, S. A., & Mashtoly, T. A. (2022). Influence of non-thermal plasma on the quality and nutritional content of palm dates. *Applied Sciences*, *12*(17), 8587. https://doi.org/10.3390/app12178587

Maghsoudi, H., Balvardi, M., Ganjovi, A., & Amir-Mojahedi, M. S. (2023). Investigating the effect of cold plasma on some chemical properties of date fruits (Phoenix dactylifera L.). *Biomechanism and Bioenergy Research*, *2*(1), 56–67.

Mahmoud, R., & Abdel-Khalik, A. R. (2022). Comparison between two physical methods to control the stored dates fruit mites, Tyrophagus putrescentiae (Schrank) and Rhizoglyphus robini Claparede (Astigmata: Acaridae). *Egyptian Academic Journal of Biological Sciences, B. Zoology*, *14*(1), 149–158. https://doi.org/10.21608/EAJBSZ.2022.228058

Mahomoodally, M. F., Khadaroo, S. K., Hosenally, M., Zengin, G., Rebezov, M., Ali Shariati, M., ... & Simal-Gandara, J. (2023). Nutritional, medicinal and functional properties of different parts of the date palm and its fruit (Phoenix dactylifera L.)–A systematic review. *Critical Reviews in Food Science and Nutrition*, 1–56. https://doi.org/10.1080/10408398.2023.2191285

Mai-Prochnow, A., Clauson, M., Hong, J., & Murphy, A. B. (2016). Gram positive and Gram negative bacteria differ in their sensitivity to cold plasma. *Scientific Reports*, *6*(1), 38610. https://doi.org/10.1038/srep38610

Maqsood, S., Adiamo, O., Ahmad, M., & Mudgil, P. (2020). Bioactive compounds from date fruit and seed as potential nutraceutical and functional food ingredients. *Food Chemistry*, *308*, 125522. https://doi.org/10.1016/j.foodchem.2019.125522

Misra, N. N., Yadav, B., Roopesh, M. S., & Jo, C. (2019). Cold plasma for effective fungal and mycotoxin control in foods: Mechanisms, inactivation effects, and applications. *Comprehensive Reviews in Food Science and Food Safety*, *18*(1), 106–120. https://doi.org/10.1111/1541-4337.12398

Moreno, C., Andrade-Cuvi, M. J., Zaro, M. J., Darre, M., Vicente, A. R., & Concellón, A. (2017). Short UV-C treatment prevents browning and extends the shelf-life of fresh-cut carambola. *Journal of Food Quality*, *2017*, 2548791. https://doi.org/10.1155/2017/2548791

Mrabet, A., Hammadi, H., Rodríguez-Gutiérrez, G., Jiménez-Araujo, A., & Sindic, M. (2019). Date palm fruits as a potential source of functional dietary fiber: A review. *Food Science and Technology Research*, *25*(1), 1–10. https://doi.org/10.3136/fstr.25.1

Mtaoua, H., Sánchez-Vega, R., Ferchichi, A., & Martín-Belloso, O. (2017). Impact of high-intensity pulsed electric fields or thermal treatment on the quality attributes of date juice through storage. *Journal of Food Processing and Preservation*, *41*(4), e13052. https://doi.org/10.1111/jfpp.13052

Muhialdin, B. J., Filimonau, V., Qasem, J. M., Ibrahim, S. A., & Algboory, H. L. (2022). Traditional fermented foods and beverages in Iraq and their potential for large-scale commercialization. *Journal of Ethnic Foods*, *9*(1), 18. https://doi.org/10.1186/s42779-022-00133-8

Naliyadhara, N., Kumar, A., Girisa, S., Daimary, U. D., Hegde, M., & Kunnumakkara, A. B. (2022). Pulsed electric field (PEF): Avant-garde extraction escalation technology in food industry. *Trends in Food Science & Technology*, *122*, 238–255. https://doi.org/10.1016/j.tifs.2022.02.019

Nowosad, K., Sujka, M., Pankiewicz, U., & Kowalski, R. (2021). The application of PEF technology in food processing and human nutrition. *Journal of Food Science and Technology*, *58*(2), 397–411. https://doi.org/10.1007/s13197-020-04512-4

Ouf, S. A., Basher, A. H., & Mohamed, A. A. (2015). Inhibitory effect of double atmospheric pressure argon cold plasma on spores and mycotoxin production of Aspergillus niger contaminating date palm fruits. *Journal of the Science of Food and Agriculture*, *95*(15), 3204–3210. https://doi.org/10.1002/jsfa.7060

Palumbo, M., Attolico, G., Capozzi, V., Cozzolino, R., Corvino, A., de Chiara, M. L. V., Pace, B., Pelosi, S., Ricci, I., Romaniello, R., & Cefola, M. (2022). Emerging postharvest technologies to enhance the shelf-life of fruit and vegetables: An overview. *Foods (Basel, Switzerland)*, *11*(23), 3925. https://doi.org/10.3390/foods11233925

Pandiselvam, R., Sunoj, S., Manikantan, M. R., Kothakota, A., & Hebbar, K. B. (2017). Application and kinetics of ozone in food preservation. *Ozone: Science & Engineering*, *39*(2), 115–126. https://doi.org/10.1080/01919512.2016.1268947

Pankaj, S. K., Bueno-Ferrer, C., Misra, N. N., Milosavljević, V., O'donnell, C. P., Bourke, P., ... & Cullen, P. J. (2014). Applications of cold plasma technology in food packaging. *Trends in Food Science & Technology*, *35*(1), 5–17. https://doi.org/10.1016/j.tifs.2013.10.009

Pérez-Lamela, C., Franco, I., & Falqué, E. (2021). Impact of high-pressure processing on antioxidant activity during storage of fruits and fruit products: A review. *Molecules (Basel, Switzerland)*, *26*(17), 5265. https://doi.org/10.3390/molecules26175265

Perveen, K., & Bokahri, N. A. (2020). Comparative analysis of chemical, mineral and in-vitro antibacterial activity of different varieties of date fruits from Saudi Arabia. *Saudi Journal of Biological Sciences*, *27*(7), 1886–1891. https://doi.org/10.1016/j.sjbs.2019.11.029

Piechowiak, T., Migut, D., Józefczyk, R., & Balawejder, M. (2022). Ozone treatment improves the texture of strawberry fruit during storage. *Antioxidants*, *11*(5), 821.

Pinela, J., & Ferreira, I. C. (2017). Nonthermal physical technologies to decontaminate and extend the shelf-life of fruits and vegetables: Trends aiming at quality and safety. *Critical Reviews in Food Science and Nutrition*, *57*(10), 2095–2111. https://doi.org/10.1080/10408398.2015.1046547

Putnik, P., Pavlić, B., Šojić, B., Zavadlav, S., Žuntar, I., Kao, L., ... & Kovačević, D. B. (2020). Innovative hurdle technologies for the preservation of functional fruit juices. *Foods*, *9*(6), 699. https://doi.org/10.3390/foods9060699

Roobab, U., Abida, A., Chacha, J. S., Athar, A., Madni, G. M., Ranjha, M. M. A. N., Rusu, A. V., Zeng, X. A., Aadil, R. M., & Trif, M. (2022). Applications of innovative non-thermal pulsed electric field technology in developing safer and healthier fruit juices. *Molecules (Basel, Switzerland)*, *27*(13), 4031. https://doi.org/10.3390/molecules27134031

Sarangapani, C., Patange, A., Bourke, P., Keener, K., & Cullen, P. J. (2018). Recent advances in the application of cold plasma technology in foods. *Annual Review of Food Science and Technology*, *9*, 609–629. https://doi.org/10.1146/annurev-food-030117-012517

Sarraf, M., Jemni, M., Kahramanoğlu, I., Artés, F., Shahkoomahally, S., Namsi, A., Ihtisham, M., Brestic, M., Mohammadi, M., & Rastogi, A. (2021). Commercial techniques for preserving date palm (Phoenix dactylifera) fruit quality and safety: A review. *Saudi Journal of Biological Sciences*, *28*(8), 4408–4420. https://doi.org/10.1016/j.sjbs.2021.04.035

Silva, E. K., Meireles, M. A. A., & Saldaña, M. D. (2020). Supercritical carbon dioxide technology: A promising technique for the non-thermal processing of freshly fruit and vegetable juices. *Trends in Food Science & Technology*, *97*, 381–390. https://doi.org/10.1016/j.tifs.2020.01.025

Singh, S. P., & Golding, J. (2020). Advances in the potential use of non-thermal plasma in postharvest treatment of fresh horticultural produce. In Chris Watkins (Ed.) *Advances in postharvest management of horticultural produce* (pp. 273–296). Burleigh Dodds Science Publishing. https://doi.org/10.1201/9781003047650

Sonne, A. M., Grunert, K. G., Olsen, N. V., Granli, B. S., Szabó, E., & Banati, D. (2012). Consumers' perceptions of HPP and PEF food products. *British Food Journal*, *114*(1), 85–107. https://doi.org/10.1016/j.appet.2008.09.010

Sonntag, F., Liu, H., & Neugart, S. (2023). Nutritional and physiological effects of postharvest UV radiation on vegetables: A review. *Journal of Agricultural and Food Chemistry*, *71*(26), 9951–9972. https://doi.org/10.1021/acs.jafc.3c00481

Tang, Z. X., Shi, L. E., & Aleid, S. M. (2013). Date fruit: Chemical composition, nutritional and medicinal values, products. *Journal of the Science of Food and Agriculture*, *93*(10), 2351–2361. https://doi.org/10.1002/jsfa.6154

Trinetta, V., Vaidya, N., Linton, R., & Morgan, M. (2011). A comparative study on the effectiveness of chlorine dioxide gas, ozone gas and e-beam irradiation treatments for inactivation of pathogens inoculated onto tomato, cantaloupe and lettuce seeds. *International Journal of Food Microbiology*, *146*(2), 203–206. https://doi.org/10.1016/j.ijfoodmicro.2011.02.014

Vignali, G., Gozzi, M., Pelacci, M., & Stefanini, R. (2022). Non-conventional stabilization for fruit and vegetable juices: Overview, technological constraints, and energy cost comparison. *Food and Bioprocess Technology*, *15*(8), 1729–1747. https://doi.org/10.1007/s11947-022-02772-w

Wang, Y., Johnson-Cicalese, J., Singh, A. P., & Vorsa, N. (2017). Characterization and quantification of flavonoids and organic acids over fruit development in American cranberry (Vaccinium macrocarpon) cultivars using HPLC and APCI-MS/MS. *Plant Science*, *262*, 91–102. https://doi.org/10.1016/j.plantsci.2017.06.004

Wilson, M. D., Stanley, R. A., Eyles, A., & Ross, T. (2019). Innovative processes and technologies for modified atmosphere packaging of fresh and fresh-cut fruits and vegetables. *Critical Reviews in Food Science and Nutrition*, *59*(3), 411–422. https://doi.org/10.1080/10408398.2017.1375892

Younis, M., Ahmed, I. A. M., Ahmed, K. A., Yehia, H. M., Abdelkarim, D. O., Fickak, A., ... & Elfeky, A. (2023). Pulsed electric field as a novel technology for fresh Barhi date shelf-life extension: Process optimization using response surface methodology. *Horticulturae*, *9*(2), 155. https://doi.org/10.3390/horticulturae9020155

Zamir, R., Islam, A. B. M. N., Rahman, A., Ahmed, S., & Omar Faruque, M. (2018). Microbiological quality assessment of popular fresh date samples available in local outlets of Dhaka City, Bangladesh. *International Journal of Food Science*, *2018*, 7840296. https://doi.org/10.1155/2018/7840296

Zarbakhsh, S., & Rastegar, S. (2019). Influence of postharvest gamma irradiation on the antioxidant system, microbial and shelf life quality of three cultivars of date fruits (Phoenix dactylifera L.). *Scientia Horticulturae*, *247*, 275–286. https://doi.org/10.1016/j.scienta.2018.12.035

Zhang, Y., Zhang, J., Zhang, Y., Hu, H., Luo, S., Zhang, L., ... & Li, P. (2021). Effects of in--package atmospheric cold plasma treatment on the qualitative, metabolic and microbial stability of fresh-cut pears. *Journal of the Science of Food and Agriculture*, *101*(11), 4473–4480. https://doi.org/10.1002/jsfa.11085

Index

Note: Page locators in **bold** and *italics* represents tables and figures, respectively.

A

D

E

N

O

P

Q

R

S

T

U

V